WITH MALICE TOWARD NONE

Books by Stephen B. Oates

William Faulkner: The Man and the Artist

Biography as High Adventure: Life-Writers Speak on Their Art

Abraham Lincoln: The Man Behind the Myths

THE CIVIL WAR QUARTET:

Let the Trumpet Sound: A Life of Martin Luther King, Jr.

With Malice Toward None: A Life of Abraham Lincoln

The Fires of Jubilee: Nat Turner's Fierce Rebellion

To Purge This Land with Blood: A Biography of John Brown

Our Fiery Trial

Portrait of America (2 volumes)

Visions of Glory

Rip Ford's Texas

Confederate Cavalry West of the River

WITH MALICE
TOWARD NONE
A LIFE OF
ABRAHAM LINCOLN

STEPHEN B. OATES

HarperPerennial

A Division of HarperCollins*Publishers*

A hardcover edition of this book was originally published in 1977 by Harper & Row, Publishers, Inc.

WITH MALICE TOWARD NONE: A LIFE OF ABRAHAM LINCOLN. Copyright © 1977, 1994 by Stephen B. Oates. All rights reserved. Printed in the United States of America. No part of this book may be used or reproduced in any manner whatsoever without written permission except in the case of brief quotations embodied in critical articles and reviews. For information address HarperCollins Publishers, Inc., 10 East 53rd Street, New York, NY 10022.

HarperCollins books may be purchased for educational, business, or sales promotional use. For information please write: Special Markets Department, HarperCollins Publishers, Inc., 10 East 53rd Street, New York, NY 10022.

First HarperPerennial edition published 1994.

The Library of Congress has catalogued the hardcover edition as follows:

Oates, Stephen B.
    With malice toward none.
    Includes bibliographical references and index.
    1. Lincoln, Abraham, Pres. U. S., 1809–1865. 2. Presidents—United States—Biography. I. Title.
E457.017   973.7′092′4 [B]   76-12058
ISBN 0-06-013283-3

ISBN 0 -06-092471-3 (pbk.)

06 07 08 RRD H 40 39 38 37 36 35 34 33

For
Greg and Stephanie

Happy reading this incredible history lesson. Harry Miller

CONTENTS

ILLUSTRATIONS

The biographer's mission is to perpetuate a man as he was in the days he lived—a spring task of bringing to life again.
Paul Murray Kendall

I hope, however, that the following pages may prove to be of interest from the strictly biographical no less than from the historical point of view. Human beings are too important to be treated as mere symptoms of the past. They have a value which is independent of any temporal processes—which is eternal, and must be felt for its own sake.
Lytton Strachey

*I shall do nothing in malice. What I deal with
is too vast for malicious dealing.*
A. Lincoln to a Louisiana Unionist, 1862

PREFACE TO THE HARPER-PERENNIAL EDITION

I

Since it was first published in 1977, *With Malice Toward None* has enjoyed a critical and popular reception beyond all my expectations. It was extensively reviewed, was a Book-of-the-Month Club selection, got me on the "Today Show" with Tom Brokaw, won several awards, appeared in various foreign editions, came out in a paperback edition that was highly successful, and elicited a great many letters from readers all over the world. I like to think that readers identified with the Lincoln in my story, as I did. For behind the myths, behind the god of marble and stone, I had discovered a man of rich humanity—a principled man who understood the complexities of human nature, a self-made man who was proud of his achievements, substantially wealthy, morbidly fascinated with madness, obsessed with death, troubled with bouts of melancholia, and gifted with a major talent for literary expression.

I also like to think that readers were attracted to my empathetic narrative voice, for empathy is what calls a human being back from the darkness. In Frank Vandiver's felicitous phrase, empathy is the biographer's "quintessential quality," his "spark of creation." Empathy is not the same thing as hero worship. On the contrary, it is an attempt to understand frailty and failings in another, not hide from them. It is an effort to experience other human beings, by seeing the world from their view, feeling their feelings, and thinking their thoughts. To be properly empathetic, the biographer must be prepared to walk many a lonely mile in his subject's footsteps. Certainly I tried to do that in my pursuit of Lincoln: I walked in his footsteps in New Salem, Springfield, and Washington, D.C. I sat in his head and heart so much that I got depressed when he did; I laughed when he laughed. When I left my study after a day's writing in his world, I was stunned to find myself in the twentieth century. I hope that my readers have a similar sensation when they put the book down. If

I have done my job properly as a biographer, they will willingly suspend their knowledge of how the story ends and allow my narrative to transport them back into Lincoln's world, where they can walk with him as his life unfolds, sharing in his humanity.

In my original Preface, I claimed that I was presenting not the Lincoln of myth and legend, but the actual man of history. This was a misleading statement, for it suggested that no art was involved in my reconstruction of his life, that all I or anyone else had to do was to examine the facts critically and there "the real" Lincoln would be. When I referred to resurrecting the Lincoln of history, I did not mean that a definitive portrait could be shaped that would stand forever as the way he really was. As I've said elsewhere, historical biography is an interpretive art, not an exact science. The very materials we use to forge biography—letters, journals, diaries, interviews, recollections, and the like—were all recorded by people who filtered things through their own perceptions and sensibilities. Because biographical materials are themselves imprecise and interpretative, it is impossible for anyone to produce a definitive biography—a fixed and final portrait—of Lincoln or any other figure. As we strive for biographical and historical truth, the best we can hope for is a careful approximation of what Lincoln was like in the days he lived. *With Malice Toward None,* then, is my approximation of Lincoln, based on authenticated detail and a synthesis of Lincoln scholarship.

II

If my biography has had its share of successes, it has also been the subject of an acrimonious public debate about its integrity. In 1990–1991, a literary critic, a professor of classics, an associate professor of criminology, and two historians contended that I had plagiarized in writing *With Malice Toward None.* They cited as evidence similarities of phrases and short bits of factual matter between my account of Lincoln's early years and that in Benjamin Thomas's 1952 biography. The original accusers presented their challenge at a symposium of the Illinois State Historical Society, which had elected not to forewarn me about the nature of the complaint. According to a historian who was present, the symposium organizers hoped that the charges would get into the *New York Times* the next morning. They were "disappointed," he said, when the *Times* ignored them.*

After studying the allegations, I issued a public rebuttal, pointing out that Lincoln literature consists of a common body of knowledge about him, particularly his well-known early years, that has accumulated for more than a century

*See Jean Caldwell's report in the *Boston Globe* (April 17, 1991), 21.

and is in the public domain. If there are similarities between my book and Thomas's, I said, it is because both biographies draw from that common "text" or body of writing and information. I emphasized the point that there are also similarities of facts and phrases between Thomas's work and previous Lincoln biographies by Albert J. Beveridge and Carl Sandburg. In truth, there are textual similarities among Lincoln books as a whole, for the simple reason that they derive from and are part of a common body of recorded knowledge. In conclusion, I drew on James E. Porter's fine article on intertextuality, which appeared in *Rhetoric Review* (Fall 1986). "Examining texts 'intertextually,'" he states, "means looking for 'traces,' the bits and pieces of Text which writers or speakers borrow and sew together to create new discourse." In writing *With Malice Toward None,* to borrow Porter's words, I took "the bits and pieces" of the Lincoln biographical tradition and tried to weave it into "a new discourse."

In response to my rebuttal, twenty-three prominent Lincoln and Civil War scholars, including five Pulitzer Prize winners and one Parkman and Bancroft Prize winner, issued a public statement rejecting the allegations of plagiarism as "totally unfounded." Subsequently, the University of Massachusetts at Amherst, my employer, announced that "distinguished scholars of two departments" had examined the charges and had found them "groundless." The American Historical Association, after dragging the case out for a year and a half, also handed down a verdict against my accusers. "With the assistance of specially qualified advisors," the AHA reported later, "the Council found that plagiarism had not occurred, but rather an insufficiency of acknowledgment of one particular source" (*Perspectives: Newsletter of the American Historical Association,* September 1992). In other words, even though I had done nothing wrong, I should have included more references to Thomas's biography in my account of Lincoln's early years. I and other Lincoln scholars disputed that point, since Thomas's scholarship is cited repeatedly in my references and since Thomas's own biography has no references at all—only a general discussion of sources at the back.

The matter of attribution in popular biography is a crucial point. Because *With Malice Toward None* is a one-volume synthesis of Lincoln and Civil War scholarship and because it is directed at a general audience and not just at scholars, it does not contain the kind of copious, line-by-line footnoting associated with a scientific research paper or a historical monograph on an unfamiliar subject. This is something my accusers have persistently misunderstood. Like Thomas, I made no attempt to attribute every line of my text, or to list all the previous Lincoln biographies and histories that had likewise touched on a particular point or episode. What my biography contains are extensive lists of sources and secondary works, keyed to specific pages of text, from which I quoted directly or borrowed original findings and ideas. All told, there are thirty-seven pages of annotations at the back of this volume, compared to seventeen pages in Thomas's. By the standards of popular biography, *With Malice Toward None* has more than adequate attribution of its sources.

Much of the concern about my book has stemmed from confusion about what constitutes plagiarism. The clearest definition, the one in vogue when I wrote my biography in the early and mid-1970s, is that offered by an expert, Alexander Lindey, in his authoritative *Plagiarism and Originality* (1952). "Plagiarism," Lindey writes, "is literary—or artistic or musical—theft. It is the false assumption of authorship: the wrongful act of taking the product of another person's mind, and presenting it as one's own. Copying someone else's story . . . intact or with inconsequential changes, and adding one's name to the result constitute a simple illustration of plagiarism." I did not, of course, "copy" Thomas's story and sign my name to it. Nor did I "steal" his ideas or sentences, paragraphs, or pages, and present them as my own.

In one of the most perceptive passages of his study, Lindey warned that "any method of comparison which lists and underscores similarities and suppresses or minimizes differences is necessarily misleading." In focusing on minuscule fragments to the exclusion of everything else, my challengers ignored how different my book is from Thomas's and the other earlier biographies as well. Among other things, it is different in its interpretation of Lincoln's character and personality, of his feelings about his log-cabin origins, of slavery and race, of the Lincoln marriage, of the Lincoln-Douglas debates and the divisive politics of the 1850s, of Lincoln's views of the secessionist South, of his relationship with his generals and the so-called radical wing of the Republican party, of his Emancipation Proclamation and his changing attitudes toward blacks, and of his approach to reconstruction. It is also different in design, organization, and style. Even so, I learned a great deal from Benjamin Thomas about preparing a Lincoln biography. An outstanding scholar and life-writer, he was an inspiration to me and every other student of Lincoln who followed him. I also benefited enormously from the Lincoln biographies of my other predecessors, not only Sandburg and Beveridge, but Richard N. Current, James G. Randall, Reinhard H. Luthin, Ida Tarbell, Lord Charnwood, and William H. Herndon, and I am indebted to them all for breaking the ground ahead of me.

PART·ONE
RIVERS OF TIME

Outside of Illinois, people knew little about him. Even newspapers were conspicuously reticent about his life and background. All most could say was that he hailed from Illinois, that he had served a single term in Congress and had lost a bitter Senate contest to Stephen A. Douglas a couple of years before. And now, in the summer of 1860, he was the Republican candidate for President of the United States in what promised to be the most combustible election the Union had ever known. In the South, Democrats who understood nothing about the candidate as a man, nothing at all, castigated him as a symbol of "Black Republicanism"—a "sooty and scoundrelly" abolitionist who wanted to free the slaves and mongrelize the white race. In the North, Democratic papers disparaged him as a party hack and a political unknown who lacked the ability to serve as President. Even many Republicans were hard-pressed to talk specifically about their candidate, to sell voters on his appeal and his talents. Some party bosses mistakenly thought his first name was "Abram," and various newspapers persisted in calling him that.

"There are thousands who do not yet know Abraham Lincoln," observed Horace Greeley of the *New York Tribune,* and he called for the publication of an inexpensive biography that Republicans could read and circulate across the North. As several writers set about compiling their own profiles (one inevitably called him "Abram" Lincoln), Joseph Medill out in Chicago decided that a terse campaign biography should be prepared under the auspices of the *Chicago Press & Tribune,* Medill's influential Republican newspaper. Medill assigned senior editor John Locke Scripps to write the portrait, and Scripps caught a train for Springfield, where Lincoln lived and practiced law.

Scripps called on Lincoln in his Springfield office and found him besieged with Republican bigwigs, office seekers, and reporters (Mr. Lincoln, what are your plans? policies? Cabinet choices? Mr. Lincoln, what will you do if you are elected and the slave states secede?). It was sometime early in June, and the air was scented with summer smells; outside, hogs rooted in the dirt streets, wagons jingled by, and people strolled about the public square.

Lincoln shooed people out of his office and closed the door so that Scripps could interview him in private. The candidate was tall and melancholy, with coarse black hair, large ears, a hawkish face, and long and bony limbs. As he commented on his rise to prominence and the impending campaign, Scripps took notes and urged Lincoln to talk more about himself. "The chief difficulty I had," Scripps reported later, "was to induce him to communicate the homely facts of his youth." "Why, Scripps," Lincoln protested, "it is a great piece of folly to attempt to make anything out of my early life. It can be all condensed into a simple sentence, and that sentence you will find in Gray's Elegy, 'The short and simple annals of the poor.' That's my life, and that's all you or any one else can make out of it."

The truth was that Lincoln felt embarrassed about his log-cabin origins and never liked to talk about them. In fact, he had worked all his adult life to overcome the limitations of his frontier background, to make himself into a literate and professional man who commanded the respect of his colleagues. So if he ever discussed his boyhood or his parents, said William Herndon, Lincoln's law partner, "it was with great reluctance and significant reserve. There was something about his origin he never cared to dwell on."

Still, Lincoln conceded that it might be well for Scripps to write a brief, authorized biography, so that the public might know the essential facts of his life . . . like the fact that his first name was *Abraham,* not Abram, as he was tired of pointing out. To ensure factual accuracy, Lincoln even agreed to furnish a summary of his early life and career. As he spoke, Scripps noted, Lincoln "seemed to be painfully impressed with the extreme poverty of his early surroundings, and the utter absence of all romantic and heroic elements." At least that was how Lincoln remembered things now, as a self-made lawyer reaching for the highest office in the land. He told Scripps a few details about his ancestry, but warned that he did not want them published.

Later, after Scripps had gone, Lincoln toiled over an autobiographical sketch in which he referred to himself as "A." It was little more than an outline—as lucid, exact, and careful as he could make it, but an outline all the same. Because he could not bring himself to compose a full record of his life—that lay in uncertain fragments in a thousand letters, speeches, and newspaper clippings, in family trunks with their photographs and memorabilia, and in the diaries and recollections of his family, friends, colleagues, and adversaries. No, not a complete record of his life, because there was much about himself he would never reveal . . . much about his parents and their backgrounds and deficiencies he refused to make public for the opposition to exaggerate and use against him. No, not a complete record of his life —not so much as a glimpse of who he was, of how he had suffered and what he had come to know in the decades since he had romped and roamed and brooded and labored in the fields and creek bottoms of his youth. Still, he must have paused from time to time in his composition and thought back over those faded years, recalling some half-forgotten episode of his boyhood (a storm, a dream) and listening to the echoes of his past.

Try as he might, he could not remember much about Kentucky—and nothing at all about the log-cabin farm on the south fork of Nolin Creek, a hardscrabble place where "A." was born on February 12, 1809, christened Abraham after his grandfather. His earliest recollection—a memory swimming out of some secret river of time—was a scene at their subsequent homestead on Knob Creek, several miles to the north. The scene floated in his memory like a photograph: Thomas Lincoln, his stout, barrel-chested father, hoeing down in the creek bottom while little Abraham followed along dropping pumpkin seeds between hills of corn. And the Sunday not long after that when a storm blew up, attacking the countryside with doomsday thunder and pelting rain. And then a flash flood came roaring down the ravines, washing away the pumpkin seeds and the corn and the topsoil itself. And, too, there was the time he almost drowned in the creek as a child: little A. thrashing about in the water, a neighbor boy wading in to save his life.

Storms and deaths, his mother would say, were part of the workings of God, mysterious and incomprehensible. His thin, dark-haired mother, her eyes like pools of sadness, raised him and sister Sarah with a melancholy affection. Unable to read, she recited prayers for the children and quoted memorized passages from the family Bible. Incapable of even writing her name, Nancy Hanks Lincoln signed legal documents with her mark. Yet if she could not teach the children their letters, she did instruct them in the primitive Baptist doctrine of fatalism in which she gloried. "Nothing can hinder the execution of the designs of Providence," the creed went. "What is to be will be and we can do nothing about it." That was why Abraham was still alive, why he had not drowned that time in the creek: because it was *not his time,* because Providence had *other designs.* But with the Lincolns' third child God had decreed otherwise. The child, who had died of some unnamed affliction, was buried in a small grave visible from the Lincoln cabin.

In contrast to Nancy, Thomas Lincoln was a convivial fellow, fond of good company and a "leader of grocery-store dialogue." As a man who knew him recalled, "His chief earthly pleasure was to crack and tell stories to a group of chums who paid homage to his wit by giving him the closest attention and the loudest applause." Thomas descended from a long line of Lincolns, one of whom had migrated from England to Massachusetts back in 1637. Originally the Lincolns were Quakers, but gradually they fell away from the beliefs and habits of those high-minded folk. In successive waves, the Lincolns moved on to Pennsylvania, down to Virginia, and finally out to Kentucky where Abraham—A.'s grandfather—had been killed by an Indian. Young Abraham had heard the tale of Grandfather's death over and over, since Thomas enjoyed describing it in the evenings around a fire. He met Nancy Hanks when she was living with guardians or relatives in Washington County, Kentucky, and married her there in 1806. Although myths were to flourish about Thomas as a lazy ne'er-do-well, he was actually a hard-working carpenter and farmer who won the respect of his neighbors. He stayed sober, accumulated land, paid his taxes, sat on juries, and served on the country slave patrol. Though he came from a family of small slaveholders and undoubtedly shared the anti-Negro prejudice of nearly all whites of his generation, he came to question the peculiar institution itself. Indeed he associated with a group of antislavery Baptists in his neighborhood who had walked out of the regular Baptist church during a stormy debate over slavery; they had established a separatist church which not only renounced human bondage but eschewed all written creeds and official church organization, relying on the Bible as the sole rule of faith. In 1816 Thomas and Nancy Lincoln united with the separatist church and sang and prayed with its antislavery ministers.

Nancy, for her part, had a confused and cloudy past. In truth, her origins are so vague that one can say little about her with certainty. According to tradition, her mother was Lucy Hanks, but her father's identity remains a mystery. As a consequence, a controversy has long raged over Nancy's legitimacy, with many authorities insisting that she was born out of wedlock and others retorting that she was not. Since Nancy herself was raised mostly by her relatives or guardians, she probably knew little about her mother and even less about her obscure father. It's possible that her son grew up thinking her illegitimate (Herndon, Lincoln's law partner of later years, claimed he did), but this cannot be substantiated. In any event, Lincoln did become resolutely silent about his mother and her background. In his autobiography, he dismissed Nancy with only a brief statement that she was born in Virginia. Yet in mood and appearance he resembled her more than he did his father.

So he never knew much, perhaps never tried to know much, about his parents' ancestry, characterizing both as products "of undistinguished families." Later he claimed that because his father was "left an orphan at the age of six years, in poverty, and in a new country, he became a wholly uneducated man; which I suppose is the reason why I know so little of our family history."

He had only random images, then, of his grandfather's death, of the Knob Creek farm, of the first exhilarating brush with education—two brief sessions in 1815 and 1816 when he and Sarah could be spared from the family chores in the winter, when they walked through the scrub trees and creek bottoms over to the log schoolhouse on the Cumberland Road. Here he learned his ABC's "by littles," his teacher a fifty-two-year-old Catholic slave owner. Sister Sarah, two years older than Abraham, was a plump girl with dark hair and gray eyes, a grown-up child of nine who admonished her brother to behave himself. By now Abraham was "a tall spider of a boy," a neighbor recalled, who "had his due proportion of harmless mischief." And then came the move to Indiana that fall, the Lincolns uprooting themselves for reasons the boy only vaguely understood—an error in his father's land titles, threatened lawsuits, and something about slavery Abraham did not fully comprehend: his father announcing that they would be better off in the free territory of Indiana, where a number of Kentuckians had already migrated and where land laws and surveys were more exact than in Kentucky. So "like a piece of human flotsam thrown forward by the surging tide of immigration," as one writer phrased it, Thomas Lincoln took his family across the Ohio River to find a better life in the Indiana wilderness.

They reached their new claim sometime in December, 1816, the very month Indiana entered the Union as a free state. The claim was situated in what became known as the Little Pigeon Creek community, in a remote backwoods some sixteen miles northwest of Troy and the Ohio River. It was a wild region, an area of dense forests and grapevine thickets so entangled that travelers often had to cut their way with axes. With wolves howling in the distance and a winter fog hanging in the woods, the Lincolns threw up a three-corner shelter until Thomas could construct a permanent cabin. Here they endured their first winter in Indiana, huddled in furs around a whipping fire in the open side of the lean-to.

In February, a few days after the boy's eighth birthday, the family moved into a new log cabin, with a packed earth floor, a stone fireplace, and a loft where Abraham slept. Not long after that he stood inside the doorway and shot a wild turkey as it approached. It was a traumatic experience, for he loved birds and animals, hated killing them even for food. He never liked to hunt or fish again.

There was little time for that anyway: come spring his father put an ax in his hands and sent him out to clear fields, split fence rails, chop firewood. Help Thomas clear a farm out of the timber, establish a carpenter's trade, make the name of Lincoln mean something in the neighborhood. It seemed to the boy that he was almost constantly handling his ax, or struggling behind a plow, or thrashing wheat with a flail, or carrying grain down to the mill in Troy.

In 1817 some of Nancy's relatives—Thomas and Elizabeth Sparrow—joined the Lincolns on Little Pigeon Creek, riding a wave of immigration onto the Indiana frontier. With them came Dennis Hanks, illegitimate son of another of Nancy's aunts, a congenial, semiliterate youth of nineteen. For some reason Dennis took a disliking to Thomas Lincoln and later falsely characterized him as a slow and shiftless oaf who neglected his family. But Dennis loved Nancy, describing her as a kind, good, and affectionate woman —"the most affectionate I ever saw." Despite their age difference, he and Abraham chummed around together: on Sundays they ran, jumped, and wrestled and joined Sarah and the adults in singing "how tasteless the hour when Jesus no longer I see." Abraham tried his best to keep a tune, Dennis said, but "he never could Sing Much."

The following summer an epidemic of the dreaded "milk sick" swept through the area. Many settlers died, including Thomas and Elizabeth Sparrow, and then Nancy too fell sick and died. She was only thirty-four years old. While Thomas fashioned a black cherry coffin, the dead woman lay in the same room where the family ate and slept. Then came the funeral on a windy hill, with Thomas, Sarah, Abraham, and Dennis Hanks huddled around the grave. In subsequent years Abraham said little about his mother's death, as reticent about that as he was about her life and family background. But he once referred to her as a wrinkled woman, with "withered features" and "a want of teeth."

Dennis Hanks now moved into the Lincoln cabin and shared the loft with Abraham. Twelve-year-old Sarah tried to fill her mother's place, to make and mend clothes for the menfolk, to clean, cook, and wash for them. But it was hard without a woman, and the Lincoln homestead sank into gloom and squalor. Then in 1819 there occurred another scrape with death: a horse kicked Abraham in the head "and apparently killed him for a time," as he put it later. The boy survived, but that same year Thomas Lincoln headed back to Kentucky, returned in a couple of months with a new wife—a widow named Sarah Bush Johnston—and her three children. Though "Sally" was ten years younger than Thomas, they had known one another for almost a decade. She had married Daniel Johnston, jailer of Elizabethtown, and had made her home in the stone jail where she cooked for the prisoners and reared her children. Since her husband's death, she had lived in a modest cabin she had bought herself. Thomas found her there, proposed, paid her debts, and married her in a Methodist ceremony.

Tall, straight, and light-skinned, Sally could not write and probably could not read either. But she was loquacious and proud, she radiated sunshine into the Lincoln home, and she raised Sarah and Abraham as her own. Because of her warm and fair-minded ways, Abraham became very attached to Sally, calling her "a good and kind mother."

Thomas and Sally Lincoln were hospitable folk, eager to help newcomers and to visit with neighbors who came their way. Thanks to Sally, Thomas was his old self again, cracking jokes with visitors and expatiating on the opportunities in Indiana if a man took advantage of them. Plenty of good cheap land—the government was now selling it off at $1.25 an acre. Eventually Thomas bought one hundred acres and raised stock and grain for market. The country was filling up with settlers, most of them Southern in origin, so that by 1820 some forty families were homesteading in the Lincoln neighborhood. The richest and most prominent was James Gentry, a North Carolinian who sired a brood of eight children, amassed a large farm, and built a trading post called Gentryville. Though not so eminent as Gentry, Thomas was a popular yarn-spinner and enjoyed considerable status as a skilled carpenter, whose cupboards and furniture enriched the cabins of his neighbors.

As a tribute to his carpentry, the local Baptists chose him to supervise the construction of a meeting house on a nearby farm. After he had completed the building, Thomas and Sally united with the Pigeon Creek Baptist Church and Thomas himself became a trustee. Through his influence young Abraham got a job as sexton, charged with sweeping the place out and furnishing it with candles. The preacher was known for washing his feet and inveighing against slavery, and Abraham no doubt heard some of his antislavery sermons. Like his father, Lincoln came to oppose human bondage. But he never joined his father's church.

Thomas raised his son to be a farmer and even hired him out to other homesteaders from time to time. But Abraham disliked farm work, prompt-ing some to remark that he was "lazy, awful lazy." But others recalled that he toiled "hard and faithful" and was "mighty conscientious" about getting in a full day for twenty-five cents—which he gave his father, who legally commanded his wages until he came of age. If manual labor did not exactly excite him, Abraham nevertheless learned a great deal about the business of farming. And he became a master axman. "My how he could chop," said a companion. "His ax would flash and bite into a sugar tree or sycamore, down it would come. If you heard him felling trees in a clearing, you would say there were three men at work, the way the trees fell."

By the time he was twelve or thirteen, the boy reflected his father's conversational habit and love for anecdote, entertaining fellow hands with a procession of hilarious stories. His jokes were as raw and pungent as the frontier he lived in, and so were the tunes he sang with his comrades—raunchy ballads about one Barbara Allen or about the silk merchant's daugh-ter that left them howling with delight. Once they acquired a copy of *Quin's Jests,* a collection of lubricious repartee, and the boys giggled and whooped as Lincoln read the book aloud in the forest.

At other times Lincoln was a serious, brooding youth, with a driving passion to understand that set him apart from his friends. He got irritated when adults spoke to him in ways he could not comprehend—"that always disturbed my temper," he recalled. When Thomas chatted and argued with neighbors, Abraham was "a silent and attentive observer," Sally Lincoln related, "never speaking or asking questions till they were gone and then he must understand everything—even to the smallest thing—minutely and ex-actly—he would then repeat it over to himself again and again—sometimes in one form or another and when it was fixed in his mind to suit him he became easy and he never lost that fact or his understanding of it."

Between his eleventh and fifteenth years he went to school irregularly, attending brief sessions between winter harvest and spring plowing. All told, he accumulated about a year of formal education. These were "blab" schools he went to, so called because pupils studied aloud so that the teacher, rod in hand, could grade their progress. Lincoln came to school with an old arithmetic under one arm, dressed in a raccoon cap and buckskin clothes, his pants so short they exposed six inches of his calves. In later years he scoffed at the instruction he received in Indiana, insisting that "there was absolutely nothing to excite ambition for education." "Still somehow, I could read, write, and cipher to the rule of three; but that was all."

Well, that was not quite all. He developed an interest in poetry and even wrote some himself, recording some playful lines in a homemade copybook:

Abraham Lincoln
his hand and pen
he will be good but
god knows when

Abraham Lincoln is my nam[e]
And with my pen I wrote the same
I wrote in both hast and speed
and left it here for fools to read

He took pride in his penmanship, loved to sketch letters in the dust and snow, loved to construct words and sentences in his copybook. He became the family scribe and wrote letters not only for his parents but for neighbors as well. It was an invaluable experience, obliging the boy to see the world through the eyes of other people.

And he enjoyed reading, too, losing himself in the adventures of *Robinson Crusoe* or the selected fables of *Dilworth's Spelling-Book.* Books were rare in frontier Indiana, but Lincoln consumed the few that he found, reading the same volume over and over. He would bring his book to the field and would read at the end of each plow furrow while the horse was getting its breath; and he would read again at the noon break. Perhaps his favorite volume was Parson Weems's *Life of Washington,* which sketched the American story with a romantic brush, mythologizing the Founding Fathers as immortal statesmen who endowed America with "the Genius of Liberty." Young Lincoln was entranced with the fabled beginnings and star-flung destiny of the young Republic. And Weems's description of the Revolutionary War—especially the battle of Trenton—so thrilled the boy that he could almost hear the rattle of musketry and smell the acrid scent of gunpowder in the Indiana wind. As he labored in the fields, he dreamed of Washington and Jefferson, came to idolize them as heroic men who had shaped the course of history.

The more he read, of course, the more it fired his own ambitions. Surely there was more to life than plowing fields and chopping wood, more than living and dying here in Indiana which he thought as "unpoetical as any spot on earth." Yet this same unpoetical spot was having an ineradicable influence on Lincoln, marking and molding him in ways he could neither erase nor forget. For he came to manhood in a rural backwoods where people accepted the most excruciating hardships as commonplace. Where they went for seasons without baths, saw whole families wiped out in epidemics, endured a lifetime of backbreaking toil for the sake of raising children and getting ahead. Where people relied on corn whiskey, fire-and-brimstone revivals, political discussion, and bawdy jokes to ease the painful grind of day-to-day existence. Where they ordered their lives around a morass of superstitions, believing that one must plant and harvest according to the stages of the moon. That a dog howling in the distance meant that death would visit one's family. That anybody who brought a shovel into a cabin would go out with a coffin. That a bird lighting on a window sill—or flying inside a home— signaled the approach of sorrow. And that dreams were filled with hidden meanings, were auguries of triumphs or calamities.

And then there was the accent—a southern Indiana dialect that would barb much of Lincoln's own speech all his life. Like his neighbors, young Lincoln said "howdey" to visitors. He "sot" down and "stayed a spell." He came "outen" a cabin and "yearned" his wages and "made a heap." He "cum" from "whar" he had been. He was "hornswoggled" into doing something against his better judgment. He "keered" for his friends and "heered" the latest news. He pointed to "yonder" stream and addressed the head of a committee as "Mr. Cheermun." And he got an "eddication" in log-cabin schools and "larned" about adversity, self-reliance, and the necessity for mutual cooperation with his neighbors.

He also learned to live with death, with madness, with the bizarre and the macabre. When he was sixteen he witnessed a hideous spectacle that so affected him he later set it to verse. This was the incredible seizure of Matthew Gentry, one of James Gentry's boys, nineteen years old then and "rather bright." Young Lincoln saw it happen. He was there when Matthew became unaccountably, "furiously mad." His eyes bulged, he tried to maim himself, he attacked his father and then his mother, all the while gurgling maniacally. For days afterward Lincoln brooded over Matthew's fit. Why, for what reason, had this "fortune-favored" youth been so stricken? Lincoln lay awake at night, recalling Matthew's face, trying to understand the "pangs" that killed his mind. And though Matthew seemed less distracted as time went by, young Lincoln retained a morbid fascination with his condition. Lincoln would stand in the forest near the Gentry place, listening as Matthew "begged, and swore, and wept and prayed." And sometimes Lincoln would wake early in the morning and steal off to Gentry's house to "drink in" Matthew's plaintive song, when even the "trees with the spell seemed sorrowing angels. . . ."

At sixteen he was over six feet tall, a gangly youth with high cheekbones, dark eyes, and unruly hair. His legs gave him his height, so long that he seemed to stand on stilts. But if he looked awkward, all arms and legs, he was in actuality a superb athlete, one of the best wrestlers and fastest runners in his age group. His arm muscles were like cables, so strong that he could seize an ax by the end of the handle and hold it straight out at arm's length.

Emotionally he was given to intense oscillations of mood, was "witty, sad, and reflective by turns," as one settler recalled. Though laconic by nature, he hungered for male companionship and liked to visit Gentry's store, to quip and swap stories with the men. He knew he was "gawky" looking and so "didn't take much truck with girls," as a companion phrased it. When he was around girls, he covered up a painful shyness by acting the clown. Because of his appearance, "all the young girls of my age made fun of Abe," one woman remembered. But she was certain "Abe" didn't mind because he was such a good fellow.

Though Lincoln could be good-humored like his father, there was a permanent estrangement between them now. Dennis Hanks blamed it on Thomas, claiming that he thought Abraham was ruining himself with education and that he beat the youth for reading books. According to others, though, Sally Lincoln insisted that Abraham pursue his studies and Thomas Lincoln did not interfere. Still others maintained that it was Thomas who urged Lincoln to attend school and get his learning. "Old Tom couldn't read himself," one settler recorded, "but he wuz proud that Abe could, and many a time he'd brag about how smart Abe wuz to the folks around about." Whichever claim is true, Dennis was undoubtedly right when he declared that father and son never understood each other. Probably Thomas felt both respect and resentment for a son who read books and wrote poetry, moving toward a world of the mind Thomas could neither share nor comprehend. And young Lincoln, for his part, had considerable hostility—all mixed up with love, rivalry, and ambition—for his father's intellectual limitations. In later years Lincoln remarked that his father "never did more in the way of writing than to bunglingly sign his own name."

At the age of seventeen Lincoln left home for a time and worked at a ferry down at the confluence of the Anderson and Ohio rivers. Two years later his sister died: round, gray-eyed Sarah, who had half-mothered him all these years. He had seen her married off to Aaron Grigsby, whom Lincoln had never liked, and now on a cold January day in 1828 she lay dead from trying to bear Grigsby's child. And so another funeral, another round of prayers, another coffin: twenty-one-year-old Sarah, her stillborn baby in her arms, buried under a sandstone slab in a new cemetery by the church Thomas Lincoln had built. *For nothing can hinder the designs of Providence. Whatever will be will be.*

In April, 1828, Lincoln contracted with James Gentry to take a boatload of farm produce down to New Orleans. It was a welcome opportunity to get away from Indiana, away from the memory and sorrow of Sarah's death. Sometime that month he and Allen Gentry—one of James's sons—shoved off from Rockport and guided their flatboat down the placid waters of the Ohio. At last they came to the Mississippi and headed southward in its tempestuous currents, pulling on their extended oars to avoid snags and sandbars, passing occasional flatboats and smoking steamers. For most of the 1,200-mile journey the banks were monotonous variations of bluffs and forests, but below Natchez the trees were festooned with Spanish moss, which gave the woods an aura of ineffable sadness.

Presently they passed Baton Rouge and anchored at a river plantation, where they commenced trading part of their cargo for sugar, cotton, and tobacco. One night seven slaves from a nearby plantation attacked the flatboat, but the youths fought them furiously, hand to hand. Though both were hurt some in the melee, they drove the blacks away, weighed anchor, and set out for New Orleans as fast as they could go.

When they reached their destination, they were incredulous at what they saw: the wharves were teeming with activity, with over a thousand flatboats tied up there and whites and slaves alike stacking produce in carts and wagons, produce that came from Missouri, Illinois, Indiana, and Ohio and that would be loaded on square-rigged sailing ships bound for ports around the world. After selling off their own cargo, the boys made their way along the levees, piled high with cotton and sugar, and roamed the narrow streets of New Orleans, taking in its mixture of Southern and Old World cultures. Now they were in the French Quarter, marveling at the picturesque homes with their painted windows, iron grillwork, and second-story porches. Now they were passing the slave markets, where unhappy blacks—the men on one side and the women on the other—were being bought and sold like cattle. And then there were the sultry nights, when dives along the waterfront roared with drunken rivermen and women of the evening advertised themselves in doorways and windows, offering to help the youths lose their virtue for a drink and a fee.

Finally it was time to go, time to leave this complex and unforgettable city, and return to the prosaic farms of Indiana. But the trip home was not so mundane as the youths might have expected: they made a good part of it by steamboat, chugging upriver like a couple of young entrepreneurs. At last, three months after they had set out, they landed at Rockport and made their way back to Little Pigeon Creek, where young Lincoln dutifully gave his father the twenty-five dollars he'd earned for his labors.

Sometime that summer or winter Lincoln began hanging around the log courthouses in Rockport and Boonville. A sort of legal buff, he watched transfixed as young country lawyers wooed juries, cross-examined witnesses, delivered impassioned summations. He listened, too, as old-timers sat on the steps of the courthouses, spitting tobacco juice and discussing the latest trials and the capricious workings of the law—the verdict a jury might reach, the sentence a judge might hand down. It was all very exciting to him, a challenging new world that made his adrenaline flow. He was so thrilled, in fact, that he borrowed and read the *Revised Statutes of Indiana.* He also studied the Declaration of Independence and the Constitution, observing as he did how frequently Indiana lawyers referred to them.

In March, 1830, Lincoln's father decided to leave Indiana and migrate west into Illinois. John Hanks—a cousin on Nancy's side—had settled there and was writing so ecstatically about the prospects that Thomas's head began to swim. He was doing all right in Indiana—even had a new house under construction—but Illinois seemed too promising to resist. And both Sally and Dennis Hanks wanted to go. So Thomas sold his farm and uprooted his family again. Though Abraham was twenty-one now and entitled to keep his wages, he went along with his father. Maybe Illinois would not be so unpoetical as Indiana. Maybe he could find something there—a job, an enterprise —that would measure up to his ambition. Abraham drove one of the wagons himself, lost in reflection while the axles squealed like violins.

The Lincolns relocated in central Illinois, on a wind-swept prairie some ten miles west of Decatur. Here Abraham helped his father erect another log cabin, clear ten acres, split fence rails, and plant corn. That autumn everybody on the Lincoln claim fell sick with the ague, a malarial fever attended by flaming temperatures and violent shakes. Then in December a blizzard came raging across the prairie, piling snow high against the Lincoln cabin. Then it rained, a freezing downpour that covered the snow with a layer of ice. Now a wind came screaming out of the northwest, driving snow and ice over the land in blinding swirls. Cows, horses, and deer sank through the crust and froze there or were eaten by wolves. For nine weeks the temperature held at about twelve below zero. Settlers called it the winter of the "deep snow," the worst they had ever known.

During the slow winter days, young Lincoln mulled over with John Hanks and John Johnston—Sally's boy—a scheme that might make some money. They would hire out to a speculator named Denton Offutt, who would pay them to take another boatload of cargo down to New Orleans, where the sweltering weather would be a welcome relief from the winters of Illinois. They would join Offutt in nearby Springfield, Lincoln decided, once "the snow should go off."

When the snow melted that March, rivers overflowed and floods washed across the prairie. Though travel was perilous, Lincoln could wait no longer. He was ready to set out on his own, to head south while Thomas and Sally Lincoln moved to another homestead in Coles County. With Hanks and Johnston, young Lincoln paddled a canoe down the Sangamon River and then slogged to Springfield through mud and rivulets of water. They found Offutt at Buckhorn Tavern, swigging whiskey with an arm around an acquaintance. He was a "wild, harumscarum kind of man," settlers recalled, "who always had his eyes open to the main chance." Somehow, he told Lincoln, he had failed to secure a flatboat for the New Orleans venture. But couldn't the boys chop logs on government land and make one of their own? They left Offutt to his drinking and started fashioning a boat down on the Sangamon River. When he felt like it, Lincoln did the cooking. When he didn't, he took his meals at Caleb Carmen's cabin nearby. At first Carmen thought Lincoln "a green horn, tho after half hour's conversation with him I found him no Green Horn."

At last they completed a crude boat and charted their course to New Orleans: they would head northwest up the Sangamon and then ride the Illinois River south to the Mississippi. In late April or early May, 1831, they shoved off with a cargo of corn, live hogs, and barreled pork. Offutt was with them, forecasting a time when they would all be on easy street, eating three squares a day and swilling expensive whiskey. By now the floods had receded and the Sangamon was so low that Lincoln had to steer the flatboat carefully along the channel. Presently, some twenty miles northwest of Springfield, they came to a small village called New Salem, where a mill dam obstructed the river. They decided to run the dam and pushed the boat on faster and faster; it hit the dam and ground across, only to stick about halfway over, filling slowly with water that poured in over the down-tilting stern. As a crowd gathered on the bank to watch, the tallest flatboatman stood on the dam and tried to pry the water-laden boat over. A villager observed that he was dressed "very rough," wearing blue-jean breeches, a buckeye-chip hat, and a cotton shirt with alternate blue and white stripes.

When the raft would not budge, the tall flatboatman got an idea. He helped the crew move part of the cargo to the bank and shove the rest forward to balance the raft. Then he fetched a hand drill from the village, bored a hole in the bow to let water out, plugged the hole, and eased the boat across the dam.

The spectators were amazed at his ingenuity. And so was Offutt, who boasted that he would construct a boat with rollers on it to ride over shoals and dams. "By thunder," he declared, with Lincoln in charge "she would have to go." Actually Offutt had a better plan than that: he was certain that the Sangamon River could be navigated by steamboats and that New Salem would become a boom town. Think of the profits that could be made here in trade and speculation! Before he left, Offutt made arrangements to rent the mill and open a general store. He then hired Lincoln to work as his clerk, saying he had taken a liking to the ingenious young man and wanted to "turn him to account." Lincoln accepted the job with alacrity and high hopes: now he would be free of manual labor, would have a respectable position that required some intelligence. At the same time, Lincoln may have had his sights on something higher still. One villager, who noted the six inches between the bottom of Lincoln's pants and the top of his socks, asked what he would do if he had money. According to the villager, Lincoln said he would like to study law.

The crew climbed aboard the flatboat and moved up the river, away from the dam and the crowd and the little village on the bluff. The tall flatboatman returned some three months later and set about ingratiating himself with his new neighbors. Nothing much had happened on his second New Orleans journey; afterward he'd made a flying trip to Coles County to inform his father of his plans. And so A. was on his own at last, a "friendless, uneducated, penniless boy," as he recalled himself, who at the age of twenty-two had finally "separated from his father."

New Salem was a roaring pioneer settlement when Lincoln came there, with two saloons, a tavern, and a couple of general stores that sold everything from gunpowder to whiskey. Drunken riffraff often hung around the saloons, and speculators like Denton Offutt came and went in search of easy fortunes. At its zenith, New Salem had a population of about a hundred people, most of them Southern in background, and boasted of a justice of the peace and two physicians. In the surrounding area, farmers worked the land with crude plows, raised gaggles of children, and suffered through a myriad of afflictions from rotten teeth to chronic attacks of the ague. Many of them were reluctant to call on doctors when they were ill, for doctors prescribed savage treatments in those days: they "purged, bled, blistered, puked, and salivated" their patients and crammed pills down them as big as cherries. On Sundays people drank corn liquor and attended riotous camp meetings as settlers did back in Indiana. Peter Cartwright, the Methodist circuit rider, lived nearby and often preached at village revivals, exhorting New Salemites until they shook and howled. On holidays people whooped it up in besotted hoedowns, in boisterous foot races and wrestling matches.

Like other frontier communities, New Salem had its better sort—a handful of "respectable" folk who deplored all the violence and drunkenness that plagued the neighborhood, especially on Sundays. There was rhetorical James Rutledge, proprietor of Rutledge's Tavern on the Springfield road and father of ten children, including a daughter named Ann. He was an educated man who owned a library of twenty-five or thirty books and founded the local debating society. There was Mentor Graham, the schoolmaster, a self-educated Kentuckian and a natty dresser with a collection of fancy vests. And there was Dr. John Allen, a native of Vermont with a degree from Dartmouth Medical School, a consumptive, crippled little physician who had come west for his health. A member of the debating society and a leading "intellectual," Allen tried to reform his unruly neighbors by organizing a temperance society, but most of them went right on drinking nevertheless.

At first Lincoln circulated with New Salem's meaner sort, who liked his humor and gave him ready acceptance. He swapped jokes with the hardbitten farmers who called at Offutt's store, competed in foot races, and showed off his feats with an ax. He had his share of fights, too, lest he be branded a coward—the kiss of death on the frontier. Inevitably he found himself in a dramatic showdown with burly Jack Armstrong, the village wrestling champion and leader of the Clary Grove boys, who hung around Bill Clary's grocery and saloon over near the river. As it happened, Denton Offutt boasted one day that Lincoln, at six feet four and 185 pounds, was the best wrestler in New Salem. Bill Clary heard that and wagered ten dollars that Lincoln couldn't throw Armstrong, whereupon the two promoters arranged a match Lincoln was obliged to accept. On the prescribed day, with a large and happy crowd gathered on the bluff to watch, the two fighters stripped to their waists and approached one another on the riverbank below. They grabbed, grunted, wrenched, and struggled while the spectators cheered them on. When it seemed that Lincoln might throw the champion, the entire Clary Grove clan jumped him and drove the tall clerk back against the bluff. But Lincoln, blazing with defiance, offered to take them all on— one at a time. Impressed with his pluck, Armstrong shook Lincoln's hand, turned to the spectators, and pronounced the fight a draw.

After that Lincoln became fast friends with the Clary Grove boys, brawling and clowning around with them during his leisure hours. They came to admire Lincoln tremendously, not only for his yarns and physical strength, but for his honesty as well. Lincoln was incorruptible; he told the truth come what may, so that with him you always knew where you stood. In time the Clary Grove boys regarded him as a leader. And Lincoln basked in their friendship, in the way they looked up to him.

Among the better sort, Lincoln now had a reputation as a ruffian, though they conceded that he was a ruffian with a difference, since he was honest, literate, and sober. Yes, to everyone's astonishment, the newcomer was a confirmed teetotaler. He used to drink some back in Indiana, but whiskey, he said, left him "flabby and undone," blurring his mind and threatening his self-control. And he hated anything that threatened his self-control, feared and avoided it. He didn't care if others drank—that was their business. And he felt sorry for alcoholics, too, regarding them not as criminals—the way temperance advocates generally did—but as unfortunate people, many of them intelligent and kind, who deserved sympathy instead of vilification. As he said later, he looked forward to the day when all men could free themselves from the "tyrant of spirits," when all appetites would be controlled, all passions subdued, and *"mind,* all conquering *mind,"* would rule the world. Thus, if he chummed around with the rowdy Clary Grove clan, whose friendship meant a lot to him, he always passed the liquor by.

He avoided eligible young women, too, because he was insecure in their presence and was afraid of failure and rejection in love. But while boarding for a time at Rutledge's Tavern, he did befriend Ann Rutledge. She was nineteen then, pretty and unaffected, with auburn hair and blue eyes. In later years, some of Lincoln's associates spawned legends about him and Ann, contending that they were romantically involved and that Ann was the only woman he ever loved. In actual fact, Ann was engaged to another man. And though the man left New Salem and eventually broke off the engagement, there is no evidence that Ann and Lincoln ever had anything more than a platonic relationship. And when in time the Rutledges moved to a farm over on Sand Creek, seven miles from New Salem, Lincoln often visited them because he cherished Ann's friendship and respected her father.

Once he found acceptance in New Salem, Lincoln hungered to better himself, hungered for the education and status which Rutledge, Dr. Allen, and Mentor Graham enjoyed. During his first winter there, Lincoln told a fellow clerk he wanted to "get hold of something that was knotty." So he went off to a meeting of the local debating society, where he surprised Dr. Allen by asking for a chance to speak. What? This ruffian a debater? But Lincoln astonished the entire club by delivering a reasonable and forcible argument, speaking out in a high-pitched voice which became his trademark. He was extremely nervous, though, and didn't know what to do with his hands, finally burying them in his pockets. Still, Allen was impressed; he told his wife that there was more in Lincoln's head than jokes. He was already a good speaker, the physician said. All he needed was some culture.

Painfully aware of his want of education, Lincoln embarked on a rigid curriculum of self-improvement that earned him the respect and friendship of Graham and Dr. Allen—and that awed his unlettered friends. He studied the *Columbian Orator* and "practiced polemics" throughout the whole winter of that year. He taught himself mathematics, a subject that held a lasting fascination for him, and practiced writing out propositions in three different forms. He discussed poetry with the village blacksmith, an eccentric fellow who knew Burns and Shakespeare by heart. He borrowed *Kirkham's Grammar* from Graham and used it to polish up his English, laboring over his sentences until they sparkled and flowed.

It was politics, though, that intrigued him the most. One of the first things he'd done in New Salem was to vote in an election. Since then he'd become increasingly absorbed with politics, with the tremendous possibilities for recognition and self-advancement that politics afforded in rural Illinois. In those days a general store was the focal point of political discussion, and Lincoln spent a lot of time in Offutt's place arguing with his customers around a blazing stove. In 1832 national politics was in a state of flux and party lines were vague. The old Federalist party of Washington and Hamilton had died out years before, leaving the Jeffersonian Republicans the only national political organization. In the mid-1820s, though, the Republicans had divided into rival factions, one supporting John Quincy Adams of Massachusetts and Henry Clay of Kentucky, another backing Andrew Jackson of

Tennessee and John C. Calhoun of South Carolina. Jackson won the Presidency in 1828 and he and his followers eventually called themselves Democrats. Nevertheless, party lines remained blurred for several years, so that in 1832 voters identified themselves by the leaders they supported—they were either Jackson, Calhoun, or Clay men. The most popular national figure was President Jackson himself, a slaveholder and a man of wealth who drew the bulk of his support from Pennsylvania and the South. Though celebrated as a symbol of the Common Man, Jackson was an imperious Chief Executive who often defied Congress and waged a destructive personal vendetta against the United States Bank, which he viewed as a tool of the moneyed interests and a symbol of British influence. In terms of popularity, his chief rival was Henry Clay of Kentucky, who was building a reputation as the Great Compromiser between the slave and free states. What was more, Clay was an outspoken nationalist who advocated a protective tariff, internal improvements, and a strong national bank to weld the sections together and make America self-sufficient. As a founder and a member of the American Colonization Society, Clay also wanted to solve America's slavery problem by gradually freeing the blacks and transporting them all back to Africa.

Like his father, young Lincoln was a Clay man and "all but worshipped his name." He liked Clay's ringing nationalism. He liked Clay's economic program, whose goal of sectional interdependence and national unity appealed to Lincoln's love of logic, of symmetry and stability. And he liked Clay's stand on slavery and colonization.

In New Salem, though, local issues were more significant than national ones, and the hottest talk around the village focused on steamboats and those newfangled railroads. Which was the more profitable means of transportation? which the way of the future? Because New Salem was a river town, most villagers favored river and steamboat development. And when a small steamer actually chugged up the treacherous Sangamon, blowing its whistle as it went, rumors flew about that the river would become a busy water route. If so, it meant cheaper goods for New Salem. It meant a boom, a rise in population, a line to the outside world.

Swept up in all the talk about river improvements, Lincoln made a meticulous study of the Sangamon, noting the currents and the stages of the water. On certain days villagers who passed the mill could see Lincoln out on the river in a flatboat, gauging the depth of the water and making notations about the river's bends and snags. By now the industrious clerk had become enormously popular in New Salem, so much so that his friends thought he really ought to go into politics. Given his intelligence and his keen interest in river improvements, he might do very well indeed.

Lincoln scarcely needed any urging. In fact, he decided to run for political

office that very year of 1832. Anybody else might have sought a local position, but not Lincoln. At the age of twenty-three, he announced himself a
candidate for nothing less than the state legislature. With painstaking care,
he wrote out, revised, and polished his first political platform and carried it
down to Springfield, where it appeared in the *Sangamo Journal* of March 15,
1832.

If elected, the candidate would give his undivided support to internal
improvements. While railroad construction was a grand idea, an imaginative
idea, he thought it would cost too much. In his view, improving the Sangamon was better suited "to our infant resources," yet that could only be
achieved through "art." He had studied the river, he said, and believed that
the driftwood would have to be cleared away and some bends cut through
to expedite steamboat travel. In addition to river development, the candidate
wanted a law to curtail exorbitant interest rates, but indicated that he would
be practical: in cases where high rates seemed desirable, ways could be found,
he declared, to "cheat" the very law he favored. Beyond that, he made a
general endorsement of education, contending that all men should have
enough schooling to read the history of America and other nations as well,
so that they might appreciate the value of free institutions.

He concluded that every man is said to have his own ambition. "I have
no other so great as that of being truly esteemed of my fellow men, by
rendering myself worthy of their esteem. How far I shall succeed in gratify-
ing this ambition, is yet to be developed." He conceded that he was young
and unknown to many outside New Salem. And he had no wealthy or
popular relatives to recommend him, so that he rested his case entirely with
the independent voters of the county. If they elected him, he would regard
it as a favor. If not, "I have been too familiar with disappointments to be very
much chagrined."

In April a dispatch rider burst into
New Salem with electrifying news. The Black Hawk Indian war had flamed
up in northern Illinois, and the governor wanted volunteers at once. A
company of New Salem men mustered in the streets and rode off to war in
noisy jubilation. Lincoln was there, along with the Clary Grove clan, all
looking for excitement and extra money. Offutt was about to close his store
and chase after another speculative rainbow, so Lincoln took off on a bor-

rowed horse, planning to return in time for the August elections.

To his surprise, the company elected him as captain—a triumph of popularity that gave him intense satisfaction. In May, Lincoln's outfit joined a regular army company under Captain Zachary Taylor and headed up Rock River into northern Illinois, slogging through spring cloudbursts that turned the riverbanks to mud. When all they did was march and camp, march and camp, Lincoln's troops began to grumble. Where were all the Indians, the battles, the glory? With no Indians to shoot, they "made war" on nearby farms, confiscating pigs and chickens for their suppertime fires. Once several of them stole some whiskey from the officers' quarters and got so drunk they could not march the next morning. The officers held Captain Lincoln responsible and punished him by making him carry a wooden sword for two days.

Once their enlistments were over, most of the company went home in disgust. Lincoln, though, signed up again and spent several sweltering weeks thrashing about in swamps and sinkholes, only to come across five whites killed in a recent skirmish. "The red light of the morning sun was streaming upon them as they lay heads towards us on the ground," Lincoln recalled, "and every man had a round, red spot on top of his head" where the Indians had scalped him. It was grotesque, Lincoln said, "and the red sunlight seemed to paint everything all over."

That was the closest Lincoln ever got to combat. While his outfit meandered about in the wilderness, other troops drove Black Hawk into Wisconsin and slaughtered most of his people as they tried to escape west across the Mississippi, thus ending the so-called Black Hawk War.

In later years Lincoln joked about his follies in what turned out to be his first and only military experience. Ridiculing the tendency in his day to make a military hero out of every political candidate, Lincoln recalled his deeds of derring-do in the Black Hawk War, telling how he'd survived a good many bloody struggles with mosquitoes and led a number of dashing assaults on wild onion patches. Still, Lincoln's friends thought he was actually rather proud of his service—after all, it had given him his first experience in leading and handling men, invaluable for an aspiring politician. Also, the government eventually gave him $125 and a land claim in Iowa—a pretty fair recompense, as Lincoln would say, for battling mosquitoes and onion patches.

By late July Lincoln was back in Sanga-
mon County and campaigning hard for the legislature. He spun yarns in
country stores, pitched horseshoes with voters, and declaimed his sentiments
from boxes and tree stumps. Never had he been so exhilarated, so sure of
himself. Yet he lost the election, running eighth in a field of thirteen candidates. In the New Salem precinct, though, he polled 227 out of 300 votes
cast.

By now the flames of political ambition burned in him like a furnace. But
what to do until the next election? Offutt had left New Salem and Lincoln
was out of a job. He thought about learning the blacksmith trade. He thought
about studying law—and even bought a book on legal forms so he could
draw up deeds and mortgages for his neighbors. But wouldn't he need a
better education to become a professional attorney? Well, whatever he decided to do, he would stay here in New Salem, where people had been
generous to him. Anyway, he didn't know where else to go.

In August or September, Lincoln's fortunes—like the prairie wind—
changed again. An opportunity arose to buy out J. Rowan Herndon's interest
in a country store he operated in New Salem with William F. Berry, a
preacher's son. Though Berry was a known alcoholic, Lincoln jumped at the
chance to join the business and gave Herndon a promissory note for his
share. Later he and Berry bought out another store in New Salem, merged
their stock, and moved into the new place. Thus Lincoln became a frontier
merchant, a move that was not without political considerations since businessmen were natural community leaders.

Lincoln & Berry was a typical general store, where customers could shop
for a variety of items from bacon to calico. A tavern license taken out in
Berry's signature allowed the partners to sell limited quantities of liquor, so
that bottles of wine, rum, and French and peach brandy graced their shelves.
They also sold whiskey from a barrel at the back, allowing customers who
paid cash a free dipperful.

Instead of paying their debts, Lincoln and Berry succumbed to the go-for-
broke spirit of the age and plunged into other financial and business ventures,
assuming mortgages and giving notes they could scarcely meet. At the same
time, neither man gave the store the attention it demanded to be successful.
Berry could not resist the whiskey barrel. And Lincoln could not resist talking

politics all day with his customers. Lincoln admitted that the partners were not shrewd bargainers and did much too much business on credit. As a consequence, he declared, "they did nothing but get deeper and deeper into debt" until the store finally "winked out."

Plagued with debts, Lincoln looked about desperately for a job. Again luck was with him, for in 1833, at either his own request or that of his friends, President Jackson appointed him postmaster of New Salem at a salary of twenty-five to thirty dollars a year. Lincoln quipped that the position was "too insignificant to make his politics an objection," but still he was pleased. It was his first political appointment and could not help but enhance his prospects. Also, he could now read all the newspapers that came to Sam Hill's general store, where the post office was located.

Still, the position did not pay enough to live on, and Lincoln had to work at odd jobs and even hire out as a farmhand from time to time. Then came another bit of patronage. Thanks to the rapid influx of settlers into the Sangamon country, the county surveyor had more work than he could handle and appointed Lincoln as deputy. Lincoln knew nothing at all about surveying, so he borrowed a couple of books on the subject and put himself through a crash course. Then he bought some surveying instruments and a horse and bridle—all on credit—and set about laying off roads, school sections, and town sites. Surveying also fueled his incipient political career, because he met people all over the county, proving to some who considered him "a sort of loafer" that he could be serious and dependable. On his way out to survey some tract of land, he often put letters in his hat and delivered them to farms along the way.

By 1834 Lincoln's debts had caught up with him, ensnaring him in a web of legal tangles. When he failed to pay his notes, some of his creditors sued him and the sheriff had to levy on his personal possessions, including his horse, bridle, and surveying instruments, which were to be auctioned off. Lincoln was in a black mood, but his friends helped all they could. One put up a horse against the judgment. And a farmer named Jimmy Short bought Lincoln's other possessions at the auction and returned them to him.

But his troubles kept piling up. In January, 1835, Berry died, leaving Lincoln solely responsible for the rest of their debts—which amounted to some $1,100. It was such a huge sum for Lincoln's modest means that he called it the National Debt. Of course he could have reneged on his creditors and left the county as a lot of other bankrupts had done. But he vowed to pay every dollar he owed and spent the next decade and a half doing so. As a consequence, he earned a reputation as a man of unimpeachable honesty.

Still, all was not bleak despair in these years. In 1834 he ran for the legislature again, which cheered him considerably. By now party lines had

solidified, with Jackson as head of the Democrats and Clay as leader of the
Whigs, a coalition of Jackson and Van Buren opponents that included Southern planters as well as Yankee farmers and businessmen. In Illinois, Democratic and Whig organizations had begun to form, with alignments depend-
ing in large part on what men thought about Jackson and his crusade to
demolish the U.S. Bank. In Jacksonville that year, a tough little lawyer named
Stephen A. Douglas gave a roaring defense of Jackson which earned him the
nickname "the Little Giant." After that he set about building an inchoate
Democratic party into a cohesive political machine and eventually became its
"generalissimo."

On the county level, though, party lines still were blurred. Since 1834 was
not a Presidential election year, personal magnetism and tactical maneuvering were apt to determine who would win. This was certainly true in New
Salem, where Lincoln, though a Clay man, enjoyed bipartisan support. In his
campaign, Lincoln gave few speeches and offered no platform as he had in
1832. He simply chatted with folks around the county and drew on his
reputation as a postmaster and surveyor. On one occasion he won the votes
of thirty farm hands by cradling more wheat than any of them.

Lincoln's soft-sell campaign paid off. On election day, August 4, he placed
second out of thirteen candidates and so was one of four men elected to the
Illinois house of representatives from Sangamon County. He was only
twenty-five years old.

Among the other winners was John Todd Stuart, a young Springfield
attorney and rising Whig politician with one term in the legislature already
behind him. During the campaign, Stuart bumped into Lincoln, took a liking
to him, and advised "A" to become a lawyer because that was the surest road
to political success. Lincoln agreed. In truth, he had been considering the law
for some time now, but had balked because of his lack of education. Nevertheless, his interest in the law had grown steadily. It appears that he'd argued
a few cases before New Salem's justice of the peace, though as an amateur
he received no compensation for his work. Also, he'd served on juries down
in Springfield, something that rekindled his fascination with courtrooms and
legal processes that had begun back in Indiana. Lincoln was impressed with
the men of the Springfield bar, most of whom had never attended college
or seen the inside of a law school. They were indisputable proof that in
Illinois one could become a lawyer without formal education.

Now, fresh from his first political triumph and encouraged by young
Stuart, Lincoln decided to take up the law. Politics, of course, remained his
first and deepest love, but the law would provide him with a steady income,
give him professional status, and afford him excellent political contacts as
well. So he borrowed lawbooks from the firm of Stuart & Drummond, one

of the busiest in Springfield, and walked home to New Salem, where he "went at it in good earnest" until the legislature convened in December.

When at last it was time to go, Lincoln borrowed two hundred dollars from a friend to pay his more pressing debts and buy a tailor-made, sixty-dollar suit. After all, he had no intention of showing up in the Illinois capital of Vandalia dressed in the kind of frontier garb he'd been wearing around New Salem. On the day of departure, young Lincoln boarded a four-horse stage just as dawn was breaking over the village. He stored his bag on top and settled in for a bouncing ride down to Springfield and from there south to the capital. As the stage pulled out of the little river settlement, shattering the morning stillness with its bells, Lincoln hoped that his rustic past was truly behind him now . . . hoped that he was on his way to becoming a professional man who would never have to toil in the fields again.

When Lincoln swung down from the stage at Vandalia, he found a muddy prairie village of some eight hundred souls, with smoking cabins clustered about a public square. Across the street from the square was the statehouse, a two-story brick structure that was falling apart. When the legislature was not in session, Vandalia was a lethargic backwater known mainly for the mosquitoes and flies that plagued it in rainy seasons; the flies, people said, were big enough to kill a horse. Once the legislature convened, though, the village sprang to life, as carriages and coaches clattered through the rutted streets, politicians and their wives promenaded on the plank sidewalks, irreconcilables slugged it out in the famous "bull pen," and legislators and lobbyists got drunk together in gloomy taverns and saloons.

Lincoln registered in the same tavern as the other Whigs and shared a room with Stuart, the Whig floor leader. As politicians came and went, Lincoln observed how Stuart handled himself, how graceful and charming he was. A fellow Kentuckian, Stuart was two years older than Lincoln and enjoyed advantages Lincoln had never had. Stuart's father was a Presbyterian minister and a professor of classical languages at Kentucky's Transylvania College, and Stuart had grown up in a world of high ideas and cultivated discussion. He had graduated from college in 1826, obtained his law license, and set out on horseback to make his fortune in Illinois. While Lincoln was flatboating on the Mississippi, Stuart settled in Springfield, built up a thriving law practice, and made a name for himself in Whig political circles. Lincoln stood to learn much from him about politics. And Stuart, for his part, thought enough of Lincoln to groom him as his protégé.

When the session began that December, Lincoln took his place among the other representatives in the first floor of the statehouse; they sat at long tables with sandboxes on the floor which served as spittoons. A stove and a roaring fireplace heated the hall, and candles furnished light on cloudy days and in the evenings. Most of the legislators were professional men, all were young, few were natives of Illinois. As they debated the issues of the day, falling plaster often punctuated their orations.

As a freshman legislator, Lincoln said little and learned much that winter: he saw Whigs and Democrats alike logroll for their pet projects, saw lobbyists at work inside the statehouse and out, saw party spokesmen duel one another in rancorous speeches. Clay man that he was, he voted with a bipartisan coalition which established a state-chartered bank in Springfield and incorporated the Illinois and Michigan Canal. In all, he served on twelve special committees during the session—an above-average record for a freshman legislator. But he did his most influential work in drafting bills and resolutions for other Whigs, who could not write so lucidly or logically as he. In truth, his writing abilities earned him the most accolades in those early days in the Illinois legislature.

When the legislature adjourned in February, 1835, Lincoln pocketed $258 for his labors, rode back to New Salem in subzero weather, and resumed his legal studies with single-minded zeal. Though most students in those days read with an accomplished attorney in his office, Lincoln taught himself entirely on his own. He memorized Blackstone's *Commentaries,* Greenleaf on evidence, Chitty's *Pleadings,* and Story's *Equity,* rehearsing cases aloud, analyzing some legal point from various angles until he understood the essence of the problem. When warm weather came on, he liked to read outside, lying under a shade tree with his long legs stuck up the trunk. People often saw him walking along a road, reading and reciting from one of his books. To those around New Salem who never read anything, Lincoln seemed a bookworm, a bizarre visitation from another world. One villager, who found him on a woodpile absorbed in a huge volume, asked what on earth he was reading. "I'm not reading," Lincoln retorted. "I am studying law." "Law!" the man exclaimed in amazement. "Good God A'mighty."

On Sundays Lincoln would hike out to Mentor Graham's schoolhouse, to borrow a book or talk with the schoolmaster about things that puzzled and perplexed him. For in this period he was trying to sort out his opinions not only about the law, but about politics, too, and religion—especially religion. Frankly, he thought he might be a skeptic. He loathed all the emotionalism and fierce sectarian disputes that characterized organized religion in his day, and so he never joined a church. Still, he believed in God, believed there was a Supreme Being who endowed people with individual destinies. And he had read the Bible and was a religious fatalist like his mother. Yet he had reservations. What, for instance, was he to make of Christ? of sin and salvation? of Heaven and Hell? Well, perhaps he was a deist then. Several members of New Salem's debating club were freethinkers who eschewed churches and revivals, who read Voltaire and Thomas Paine and talked about how reason must triumph over passion. Lincoln often consorted with this group and most likely read some Voltaire and Paine himself. And he looked forward to the "fall of fury" and the end of passion, when at last the "Reign of Reason" would begin. At the same time, he retained a powerful belief in what he termed "the Doctrine of Necessity"—"that is, that the human mind is impelled to action, or held in rest by some power, over which the mind itself has no control."

Yet that troubled him. Why, toward what purpose, did God control his mind? Was it God that caused him to strive—to study, work, and get ahead —even though he was destined by that same God to die? Since death was his final destination, what was the meaning of life? The meaning of *his* life? Why should one strive when all that awaited one was a coffin? Because God willed that one should strive? Yes, but toward what end? And the more he brooded about such things, the more withdrawn he became, so lost in *"intensity* of thought," as he put it later, that he wore ideas "thread-bare" and turned them "to the bitterness of death." Obsessed with death, beset with anxieties and with questions for which there were no final answers, Lincoln would often sink into a deep depression, or "hypochondria" as they called it then, that might plague him for days.

In August, 1835, news came of a tragedy that gave him a severe case of the "hypo." Ann Rutledge was dead—dead of the "brain fever" at the age of twenty-two. Her coffin resting in the Rutledge farmhouse was reminiscent of the coffins of his sister and the mother he would never talk about. His friend Ann was gone, proof of how transient were human dreams of happiness and lasting life. For all are fated to die in the end. All are fated to die.

Even as a law student, Lincoln had business interests, a good Whig stake in society. He owned several lots in nearby Huron, acquired as remuneration for surveying the townsite. In hopes of making money and retiring some of his debts, he bought an additional forty-seven acres near Huron, plus a couple of lots down in Springfield, and promoted a pet canal project that might precipitate an economic boom in his neighborhood, whereupon he could sell his property for a profit.

In December, 1835, Lincoln was back in Vandalia, attending a special session of the legislature. By now the young legislator held fairly clear political convictions predicated on the Whig principle of order and national unity. Though a state politician, Lincoln was national in his outlook: he desired a strong federal government which had an explicit responsibility to all the people, providing a prosperous, stable economy so that everyone would have an opportunity to get ahead. Thus he favored internal improvements financed by the federal government, federal subsidies to help the states build their own canals and turnpikes, and state banks whose task was to ensure financial growth and stability. "The legitimate object of government," he later asserted, "is 'to do for the people what needs to be done, but which they can not, by individual effort, do at all, or do so well, for themselves.' "

Many of his political ideas derived from his study of the past. In between his lawbooks, he read American history again, paying particular attention to the Revolution and the Federalist era and applauding the nationalist programs of Alexander Hamilton. Thanks to the romantic histories he perused, Lincoln worshipped the Founding Fathers as apostles of liberty who'd begun an experiment in popular government on these shores, to show a doubting Europe that people could govern themselves without hereditary monarchs and aristocracies. And the foundation of the American experiment was the Declaration of Independence, which in Lincoln's view contained the highest political truths in human history: that all men are created equal and that all are entitled to life, liberty, and the pursuit of happiness. Which for Lincoln meant that men like him were not chained to the conditions of their births, that they could better their station in life and harvest the fruits of their own talents and industry. Thus he had a deep, personal reverence for the Declaration, deemed it "the sheet anchor" of American republicanism, insisted that all his political sentiments flowed from that sacred document.

In the opening day of the special session, the Democrats held the first political convention in Illinois. Led by pugnacious Stephen A. Douglas, the Democrats chose a uniform slate of candidates and drafted a platform that all Democrats were honor-bound to support or be read out of the party. Since 1836 was a Presidential election year, the Democrats also endorsed Martin Van Buren for President and prepared to battle the Whigs with a well-oiled political organization designed to ensure party regularity.

The Whigs, for their part, emphatically opposed the new convention system, disparaging it as a menace to liberty and republican government. Young Lincoln was in the forefront of the anticonvention fight, often serving as floor leader when Stuart was politicking in the halls or out in the taverns. In their speeches and resolutions, the Whigs vowed to retain the old system whereby anyone could run for office simply by announcing his candidacy.

Beyond the convention fracas, the legislators enacted more bills related to canals and railroads, then adjourned in February, 1836, so they could go home to campaign. Up for re-election, Lincoln blamed the "locos"—his favorite name for Democrats—for all that was wrong with the country. When an anonymous circular accused him of unscrupulous activities, Lincoln promptly distributed his own anonymous handbill. In it he upbraided his opponent as "a *liar* and *scoundrel*" and promised to "give his proboscis a good wringing" if he made himself known. On the campaign trail, Jack Armstrong and the Clary Grove boys sang Lincoln's praises and helped keep order at his political rallies. At New Salem, Petersburg, and Springfield, Lincoln glorified the national Whig program. What did he think about universal manhood suffrage? Well, he wasn't for that, he said. But he would allow all whites to vote who paid taxes and bore arms—and that included women, too. For he didn't agree with the practice in his day of denying political rights to white females. Since the national Whigs were split over Daniel Webster and Hugh White for President, whom did Lincoln support? "If alive on the first Monday in November," he informed the *Sangamo Journal,* "I shall vote for Hugh L. White for President."

In the state elections, the Whigs swept Sangamon County, where Lincoln led a list of seven victorious Whigs elected to the state house of representatives. A Whig also won a senate seat and another was a holdover there, so that the Whigs controlled the entire Sangamon delegation. Because the average height of the Whig delegates was six feet, they became known as the Long Nine, with Lincoln, at six feet four, ranking as the longest of them all.

Elsewhere, though, the Whigs suffered disastrous losses, as a Democrat whipped Stuart in his bid for Congress and Stephen Douglas led a crowd of cheering Democrats to power, winning undisputed control of the legislature.

And in the Presidential contest in November, Van Buren carried Illinois and the nation at large, so that the divided and outgunned Whigs had to endure another four years under Democratic rule. The message seemed clear to Whigs who would listen: either adopt the convention system and ensure party regularity like the Democrats, or remain permanently out of power.

Lincoln, meanwhile, was well on his way to realizing a cherished goal. In March, 1836, he took his first step toward becoming a lawyer when the Sangamon County Court registered him as a man of good moral character. Afterward, between speeches on the political stump, he crammed hard for his bar exams, an oral grilling in which practicing attorneys would interrogate him on technical points of the law and legal history. At last he got up his courage and took the exams, sailed through without mishap, then treated his examiners to dinner according to the custom of the day. On September 9, 1836, he received his law license and went right to work on his first case, a complicated action involving three related suits over disputed oxen and farmland.

There were other portents of change that autumn. One day Mrs. Bennett Abell, whom Lincoln described as "a good friend of mine," made a weird proposition. Did Lincoln recall her sister Mary Owens, who'd visited New Salem three years ago? Yes, he remembered her. She was a couple of months older than he, came from a good Kentucky family (a better and more refined family than his had been), and was pretty and coquettish. As it happened, Mrs. Abell planned to visit Kentucky and offered to bring Mary back if Lincoln would marry her. What did Lincoln think? Well, he said with a shrug, he guessed he would give it a try. Though deep down he was afraid of marriage, he cracked that he considered Mary intelligent and agreeable enough "and saw no good objection to plodding life through hand in hand with her."

If he thought Mrs. Abell was joking, he was wrong. She went off to Kentucky and returned with Mary in tow. Lincoln blanched when he saw her, for he thought she'd grown stout and lost her prettiness. Still, he'd made a promise. So they embarked on a strained, formal courtship filled with sighs and silences. In their notes and letters, there was not a single mention of physical passion or even a kiss; rather they referred to one another as "my friend." Mary of course blamed their troubles on Lincoln, whom she considered remote and even ungallant. Once when they went horseback riding with other young couples, they came to a stream and all the other men helped their women across. But not Lincoln: he rode on alone and left Mary to fend for herself. She was miffed. Frankly, she thought he had terrible manners. And he was moody, too, and seemed never to have anything to say that was light and fun and tender. He never said much at all.

Still, they courted through the autumn and evidently had some vague understanding to marry. Certainly the gossips around New Salem whispered that they were engaged. But when Lincoln left for the legislature that December of 1836, all they decided on explicitly was that they would write.

L incoln was ill when he checked into his room in Vandalia and he remained sick through the opening skirmishes of the session. Alone in his room one December day, with a cold and windy rain spattering against his windows, he wrote Mary Owens a long letter by candlelight. He'd written her before, but she hadn't replied and he was upset about it. In truth, now that they were apart he realized how much he cared for Mary. He felt mortified, he said, every time he called at the post office but found no letter from her. "I dont like verry well to risk you again," he wrote now. "I'll try you once more anyhow." He confessed that he'd been ill and that this, along with "other things I can not account for, have conspired and have gotten my spirits so low, that I feel that I would rather be any place in the world than here. I really can not endure the thought of staying here for ten weeks. Write back as soon as you get this, and if possible say something that will please me, for really I have not [been] pleased since I left you. This letter is so dry and [stupid] that I am ashamed to send it."

Troubled about Mary, feeling ill and dispirited, Lincoln flung himself into Whig party caucuses, trying to revive himself in hard work. In Whig strategy sessions, Lincoln got to know Ninian Edwards of Springfield, a vain, imperious individual who owned a mansion on a hill and was a leading member of Springfield high society. "Constitutionally an aristocrat," as one fellow said, Edwards hated Democrats "as the devil is said to hate holy water." Lincoln also developed close ties with state senator Orville Browning of Quincy, a suave, robust, conservative Whig who was almost alone in his opposition to internal improvements. Browning was married to a tall, striking woman named Elizabeth, whom Lincoln was soon to befriend.

As the session progressed, Whigs and Democrats alike trumpeted the glories of internal improvements, prophesying a golden new era of boom and prosperity should the legislature enact a system of public works as other states had done. Internal improvements enjoyed wide popular support in Illinois, for it was a time of tremendous growth and expansion there, of feverish speculation, get-rich-quick schemes, and economic panaceas. As snowstorms raged in January and February, 1837, a bipartisan coalition—including Lincoln and archrival Stephen A. Douglas—enacted a mammoth internal improvements bill to meet the public demand. The measure appropriated ten million dollars to subsidize a network of railroads, canals, and turnpikes around Illinois that would surely usher in an economic millennium. When the bill passed, people in Vandalia danced in the icy streets, ignited bonfires, and shot off fireballs against the sky. "All was joy! joy!" reported a newspaper correspondent, and Whigs and Democrats competed with one another in wild enthusiasm.

On the issue of state banks, though, the bipartisan coalition began to crack. When Whigs and pro-bank Democrats enacted bills that strengthened the state bank in Springfield and made it the state's fiscal agent, a faction of hard-line Jacksonian Democrats rose up in arms. Led by Stephen A. Douglas, the Little Giant, they assailed the bank as a tool of elitist financiers—mostly Whigs—and pronounced it an unconstitutional monstrosity which ought to be investigated and dismantled.

Incensed at their accusations, Lincoln took the floor and gave a ringing defense of the bank as a fair, constitutional, and indispensable institution. And he flayed away at the anti-bank maneuvering: "Mr. Chairman, this movement is exclusively the work of politicians; a set of men who have interests aside from the interests of the people, and who, to say the most of them, are, taken as a mass, at least one long step removed from honest men. I say this with the greater freedom because, being a politician myself, none can regard it as personal."

Meanwhile Lincoln and his colleagues had raised another controversial issue—the relocation of the state capital. Since Illinois was booming, migrants from New York and New England came streaming into the central counties in search of a better life, mixing with older settlers there—mostly from the South—to produce a proverbial melting pot in the Illinois heartland. Because the population in Illinois was shifting from the south northward, phlegmatic Vandalia, situated in the south, seemed wretchedly inadequate as the seat of government. Many politicians, Lincoln and the Long Nine chief among them, wanted to transfer the capital up to Springfield, which was strategically located in central Illinois and was growing rapidly in population and refinement.

Accordingly, Lincoln and the Sangamon Whigs introduced a relocation bill —and undoubtedly used pressure and a little bargaining to gain votes. But available records do not support the widespread belief that Lincoln and the Long Nine, out to make Springfield capital come what may, logrolled left and right throughout the session, promising to support this or that public works project in exchange for votes in favor of relocation. Lincoln, for his part, supported internal improvements not because of cynical bargains over Springfield, but because he believed in public works as a matter of deep conviction. And in any case, the Long Nine were hardly united enough on internal improvements to logroll as flagrantly as most writers have contended. Moreover, a few Whigs who voted against the public works bill now turned around and supported relocation; obviously they hadn't made any deals with anyone. Orville Browning, who resolutely opposed internal improvements, was the man who guided the relocation bill through the senate.

As it turned out, Springfield was not so unpopular that it required a great deal of politicking to get it selected as the seat of government. In fact, cities from Shawneetown in the south to Chicago in the north all hailed Springfield as the logical place for the capital. Vandalians, of course, hotly opposed relocation and they and their allies fought bitterly against it, but Browning, Lincoln, and their colleagues sold the legislature on the merits of Springfield. On February 28, six days after the senate ratified the internal improvements bill, a joint session designated Springfield as the new capital, with the government to move there officially in 1839.

That evening Springfield supporters held a victory celebration in Ebenezer Capps's tavern and passed out free champagne, cigars, oysters, almonds, and raisins to virtually the entire legislative contingent. Lincoln was on hand, reveling in all the camaraderie even if he didn't drink. By night's end the politicians had consumed eighty-one bottles of champagne, which Orville Browning, who got much of the credit for relocation, paid for out of his own pocket.

The session of 1836 and 1837 was also significant in another respect: back in December, the governor received memorials from several Southern states, whose spokesmen were extremely upset about a small but vociferous abolitionist movement now sweeping the North. The memorials called on states like Illinois to muzzle the abolitionists lest they ignite a sectional powder keg over slavery that could wreck the Union.

In a broader sense, these memorials grew out of a train of ominous events that began back in 1800, when Virginians uncovered the Gabriel slave conspiracy in Richmond. Ever since then, Southerners had become increasingly apprehensive about slave unrest and alleged outside Yankee agitation.

In the 1820s a series of insurrection panics rocked the Deep South, especially the South Carolina tidewater, where blacks outnumbered white people. Then came 1831 and Nat Turner's apocalyptic slave rebellion in southeastern Virginia, an explosion of black rage that shook the entire South to its foundations. Afterward the Southern states enacted tough new slave codes and strengthened their patrol and militia systems to guard against another Nat Turner uprising. At the same time, Southerners searched about desperately for somebody to blame for Turner besides themselves. For them it was no mere coincidence that the revolt came at a time of rising abolitionist militancy in the North. In fact, just six months before the revolt, William Lloyd Garrison began publishing the *Liberator* in Boston, demanding in bold, uncompromising editorials that the slaves be immediately and unconditionally emancipated. Southerners, of course, insisted that Garrison's rhetoric had incited the Turner outbreak, insisted that all abolitionists were bloodthirsty lunatics who wanted to obliterate the South in a carnage of racial violence. While this was hardly true (Garrison and most other abolitionists were Christian pacifists), anxious Southerners believed what they needed to believe. As a consequence, Northern abolitionism and slave rebellions were inextricably linked in the Southern mind.

Determined to protect their slave regime at all costs, Southerners embarked on a rigid program of internal censorship, military preparedness, and resounding proslavery propaganda designed to counter the arguments of the abolitionists. To prevent any national emancipation law (and to rally support back home), Southern leaders in Washington sought to squelch antislavery protest and to control and manipulate the federal government itself. In addition, Southern leaders defended the institution as an indispensable means of race control and warned that whites in the slave states would never liberate all their hundreds of thousands of Negroes, because in Southern minds that would lead to social chaos and racial catastrophe. When the abolitionists formed the American Antislavery Society and set out to convert Northern opinion to abolitionism, Southern whites howled in rage and despair, insisting that slavery, the foundation of their "way of life," could not be subverted without drenching the country in blood.

So it was that several Southern states sent their antiabolitionist memorials to Illinois and other Northern states, beseeching them to suppress abolitionist protest if the nation itself were to survive. At the end of December, 1836, the governor forwarded the memorials to the legislature, where Democrats and Whigs alike were extremely sympathetic. In January, 1837, the legislature adopted resolutions which castigated the abolitionists as a dangerous and reprehensible menace to the Union, affirmed the "sacred" Constitutional right of Southerners to own Negroes, pronounced slavery a state institution which the federal government was powerless to molest, and declared that slavery could never be removed in Washington, D.C.

These resolutions reflected the opinions of the vast majority of Northern whites, who regarded abolitionists as sinister agitators out to mongrelize the white race. Since the North was also a white supremacist society, most whites there were perfectly content to leave slavery alone where it already existed. Many of them may have opposed slavery in the abstract, but most rejected actual emancipation—unless accompanied by wholesale colonization—lest abolition result in a massive exodus of Southern Negroes into the North. To control Northern "free" blacks, most Yankee states outside of Massachusetts had enacted "black laws"—prototypes of the Jim Crow laws of later years— which arbitrarily relegated Negroes to the bottom of society.

Illinois was no exception. If there were "islands of abolitionism" in central and northern Illinois, where antislavery New Englanders had resettled, most Illinois whites were hostile to abolition, for fear that it would bring thousands of blacks flocking into the state from neighboring Missouri and Kentucky. Anti-Negro to the core, Illinois had its set of "black laws," too, which discriminated against Negroes in a host of ways. By law, blacks could not settle in Illinois without a legally stamped certificate of freedom. (Though the measure was seldom enforced, Illinois later voted overwhelmingly to prohibit any more Negroes from entering the state.) Moreover, blacks there could not vote, hold political office, sit on juries, give testimony against whites, or attend schools, even though Negroes had to pay taxes like everyone else. Thanks to white prejudices, Negroes were lucky even to get menial jobs and suffered from a number of other discriminatory practices. Although the state constitution outlawed slavery, Illinois retained a system of "voluntary" servitude whereby hapless blacks could sell themselves as "indentures" for twenty years. Many Negroes were so destitute that they had little choice but to become "voluntary" slaves just to eat and have a place to sleep. Others, unable to read or write, could be tricked into signing an "X" on a contract whose terms they could not understand.

So given the racial attitudes and antiabolitionist sentiments that prevailed in Illinois, most state politicians gladly endorsed the proslavery resolutions ratified by the legislature that January of 1837. In the house the official tally was seventy-seven in favor of the resolutions and six opposed.

Lincoln was among those who voted no, and his vote is significant because it was the first time he'd publicly recorded his stance on slavery. Later he claimed that he'd always hated the peculiar institution—as much, he thought, as any abolitionist. As a boy, he'd heard his father and several antislavery preachers denounce human bondage, and he'd grown up and entered politics thinking it wrong. Yet because of all the racial prejudice that existed in

Illinois, he'd been extremely careful in what he said about slavery, abolitionists, and the position of the free Negro in American white society. In fact, in the legislative session of 1835–36, he'd voted in favor of restricting the suffrage to whites only. Public opinion was almost universally against political rights for black people, and young Lincoln, who had elected to work within the system, was not about to ruin his career by supporting Negro suffrage. Nor was he going to get himself branded an abolitionist, because in Sangamon County that would be certain political suicide.

And anyway Lincoln didn't agree with the tactics of the abolitionists. He thought them much too strident, too uncompromising, too obsessed with damning Southerners as unregenerate sinners. Screaming at misguided people, Lincoln believed, was not the way to correct their wrongs. As he put it later, you won people to your side through "persuasion, kind, unassuming persuasion," making friends with them, appealing to their reason, gently telling them that they were only hurting themselves by their follies. For it was "an old true maxim," Lincoln contended, "that a drop of honey catches more flies than a gallon of gall." But if you assailed, damned, and vilified the misled, they would shut you off and lash back. Thus Lincoln thought the abolitionists were only polarizing the country, were exacerbating racial tensions in the North and driving Southern whites to defensive and potentially dangerous extremes. In doing so, he believed, the abolitionists were setting back the clock of emancipation by fifty years. For Lincoln persuaded himself that if slavery were kept in the South and left alone there, time would somehow solve the problem and slavery would ultimately die out. And he believed that the Founding Fathers had felt the same way, that they too had expected slavery to perish someday. Why else had they outlawed the international slave trade, excluded slavery from the old Northwest territories, if not to restrict the growth of slavery and place it on the road to extinction?

Still, slavery gnawed in him. He realized that the institution was a blight on America's experiment in popular government, realized what a profound moral contradiction it was that slavery should exist at all in a self-proclaimed free and just republic, one founded on the enlightened ideals of the Declaration of Independence. He who regarded the Declaration as "the sheet anchor" of American liberty understood only too well how human bondage mocked that noble document. At the same time, as a student of the law and the Constitution, Lincoln conceded that slavery was protected by a veritable web of constitutional, legal, and political safeguards.

So what to do? One bided one's time. One believed that slavery would eventually disappear, trusting to the future to square America with her own ideals. Meantime, one could reject as a matter of principle such proslavery resolutions as his legislative colleagues adopted. Lincoln could have let it go at that, but a mere vote against those resolutions was not enough. He wanted to make it clear that one could object to slavery and yet not be an abolitionist in the manner of Garrison and his kind. Therefore, after the legislature had enacted the public works program and removed the capital to Springfield, Lincoln and Dan Stone—a Sangamon Whig who shared many of his views —composed an official protest against slavery and had it recorded in the House Journal on March 3, 1837. In the protest, they decried human bondage as a bad and unjust policy—and yet condemned the abolitionists for only compounding the evil. They agreed that Congress had no constitutional authority to interfere with the institution in the Southern states, but insisted that if the voters in the national capital approved, then slavery could be legally extinguished there.

So at the age of twenty-eight, Lincoln made his first official statement about slavery, the most inflammable issue of his generation. In his 1860 autobiography, he asserted that "A" 's protest "briefly defined his position on the slavery question; and so far as it goes, it was then the same that it is now."

By the time the legislature adjourned in March, having enacted some four hundred bills into law, Lincoln had completed the final step toward becoming an attorney. On March 1 the clerk of the Illinois Supreme Court enrolled his name as a lawyer. At last he was a professional man, and he had done it almost entirely on his own.

He had reached another milestone as well. Young Stuart, who'd come down to Vandalia and lobbied for Springfield as capital, invited Lincoln to join his law firm there. Lincoln happily accepted and decided to make Springfield his home. After all, the new capital offered splendid professional opportunities, and all his new friends there wanted him to come. Anyway, he'd outgrown bucolic New Salem; it was too small and provincial for him now. So in April, 1837, he rode home to New Salem for the last time, packed his things in his saddlebags, and set out for Springfield on horseback, his legs so long that his feet almost touched the ground. By now the winter had lifted and so had Lincoln's spirits, even though he was still confused about Mary Owens. Out on the prairies the sunflowers were in bloom.

So it was that Lincoln left New Salem as he'd come: in the aftermath of a terrible winter, a winter of snow and ice storms almost as severe as that of 1830 and 1831. Behind him the village that had given him his start in politics was dying out now. The Sangamon River had proved too treacherous for steamboat navigation, and villagers were moving on to Petersburg, a bustling new trading center several miles to the northwest. By 1840 New Salem was a ghost town, a cluster of windy and uninhabited cabins on the bluffs above the Sangamon.

PART·TWO

WHY SHOULD THE SPIRIT OF MAN BE PROUD?

To be honest, Lincoln found that living in Springfield was "rather a dull business after all, at least it is for me." In many ways Springfield was still a crude frontier town: hogs rooted in the dirt streets, and stables and neglected privies gave off a pungent stench when the wind was up. After rainstorms lashed the countryside, Springfield became a huge mudhole, with rancid ponds, impassable streets, and oozing walkways. In the dry seasons, though, the tall grass out on the prairie became dangerously inflammable, and prairie fires often raged out of control, filling the air with ash and smoke and lighting up the sky at night with a dazzling glow.

Nevertheless, Springfield was the most civilized place Lincoln had ever lived in. It had a population of two thousand people and boasted of its share of elegant homes and debonair couples, who flourished about in carriages and made Lincoln feel lonely and destitute by comparison. There were several private schools, an academy, and an active lecture circuit that brought in such famous people as Daniel Webster and Martin Van Buren. A Young Men's Lyceum and a Thespian Society conducted stimulating debates, and six churches overflowed on Sundays with fastidious gentlemen and fluttering ladies. Iconoclast that he was, Lincoln stayed away from church because he was certain he would not behave himself.

A month after settling in Springfield, in May of 1837, Lincoln wrote Mary Owens, who was back in Kentucky now, waiting for Lincoln to clarify their relationship. "Friend Mary," he said, "I am quite as lonesome here as [I] ever was anywhere in my life." Only one woman had spoken to him—but only because she couldn't avoid it. Still, lonely or not, Lincoln had thought over his promise to wed Mary and decided to let her out if she wanted. After all, he was very poor and if they married she would have to live in unaccustomed poverty while other people their age went about in carriages. He wanted Mary to be happy. He himself would be happier with her than to be alone, but he asked her to think it over before throwing in with him. If she liked, they could still get married. But his frank opinion was that she "better not do it" because of the hardships she would have to suffer. He would await her decision.

That August Mary visited Springfield. Lincoln brooded and said little, and when she left he was more confused than ever. Had she misunderstood his intentions? Did she think him indifferent? insincere? He wrote her a tortured, befuddled letter in which he presented his "case," his "plea." "I want in all cases to do right, and most particularly so, in all cases with women. I want, at this particular time, more than any thing else, to do right with you." For her part, she could dismiss him from her thoughts, forget him. But she mustn't think he wanted to cut off their "acquaintance," because he didn't. He would leave it up to her whether to stop or continue seeing one another. If she felt bound to him by any promise, he now released her from all obligations—if that was what she wanted. "On the other hand, I am willing, and even anxious to bind you faster, if I can be convinced that it will, in any considerable degree, add to your happiness. This, indeed, is the whole question with me. Nothing would make me more miserable than to believe you miserable—nothing more happy, than to know you were so." But "if it suits you best not to answer this," he concluded, then "farewell—a long life and a merry one attend you."

Offended by his tone and his entire behavior, Mary never replied, thus terminating their relationship. Later she remarked that Lincoln "was deficient in those little links which make up the chains of a woman's happiness." And Lincoln? Supposedly he told Mrs. Abell that Mary was "a great fool for not marrying me." And several months after their breakup, in a letter to Mrs. Browning on April Fool's Day, he penned a cruel and merciless satire about Mary—and about himself as well. He described how he'd promised Mrs. Abell to wed Mary and how shocked he'd been when he saw her again. In truth, Mary reminded him of his mother, "not from withered features, for her skin was too full of fat" to resemble his mother, but from Mary's want of teeth and "weather-beaten appearance in general." Obviously this rotund and wretched woman was desperate for a husband, which was why she came back to New Salem with Mrs. Abell. Still, Lincoln had made a promise and gallantly stuck by it. After much procrastination, he got around to proposing, but she stunned him, mortified him "almost beyond endurance," by turning him down. "My vanity was deeply wounded," he wrote, turning his pen on himself now, "that she, whom I had taught myself to believe no body else would have, had actually rejected me with all my fancied greatness; and to cap the whole, I then, for the first time, began to suspect that I was really a little in love with her. But let it all go. I'll try and out live it. Others have been made fools of by the girls; but this can never be with truth said of me. I most emphatically, in this instance, made a fool of myself. I have now come to the conclusion never again to think of marrying; and for this reason; I can never be satisfied with any one who would be block-head enough to have me."

On the day Lincoln arrived in Springfield, the *Sangamo Journal* carried an announcement that Stuart & Drummond had dissolved and that Stuart would now practice law with "A. Lincoln." Their office was situated in Hoffman's Row across from the public square and was outfitted with "a small dirty bed, one buffalo robe, a chair and a bench," and a feeble bookcase containing legal volumes. For the next four years the partners got along amicably enough, as Stuart educated Lincoln in the practical points of courtroom law and Lincoln composed pleas and pleadings in expert fashion. They enjoyed an extensive practice—and once handled sixty cases in a single term of the Illinois circuit court, which met in a store beneath their second-story office. While most of their business involved trivial matters, the partners got their share of dramatic criminal cases. In fact, it was in a murder trial that Lincoln won his first courtroom victory: he helped secure the acquittal of a Democrat who'd shot a rival in self-defense.

After a long day in court, Lincoln would walk home to Joshua Speed's General Store, where he and Speed lived together. When Lincoln first came there looking for a place to stay, Speed was astonished at his sadness. "I never saw so gloomy and melancholy a face in my life," he declared. But Speed liked the strapping attorney and invited him to share a double bed in the upstairs room, only to burst out laughing when Lincoln dumped his saddlebags on the floor and grinned. "Well, Speed, I'm moved."

A bachelor, too, Speed was a brooding, hefty Kentuckian who soon became Lincoln's "most intimate friend." They not only slept together, but confided in one another about their fears, their feelings, and their problems with women. In the evenings, they often met with other young bachelors at the back of the store, where they vied with one another in spinning bawdy tales. At such times, recalled one fellow, Lincoln's "wit and humor boiled over." On other evenings, Lincoln stayed to himself and read poetry—some Burns, Byron, and Shakespeare—or polished up his law. "By nature a literary artist," as one writer has stated, Lincoln also studied how poets and orators expressed themselves, noting the way they turned a phrase and used a figure of speech, admiring great truths greatly told.

By now Lincoln was thoroughly addicted to newspapers and read them more than anything else. In those days newspapers were largely mouthpieces for political parties and thus were intemperate, belligerent, and fanatically partisan. As a consequence, Lincoln had to examine a number of different journals—Northern and Southern, Whig and Democrat—in order to understand rival points of view and get all the news about the events and issues that divided Americans of his time.

One thing he read about troubled him deeply—an escalating rate of violence all over the country, particularly against those hapless scapegoats of Jacksonian America—abolitionists and Negroes. By 1837, various abolitionist societies were at work in the North, trying to convert people through nonviolent moral suasion, and Southerners were responding with a fusillade of murderous threats. North Carolina and Georgia each put a price of five thousand dollars on the head of William Lloyd Garrison; communities in Louisiana and Alabama offered fifty thousand dollars for New York abolitionist Arthur Tappan, dead or alive; and politicians in Mississippi and Virginia vowed to kill any abolitionist who came to their states. In the North, race riots flared up in Boston, New York, Philadelphia, Cleveland, and Cincinnati, where whites of all classes—from factory workers to gentlemen of property—mobbed abolitionists and set fire to black sections of town. In Saint Louis, a mob seized a young black sailor who had killed a lawman in a scuffle, tied the Negro to a tree, and burned him to death. And right here in Springfield, whites threatened to hang a minister suspected of abolitionist tendencies, and they adopted resolutions which branded abolitionists as "dangerous," "designing," and "ambitious" men. Then in November, 1837, occurred the most publicized atrocity of all. In Alton, Illinois, a mob shot and killed Elijah Lovejoy, an abolitionist editor, and dumped his press into the swirling waters of the Mississippi.

Lincoln was distressed about all the riots and killing. Was the country on the verge of anarchy? of mob despotism? On January 27, 1838, he stood before the Young Men's Lyceum in Springfield and delivered a spirited oration about violence in America, "this land so lately famed for love of law and order." He began with a hymn to the American system of government, which guaranteed more civil and religious freedom than any other in history. Yet there was an "ill-omen" among us, Lincoln declared, a spiraling savagery that menaced the very foundations of the Republic. The newspapers crawled with reports of outrages, some committed by "pleasure hunting masters of Southern slaves," others by antiabolitionist Northerners. He cited a lynching in Vicksburg and the incineration of the Negro sailor in Saint Louis, and denounced such acts as "dangerous in example, and revolting to humanity." And though he did not mention Lovejoy by name, the specter of the slain editor loomed over his audience. Yes, Lincoln warned, the spirit of the mob was abroad in the land; and once murderous passions were unleashed, mobs were apt to terrorize the entire country, burning innocent and guilty alike, until all the walls erected to defend the people were obliterated. When that happened, when "the vicious portion of population shall be permitted to gather in bands of hundreds and thousands, and burn churches, ravage and rob provision stores, throw printing presses into rivers, shoot editors, and hang and burn obnoxious persons at pleasure, and with impunity; depend on it, this Government cannot last."

And that would be an incalculable disaster, an irreparable blow to human liberty. For the Founding Fathers had staked their lives—and their destinies —on "an undecided experiment" in popular government, to show a skeptical world that people could govern themselves. The Fathers succeeded, too: they harvested the fields of glory and passed on the fruits of that harvest to Lincoln's generation. And it was up to Americans of his time, through an abiding respect for the law, to defend popular government from the tyranny of the mob.

In the history of nations, he went on, there was a time for passion and a time for restraint and rational debate. America had her season of passion in the Revolution, which swept up and utilized man's basest nature—"the jealousy, envy, and avarice, incident to our nature"—to create out of the blood and fire of war a nation based on equal rights and self-government. And now, in order to preserve and consolidate their experiment, Americans no longer needed passion. What they needed was "Reason, cold, calculating, unimpassioned reason." And in the spirit of reason, let them now reassert their faith in the American system. Let them "swear by the blood of the Revolution" to respect and obey the laws and stop all the violence and killing. Let reverence for the law be sung by every mother to her child, "let it be preached from the pulpit, proclaimed in legislative halls, and enforced in courts of justice. And, in short, let it become the *political religion* of the nation; and let the old and the young, the rich and the poor, the grave and the gay, of all sexes and tongues, and colors and conditions, sacrifice unceasingly upon its altars."

Thanks to oratory like that, Lincoln rose rapidly in Springfield's Whig circles as a gifted and eloquent speaker. By 1839 he was a member of the Whig "Junto" in Springfield, a powerful clique of young lawyers who were influential in the choice of Whig candidates all about the state. Stuart was a member, and so were Stephen T. Logan and Edward Baker, the latter a small, sprightly Englishman who became another of Lincoln's friends. Lincoln, for his part, was proud to be a member of the Young Whigs: he strutted about "in high feather" with them and attended their clubbish meetings in the office of the *Sangamo Journal,* owned and operated by that irrepressible Whig Simeon Francis.

Their archrivals were Stephen A. Douglas and his "claque" of Young Democrats, who gathered at the *Illinois Republican* (later the *Illinois State Register*), a Democratic journal. In his mid-twenties, three years younger than Lincoln, Douglas was a cocky little dynamo, with stubby legs, a massive head, bushy eyebrows, a booming voice, and a hell-for-leather style. Vermont-born, he'd come west and settled on the Illinois prairie, where he taught school near Jacksonville before becoming a lawyer and going into politics as a dedicated Jacksonian who hated state banks. More than anything else the Little Giant loved a good political brawl, the give-and-take of floor battles, and two-fisted oratory. He drank hard, chewed cigars, and argued and fought so tempestuously that he reminded people of a pugilistic bantam rooster. Admitted to the Illinois bar in 1834, Douglas rose swiftly in politics: he served as state's attorney general and a state legislator, and in 1837 became register of the Springfield Land Office. In all the political battles down in Vandalia, Douglas had become a man of the political arena, an improviser and tough little pragmatist who admitted nothing and stayed constantly on the offensive. Next to Lincoln, Douglas at five feet four looked like a dwarf, but his stentorian voice and menacing frowns made up for his size. Though Lincoln and the Young Whigs laughed Douglas off as "a small matter," the Little Giant was a leading Democratic boss who'd shaped his party into a cohesive political machine. Already he was engaged in a driving, headlong pursuit of political power that was to last all his life.

In Springfield, Douglas's Young Democrats and the Young Whigs were locked in a blood feud, competing in everything from fiddle dancing to politics. The two groups smeared one another without mercy, belching out accusations of "unprecedented" dishonesty and "dastardly" bargains. Lincoln, who was known for his literary talents, published in the *Sangamo Journal* a number of hilarious parodies of the Democrats, signing them variously as "Madison," "Citizen of Sangamon," "John Blubberhead," and "Sampson's Ghost." The Democrats responded in kind, until both sides were slinging mud—and much else—in a torrent of mutual abuse. In one memorable contest for county probate judge, the two cliques waged a distempered newspaper war against one another, with Lincoln writing many of the Whig assaults. When the Democratic candidate accused the Whigs of conspiring to destroy him, Lincoln penned a slashing letter to the editor which dismissed the Democrat's claims as "false as hell" and called him "a liar" and not a lawyer. The journalistic vendetta raged on for weeks, illustrating Lincoln's penchant in these years for polemics and parodies at somebody else's expense. Still, he was hardly the only Springfield politician who wrote such things.

In 1838 Lincoln won re-election in Whiggish Sangamon County, and Stuart narrowly defeated Stephen A. Douglas in a rough-and-tumble campaign for Congress marred by an actual fight between the two candidates: Stuart grabbed Douglas in a headlock and dragged him around a Springfield market, and Douglas bit Stuart's thumb so severely that it left a scar. When Stuart went off to Congress, leaving his partner to tend their law firm, Lincoln turned to a new page in their account book and scrawled across the top: "Commencement of Lincoln's Administration."

In 1837 a terrible panic shook the U.S. economy with volcanic fury, toppling banks and businesses alike and causing financial ruin everywhere. Unable to raise money, state governments in the South and Midwest repealed their internal improvement programs and defaulted on their loans. In Illinois, the state bank in Springfield suspended specie payment as other state banks had done, which meant it would no longer redeem bank notes with gold or silver. With money increasingly scarce, the legislature met in three successive sessions from 1837 to 1840 and debated what to do with the public works projects then under way. Lincoln, serving as Whig floor leader in the house, joined the majority in voting not merely to retain the internal improvements system, but to enact additional canal and railroad projects. This was not so imprudent as many have charged, since the programs would provide employment and stimulate the state's stricken economy. By now Lincoln was a strong Whig boss and even ran for speaker of the house, but lost to a Democrat he castigated "as not worth a damn."

By the winter of 1839–40, when the legislature convened in Springfield for the first time, the state debt stood at $10 million, with an annual interest of $600,000 the government could scarcely pay. Meeting in churches until the new statehouse was completed, the legislature roared with motions, resolutions, and amendments regarding the financial crisis. Lincoln, for his part, steadfastly refused to abandon the internal improvement system, which he was certain would benefit workers, farmers, and businessmen alike. He was pledged to public works, he declared, and would have his limbs ripped off before he would violate that pledge. In addition, he helped renew the charter of the Springfield state bank, doing so partly out of principle and partly out of loyalty to Whig chums who were heavily indebted to that institution.

Still, Lincoln had an idea how to raise revenue to finance internal improvements. Have Illinois buy up twenty million acres of federal land within her borders—at a price of twenty-five cents an acre—and then sell it off at the federal rate of $1.25 an acre and turn a handsome profit. It was an intriguing proposal, which the legislature warmly adopted, but nothing ever came of it in Congress.

After weeks of wrangling, the legislature finally voted to continue the public works now under construction, but not to add any new ones. To raise money, the legislature also enacted a general property tax, which Lincoln gladly endorsed because it took from the *"wealthy* few" and not the "many poor." Frankly he didn't care whether the rich disliked the tax or not since they were too small a class to carry elections.

With crucial state and national elections approaching in 1840, Lincoln and his friends tried to bring some order and organization to the state Whig party, lest the Democrats trounce them again. In October, 1839, the Whigs held a convention in Springfield and endorsed William Henry Harrison, military hero of the War of 1812, for President. They also established a Whig State Central Committee, consisting of Lincoln and four close friends. Since Lincoln and Stuart were both up for re-election, Lincoln was determined that his party should crush Douglas's people through Whig solidarity. The central committee even published a Whig campaign circular called "Old Soldier," and Lincoln himself conceived a top-secret plan that organized the Whigs down to the precinct level, instructing county committees all about the state to woo "THE DOUBTFUL VOTERS" and get all Whigs to the polls on election day.

And so all across the Union the campaign of 1840 was on, with both parties largely ignoring the issues and appealing to popular passions. Out to demonstrate that theirs was the true party of the people, the Whigs branded Democrats as a gang of elitists and portrayed Van Buren as a mustached patrician who dined like a king. At the same time, Whigs celebrated Harrison as a man of the common folk who would be content with a log cabin and plenty of hard cider. Thus they called it the log-cabin-and-hard-cider campaign, as Whigs disseminated log cabin badges, sang log cabin songs, formed log cabin clubs, and handed out log cabin jugs filled with hard cider.

The campaign was as noisy in Illinois as everywhere else, complete with parades, bands, glee clubs, barbecues, and flatulent oratory. In Springfield, Lincoln and Douglas squared off in their first public debates, setting a style of political rivalry that would last for twenty years. The debates had begun back in November, 1839, when the two got into a heated discussion in back of Speed's store: *I say the state bank is necessary! I say it's a monstrosity!* Douglas suggested that they argue in public, and so for three days the Little Giant and the tall Whig harangued one another before enchanted crowds. The debates spilled over into the legislature, where Douglas punched away at the Whigs' economic dogmas and held them responsible for all financial woes. Lincoln struck back hard. On the day after Christmas, 1839, he defended the now moribund U.S. Bank and pummeled the Democratic administrations of Jackson and Van Buren for bringing on economic chaos, national bankruptcy, and unbridled corruption. "I know," he cried, "that the great volcano at Washington, aroused and directed by the evil spirit that reigns there, is belching forth the lava of political corruption, in a current broad and deep, which is sweeping with frightful velocity over the whole length and breadth of the land." Such empurpled rhetoric was uncharacteristic of Lincoln, but the Whigs loved it: the state committee even published the speech as a campaign document; and in August, 1840, the *National Intelligencer* in Washington quoted from Lincoln's address, thus giving him his first national recognition.

As the campaign of 1840 roared on, Lincoln and Douglas debated one another across Illinois. In one encounter, Lincoln mimicked another Democratic rival, imitating his accent, walk, gestures, and facial expressions, and sending the crowd into paroxysms of laughter. That sort of thing enraged Douglas's paper, the *Illinois State Register,* which berated Lincoln for his "assumed clownishness" and "game of buffoonery," which won nobody to his side, nobody at all. The next day Lincoln apologized if he'd offended anyone.

When the two sides grew tired of political smears, they resorted to racial smears. Because of Lincoln's dark complexion, the *Register* hailed him as the lion of the Whig tribe who came originally from Liberia. In reply, Lincoln's paper—the *Sangamo Journal*—declared that Douglas was "clothed with sable furs of Guinea—whose breath smells rank with devotion to the cause of Africa's sons—and whose very trail might be followed by scattered bunches of *Nigger wool.*" When Douglas accused Harrison of being soft on abolition-ism, Lincoln retorted that it was Van Buren who was the abolitionist. To prove it, Lincoln produced a Democratic campaign biography and an-nounced that Van Buren had once voted for Negro suffrage in New York. *Lies,* Douglas cried, *all lies.* But at a subsequent debate Lincoln exhibited a letter from Van Buren admitting that the biography was correct. In an explosion of rage, Douglas grabbed the book and flung it into the audience.

All that summer, Lincoln rode through rainstorms and blazing sunlight to attend Whig rallies. "He stumped all the middle and lower part of the state," recalled a friend, "traveling from the Wabash to the Mississippi in the hot month of August, shaking with the ague one day, and addressing the people the next." But all his efforts were in vain: in the state elections, the Demo-crats once again won control of the Illinois legislature. And in November Van Buren carried Illinois by a slim margin even though Harrison won the Presidency. In Sangamon County, Lincoln was re-elected for a fourth term, but he placed last of five Whigs sent to the house—the poorest he'd run since 1832.

Even so, Lincoln now enjoyed a hard-won reputation as a promising politician and a gifted young lawyer. In court, he held his own against the best legal talent in Illinois. And in January, 1840, he appeared in his first case in the Illinois Supreme Court, the most prestigi-ous tribunal in the state.

As Lincoln's friends observed, he was successful in all phases but one—his romantic life. Since the government had moved to Springfield, bringing in a train of eligible bachelors, marriageable young women had streamed into town, all looking for husbands. Two of the leading socialites were Ninian and Elizabeth Edwards, whose hilltop mansion served as a meeting place for young couples, a frequent site for picnics and cotillions.

Stuart was a cousin of Elizabeth's and often attended social functions at the Edwards home. And Lincoln sometimes came along, although he was ex-tremely ill at ease around coquettish young women: he did not understand them, felt stupid in their presence, was sure they all thought him ugly, and sat with his hands hidden under his knees. Though he was the longest of the Long Nine, he once said, no woman ever found any charm in that asset. Except for his friend Ann Rutledge, Lincoln's closest female relationships thus far had been with married women who tended to mother him.

But in December, 1839, that suddenly changed. Between party maneuvers and political debates, he attended a cotillion party in Springfield. Speed was there. And so were Douglas and other young men about town. In the course of the dancing, Lincoln met a young woman named Mary Ann Todd, who was Elizabeth Edwards's younger sister. At first Lincoln did not know what to make of Mary. She was witty and vivacious. She seemed interested in him. She could talk about poetry and politics. Why, yes, Mr. Lincoln, she was a fiery little Whig herself. She even knew Henry Clay. You know *Henry Clay?* Yes, he was a friend of her father's. And when she smiled, Lincoln was moved in spite of himself.

He wanted to call on her, but couldn't without an official invitation from the aloof and critical Edwardses. Luckily for him, Speed was a favorite of theirs and evidently secured permission for Lincoln to see Mary. One visit led to another, until at last Lincoln was a regular caller at the Edwards mansion. He and Mary would sit on a horsehair sofa in the parlor, Mary flirtatious and talkative about whatever came to mind. Elizabeth Edwards, who watched them in her home, didn't like Lincoln and said so. "He would listen and gaze on her as if drawn by some superior power," she huffed. "He never scarcely said a word," because he "could not hold a lengthy conversation with a lady—was not sufficiently educated and intelligent in the female line to do so."

Yet Mary admired him. She found him refreshing and attractive in his own way. Behind his terrible shyness she saw an intense and supremely ambitious man who would surely become a success. And in spite of his anxieties, Lincoln felt a growing affection for Mary, too. Physically, she was five feet two and fashionably plump, with a turned-up nose that Lincoln liked. She had a prominent chin, which, when she wasn't smiling, made her little mouth seem bitter, as though she were sucking a lemon. Around Lincoln, though, she smiled most of the time. She was twenty-one now, almost ten years younger than he, and came from the distinguished Todd family down in Lexington, Kentucky. The fourth of seven children, Mary had been a sensitive little girl, pampered and loved by her parents, especially her mother. But her mother died when she was six, a traumatic experience that left scars in Mary, who described her subsequent years as desolate ones. Her father married again and had nine more children by his second wife. For her part, Mary was hostile to her stepmother and felt neglected, unloved, in so large and boisterous a family. As she grew up, she developed into a fiery, impulsive girl with a stubborn streak and an insatiable passion for pretty clothes. At times she tried to be "sweetly hypocritical" like other well-raised Southern girls, but at other times she was fiercely independent and said what she thought—often with a biting sarcasm. But when she was frustrated, when things went wrong and she couldn't cope with them, she wept uncontrollably and threw temper tantrums that became legendary among her family and friends.

Her father, Robert Smith Todd, was an eminent Whig banker in Lexington, and Mary idolized him and loved the world of politics and power in which he moved. Todd entertained extensively, and Mary, bright and politically aware, circulated with governors, congressmen, and United States senators. She became an ardent admirer of Senator Henry Clay, describing him as "the handsomest man" in Lexington.

In those days Lexington was called the "Athens of America," celebrated for its culture and refinement. Yet it was also a slave town, with auction blocks and rancid slave jails. From her bedroom Mary could see manacled Negroes—women and children included—being driven down the dusty streets. Yet she accepted slavery as simply the way things were in her world. And anyway her father owned slaves himself: Negro servants tended the Todd family, and "Mammy" Sally had cared for Mary since she was a baby, had fussed over her, given her confidence, and made her fear the "debil." In addition, she'd taught Mary—a secret between them—that it was all right to help fugitive slaves on their way to Canada: clandestine instructions, a bag of food. Yet Mary did not question Sally's own bondage or that of the other Todd "servants." If they were turned out in the streets, what would they do? Where would they go? Mary thought them better off enslaved.

In 1832, the year Lincoln entered politics, Mary enrolled in a fashionable women's academy. Here she studied English literature, acquired a reading knowledge of French, learned the rules of etiquette, mastered all the dances —the waltz, the polka, galop, and schottische—and learned to flutter a fan as well as any girl. When she left the finishing school at the age of seventeen, she was fully trained for marriage, the central goal in life for young ladies of that Victorian era.

To find a husband, Mary moved up to bustling Springfield, to live with Ninian and sister Elizabeth and flourish about in Springfield's "gay" society, which included three of her cousins—Stuart, Logan, and John J. Hardin. Attracted to all the "glitter, show and pomp and power," as she phrased it, Mary became a conspicuous member of the Edwards clique—comprising Springfield's prettiest and most marriageable "belles"—and attended all the dances around town, gliding gracefully about on the arms of eligible bachelors. She was "the very creature of excitement," said one young fellow. How she loved to dance—it was "poetry in motion," she sighed—and talk and picnic. She even courted Stephen A. Douglas some, but he never asked for her hand and she wasn't sorry. "I liked him well enough," she recalled, "but that was all." Clever and flirtatious, she was doing in Springfield high society exactly what she was trained to do. Her friends described her as fun-loving but temperamental—now utterly charming, now curt and sharp-tongued. "Her face," declared a nephew, "was an index to every passing emotion."

Her best friend was Mercy Ann Levering, who lived with a married brother next door to Mary and who courted and eventually married James C. Conkling, dapper and dashing, a Princeton graduate and rising young Whig attorney in Springfield. In the winter and spring of 1840, the two girls talked about their mutual romances, and Mary revealed her interest in Lincoln, who called her "Molly" and was visiting her regularly now.

By the late summer of 1840—the summer of the obstreperous log cabin campaign—Lincoln and Mary Todd were "a-courtin'," as they said in Illinois. Mary found in Lincoln "the most congenial mind she had ever met," took a keen interest in his political work, and felt a deep affection for this tall attorney who was unlike any man she'd ever known. For his part, Lincoln still had gnawing doubts about himself and his paltry education and low-class background when compared to Mary's. Nevertheless, he found her entirely free of snobbishness—she cared about him, not his background—and so they talked about marriage. In December, 1840, they reached an "understanding" and evidently became engaged.

At this point Ninian and Elizabeth Edwards stepped in and tried to break up their romance. For Mary to see Lincoln was one thing. But to marry him was quite another. And the Lexington Todds emphatically agreed. After all, Mary came from a proud and educated family. By contrast, Lincoln was from "nowhere" and his future was "nebulous." As a consequence, said the imperious Edwardses, he was no longer welcome at their home. All of which upset Mary, who complained bitterly to Mercy about "the *crime* of matrimony."

Lincoln was devastated. The hostility of the Todd and Edwards families—especially Elizabeth—caused incalculable pain in one so insecure about himself and so resentful of his own family that he hadn't visited his father in over nine years. One of Lincoln's greatest sorrows—from his view—was that he'd worked himself to the bone for recognition and success and yet had carried a social albatross about his neck: the lack of family respectability. No wonder he never discussed his background in Springfield high society. What could he say in the presence of Elizabeth and Ninian Edwards? That his father was all but illiterate? That the origins of his real mother were obscure?

As winter came on, Lincoln sank into a profound depression about the way

Mary's people had rejected him. In despair, he remarked that one *d* was enough to spell God. But it took two *d*'s to spell Todd.

D epressed about "Molly" and exhausted from his political campaigns that year, Lincoln dragged himself over to the Presbyterian church for the legislative session of 1840–41. Melancholy or not, he fought hard to save the state bank and its various branches, which the Democrats sought again to eradicate, this time for suspending specie payments. Thanks to Lincoln, the legislature authorized the bank's move and the institution continued a precarious operation. At the same time, Lincoln told the entire House that he felt his share of responsibility in the "present crisis" regarding internal improvements and the state debt, which now stood at a staggering $13,644,000. Declaring that Illinois must pay every dollar it owed, Lincoln helped enact emergency measures to raise money and pay the interest on the state debt. In addition, he introduced a resolution to improve the quality of instruction in Illinois schools—one of the few times he showed any initiative in this area. At his recommendation, the legislature adopted a new public education bill, which required that teachers pass an examination administered by district trustees and obtain a certificate of qualification before they could appear in public classrooms.

But in January, 1841, Lincoln went through a terrible emotional crisis that left him ineffective for the rest of the session. On "the fatal first of Jany.," as he called it afterward, Lincoln broke off his engagement with Mary Todd. He did so because the opposition of her family had shattered whatever confidence he'd mustered about marriage and intimacy with a woman. Stricken with self-doubts, certain that he was indeed inferior and incapable, that he could neither support a wife nor make her happy, Lincoln apparently told Mary he could not go through with the marriage and asked her to release him, which she gracefully did. She was not bitter about it and sympathized with his anguish, but still she was hurt. He knew it and told Speed that her unhappiness "killed" him. In his misery, Lincoln hated himself for not keeping his resolve, "the only, or at least the chief, gem of my character." Now he had lost that, too, and felt worthless and dead inside.

A few days after the breakup with Mary, Lincoln filed divorce papers for a client named Ann McDaniel, which didn't help his mood any. In truth, everything seemed to be coming apart: Speed had recently sold his store and

Lincoln had been forced to take another room by himself. Now Speed was planning to move back to Kentucky, and Lincoln was distressed about that, too. His closest friend, his engagement, his confidence—all were gone or going. As his melancholy grew worse, he called on Dr. Anson Henry, who diagnosed his trouble as "hypochondria" all right, a deep depression brought on by extreme anxiety, overwork, and exhaustion. For a week in mid-January, Lincoln lay in his boarding room in acute despair, with snow and rain lashing his windows. He couldn't sleep, and his insomnia and weariness only aggravated his depression, which in turn made him toss all the more at night.

After a week of "making the most discreditable exhibition of myself in the way of hypochondriaism," Lincoln forced himself back to the legislature, but he was too dispirited to do much. Frankly his colleagues were shocked at his appearance. He looked emaciated, said James Conkling, "and seems scarcely to possess strength enough to speak above a whisper." On January 23 Lincoln wrote Stuart and confessed that his mind was in such a "deplorable state" that he couldn't summarize the news. He couldn't attend to business either. "I am now the most miserable man living," he sighed. "If what I feel were equally distributed to the whole human family, there would not be one cheerful face on the earth. Whether I shall ever be better I can not tell; I awfully forebode I shall not. To remain as I am is impossible; I must die or be better."

By now Springfield hummed with talk about Lincoln's condition. One acquaintance heard that he'd thrown "two cat fits, and a Duck fit since we left." And another said he'd gone "crazy for a week or so." "Crazy as a loon," growled Ninian Edwards. Lincoln "went crazy," Elizabeth Edwards contended, "because he wanted to marry and doubted his ability and capacity to please and support a wife."

"You see by this," Lincoln wrote Stuart on February 3, "that I am neither dead nor quite crazy yet." In truth he was feeling a little better now and had been attending the legislature more or less regularly for the last week—for work, however mundane, was all he had to cling to. He mentioned that the Democratic legislature had passed a controversial court-packing bill, which raised the number of Supreme Court judges from four to nine. Lincoln predicted that the court would now become "a loco concern," and he was right. The Democratic governor promptly appointed five Democrats to the high tribunal—among them young Stephen A. Douglas. Lincoln witnessed the stormy debates over the bill, but was still too depressed to speak out, thus depriving the Whigs of one of their most effective voices.

When the legislature adjourned in March, 1841, Lincoln's legislative career came to an end. He was now thirty-two years old. Though he remained on the Whig State Central Committee, the Sangamon Whigs did not renominate him for the legislature and he was tired of it anyway. In his four terms there, he'd learned a lot about practical politics, served as an above-average floor leader, proved himself an articulate debater, and exhibited a first-class talent at drafting bills. Yet he was not a creative lawmaker and seldom initiated original legislation. Still, he was an honest politician—even his opponents conceded that—and he could be unbending when it came to his principles, as evidenced by his recalcitrant stand on internal improvements.

In the end, the public works program he supported proved too costly to continue: the state not only defaulted on its debt—which almost ruined its credit rating—but dissolved Lincoln's cherished state bank and canceled most of the unfinished public works, leaving farmers to raise corn on abandoned railroad gradings.

By the spring of 1841, a lot of people in Springfield thought Mary Todd had jilted Lincoln for another man. In truth, one Edwin Webb, a middle-aged widower with two children, was trying to woo her now, but Mary's thoughts were about Lincoln. She wrote Mercy Levering that he "deems me unworthy of notice, as I have not met *him* in the *gay* world for months . . . yet I would that the case were different, that he would once more resume his Station in Society, that 'Richard should be himself again,' much, much happiness would it afford me." As for the *"winning widower,"* Mary could never give her heart to him. For one thing, there was the "slight difference of some eighteen or twenty summers in our years." For another, there were his children, those "two *sweet little objections."*

"Poor A," Mercy wrote James Conkling. "I fear his is a blighted heart! perhaps if he were as persevering as Mr. W—— he might finally be success-ful." Conkling was certain Lincoln now realized that love was "a painful thrill," but that not to love was even more painful. "I suppose he will now endeavor to drown his cares among the intricacies and perplexities of the law."

That was exactly what Lincoln did. He flung himself into a flurry of law cases—and also changed business partners, which helped him mend a little. He and Stuart both were too involved in politics to carry on a successful practice; and besides, with Stuart off in Washington much of the time, Lincoln had to do most of the firm's bookkeeping and clerical chores, which he detested. In May, 1841, they dissolved their firm by mutual consent, and

Lincoln formed a new partnership with Stephen T. Logan, a fellow member of the Whig Junto.

Logan was a shriveled little man with frizzled red hair and a shrill voice. He scarcely ever varied his dress, which consisted of a twenty-five-cent straw hat, baggy pants, and an oversize coat. Because he was a compulsive whittler, the sheriff kindly provided him with white plank shingles and a wool sack when court day began. Though bizarre and cantankerous, Logan proved an excellent partner for a lawyer on the climb. As an attorney, he was precise and meticulous, with a broad command of both the practical and philosophical sides of the law. He'd practiced ten years in Kentucky, come to Springfield in 1833, served a couple of years as a circuit court judge, and become the undisputed leader of the Sangamon bar.

Because of their industry and punctiliousness, Logan and Lincoln built theirs into one of the busiest firms in Springfield and all but monopolized litigation in the Illinois Supreme Court. In time Logan considered Lincoln "a pretty good lawyer" and taught him a great deal about the importance of painstaking preparation, of compiling an exact and thorough brief before defending a client in court.

Still, Lincoln suffered from recurring bouts of the hypo. So in August, 1841, he took a much-needed vacation and visited Joshua Speed at his parents' Kentucky plantation. Here Lincoln got his first taste of real luxury, as Negro servants brought him coffee in bed and served him dishes of peaches and ice cream. And he chatted amiably with Speed's mother and kidded around with his half-sister until she and Lincoln became "something of cronies."

Alone with Speed, Lincoln opened up about Mary and his depression, and Speed confided in Lincoln about his own romantic troubles. As it happened, Speed was courting a young woman named Fanny Henning, but should he marry her? Could he make her happy? Was he capable of love? What if they married and then she should die? Why was he obsessed with her death? Was he really not in love with her then? In full sympathy with Speed's forebodings, Lincoln went with him to see Fanny—and found her "one of the sweetest girls in the world" with "heavenly *black eyes.*" She had only one flaw, Lincoln decided, and that was "a tendency to melancholly." He added: "This, let it be observed, is a misfortune not a fault."

Still trying to sort out his feelings, Speed accompanied Lincoln back to Springfield. As they sailed down the Ohio on a river steamer, Lincoln studied a coffle of twelve slaves on board, all chained together, he said, "like so many fish upon a trot-line." The blacks, Lincoln learned, had been torn from their families and friends in Kentucky and were on their way to the Deep South. The more he watched them, the more he brooded over their condition. Here they were, being sold "into perpetual slavery where the lash of the master is proverbially more ruthless and unrelenting than any other where." Yet to Lincoln they seemed happier than anybody on board. One slave played the fiddle and the others danced, sang, joked, and played card games. Lincoln decided that here was an example of how " 'God tempers the wind to the shorn lamb,' or in other words, that He renders the worst of human conditions tolerable, while He permits the best, to be nothing better than tolerable." Even so, the sight of those chained Negroes troubled him. Years later he asserted that the spectacle "was a continual torment to me" and that he saw something like it every time he touched a slave border. Slavery, he said, "had the power of making me miserable."

Back in Springfield, Lincoln had much work to do in the circuit court. Speed remained with him until January, 1842, then headed home more or less resolved to marry Fanny. For several weeks he and Lincoln exchanged intimate letters, with Speed fretting about his "intense anxieties" and Lincoln offering encouragement. Speed, he said, you are my everlasting friend and I feel your pain and suffering as though they were my own. Everybody worries about marriage, Speed, but the "painful difference" between you and the mass of the world is your "nervous debility." But you'll overcome it. After you're married, your nerves may still fail you now and then, but once you've got them straightened out, you'll be over that problem. In any case, you are clearly in love with Fanny and should marry her. Your fear that she is destined for an early grave is evidence of your love, Lincoln insisted, because if you didn't love her, you wouldn't be so apprehensive about her death. Lincoln was sorry to dwell on this, but "you know the hell I have suffered on that point, and how tender I am upon it." Still, in counseling Speed, Lincoln helped assuage his own anxieties. He was now "quite clear of hypo," he told his friend, and felt like "quite a man."

With Lincoln's reassurances, Speed went through with his marriage and then wrote Lincoln that most of his nervous fears had been groundless. "I tell you, Speed, our *forebodings,* for which you and I are rather peculiar, are all the worst sort of nonsense," Lincoln rejoiced. When Speed wrote again and said he was happier than he'd ever hoped to be, Lincoln replied: "I am not going beyond the truth, when I tell you" that "your last letter, gave me more pleasure, than the total sum of all I have enjoyed since that fatal first of Jany. '41."

Although Lincoln avoided Mary, he thought about her constantly. He knew he'd made her sad and that "still kills my soul." He could not but reproach himself for even wanting to feel happy when she was otherwise. Speed, now advising Lincoln, warned him either to wed Mary or forget her. Lincoln agreed. "But before I resolve to do the one thing or the other, I must regain my confidence in my own ability to keep my resolves when they are made." Until he did, he could not trust himself in love. Meanwhile, he was extremely pleased for Speed and Fanny. "I always was superstitious," he wrote his friend; "and as part of my superstition, I believe God made me one of the instruments of bringing your Fanny and you together, which union, I have no doubt He had fore-ordained. Whatever he designs, he will do for *me* yet."

As the summer of 1842 came on, friends assured Lincoln of Mary's continued affection for him. Finally Mrs. Francis, wife of the *Journal* editor, invited them separately to her house and persuaded them to "be friends." Encouraged by Speed's success, Lincoln started seeing Mary again, meeting her secretly at the Francis home so that Ninian and Elizabeth wouldn't find out.

As they carried on a clandestine courtship, Lincoln and Mary conspired to poke some fun at the Democrats—and singled out state auditor James Shields, an outspoken little Irishman Lincoln thought was pompous, for some clever Whig ridicule. In August and September, there appeared in the *Sangamo Journal* a series of anonymous letters, narrated by a fictitious country lady named Rebecca, which not only derided Shields's financial policies, but questioned his honesty and mocked his physical courage as a man. With barely restrained giggles, Mary and a girlfriend wrote one or more of these sarcastic compositions, and Lincoln contributed one that called Shields "a conceity dunce" and "a liar as well as a fool."

Shields was transported with rage. He burst into Francis's office and demanded to know who had written those letters. Francis went to see Lincoln, who gallantly covered for the young women. "Tell Shields I am responsible," Lincoln said. Determined to defend his honor, Shields sent Lincoln an ominous demand: either he confess his authorship of the offensive letters and retract them in full, or face "consequences which no one will regret more than myself." Lincoln retorted that Shields's note contained so many assumptions and threats that he would give no further reply. But Lincoln added that the consequence Shields alluded to would be "as great [a] regret to me as it possibly could to you."

Springfield soon buzzed with excitement, as rumors flew about of an impending duel. By now Lincoln was mortified about the whole affair; he didn't want to fight Shields, who was a smaller man than he, and yet he wouldn't back away from a challenge either, lest that damage his own honor. In his dilemma, Lincoln on September 19 sent Shields an explanation: he would confess to writing only one of the Rebecca letters and would assure Shields that it was intended "wholly for political effect" and not in any way to defame his character. If Shields found this unacceptable and insisted on a duel, then Lincoln stipulated the following conditions: they would stand face to face in an eight-foot circle and fight it out with cavalry broadswords. This was so absurdly grotesque that Lincoln couldn't have been entirely serious. As another biographer has suggested, Lincoln was no doubt trying to show his opponent how ridiculous their confrontation had become.

Their seconds now took over and arranged for them to fight a duel over in Missouri, where that dubious activity was still legal. As it happened, the duel site was situated across the river from Alton, Illinois, where Elijah Lovejoy had been murdered—an outrage that had helped inspire Lincoln's antiviolence speech of 1838. On September 22 Lincoln and Shields turned up on the Missouri border with a retinue of seconds. But at this juncture their friends interceded and persuaded a smoldering Shields to accept Lincoln's explanation, thus reconciling their differences.

So ended what one writer has called "the most lurid personal incident in Lincoln's entire life." Afterward, he was so ashamed of the affair that he would never talk about it. He never published any more cruel and anonymous satire either, for he'd finally learned that it could hurt people and lead to embarrassing and potentially dangerous repercussions. Moreover, he avoided even the threat of fighting to uphold his honor. He deplored violence anyway. When someone insulted him or tried to pick a fight, Lincoln just laughed at the man and walked off.

Ironically, though, the Shields episode brought Lincoln and Mary closer together. Mary thought him terribly chivalrous in covering for her, and her admiration did much to restore his self-esteem. On October 5 he wrote Speed again. "Are you now, in *feeling* as well as *judgement,* glad you are married as you are?" Yes, Speed replied, he was really glad. Convinced that he too was capable of love, Lincoln threw his doubts to the wind and asked for Mary's hand. They set November 4, 1842, as their wedding day.

Ninian and Elizabeth, of course, knew nothing about their plans. In fact, Lincoln and Mary refrained from telling them until the very morning of the wedding. At first they were both quite upset, but finally relented when they saw how determined Mary was. Well, they sighed, if there must be a wedding, it would happen here in their home.

Later, as Lincoln blackened his boots and dressed for the ceremony, a young fellow entered his room and asked where he was going. Lincoln cracked, "To hell, I reckon." Which was his way of fighting back anxiety. That and reassurances like those he'd given Speed: your nerves may still fail you on occasion, but once "you get them fairly graded now, that trouble is over."

That evening, with rain pelting the Edwards mansion, Lincoln and Mary Todd—aged thirty-three and twenty-three respectively—stood before an Episcopal minister in the parlor, with a small group of friends in attendance. Lincoln seemed pale and nervous as he exchanged vows with Mary. But with a rising hope that his old trouble would soon be over, that he would be happier or at least *"less miserable"* living with Mary than alone, Lincoln took her hand now—took the hand of this woman who had stood by him through all his private anguish, who had assured him that he was worthy and tender enough to merit her affection—and he gave her a wedding ring with the inscription "Love Is Eternal."

Later that night, the newlyweds drove their carriage through a blinding rain and came at last to the Globe Tavern, where Lincoln had rented a single room as their home. A few days afterward Lincoln wrote an acquaintance that nothing was new "except my marrying, which to me, is a matter of profound wonder."

Lincoln's marriage did nothing to reconcile him with his own family, not one of whom had been invited to the wedding. In fact, Thomas and Sally Lincoln knew nothing about it for some time, finding out only when Lincoln, in Coles County on business, dropped by for a brief visit and mentioned that he was married now.

He and Speed also drifted apart. Lincoln wrote that "you and your Fanny" and "I and my Molly" should get together, but apparently they never did so in these years. Lincoln was sorry that they were letting "a friendship, such as ours, to die by degrees," but he and Speed were moving in separate worlds now.

Because his New Salem debts still drained away much of his earnings, Lincoln and Mary spent their first year in the Globe Tavern, where room and board cost them only four dollars a week. It was a rough time for Mary, who was not accustomed to such conditions and fretted that their wealthier friends were snickering about them. Also, a couple of months after their wedding, they discovered that she was pregnant. She could have had the child in the

Edwards mansion, but Mary was too proud and stubborn to budge from her Globe Tavern room. On August 1, 1843, she gave birth to a son, whom they named Robert Todd after her father.

Ashamed of their living conditions, Lincoln scrimped and toiled in a relentless effort to get ahead, putting in long hours at his office and building up a sizable practice out riding the circuit. At last he had enough money to move out of the noisy tavern and into a rented cottage. Then in January, 1844, he paid out $1,200 for a lot and a house, situated on the corner of Eighth and Jackson streets, within walking distance of his downtown office. It was a comfortable frame home, one and a half stories, with a shade tree in front and a stable and privies out in back. Now he and Mary could entertain their friends without embarrassment and "Bobby" had a yard to play in. Two years later Mary gave Lincoln another son, whom they christened Edward after Lincoln's friend Edward Baker. Lincoln loved to romp with his boys, to lie on the hall floor and tickle and toss them in the air. He spoiled them with unabashed pride, chuckled at their mischief, never took his hand to them.

By the mid-1840s, Lincoln was making over $1,500 a year—a good income in those days—which eased their financial worries. He hired a maid to help Mary and free her to shop and socialize more. And in time he managed to retire the "National Debt" that had plagued him since the 1830s.

By now the Lincolns had learned to live with one another and to cope with the pressures of marriage, but that didn't come easily. Of the two, Lincoln may have adjusted more readily to married life, because he was used to hardship and responsibility. In fact, he derived tremendous satisfaction from being a husband and father—it gave him a sense of real worth and security for the first time in his life. Mary, on the other hand, was shocked at the realities of marriage and utterly unprepared for the demands it placed on her. After all, she'd grown up in an aristocratic family, accustomed to material luxury and to servants who waited on her. Though she could stitch and entertain, neither her family experience nor her finishing school had taught her what to expect once the wedding bells were over—had trained her to keep house, fix meals, watch her budget, have babies, wash and change their clothes, stay up all night with a sick and wailing child, contend with a moody, hard-working husband. Lincoln was considerate about the children and often took care of them himself. But they both suffered until Lincoln could afford to hire a woman to help. "Why is it," Mary once asked Mercy Levering, "that married folks always become so serious?" Now at last she knew.

Though on the surface Mary seemed a blunt and willful woman, she was extremely sensitive and suffered from deep insecurities which marriage and motherhood only aggravated. Thanks to those tough Globe Tavern days, she became terrified of poverty, obsessed with it. To her, poverty was a form of death, a strangling void which deprived her of all her material values—her love of beautiful things, her desperate need to wear fashionable clothes and to receive compliments about her appearance. In truth, once the Lincolns enjoyed financial security, Mary spent lavishly on herself, her family, and her home. Yet her spending sprees left her guilt-stricken and afraid. Would her splurges send them spiraling into debt and destitution again? Would they wind up back in that cramped little Globe Tavern room, with its creaks and noises and Bobby's crying through the night, all the while the richer people whispered about her behind her back? To compensate she would become as parsimonious as a miser, haggling over prices for incidental things. Then she would turn around and go off on another shopping binge, only to fret about poverty again. It became a recurring syndrome for Mary, who, even during Lincoln's most lucrative years, was certain that they were on the brink of privation.

There were other fears, too. When Bobby would wander off, Mary would become hysterical and would send a messenger to fetch Lincoln from his office. But the boy usually turned up by the time Lincoln reached home. Once when Robert accidentally swallowed some lime, Mary lost all control and screamed over and over, "Bobby will die!" "Bobby will die!" until Lincoln convinced her that Robert would be all right. And then there were the thunderstorms. They terrified her and brought on blinding headaches that sent her to bed for days at a time. When this happened, Lincoln tended to her every need, lavished affection on her, and she reveled in all his attention. Only with his care did she gradually recover from her headaches. Lincoln, for his part, knew that Mary loved him to dote on her, call her "little woman" or "my Molly," tease and pamper her in a fatherly, gentle way. She was pleased, her friends claimed, when he called her "my child wife."

A neighbor sympathized with Mary's insecurities and intense need for affection. After all, Lincoln's law practice and political activities took him away from home for long stretches of time, leaving Mary to run the house and shepherd the boys alone. He rode the circuit twice a year, which meant that he was gone three months at a time, six months out of every twelve, and his long absences troubled Mary. But his circuit practice was much too profitable and rewarding for Lincoln to abandon or even curtail, and Mary never asked him to do so. Besides, she needed the money his practice brought in, realized where they would be without it.

Other conflicts derived from their different habits and temperaments. If Mary liked a good argument to get everything out in the open, he often withdrew at the first sign of a quarrel, for he hated fussing, avoided it whenever he could. And frankly there were other things about Lincoln that irked Mary. There was his practice of answering the door himself, often in his stocking feet, instead of leaving it to the maid. He also liked to lie on his back in the hallway, resting his head against an upside-down chair, and read his newspapers aloud. And there was his carelessness of dress, which, if scarcely so extreme as legend claims, was still a problem until Mary taught him how to match his clothes and improve his appearance. Then there were his mood swings, his habit of withdrawing into himself, of being glum and remote when she wanted to talk. She did not understand his hypo any more than his friends did and was irritated by his spells of abstraction. They might come on at the dinner table, where he would stare off into space, impervious to conversation and Mary's glances. Or he would go off and sit in his rocking chair, immersed in himself as he mulled over some law case or the state of the Union, mulled over the meaning of life and the inevitability of death, his death and that of his wife and children, until he would shake such thoughts away and pull himself back to his house, this room, his playing sons, his anxious wife. Once a spell even came over him while he pulled one of his boys in a wagon. Lost in thought, he tugged the wagon over an uneven plank sidewalk and the child fell off. But Lincoln was oblivious to the fallen boy and went on with his head bent forward, hauling the empty wagon around the neighborhood.

Inevitably, as in any marriage, the Lincolns had their conjugal spats, especially when Lincoln was too melancholy or Mary became frustrated and "got the devil in her," as a neighbor recalled. Then they would both lose their tempers and have a pretty good row. Still, they didn't quarrel very often— and always made up when they did. In truth, the Lincolns enjoyed a relatively stable marriage, with a physical need and a mutual respect for one another which transcended their differences. Lincoln, for his part, understood Mary better than anyone, loved her in spite of her flaws, shielded her from criticism, and remained thoroughly loyal to her as a husband. In turn, Mary could be tender to him, extremely tender. And no matter how lonely she was with him gone so much of the time, she shared his love of politics, wanted him to succeed, and was fiercely proud of him. Once when somebody compared him to Stephen A. Douglas, Mary declared that "Mr. Douglas is a very little, little giant by the side of my tall Kentuckian, and intellectually my husband towers above Douglas just as he does physically." Because of all the ways he cared for her, Lincoln was everything to Mary—"lover—husband—father, all."

When some of Mary's relatives visited her, they wanted and expected to find her miserable with Lincoln. Instead the Lincolns talked affectionately about their sons and about one another. At last the Todds and Edwardses came around and gave the marriage their official sanction. But the lordly Todds never fully approved of Lincoln and did little to help him politically. In their presence, he always felt he had to prove himself.

In the spring of 1843, during those lean Globe Tavern days, Lincoln set his political sights on the United States Congress: with his romantic life straightened out, his ambition for political office was burning at white heat again. In the Whiggish Seventh Congressional District, all he had to do was secure the Whig nomination to win the election, but there lay the rub. For two other young Whigs were aiming for the nomination, and Lincoln would have to campaign with all his talents to beat them. First was Mary's cousin John J. Hardin, thirty-three, Kentucky-born, son of a U.S. senator, college graduate, prominent attorney, and the favorite to win at the Congressional convention in Pekin that spring. Second was Lincoln's friend Edward Baker, thirty-two, a state senator now and a dramatic orator who liked to appear on the platform with a pet eagle chained to a ring. When Baker described how low the Republic had sunk under the Democrats, the eagle would lower its head and droop its wings. But when Baker thundered about how the Whigs would restore America to glory, the eagle would spread its wings and scream, much to the delight of Baker's crowds.

Faced with formidable competition, Lincoln set about early in March lining up support and proving himself a loyal party man, a leader and a winner. He even drafted a Whig campaign circular in which he gave a powerful endorsement of the convention system, insisting that whether it was right or wrong was no longer the issue. Since Douglas's Democrats were using the system to pulverize Whigs across the state, Lincoln pronounced it "madness in us not to defend ourselves with it." The Whigs must stop fighting among themselves and understand the maxim that "union is strength," that "a house divided against itself cannot stand."

With that, Lincoln battled Baker for control of the Sangamon delegation, the first step to victory at Pekin. Yet the Sangamon Whigs chose Baker over Lincoln, and he was deeply disappointed. How had he lost to Baker? Why? Lincoln confided in friends that some people attacked him—*him* of all people —as "the candidate of pride, wealth, and aristocratic family distinction." Also, a sortie of county churchmen opposed him because "I belonged to no church, was suspected of being a deist, and had talked about fighting a duel." In addition, a year ago he'd given an address on temperance, calling for reformers to stop castigating and vilifying drunkards and treat them as generous, sensitive folk who were simply unfortunate. As a consequence, many church and temperance people accused Lincoln, a teetotaler, of favoring drunkenness.

So he lost the nomination—lost his bid for Congress. Or had he? Though he'd been selected as a member of the Sangamon delegation and was honor bound to vote for Baker, what if the Pekin convention ended in a deadlock between Baker and Hardin? Desperate for the nomination, Lincoln now maneuvered to get himself supported by the Menard County delegation, so that in case of a deadlock the convention would swing to him.

But it was no use. At Pekin, Hardin had all the votes he needed to win. For the sake of party harmony, Lincoln withdrew Baker's name and threw his support to Hardin even though Lincoln disliked him as a man. Hardin thus secured the nomination and ensured himself of victory in the coming election. But with an eye on the future, Lincoln engineered the so-called Pekin agreement, which embodied a rotation plan to guarantee party unity —and Lincoln's own advancement. As Lincoln understood the arrangement, he and Baker promised to back Hardin in the present contest and Hardin agreed to step aside for them in subsequent elections. Thus Baker would automatically get the nomination in 1844 and Lincoln in 1846. This, said Lincoln, would avoid destructive party infighting and establish the principle that "turn about is fair play."

Now all Lincoln could do was wait. In August Hardin won a Congressional seat as expected, but so did Stephen A. Douglas, who whipped Lincoln's friend Orville Browning over in another Congressional district. At last 1844 came on, a Presidential election year featuring Henry Clay of the Whigs against James K. Polk of the Democrats, the latter running on a platform of militant expansion in Oregon and Texas. A third candidate, James G. Birney, carried the standard of the antislavery Liberty party. Always an indefatigable Whig campaigner, Lincoln stumped Illinois for Clay, giving Whig litanies in defense of protective tariffs and national banks.

In the summer Lincoln rode horseback into Indiana, where he addressed Clay rallies at Vincennes and Rockport. Then he went on to Gentryville and the neighborhood where he'd spent his boyhood—a neighborhood he hadn't seen in fourteen years. He made his way over sandy roads he used to walk on as a youth, passing familiar homesteads and dense forests with their cawing birds and tangled vines. He felt pensive and poetic. In truth, nostalgia eased his resentments for this region that had marked and shaped him in such painful ways. He visited the old home place near Little Pigeon Creek, and the sight of "my childhood home" caused such a rush of feelings and memories in him that he later set them to verse. He walked the "old field" where he'd toiled and hated to toil for nine long years, walked "the very spot where grew the bread that formed my bones," until he felt a part of the soil here, the gusting wind and sunlit land. He thought of his mother and sister and perhaps even visited their graves, recalling "things decayed and loved ones lost," rising in his memory like "dreamy shadows." As he traveled around the area, he looked up old friends who'd seen him and his family off that day they had left for Illinois. But his friends had changed now, grown up, "and half of all are dead." In successive tableaus (a lane, a cabin, a store, a field), he listened as the survivors recounted the deaths of their kinfolk, until to Lincoln "every sound appears a knell, and every spot a grave." "I range the fields with pensive tread, And pace the hollow rooms; And feel (companions of the dead) I'm living in the tombs."

He called at James Gentry's place and found Matthew languishing in "harmless insanity." Poor Matthew. With his reason destroyed, he was locked "in mental night," condemned to a living death, spinning out of control in some inner void. Lincoln recalled the day Matthew became deranged, when he attacked his parents and terrorized the neighborhood until some men tied him down. But gradually he'd settled into a silent madness, mute and lost. Lincoln went away thinking him more the cause now than the subject of woe.

Lincoln headed back to Illinois in a somber mood, riding out of his boyhood back into the season of his middle years—back to the frame house on the corner in Springfield, back to his wife and sons. And back, too, to the law office and Stephen Logan (who could be quarrelsome sometimes, overbearing sometimes), back to politics, Whig committee meetings, the Presidential election. In November, Clay won Whiggish central Illinois, but Polk and the Democrats carried the state. Polk won the Presidency, too, although the popular vote between him and Clay was extremely close. At that, the outgoing Tyler administration, interpreting Polk's victory as a popular mandate for expansion, secured the annexation of slaveholding Texas—the very thing both the Whig and Liberty parties had tried to block.

For Lincoln, the election of 1844 demonstrated with blazing clarity the folly of splinter party movements. And in 1845 he wrote a friend what he thought about them. The Liberty men—or "Whig abolitionists," as Lincoln termed them—fought against the annexation of Texas because it was a slave place; Clay also opposed annexation; yet the Liberty men refused to endorse him because he was a slaveholder and therefore drew crucial votes away from Clay in a hopeless third-party cause. And the consequence was annexation— what Clay and the Liberty men had both striven to avoid. The lesson was clear: they should have voted for Clay even if he was a slave owner. That would not have been evil, because a tree was known by the fruit it bore. "If the fruit of electing Mr. Clay would have been to prevent the extension of slavery, could the act of electing have been *evil?*"

Anyway Lincoln wasn't all that concerned about Texas. Annexing it would not "augment the evil of slavery," because slavery already existed there. One must be realistic about the slavery issue. And realistically this was Lincoln's stand: "I hold it to be a paramount duty of us in the free states, due to the Union of the states, and perhaps to liberty itself (paradox though it may seem) to let the slavery of the other states alone; while, on the other hand, I hold it to be equally clear, that we should never knowingly lend ourselves directly or indirectly, to prevent that slavery from dying a natural death." Which he was sure it must do if confined to the South, as he thought the Founding Fathers had so confined it. He added: "Of course I am not now considering what would be our duty, in cases of insurrection among the slaves."

A year after his Indiana visit, Lincoln was still in a "poetizing mood" about the mixture of feelings his boyhood home had stirred in him. With images of Indiana lingering in his memory, he came across in some newspaper a poem he'd first seen about fourteen years ago. And the poem's mournful lines now played on Lincoln's mood, touching in him chords of deepening sadness and human fatalism. The poem was published anonymously; Lincoln never knew the author was the Scotsman William Knox. Beginning with the line, "Oh, why should the spirit of mortal be proud?" the poem borrowed its themes from Job and Ecclesiastes —themes which stressed the transitory nature of human life, suggesting that each generation undergoes the same experiences as its forebears, only to die as all previous generations have died. So why indeed should the spirit of man be proud?

The poem was called "Mortality" and Lincoln came to love it more than any other. "I would give all I am worth, and go in debt," he wrote a newspaper editor, "to be able to write so fine a poem as I think that is." He knew the lines by heart and would recite them over and over when lost in one of his spells of abstraction. "'Tis the wink of an eye, 'tis the draught of a breath, From the blossoms of health, to the paleness of death. From the gilded saloon, to the bier and the shroud. Oh, why should the spirit of mortal be proud!"

In the spring of 1846, Lincoln wrote some verse with similar themes, a poem about his youth and all the scenes, inhabitants, fields, and graves he'd seen on his trip to Indiana. It was a difficult poem for him to write, especially the stanzas on Matthew Gentry, but he stayed at it, revising and polishing the lines until he had them right. In the final version, he divided the poem into two cantos—one of ten stanzas about his boyhood home and another of thirteen stanzas about Matthew. The latter canto is especially revealing, for Lincoln was fascinated with madness, troubled by it, afraid that what had happened to Matthew could also happen to him—his own reason destroyed, Lincoln spinning in mindless night without the power to know.

> But here's an object more of dread
> 　Than ought the grave contains—
> A human form with reason fled,
> 　While wretched life remains. . . .

> I've heard it oft, as if I dreamed,
> 　Far distant, sweet, and lone—
> The funeral dirge, it ever seemed
> 　Of reason dead and gone. . . .

> O death! Thou awe-inspiring prince,
> 　That keepst the world in fear;
> Why dost thou tear more blest ones hence,
> 　And leave him ling'ring here?

Lincoln sent both cantos of his poem, "or doggerel, or whatever else it may be called," to the editor of the *Quincy Whig,* who with Lincoln's consent published the verse anonymously under the title "Return," with two subtitles, "Reflection" and "The Maniac."

Back in December, 1844, Logan and Lincoln dissolved their law firm by mutual agreement. Logan desired to go into business with his son, and anyway Lincoln was tired of being a junior partner. At thirty-five, with a Congressional seat awaiting him, Lincoln wanted to run his own firm, be his own boss.

He surprised his colleagues, though, when he took young William H. Herndon as his junior associate. Nine years younger than Lincoln, "Billy" had been studying law with Logan & Lincoln and had just now received his license. Some wondered why Lincoln should choose a tyro like him when he could have formed a partnership with any number of older and more accomplished attorneys, but Lincoln knew what he was doing. A careless records keeper himself, he thought Herndon "had a system and would keep things in order"—that is, would be a reliable and systematic office lawyer. Also, since he was young and inexperienced, he wouldn't contest Lincoln's decisions, wouldn't argue with him about which cases to accept. And there were political reasons, too. Springfield Whigs were now divided loosely along generational lines: the older group consisted of successful men like Hardin, Stuart, Edwards, and Lincoln. The "Young Turks," Herndon included, were "shrewd, wild boys about town" who demanded a hand in shaping state Whig policy. Since Lincoln was scheduled to run for Congress in 1846, he wanted the backing of the younger Whigs and believed Herndon would be influential with them.

Lincoln had first met Billy back in Speed's store, where Herndon had worked as a clerk. A native of Kentucky, he attended Illinois College for about a year and eventually married a shy and reticent woman who was his exact opposite. A nervous, windy fellow, Herndon stepped about in fancy clothes, a big silk hat, kid gloves, and patent leather shoes. He was thin, stood about five feet nine, and had raven hair and black eyes. But the most memorable thing about him was his nonstop chatter, his effusive philosophizing on everything from metaphysics to sex, science, and phrenology. An intellectual gadfly, he had answers to everything and possessed an impressive library that comprised the works of Hegel, Kant, and Francis Bacon. He regarded himself as an expert psychologist and bragged about his "dog sagacity" and "mud instinct," which enabled him to divine other men's inner secrets. Yet Herndon's knowledge of people derived almost entirely from books, and he was a man of striking contradictions. Though he considered himself an intellectual, he often consorted with rowdies and hooligans in Springfield's saloons. A man of causes and a leader in the local temperance movement, he was still a hard drinker and in his later years became an alcoholic. And though he chose the law for a career and worked hard at it, he felt out of place in the legal world. "If you love the stories of murder—rape—fraud," he once snorted, "a law office is a good place." On another occasion he exploded, "I *hate* the Law: it cramps me: it seems to me priestly barbaric. . . . I say I hate the Law."

He liked working with Lincoln, though, and it became the focal point of his life. As Lincoln's junior partner, he performed only routine chores at first —he *"toated books"* and *"hunted up authorities"* and tagged along when Lincoln went to court for his clients. When Lincoln rode the circuit, Herndon remained in Springfield, tending the office and gathering data for a pending case. But after their first year together, Herndon became an accomplished lawyer in his own right, sought out cases himself, and defended people in court. As he faced a jury, he often cursed, flung his coat to the floor, shouted, and wept—you won cases, he told his nephew, by making jurors cry. Because he and Lincoln both had an enormous capacity for work, they were soon handling over one hundred cases a year and drawing business from as far away as Peoria.

Moving from one office to another, Lincoln & Herndon finally settled on the second floor of a brick building across from the public square. It was a dreary place with a couple of dusty windows that overlooked an alley. The furniture consisted of several cane-bottom chairs, a couple of desks, a bookcase stuffed with Herndon's volumes, and an old sofa where Lincoln liked to prepare his briefs. Neither man bothered to sweep the floor, so that little mountain ranges of dust rose in the corners and along the walls. Clients were apt to step on cherry and orange seeds, too, since Lincoln liked to munch on fruit for lunch. According to a law student, the office dirt was so fertile that some seeds fell to the floor and subsequently sprouted.

Though Herndon tried to keep their accounts straight, he proved as careless and haphazard as Lincoln, and the office was soon in supreme disarray, with papers thrown in drawers, piled in boxes, bunched in shapeless heaps on desks and tables alike. Lincoln even stored papers in his stovepipe hat, "an extraordinary receptacle," Herndon asserted, which served as "his desk and memorandum book." With the office in such chaos, the partners lost things constantly, and visitors often found them thrashing about in search of fugitive documents. In the litter on top of his desk, Lincoln kept a special bundle with a string around it and a notation: "When you can't find *it* anywhere else look into this."

Because the partners worked long hours together in close quarters, they came to know one another's every habit, every whim. For his part, Herndon was fascinated with Lincoln, observed his ways and moods, and haughtily presumed that he understood Lincoln better than Lincoln understood himself. He saw into Lincoln's "gizzard," Herndon claimed. Still, Lincoln had his quirks and could be a nuisance. When he arrived in the morning, for example, he would sprawl across the sofa, spread his long legs across two or three chairs, and read his newspapers aloud. Lincoln did this, he explained, because it brought two senses into play at once. But his reading so annoyed Herndon that he sometimes had to leave the room. Herndon was also irritated because Lincoln seldom even looked at books any more. A voracious reader himself, Herndon would foist his current volume on Lincoln, insisting that he study it because it had "all the answers," the ultimate truth. Lincoln would glance through a couple of pages, toss the book aside, and tell Herndon to summarize it for him. Galled, Herndon decided that Lincoln was "obtuse" and assured himself that when it came to literature or philosophy he could think circles around the older man. And then there were Lincoln's anecdotes. Herndon found nothing funny about them. "I have heard him relate the same story three times within as many hours," he sniffed. But worst of all were Lincoln's impish sons. When cases piled up and the partners had to work on Sundays, Lincoln often brought his boys to the office and let them "gut the room" while he reflected and labored on a case. With Herndon looking on in horror, they pulled books down from the shelves, "rifled drawers and riddled boxes," stabbed gold pens against the stove, overturned inkstands, and scattered letters over the floor and danced across them, all the while Lincoln worked away unperturbed. Herndon was aghast. Frankly he wanted to grab those "brats" and "wring their little necks." But he tolerated their deviltry and kept his mouth shut out of respect for Lincoln.

There was more, though, than respect. His irritations and criticisms aside, Herndon idolized his partner, referred to him always as "Mr. Lincoln," and absorbed himself in Lincoln's ambitions. Deep down, he loved the man. "He moved me by a shrug of his shoulder," Herndon sighed. "He was the great big man of our firm and I was the little one. The little one looked *naturally* up to the big one."

Herndon meant a lot to Lincoln, too. If he could be pompous and grumpy sometimes, Billy was nevertheless a thoroughly loyal and trustworthy man. Once after a bitter political defeat, Lincoln said "I expect everybody to desert me now—except Billy Herndon." In time Lincoln came to regard him as a son, overlooked his faults, and stood by him always. One time, when Herndon and his drinking companions were arrested for breaking out some tavern windows, Lincoln came down to the jail in the gray of the early morning, bailed Herndon out, and paid his fine. After that, some of Lincoln's associates urged him to dump Herndon and find a more responsible partner. "I know my own business," Lincoln snapped. "I know Bill Herndon better than anybody" and "I intend to stick by him."

Even so, Herndon could never come to socialize in the Lincoln home, because Mary had nothing but contempt for him. According to Herndon's side of the story, he'd met Mary at a ball and was so enthralled with her graceful ways that he compared her to a serpent. He meant it as a compliment, but Mary disliked him in any case and took his remark as an outrageous and unforgivable insult. Consequently, she would never invite him to her home for social functions, thought him a reprobate the way he boozed and caroused around, and objected to Lincoln's association with him. She was even more incensed when Lincoln elected to split fifty percent of the firm's income with the younger man. And Herndon, in turn, came to hate Mary for rejecting him, until in his eyes she became a hateful bitch (a view he would later popularize in his own writings about Lincoln). As Mary's most recent biographers have said, she and Herndon waged a lifelong war of attrition against one another—a bitter vendetta in which each tried to discredit and defame the other. And Lincoln, the man they both loved, was the prize they fought over.

In the summer of 1845, Edward Baker, now serving his term in Congress, assured Lincoln that it was his turn next and that Baker would step aside. But not so with Hardin. On the contrary, Hardin intended to run for the Whig nomination again, and Lincoln was indignant, protesting that Hardin was deliberately violating the Pekin arrangement of 1843.

When Hardin refused to back out, Lincoln launched a drive to take the nomination away from him and secure what was rightfully his. On the circuit that fall, Lincoln collared local Whig leaders and promoted his candidacy in county seats around the Congressional district. He argued, "I want a fair shake and I want nothing more," and reminded Whigs that the Pekin agreement had established the principle of "turn about is fair play." And now it was his turn. He did not belittle Hardin; no, "nothing deserves to be said against him," but he did alert his supporters around the district to watch for "a 'moccasan track' as Indian fighters say" and keep him informed. By astute moves, he gained control of the Whig press, too, pulling four journals into his camp while only one remained in Hardin's. By year's end Lincoln commanded most of the Whig convention delegates, in what turned out to be the shrewdest campaign he'd ever run.

Outfought and outflanked, Hardin pronounced himself "mortified" and accused Lincoln of dirty politics, of "maneuver," "management," and "combination." Nonsense, Lincoln retorted. I'm entitled to the nomination on grounds of principle, and most Whigs know it. You, Baker, and I agreed to rotate the Congressional seat and now it's my turn. Lincoln bemoaned "the utter /injustice" of Hardin's imputations, said he was having a tough time keeping his patience, said he was the one who should be mortified. Still, Lincoln didn't mock or denounce Hardin as he used to treat his political rivals. Instead he appealed to Hardin's honor, contending that surely he didn't mean to be ungenerous or unjust, that he would "yet think better and think differently of this matter."

In February, 1846, Hardin withdrew from the race. And on May 1 the district convention at Petersburg nominated Lincoln by acclamation. At that, Lincoln took to the hustings against his Democratic opponent, the Methodist circuit rider Peter Cartwright.

As their campaign got under way, the Mexican War broke out over the disputed Texas boundary, and both national capitals rang with jingoistic oratory and appeals to national pride. President Polk, to be sure, branded Mexico as the aggressor and insisted that the United States had declared war entirely for self-defense and national honor. At the same time, newspapers and politicians alike fanned the flames with hate-Mexico diatribes. And some blew the trumpets of "Manifest Destiny," proclaiming that the United States, with its superior institutions and way of life, had a God-given right to wrench the Southwest from Mexico and to rule this continent. In Congress, Stephen A. Douglas not only defended the war in patriotic flourishes, but condemned those who opposed it as "traitors in their hearts." Not to be outdone, Whig Edward Baker, dressed in full military regalia, gave an impassioned war speech in the House, then resigned to lead the Fourth Illinois off to certain military glory. In all directions volunteers headed for the front to the blare of bugles and tuck of drums, singing "Away You Rio" and vowing to defend the flag.

In Illinois, the martial spirit blazed across the prairies from Chicago in the north to Cairo in the south. Hardin and other politicians all set out for Mexico with Illinois commands. In Springfield, candidate Lincoln attended a giant war rally and joined with other luminaries in delivering "warm, thrilling and effective" speeches that called for unity against Mexico. In private, however, Lincoln may have had misgivings about the war and Polk's defense of it. But caught in the middle of a campaign in superpatriotic Illinois, with the war spirit flaming across the Seventh Congressional District, he held his tongue about whatever doubts he had and drifted with the winds of popular opinion.

As it developed, the contest against Cartwright was no contest at all. In truth, the circuit rider proved such a lackluster campaigner that Democratic bosses wrote him off and conceded the election to Lincoln. But Cartwright held on to the end, harping away on his favorite theme—that Lincoln was an infidel. In reply, Lincoln circulated a handbill—it was signed—in which he discussed his religious views, the only time he ever did so in public. He confessed that he wasn't a church member, but argued that this didn't make him an infidel. On the contrary, he believed in a Supreme Being and had never denied the truth of the Scriptures. But with an eye on the electorate, he said he wouldn't support an atheist for public office, since no man had the right to injure community feelings and morals.

On election day, August 3, 1846, Lincoln carried eight of eleven counties and polled 6,340 votes to Cartwright's 4,829, an unprecedented majority in the district. But because the thirtieth Congress would not begin until December 6, 1847, Lincoln had over a year to wait before he could assume office. He wrote Joshua Speed that his election to Congress "has not pleased me as much as I expected," but didn't explain why.

W hile Lincoln practiced law in Springfield and studied his newspapers, the Mexican War raged into its second year, with Whig generals Zachary Taylor and Winfield Scott winning spectacular victories in Mexico. Once the guns had begun to roar, though, many Whigs had come out in open opposition to the war, denouncing it as a sinister move on the part of the Polk administration to seize Mexican territory. Polk, of course, had sent Congress a war message that laid all the blame on Mexico, insisting that Mexican troops had invaded Texas and shot American soldiers there in a deliberate attempt to foment hostilities. But Polk and his allies on Capitol Hill had rammed the message through Congress without granting the Whig opposition time to study the document or engage in prolonged debate. When they examined what had happened on the Texas border, many Whigs cried out that Polk had deceived them and accused his administration of starting the war. As the Whigs saw it, the President had ordered American troops into the disputed territory between the Nueces River and the Rio Grande—a region claimed by Texas and Mexico—for the explicit purpose of provoking an incident. Predictably, Whig critics said, Mexican soldiers had fired on the Americans, thus giving Polk a pretext for war so he could wrench New Mexico and California away from Mexico by force of arms.

Some Northern Whigs went even further than that: since the South tended to control the Democratic party, they branded the war a Southern plot to grab new lands for slavery expansion. Joshua ("Old War Horse") Giddings of Ohio said so. So did John Quincy Adams and other "Conscience Whigs" of Massachusetts. With some Northern Democrats also worried about slavery extension, Pennsylvania's David Wilmot introduced the momentous Wilmot Proviso—this in the summer of 1846—which would exclude Negro slavery from all territory acquired from Mexico and save that virgin soil for free white labor, for "the sons of toil of my own race and color," as Wilmot phrased it. The Proviso became a symbol and a rallying cry for Northern free-soilers—a loose coalition of Whigs and Democrats who were content to leave slavery alone where it already existed, but were determined to keep it out of all lands won from Mexico. In the ensuing debates, proslavery Southerners roared in protest, calling the Proviso unconstitutional, calling it treason. Other congressmen invoked the spirit of the 1820 Missouri Compromise, which had drawn an imaginary line across the old Louisiana Territory —at the latitude of 36°30'—and decreed that slavery could exist below but not above that line. Would it not be fair to extend the line to the Pacific? But free-soilers blocked the move. And the Wilmot Proviso failed in the Southern-controlled Senate, thus leaving the status of slavery in captured Mexican territory unresolved.

Although Northern and Southern Whigs were divided over the slavery question, they were fairly united in their hatred of Polk. Not only had he been treacherous about the war, the Whigs said, but he'd also vetoed internal improvement legislation. To drive him out of office, the national Whigs moved to expose and disparage him for launching a needless war for territorial conquest. Then in 1848 they could defeat the Democrats and put a Whig in the White House.

That was the national Whig position in November, 1847. And it was now the position of Congressman Lincoln as well. On the way to Washington with Mary and his sons, he stopped off in Lexington, Kentucky, to visit with Mary's family. While there, he attended a Whig rally where Henry Clay— Lincoln's political idol—rehearsed all the Whig objections to the war, condemning it as one "of unnecessary and offensive aggression." Though most Whigs in Illinois still favored the war, Lincoln agreed with Clay and thus aligned himself with the Whigs in Washington. Lincoln also liked what Clay said about the slavery problem. Though a slave owner himself, Clay asserted that "I have ever regarded slavery as a great evil." And he endorsed gradual, voluntary emancipation and colonization as the best solution to the race issue in white America.

Late in the evening of December 2, the Lincolns finally arrived in Washington, a Southern city of some 34,000 souls, including several thousand slaves. Eventually they found lodging in Mrs. Spriggs's boardinghouse, across from the Capitol, whose wooden dome they could see through the December trees. Mrs. Spriggs's was somewhat inaccurately known as "the Abolition House," because free-soil, anti-Southern Whigs like Joshua Giddings resided there. At meals Lincoln either ate in moody silence or entertained the other congressmen with a story or two, his chin in his hands and his arms propped up on the table.

While he toiled in the Capitol, Mary felt increasingly out of place in Washington. For one thing, nobody made much fuss over the wife of a freshman congressman, and Mary thought she was being snubbed. For another, she was disgusted with all the licentious behavior that went on in Washington, where saloons and brothels attracted congressmen who'd left their wives at home. As winter came on, Mary confined herself to their room, which in turn only recalled those terrible Globe Tavern days. In all the tension, she placed demands on Lincoln that hindered him in "attending to business" and evidently they quarreled some. After three months she packed her bags and headed for Kentucky with the boys, leaving Lincoln to tend his business alone.

His work exhausted him. Requests for public office swamped his desk and he labored assiduously in answering them. He performed countless favors for his constitutents, ran errands and sought patents for them. In Congress, meanwhile, he watched and learned what he could from the national politicians who were now his colleagues. Sitting on the back row on the Whig side of the House, he was distinguished at first only by his height and the fact that he was "the lone Whig from Illinois." Of all the Whigs from North and South, he came most to admire Alexander H. Stephens of Georgia, calling him that "little slim, pale-faced, consumptive man" whose eloquence brought tears to Lincoln's "old withered, dry eyes."

As always, Lincoln proved himself an exemplary party man, serving dutifully on committees, seldom missing a roll call, never skulking a vote, and giving orthodox Whig speeches on tariffs and internal improvements. But Lincoln's chief effort was his outspoken stand against the Mexican War. As part of Whig floor strategy to fasten war guilt on Polk, Lincoln rose on December 22 and offered a series of resolutions, cast in the form of lawyer's interrogatories, demanding that the President reveal the exact "spot" where American blood had first been spilled. When Polk did not reply, Lincoln again took the floor and amplified his spot resolutions. The date was January 12, 1848.

"Mr. Cheermun," he sang out in his Indiana dialect, his head and shoulders tilted forward, "the spot where the war began" was crucial in determining who in fact had started it. In Lincoln's judgment, Polk's entire policy on this question was one of "the sheerest deception," part of a "design" to cover up the truth. True, Mexicans had fired on American troops in the disputed

Rio Grande country, but who really owned that country? The fact that Texas claimed it was meaningless, Lincoln said. Mexico also claimed it, so that it was just one claim against another. What one must do, he contended, was to get in back of the claims and find out which side had actual jurisdiction over the contested area. And in Lincoln's view Mexico did. Why so? Because the region was inhabited by Mexican people who'd been born there when it was indisputably part of Mexico and who'd never submitted to the civil rule of Texas or that of the United States. Therefore Polk had ordered American soldiers into a peaceful Mexican community in a deliberate effort to provoke a war. Consequently, his efforts to blame the conflict on Mexico amounted to "the half insane mumbling of a fever-dream," the product of "a bewildered, confounded, and miserably perplexed man" whose mind had been "tasked beyond it's power."

It was Lincoln's major speech of the session, a supremely partisan oration designed to bait and needle Polk into a response. But the President ignored the address and let his surrogates in Congress—prowar Democrats from the South—reply to Lincoln in condescending verbal salvos. For some reason, not a single Whig rose to defend him. And that was not the only irony. In his speech, Lincoln had derided Polk for failing to end the war. Yet just over two weeks later a peace treaty signed by Mexico arrived in Washington, and an unhappy Senate eventually ratified it. According to the terms, the United States paid Mexico fifteen million dollars and assumed her debts to American citizens, in exchange for the Rio Grande boundary, California, and the New Mexico Territory. America now had a transcontinental empire and "Mr. Polk's war" was over.

But Lincoln's troubles were not. Back in chauvinistic Illinois, Democratic papers ranted at him as an Illinois Benedict Arnold for his "treasonable assault" against Polk and the war. Lincoln's old adversary, the *Illinois State Register,* dubbed him "spotty Lincoln" and predicted that he would die of the spotted fever and go down in history as "the Ranchero Spotty of one term." Though a small group of Illinois Whigs now agreed with Lincoln and applauded his speech, many others were upset with him. Some gloomily predicted that Lincoln's antiwar stand would throw the Seventh Congressional District into the "loco" camp. And Billy Herndon, who sent Lincoln "back home" advice from Springfield, considered his partner frightfully misguided. Herndon wrote that he not only approved of the war, but thought Polk had a right to dispatch troops to the Rio Grande and guard against Mexican invasion.

Lincoln shot letters back in his own defense. He refused to believe the Seventh District would swing to the "locos" on his account—that was sheer nonsense. And Herndon—how could Billy write such things? Lincoln had spoken out against the war as a matter of conscience, he informed his partner. In blaming the war on Mexico, the Democrats had given him a clear choice —either "tell the *truth* or tell a *lie*" ("a foul, villainous, and bloody falsehood," as Lincoln told another Illinois colleague). And Lincoln went for telling the truth. Moreover, he categorically rejected Herndon's argument that a President could order an invasion of another country to prevent an aggressive act. That, said Lincoln, would allow an American President "to make war at will."

The criticism from back home gave him a mild attack of the hypo. He was tired—tired of writing letters, tired of handling requests for office. He was lonely for Mary and in April he wrote her so. "I hate to sit down and direct documents," he told her, "and I hate to stay in this old room by myself." He mentioned that he'd had a "foolish" dream about Bobby, but her recent letter assuring him that both boys were fine made him feel better. Did they get the "little letters" he'd sent them? "Don't let the blessed fellows forget father." And Mary, you say you are free of headache now? "That is good —good—considering it is the first spring you have been free from it since we were acquainted." Then he teased her a little. "I am afraid you will get so well, and fat, and young, as to be wanting to marry again."

Beyond Lincoln's windows, spring had come to Washington, with flowers blooming in the dooryards and the trees turning green again. Though lonesome and blue, he flung himself into Whig campaign work, for this was a Presidential election year and there was much to be done. By the rotation principle, Lincoln would not run for re-election this year (his present term would expire in 1849), but would step aside for Stephen Logan, his former law partner, who was stumping hard back in Illinois. Accordingly, Lincoln remained in Washington, devoting his considerable energies to the Presidential contest. As it turned out, his candidate was not Henry Clay but Zachary Taylor. While Clay was still his idol, Lincoln agreed with Whig bosses that Taylor stood the best chance of winning the White House because he was a war hero. Even so, Taylor faced a lot of hot opposition from Northern liberal Whigs. After all, he was a crusty, contentious Southerner who owned one hundred slaves and who embodied the military spirit in its most tenacious form. In addition, he seemed cheerfully ignorant about political affairs and had no discernible position on the significant questions of the day. Nevertheless, Whig bosses saw a winner in "Old Rough and Ready," and they (and Congressman Lincoln) were willing to sacrifice strict adherence to principle in order to gain the Presidency. Lincoln went for Taylor, he wrote a fellow Whig, not because Taylor would make a better Chief Executive than Clay, but because he was the only Whig who could defeat the hated "locos." "Mr. Clay's chance for an election," Lincoln contended, "is just no chance at all."

In his anxiety to get Taylor nominated, Lincoln attended the Whig National Convention in Philadelphia even though he was not a delegate. The party bosses got Taylor nominated on the fourth ballot and then offered him to the American electorate without a platform, which was one way to avoid controversy. Taylor's nomination, though, left the Whig house in an uproar, as New England Conscience Whigs cried out that the party had thrown principle to the winds and that they would never support a slave owner for President. Still, Lincoln discounted such dissidence and predicted "a most overwhelming, glorious, triumph" for the great Whig cause.

The Democratic candidate that year was Lewis Cass of Michigan, a dull party pro and a veteran of the War of 1812. To solve the divisive issue of slavery in the territories, Cass evoked the principle of popular sovereignty, which would let the people in each territory decide whether to have slavery or not. Stumping for Cass that summer was Stephen A. Douglas, now serving as a U.S. senator and enjoying a meteoric rise to power that made him one of the country's leading Democrats. In addition to the Whigs and Democrats, there was a new party in the field now—the Free-Soilers, made up of mutinous Whigs and Democrats, who ran former President Martin Van Buren as their candidate and boldly demanded that slavery be excluded from the territories and that no more concessions be made to the South.

As the campaign progressed, Lincoln toiled away in the Whig national headquarters, mailing out campaign documents and writing Whigs in New York, Pennsylvania, and Illinois to stand by old Taylor. He beseeched Billy Herndon to organize a Rough and Ready Club among Springfield's Young Turks and "let every one play the part he can play best—some speak, some sing, and some holler." In Congress that sweltering July, Lincoln defended Taylor in a half-hour speech, gesturing as he paced up and down the aisles. Lincoln insisted that the general had unimpeachable Whig credentials, insisted that he stood right on currency, the tariff, and internal improvements. What would Taylor do about the Wilmot Proviso? Frankly Lincoln didn't know. As a Northern and Western man whose constituents opposed slavery expansion, Lincoln favored the Proviso and hoped and believed that Taylor would as well. And Cass? If he were elected, Lincoln warned, slavery would flood unchecked to the Pacific. And you Democrats, he exclaimed, you accuse the Whigs of hooking on to Taylor's military coattails. Well, look what you did to Andrew Jackson. "Like a horde of hungry ticks you have stuck to the tail of the Hermitage lion to the end of his life." And now they were doing the same with "General" Cass, drumming him up as a military hero, too. Thereby hangs another "tail," Lincoln deadpanned, and he gave an outrageous spoof of Cass's berry-picking military adventures in the War of 1812 that kept the Whigs "in a continuous uproar." Cass's heroics, Lincoln went on, reminded him of his own exploits in the Black Hawk War, when he "had a good many bloody struggles with mosquitoes." But he hoped the Democrats wouldn't make fun of him, as they had with Cass, by writing him into a military star. . . . But in a moment he was serious again. The Democrats, he said, accused the Whigs of a stunning contradiction in denouncing the Mexican War and yet running a famous general of that war for President. Lincoln denied that there was a contradiction. The Whigs could assail Polk for a wicked, unconstitutional act in starting the conflict and yet praise those Whigs who fought for the flag once the war was under way. And Whigs had served gallantly—John Hardin had fallen at Buena Vista, and Clay and Webster had each lost a son. Such men deserved to be celebrated for standing by the country in an unhappy cause.

In September Lincoln stumped New England—perhaps at the request of the Whig National Committee—where he championed General Taylor and strove to prevent Whig defections to the Free Soil party. He rode slow, rattling trains across Massachusetts, speaking at Worcester, New Bedford, Boston, Lowell, Cambridge, and then Boston again. Always his message was the same: Conscience Whigs and abolitionists alike must shun another hopeless third-party movement and support the regular Whig ticket. The regular Whigs were also against slavery expansion. Yet if Massachusetts Conscience Whigs abandoned Taylor and wasted their ballots on the splinter Free Soil party, they would in effect be voting for Cass and so for slavery extension —the very thing they were all against.

On September 22 he appeared in Boston's Tremont Temple, where he shared the platform with William H. Seward, a leading Whig and former governor of New York, conspicuous for his beaked nose and ever-present cigar. Lincoln was impressed with Seward, who defended Taylor and also addressed himself to the slavery issue. The day would come, Seward declared, when the American people would expunge slavery—and they would do so through a combination of moral force and justice to Southerners, who would be reimbursed for their loss. But Seward warned that if all antislavery men crowded around the Free Soil banner, they would leave the two major parties—and the Union itself—controlled entirely by Southerners.

After the meeting, Lincoln went over to Seward and told him: "Governor Seward, I have been thinking about what you said in your speech. I reckon you are right. We have got to deal with this slavery question, and got to give much more attention to it hereafter than we have been doing."

In October he was back in Illinois with his family, speaking at various Whig rallies and stoutly defending his Mexican War stand. He was troubled about the recent state elections, though, for contrary to his own forecasts, the Whigs sustained a shocking defeat in Lincoln's own Congressional district, where the Democratic candidate slipped past Logan by just over one hundred votes. While the election may have reflected a popular repudiation of Lincoln's antiwar position, there were other reasons for Logan's defeat. For one thing, his opponent had a distinguished war record, which the Democrats played up for everything it was worth. For another, as Lincoln pointed out, Logan was a grouchy and unpopular man whom many Whigs refused to endorse. Lincoln, of course, was unhappy about the election, but insisted that the district would "right itself" in November and go for Taylor by twenty to one.

As it turned out, Taylor did triumph in Lincoln's district—and went on to win the Presidency by carrying over half of the fifteen slave states and almost half of the fifteen free ones, including Massachusetts. But to Lincoln's dismay Illinois went for Cass and remained a "loco" state.

That December Lincoln returned to Washington alone, to attend his last Congressional session. This time the major issue was slavery, so that Lincoln had plenty of opportunity to deal with that vexatious question, as he'd told Seward they must. From December to March, both houses rang with debates over the status of slavery not only in the territories but here in Washington as well. When Northern free-soilers resolved to prohibit slavery in New Mexico and California, proslavery Southerners protested in a storm of speeches and counterresolutions. Gaunt and icy-eyed John C. Calhoun, senior senator from South Carolina, argued that Southerners had a right to take their chattel into all federal territories, and he and his associates moved to extend the federal Constitution and federal laws—which protected slavery—over both New Mexico and California.

Arrayed between the free-soil and proslavery camps were Senator Douglas and his Northern Democratic followers. As chairman of the powerful Committee on Territories, Douglas was striving to organize the West—including the new lands acquired from Mexico—as rapidly and efficiently as possible. But the fight over slavery in the territories kept getting in his way. So the Little Giant looked about for expedients that would stop the controversy so he could get on with his "Western Program," which called for extensive railroad building, homestead legislation, and river and harbor improvements that would make the United States a mighty empire. When free-soilers rejected the extension of the Missouri Compromise line, Douglas turned to

Cass's doctrine of popular sovereignty as the answer he was looking for, as a useful device that would take the slavery question out of Congress—where it only kept everybody in an uproar—and banish the accursed problem to the frontier. Then in his view Congress could concentrate on more important things, such as enacting his Western Program.

For now, though, neither proslavery nor free-soil forces would accept popular sovereignty, and Congress degenerated into a confusion of voices, bills, motions, and resolutions concerning slavery in the territories. In later years, Lincoln claimed that in the session of 1848–49 he voted for the Wilmot Proviso or the idea behind it at least forty times—which was an exaggeration, but he was a thoroughly committed free-soil man. Yet he did not participate in the fiery oratory that surrounded slavery in the Far West, because he was shocked and aggrieved at how slavery could arouse such passions, such menacing threats, among national leaders he'd assumed were fairly reasonable men.

Apart from the Wilmot Proviso, Lincoln tended to be inconsistent in his votes on slavery. In the first session, he'd always favored receiving and considering abolitionist petitions, because he agreed with John Quincy Adams that Americans had a right to address their representatives. In fact, Lincoln once introduced an antislavery petition himself. But in the current session he sided now with the Southerners, now with Joshua Giddings and the Conscience Whigs, on technical motions involving runaways and other slavery matters. Moreover, when Giddings and his Whig associates moved to abolish the slave trade in Washington, Lincoln voted both for and against the consideration of their resolution. His "yes" and "no" stance suggests that he was uncertain as to how the issue should be approached, especially since the slave trade resolution brought hot and furious denunciations from Southern Whigs and Democrats alike. Lincoln had fraternized with Southern Whigs and sympathized with their apprehensions about the peculiar institution. Yet he personally detested the existence of slavery in the District of Columbia; he agreed with Giddings—his messmate at Mrs. Spriggs's—that the buying and selling of human beings in the United States capital was a national disgrace, a curse on the American experiment which Old World diplomats were quick to observe. In fact, from the windows of the Capitol, Lincoln could see the infamous "Georgia pen"—"a sort of negro-livery stable," as he described it, "where droves of negroes were collected, temporarily kept, and finally taken to Southern markets, precisely like droves of horses."

The spectacle offended him, just as the sight of those manacled slaves on the Ohio River had offended him back in 1841. Finally he stopped vacillating about slavery in the District and resolved to do something about it, evidently with the support of Giddings and his colleagues. In January, 1849, Lincoln went even further than the slave trade resolution: he informed the House that he would introduce a bill to abolish slavery itself in the District of Columbia. Since 1837 he'd contended that this could be done constitutionally. And now he was ready to offer legislation embodying the principles of gradual and compensated emancipation—henceforth Lincoln's favorite solution to the slavery problem. His projected bill would emancipate all slave children in the District after January 1, 1850, and would compensate their owners. The bill, though, would have to be ratified by District voters before it could take effect. Another provision required Washington and Georgetown authorities to send fugitive slaves back to their masters (a provision which enraged abolitionist Wendell Phillips, who later referred to Lincoln as "that slave hound from Illinois"). Lincoln announced that about fifteen of Washington's leading citizens had approved of his plan, whereupon his Southern opponents shouted out, "Who are they?" "Give us their names."

But Lincoln didn't answer—and he never formally offered his emancipation bill either. He never fully explained why. It's possible that Giddings quarreled with him about the fugitive slave provision—for Giddings hated to return runaways—and withdrew his support. In later years, all Lincoln said was that his former backers abandoned him and so he dropped his plan because he had little personal influence.

But there was clearly more to it than that. For Southerners were adamantly hostile to his bill, viewing it as one of a mounting number of Yankee assaults against slavery—first in the territories and now in the national capital—that seemed part of a sinister design to clear the way for a national abolition law, one that would plunge the South into racial chaos. Some Southerners were so upset about all this that they threatened to boycott Congress. Others lined up with John C. Calhoun, who composed a ringing "Address" to the Southern people about the present crisis. In his address, Calhoun asserted that there were Northern bills in Congress now—particularly that of "a member from Illinois"—which portended doom and disaster for slavery and the entire Southern way of life based on that institution. Then Calhoun got down to the basic issue here—the question of race. "To destroy the existing relation between the free and servile races at the South would lead to consequences unparalleled in history. They cannot be separated, and cannot live together in peace, or harmony, or to their mutual advantage, except in their present relation." Should the blacks be emancipated, Calhoun warned, the result would be "the prostration of the white race," because the liberated Negroes would gain the right to vote and hold political office, would form political alliances with Northerners, and would control the patronage in the Southern states, thus rendering that hapless region the "permanent abode of disorder, anarchy, poverty, misery, and wretchedness." How to avoid such desolation? such calamity? The Southern people must unite, Calhoun declared, and show the North that they were determined to maintain white superiority in their region come what may. This would cause the North to pause and calculate "the consequences," and then to adopt a more cordial policy toward slavery that would ease sectional tensions.

Southern resistance like that clearly made Lincoln stop and listen to the advice of moderate and conservative Whigs—that his abolition bill would only drive Southerners to more dangerous threats. However much he believed in his bill, he would never get it adopted in the present Congress anyway. And so when his own backers left him, he did indeed give the matter up. Then he sat in silence again as the debates over slavery roared on to the end of the session. In the House, as he looked glumly on, Congressmen shouted and gestured at one another, and two of them even got into a fistfight. Other politicians showed up in the House and the Senate too drunk to fight or shout. Senator Sam Houston of Texas, who was accustomed to frontier rowdiness, was nevertheless appalled at all the riot and drunkenness that distinguished the dying moments of the Thirtieth Congress. In the end Congress enacted none of the antislavery or proslavery bills, failed to provide California and New Mexico with any sort of territorial governments, and adjourned in hopeless deadlock early on Sunday morning, March 4, 1849. For Lincoln, the session had been a disquieting lesson on how "the distracting question of slavery," as he put it, could poison all reason and drive white Americans to the very brink of violence.

On Monday Lincoln dutifully attended Taylor's inauguration and the inaugural ball. Then he spent the next couple of months corresponding with office seekers and compiling recommendations for federal jobs. At last he submitted his list to the Taylor administration, including those Illinois Whigs he thought should be considered for the Cabinet. Since Lincoln had campaigned hard for Taylor, he believed the President owed him and the state of Illinois the honor of a Cabinet post.

But Taylor ignored his requests, and in May Lincoln went home in despair. "Not one man recommended by me has yet been appointed to any thing, little or big, except a few who had no opposition," he complained. By now only one post of any consequence remained unfilled—the commissioner of the General Land Office in Washington—and Lincoln went all out to get it for Cyrus Edwards, Ninian's brother, and preserve at least that political crumb for his state. But then Edward Baker promoted another Illinois Whig for the job, and the two candidates and their supporters began to fight like caged crabs. In the midst of all the bickering, Lincoln pleaded for unity lest Illinois lose the post entirely. But because "I must not only be chaste but above suspicion," he stuck by his candidate Edwards.

Then the fight thickened. Justin Butterfield, a Chicago Whig who had endorsed Clay and fought against Taylor "to the bitter end," now announced for the Land Office job. Lincoln was irate, crying that Clay men had no right to the patronage, crying that federal jobs should go to loyal Taylor men like Cyrus Edwards. At this juncture Lincoln's friends begged him to seek the commissionership himself, on the grounds that he alone could save the post for Illinois. But Lincoln had promised to support Edwards and refused to abandon him.

When Butterfield's appointment seemed imminent, Lincoln's friends interceded with President Taylor and argued that Lincoln was the only man for the job, that Butterfield's appointment would ruin the Whigs in Illinois. At that Taylor agreed to postpone his decision and give Lincoln three weeks to collect a dossier. Only when informed that the race was now down to him and Butterfield did Lincoln decide to seek the position for himself. That was on June 2. Then he set about frantically getting Whigs to write him letters of recommendation.

Now it was Edwards's turn to be irate. He accused Lincoln of violating his pledge, of selling Edwards out. He even sent a letter to Washington that damned Lincoln and praised Butterfield. Edwards "is angry with me," Lincoln groaned, and "is wronging me very much." But what could he do? Stand by and watch an ingrate like Butterfield walk off with the last political plum? Still, if Lincoln lost Edwards's friendship because of this patronage thing he would be very disheartened.

Finally Lincoln got his dossier together and mailed it off to Washington. As he awaited the outcome, Secretary of the Interior Thomas Ewing evidently tampered with Lincoln's papers, arranging his and Butterfield's dossiers so that documents against Lincoln were in prominent display. Ewing recommended Butterfield for the commissionership and Taylor accepted Ewing's choice.

Lincoln was crushed when he received the news. He complained that it was just like 1840, when Harrison won the White House and then appointed "a set of drones, including the same Butterfield, who had never spent a dollar or lifted a finger in the fight."

To make amends, the administration offered to appoint Lincoln secretary of the Oregon Territory, a lusterless post which he rejected. After that somebody in Washington—maybe old Taylor himself—felt sorry about Lincoln's treatment, and in late September the administration offered him the governorship of Oregon, a fairly prestigious job that paid three thousand dollars a year. Yet Lincoln turned it down as well, for reasons that were "entirely personal," said his friend Dr. Henry. The reasons weren't difficult to figure out. A remote satellite beyond the Rockies, Oregon must have seemed like political exile to Lincoln, who was already well established in Illinois. And Mary apparently opposed the move, too. Her father had died in July; little Eddie was in delicate health; and she simply wasn't up to a long, hazardous journey to the Far West.

Thus Oregon was out. And so was another term in Congress—"I neither seek, expect, or deserve" a seat in the next Congress, Lincoln informed a supporter. He wasn't interested in the state legislature either. So he returned to his law practice with no idea what his future in politics might be.

If Lincoln was disillusioned with public office, in November there came a final blow. He received his dossier back from Washington, only to find that a couple of letters were missing. He wrote Ewing about them, Ewing wrote something in reply, and then Lincoln mysteriously dropped the matter. Though he grumbled to a federal employee about the "villainy" and "original villain" in the land office struggle, he remarked ironically that "my great devotion to Gen: Taylor personally; and, above all, my fidelity to the great Whig cause, have induced me to be silent; and this especially, as I have felt, and do feel, entirely independent of the government, and therefore above the power of it's persecution."

PART · THREE

ON THE PILGRIMAGE ROAD

"Upon his return from Congress," Lincoln wrote in his autobiography, "he went to the practice of the law with greater earnestness than ever." As he did so, attorney Lincoln began a self-imposed exile from political office, trying to take stock of himself, sort out his ideas, figure out where he wanted to go politically.

Not long after he'd resumed his law business, Eddie Lincoln fell gravely ill. For nearly two months the Lincolns nursed the boy day and night, but it was no use. On the morning of February 1, 1850, Eddie died. Lincoln's boy was gone—only four years old, dead before he'd hardly begun—and Lincoln did not think he could bear the hurt inside. And so there was another dirge, another coffin: his own son lowered into a cold and indifferent grave. Yet in all his anguish he found escape in work—escape in the troubles and despair of others. But what was Lincoln to do about Mary? On the day Eddie died she collapsed in shock. And for weeks afterward she stayed in the bedroom and wept. Only Lincoln could get her to eat anything. Only he could get her to come out. And when she finally did emerge to manage their home again, she was more anxious and more insecure than ever, given to sudden explosions of temper against their servant girl, against Robert, and against Lincoln himself.

Lincoln commiserated with Mary, tried to comfort and understand her, but he realized he couldn't really help his wife. He watched, a brooding observer, as she reached out finally to religion . . . reached out to Dr. James Smith, pastor of the First Presbyterian Church, for the strength to continue. Over the years Mary had drifted away from religion, but she returned now because only Dr. Smith's reassurances that Eddie was in Heaven, that God loved and cared for him, could ease her grief. Lincoln appreciated her needs and even rented a family pew in the church so that Mary could attend services and sing and pray as a member. Lincoln attended some himself, befriended Dr. Smith, welcomed him to the house and thanked him for helping Mary, even read a book the pastor had written against skepticism. But Lincoln refused to let Dr. Smith convert him and declined to join Mary's church.

In his office, meanwhile, Lincoln followed political developments as reported in the *Congressional Globe* and the *Illinois State Journal* (formerly the *Sangamo Journal*), followed the sectional storm whipped up over California's bid for statehood, a storm that almost blew the Southern states out of the Union until the Compromise of 1850 brought a temporary lull in sectional hostilities over slavery. Lincoln studied the series of compromise measures which emerged from Congress that momentous year—laws that not only admitted California into the Union as a free state, but organized the New Mexico and Utah territories on the basis of popular sovereignty, leaving it up to the legislature of each territory whether to legalize or prohibit human bondage. Congress also abolished the slave trade (but not slavery itself) in the national capital and enacted a stringent new fugitive slave law. The latter may have pleased slave-owning Southerners, but it elicited cries of outrage from black and white abolitionists, who could not bear the sight of runaways being arrested in Northern communities and sent in chains back into slavery.

In July, 1850, Zachary Taylor died of "cholera morbus," leaving Millard Fillmore as the new President. With the nation in mourning, Lincoln turned up in Chicago and gave a eulogy to the old general, with Eddie's death much on his mind. "The death of the late President may not be without its use," Lincoln said, "in reminding us that *we*, too, must die. Death, abstractly considered, is the same with the high as with the low; but practically, we are not so much aroused to the contemplation of our own mortal natures, by the fall of *many* undistinguished, as that of *one* great, and well known, name. By the latter, we are forced to muse, and ponder, sadly. 'Oh, why should the spirit of mortal be proud.' " And he quoted from his favorite poem, "Mortality."

> *So the multitude goes, like the flower or the weed,*
> *That withers away to let others succeed;*
> *So the multitude comes, even those we behold,*
> *To repeat every tale that has often been told.*
>
> *For we are the same, our fathers have been,*
> *We see the same sights our fathers have seen;*
> *We drink the same streams and see the same sun*
> *And run the same course our fathers have run. . . .*
>
> *They died! Aye, they died; we things that are now;*
> *That work on the turf that lies on their brow,*
> *And make in their dwellings a transient abode,*
> *Meet the things that they met on their pilgrimage road.*

Yea! hope and despondency, pleasure and pain,
Are mingled together in sun-shine and rain;
And the smile and the tear, and the song and the dirge,
Still follow each other, like surge upon surge.

'Tis the wink of an eye, 'tis the draught of a breath,
From the blossoms of health, to the paleness of death.
From the gilded saloon, to the bier and the shroud.
Oh, why should the spirit of mortal be proud!

By autumn time, the Lincolns had learned to live with Eddie's death, Mary finding what solace she could in religion and Lincoln in his professional work. But what helped them most of all was that Mary was pregnant again. On December 21, 1850, only eleven months after Eddie's funeral, she gave birth to another son, whom they named William Wallace. As Lincoln teased and tickled the baby, Mary assured him that Willie was truly "her comfort."

A song . . . followed by another dirge. In late 1850 Lincoln's father lay dying in a farmhouse in Coles County. John Johnston—Sally's boy—wrote Lincoln semiliterate letters about Thomas's condition, saying he would not recover and wondering why Lincoln refused to reply. "Because," Lincoln finally informed Johnston in January, 1851, "it appeared to me I could write nothing which could do any good." And Lincoln couldn't visit his father either, since Mary was ill and he had pressing business commitments. But he did send a final message to the dying man, to that father who was a stranger to him, who perhaps hadn't been positive and close and understanding and intelligent enough in raising his only son. "Tell him," Lincoln wrote Johnston, "to remember to call upon, and confide in, our great, and good, and merciful Maker; who will not turn away from him in any extremity. He notes the fall of a sparrow, and numbers the hairs of our heads; and He will not forget the dying man, who puts his trust in Him. Say to him that if we could meet now, it is doubtful whether it would not be more painful than pleasant; but that if it be his lot to go now, he will soon have a joyous [meeting] with many loved ones gone before; and where [the rest] of us, through the help of God, hope ere-long [to join] them."

On January 17, 1851, Thomas Lincoln died. Lincoln did not attend the funeral.

Yet he worried about his stepmother, because John Johnston—who in Lincoln's opinion hadn't done a day's work in his life—wrote Lincoln that he was "broke" and "hard pressed" in Illinois and wanted to sell the Coles County farm and move to Missouri. Afraid he would leave Sally destitute, Lincoln replied: You say you are broke and can't get along. Well, why? It's because you are lazy, because you have not bothered even to put in a crop, because you "have *idled* away all your time." All this whining about not getting along is nonsense. *"Go to work* is the only cure for your case," Lincoln said. Moreover, he would not let Johnston sell all of "Mother's" share of the farm. "The Eastern forty acres I intend to keep for Mother while she lives —if you *will not cultivate it;* it will rent for enough to support her—at least it will rent for something. Her Dower in the other two forties, she can let you have, and no thanks to [me]." And so he made certain that Sally retained enough of the family farm to live on, and he even went by to see her when business called him to Coles County. But Lincoln still had little to do with the rest of his relatives on his side of the family. And he allowed Nancy Lincoln's grave in Indiana to remain without a monument of any kind.

Nevertheless, all the deaths and un-certainties of these years brought him closer to his own wife and sons. He tried to enjoy them more during their time together—their meals and family outings. The Lincolns had a third son now—born on April 4, 1853, and named Thomas after Lincoln's father. Because the boy had a large head and squirmed like a tadpole, they called him Tad or Taddie. Due perhaps to a cleft palate, Tad had a speech impediment which caused him to mix his words in a lisp—but this only endeared him all the more to Lincoln. An impulsive, uninhibited child whom Mary regarded as "her little *sunshine,"* Tad could throw monumental temper tantrums when he failed to get his way. But Lincoln laughed at his outbursts, holding him at arm's length as the boy squealed and tried to kick him.

The Lincolns strove to be understanding parents rather than disciplinari-ans. In fact, at a time when parents and teachers alike thrashed children severely, the Lincolns were remarkably permissive—too permissive in the eyes of several friends, who thought the Lincoln boys spoiled little hellions. But the Lincolns ignored such criticism. They listened to their children, took an interest in their worlds, boasted of their achievements. Willie became a

serious, precocious boy who liked to read books and memorize railroad timetables. Eventually he could conduct an imaginary train all the way from Chicago to New York without a mistake—an accomplishment both Lincolns could not praise enough. And they encouraged their sons to have fun as well. Mary held lavish birthday parties for them and often joined in the merriment herself—and dignity be damned. Lincoln, in turn, invited the boys to ask questions and satisfy their youthful curiosity. As Willie and Tad grew older, he took them for walks in the country, explaining things carefully as they made their way through wild flowers and wind-tossed trees.

Yet Lincoln was never so close to his oldest boy, Robert Todd. In truth, there was an estrangement between them reminiscent of the coldness between Lincoln and his father. In later years, Robert hinted at the reason for this: "My Father's life was of a kind which gave me but little opportunity to learn the details of his early career. During my childhood & early youth he was almost constantly away from home, attending courts or making political speeches."

Left alone with his mother most of the time, Robert developed into an aloof, chilly introvert. "He is a Todd and not a Lincoln," Herndon caustically remarked. "Bob is little, proud, aristocratic, and haughty, is his mother's 'baby' all through." Years older than Willie and Tad, he seemed jealous and resentful of the way Lincoln fawned over them, which only added to Robert's alienation. And then there was his visual trouble: one eye turned inward so that his schoolmates called him "cockeye," which humiliated him. And so he grew up in his mother's shadow, injured, touchy, and remote.

In his way Lincoln cared for his oldest-born and worked to give him the education Lincoln never had. He and Mary put Robert in a private school in Springfield and later sent him to another in Jacksonville to prepare him for college.

As Lincoln concentrated on his law practice in the early 1850s, he rose to the top of the legal profession in his state. He earned a reputation there as a first-rate general lawyer and litigation man who could woo judge and jury alike—and do so in the same lucid style that distinguished his best political oratory. Because he was talented and because he defended people of all classes with equal care and industry, he became one of the most sought-after attorneys in central Illinois.

On a typical workday in the early 1850s, attorney Lincoln rose at dawn and went outside to feed and curry his horse and cut his own firewood. At eight he would breakfast with his family, picking at an egg and perhaps some fruit —he always ate lightly. And then he would set out for his law office, dressed in a plain linen suit and a silk stovepipe hat that made him appear even taller than he was. While legends were to grow up about him as an ill-clothed slouch, "Mr. Lincoln dressed as well as the average Western lawyer of his day," a fellow attorney declared. "I do not think he gave much time to the tying of his necktie, and he could not have been said by his best friends to be much of a dude, but he was always respectably dressed." Now in his forties, he stepped along in a flat-footed gait, slightly stooped, his head bent forward and his hands clasped behind him. Though he called himself an old man now, Lincoln still had a young and clean-shaven face, with a swarthy outdoors complexion and luminous gray eyes. He continued to think he was a homely man, but no longer got so depressed about it. Now he had a rare ability to laugh at himself and liked to quip that if he ever met anybody uglier than Lincoln, he would shoot the wretch and put him out of his misery.

On the way to work, he stopped from time to time to chat with friends and fellow lawyers. He commanded their respect, for he had a reputation as a "lawyer's lawyer"—a knowledgeable jurist who argued appeal cases for other attorneys. They all called him Lincoln or Mr. Lincoln, and he also referred to most of them by their last names. Judge David Davis was just Davis. Stephen Logan was Logan and Leonard Swett was Swett. Nobody called him Abe—at least not to his face—because he loathed the nickname. It did not become a self-made man whose ambition for high station in life, as Herndon put it, was "a little engine that knew no rest." Though a man of great moral integrity, Lincoln took pride in his status as a professional lawyer and politician, and he not only liked money but used it to measure his worth. In the mid-1850s he earned around five thousand dollars a year from his law practice—a high income in his day. By 1860 Lincoln had more than fifteen thousand dollars invested in real estate, in mortgages, interest-bearing notes, bank accounts, and insurance policies. In contrast to the legend, which portrays him as a kind of homespun Socrates who disdained material rewards, the real Lincoln was a man of substantial wealth.

It was usually nine o'clock when Lincoln reached his office building and entered where the sign of LINCOLN & HERNDON swung on rusty hinges over the street. Then up the stairs to the second floor and down a gloomy hallway to his two-room office. Herndon was already at his desk, trying to sort out their chaotic daybook or perusing the latest copy of the *Atlantic Monthly* or the *Westminster Review,* to which he subscribed. First the partners reviewed the day's schedule. Then Billy was off to the Supreme Court library to do

research on a pending case, and Lincoln was busily interviewing clients, advising them to settle their troubles out of court if possible. But if adjudication were necessary, he was scrupulously fair in establishing his fee, content to make his money through the sheer volume of his practice. Contrary to myth, he took cases as they came to him regardless of the justice involved. And once he accepted a retainer, as one writer has stated, Lincoln played for keeps, leaving no stone unturned in representing a client. To ensure accuracy, he took down a declaration in the client's presence, writing in a prose so clear and precise that even the uneducated could understand. Then he would file the declaration and other legal documents with the clerk of the appropriate court, and the case would be listed on the judge's docket for the ensuing term.

When there was a lull in business, Lincoln would lie on the sofa as always and study the newspapers. Or if Herndon was back from the library, Lincoln would ask what he was reading now, relying on Herndon to keep him informed. "Billy," Lincoln would drawl, "tell us about the books." Whereupon Herndon would plunge into a nervous monologue about his latest reading list, jumping from history books to works on phrenology, science, and the law. But when Herndon took off on some "deep volume," like the writings of Immanuel Kant, Lincoln would wave it away as indigestible. Then they would talk about politics or people, with Herndon raving about the "God damn Irish" who did nothing but rob people and vote for Democrats. Herndon noted that Lincoln had no ethnic prejudices, "preferring the Germans to any of the foreign element, yet tolerating—as I never could—even the Irish."

Though Lincoln immersed himself in work, he always found time for friends and colleagues who dropped by the office to visit, share political news, or discuss law cases. In their conversations, Lincoln proved himself a supremely logical and analytical man with a talent for searching out the "nub" of a question. Yet at other times he spoke of superstitions, of signs, visions, and troubling dreams. And that was not his only contradiction. If he seemed friendly on the surface, seemed to enjoy people and companionship, he revealed nothing about his inner feelings, not even to his closest friends. If he'd once confided in Joshua Speed, he kept his burdens to himself now. Herndon thought him the most "shut-mouthed" man who ever lived. And others agreed. "He always told only enough of his plans and purposes to induce the belief that he had communicated all," recalled a colleague, "yet he reserved enough to have communicated nothing." Said another: "He made simplicity and candor a mask of deep feelings carefully concealed."

Even as he grew older, Lincoln continued to suffer from the hypo, from spells of melancholy that troubled his friends and associates. In the midst of conversation, they observed, he would slip away into one of his moody introspections, lost in himself again as he stared absently out the unwashed windows of his office, brooding over untold thoughts and secret storms, until he who viewed each human life as a pawn in the hands of an unknowable God, as a doomed and fleeting moment in a rushing ocean of time, would start muttering the lines of "Mortality." As his colleagues looked on in worried astonishment, his face would become so despondent, his eyes so full of anguish, that it would hurt to look at him.

But abruptly, "like one awakened from sleep," Lincoln would join his visitors again—his mood swings were startling—and joke and quip with them until laughter lit up his cloudy face. For humor was his opiate—a device "to whistle down sadness," as a friend said. Some of Lincoln's stories were quaint anecdotes which illustrated some point. Others were mindless rib-ticklers, like the one about the man in an open carriage who got caught in a nighttime downpour. As Lincoln repeated the yarn, the traveler was out on a lonely country road when the storm hit, and as he passed a farmhouse a drunk fellow stuck his head out a window and shouted, "Hullo! Hullo!" The traveler stopped his buggy and asked what the drunk wanted. "Nothing of you," the man replied. "Well," the traveler exclaimed, "what in damnation do you yell hullo for when people are passing?" "Well," the drunk retorted, "what in damnation are you passing for when people are yelling hullo?"

Still other Lincoln tales were pungent and downright bawdy. An example was the piece of foolery called "Bass-Ackwards" which he handed a bailiff in Springfield one day: "He said he was riding *bass-ackwards* on a *jass-ack,* through a *patton-cotch,* on a pair of *baddle-sags,* stuffed full of *binger gred,* when the animal *steered* at a *scump,* and the *lirrup-steather* broke, and throwed him in the *forner* of the *kence* and broke his *pishing-fole.* He said he would not have minded it much, but he fell right in a great *tow-curd;* in fact, he said it give him a right smart *sick* of *fitness*—he had the *molera-corbus* pretty bad. He said, about *bray dake* he come to himself, ran home, seized up a *stick* of *wood* and split the *axe* to make a light, rushed into the house, and found the *door* sick abed, and his *wife* standing open. But thank goodness she is getting right *hat* and *farty* again."

When it was court time, Lincoln would head across the street with his colleagues, engaging them in repartee or perhaps chatting about some legal point. In court, Lincoln was deadly serious when it came to winning his case. He took great care in selecting a jury—and once compiled a three-page, annotated list of prospective jurors for a murder trial. Though an eminently shrewd and practical attorney, Lincoln had his superstitious biases: he contended that fat men were ideal jurors because he believed them jolly by nature and easily swayed. He rejected people with high foreheads (he thought they had already made up their minds) and considered blond, blue-eyed males inherently nervous and apt to side with the prosecution in murder cases. Once a trial was under way, Lincoln proved himself a formidable jury lawyer. Eschewing "sickly sentimentalism" and "rhetorical display," said a newsman who watched him in action, Lincoln employed masterful ingenuity in cross-examining witnesses and addressed the jury with a mixture of logic, force, and wit, weaving the smallest details into a convincing argument.

As an all-purpose attorney, Lincoln argued cases that ranged across the entire legal spectrum, from divorce, murder, and rape cases to contests involving disputed wills, maritime law, the right of way of railroads, actions for injunctions, foreclosures, debts, trespass violations, slander suits, and patent infringements. And in criminal cases he could "stoutly argufy" on either side of the law, serving one day as defense counsel and another day as court-appointed prosecutor. For example, Lincoln once prosecuted a man charged with raping a seven-year-old girl and got him imprisoned for eighteen years. In another contest, though, Lincoln represented a farmer who had shot two boys—one seriously—for raiding his watermelon patch. Lincoln secured the man's acquittal, but had to sue to collect his fee.

Lincoln also defended both sides in fugitive slave cases, which illustrates the essentially pragmatic approach to the law he and most other attorneys adopted. In the case of *Bailey* v. *Cromwell,* tried in 1841 in the Illinois Supreme Court, Lincoln won the freedom of an indentured Negro girl sold by one white man to another. Lincoln persuaded the court that it was illegal to sell a human being in Illinois, inasmuch as slavery was prohibited there both by the Northwest Ordinance of 1787 and by the Illinois constitution. The case established the enlightened principle that in Illinois every person was free regardless of color and that the sale of a free person was illegal.

Yet in the Matson slave case of 1847, tried in a dreary courthouse in Charleston, Illinois, Lincoln turned around and defended a Kentucky slave owner out to retrieve a slave family that had run away from him. As it happened, the blacks were part of a slave coffle the man had brought from his Kentucky plantation, to work a tract of land he owned in Illinois. The case was freighted with ironies, since the opposition advanced the identical arguments Lincoln had made in *Bailey* v. *Cromwell,* but somehow neglected to mention the case. Lincoln, on the other hand, declared in the strongest possible terms that his client had brought the slaves to Illinois strictly on a temporary basis, that he planned to take them back to Kentucky after their work was done, and that Illinois law consequently did not apply to them. This was cold and brutal logic, demonstrating how attorney Lincoln could set aside his personal convictions—for he claimed to hate slavery—and go all out to win for a client, even if that meant sending a family back into bondage.

The court, though, ruled against him and set the blacks free. Yet it failed to cite the precedent Lincoln had helped establish in *Bailey* v. *Cromwell.*

Lincoln's varied legal practice took him into courts at all levels, from the United States Supreme Court itself—where he appeared twice in technical cases—down to the Eighth Illinois Circuit Court, where he did his bread-and-butter work. Presiding over the Eighth Circuit was Judge David Davis of Bloomington, a rotund Whig magistrate who sported an immaculately clipped beard and made a fortune in real estate and the merchandising business. Twice a year—in the spring and again in the autumn—Davis's court opened in Springfield, in Sangamon County, and then made "the swing" around the circuit, a sprawling judicial domain that comprised fourteen counties and twelve thousand square miles in the rural heartland of Illinois. For people who lived in remote neighborhoods, Davis and his cavalcade of lawyers provided a rare source of entertainment, like a traveling minstrel show. Thus when Davis opened court in some backwater county seat, farmers and villagers alike packed the courthouse and thrilled at all the human drama that unfolded there.

Besides the state's attorney, Lincoln was the only lawyer who rode the entire circuit for both sessions. He did so because he found the circuit emotionally and professionally indispensable. For one thing, all the traveling gave him plenty of time to be alone, to think about things without family interruptions. Riding in his horse-drawn rig, with a carpetbag containing his legal papers and an extra shirt, Lincoln was fascinated by the wide and silent prairies, the moonlit nights. When the weather was bad, though, he had to battle violent winds, thunderstorms, and blowing sleet and snow, making his way over muddy roads to the next county seat, only to put up for the night in some crude hostelry where he and his peripatetic colleagues slept two in a bed, three and four beds to a room. Yet he never complained about his hardships and seemed genuinely happy on the circuit, "as happy as *he* could be," said Judge Davis, "and happy no other place."

Also, Lincoln formed close personal and professional ties with prominent Whig jurists on the circuit. There was Jesse W. Fell, founder of the Bloomington *Pantagraph,* one of the leading Whig papers in Illinois. There was Leonard Swett, a famous criminal lawyer who worked with Lincoln in various rustic courthouses. There was Jesse Dubois, "a slim, handsome young man," Lincoln said, "with auburn hair, sky blue eyes, with the elegant manners of a Frenchman." There was Henry C. Whitney of Urbana, who was heavily involved with Lincoln in circuit-court litigation, often shared a bed with him, and left a vivid portrait of Lincoln's nomadic circuit life: of the rank and smoky taverns where the attorneys took their meals, of the roaring hotel fireplaces where they congregated for protracted political discussions. And then there was Judge Davis himself, who became an affectionate friend and invited Lincoln to stay at his home when court opened in Bloomington. Because of his weakness for good food and imported wine, Davis eventually swelled to three hundred pounds and required a two-horse rig to transport him around the circuit. When Lincoln wasn't traveling alone, he often rode alongside Judge Davis, who, like Billy Herndon, knew Lincoln "so well" and yet found him the most inscrutable man he'd ever met.

Finally there was Ward Hill Lamon of Danville. Lincoln had a standing business association with Lamon and numerous other local attorneys, who engaged Lincoln as co-counsel in cases they had drummed up. In fact, that was how traveling attorneys like Lincoln obtained much of their circuit practice. As they worked together in Danville, in Vermilion County, Lamon came to idolize "Mr. Lincoln," and Lincoln was fond of "Hill," too. Born twenty miles from Harpers Ferry, Lamon resided in northern Virginia until he was twenty-one, when he headed west and settled here in Illinois. A big, gruff fellow, Hill was a legendary boozer who spent much of his time in the saloon under his office, where he sang lewd and comic songs and got into brawls. Yet Lincoln accepted Lamon for what he was and never begrudged him for his drinking. In fact, Lincoln liked to quote what a Kentuckian once told him, that "its been my experience that folks who have no vices have generally very few virtues."

Back home from the circuit, Lincoln carried on a sizable practice in the federal courts, where he was as lucid and logical as always, appealing to a jury's sense of fair play and reducing complex issues to the simplest propositions—always his forte. At this time, the federal courts were swamped with patent suits, largely because rival manufacturers, striving to meet agricultural demands, produced their own reapers and other modern machinery with a cheerful disregard for patent rights. Lincoln always got his share of patent cases, for he was entranced with inventions and mechanical gadgets and could never resist a contest concerning them. In 1855 he was involved in the celebrated McCormick Reaper case tried in Cincinnati, where he was supposed to serve as co-counsel with such Eastern lawyers as Peter H. Watson of Washington and rude and supercilious Edwin M. Stanton of Pittsburgh. As it turned out, both men disparaged Lincoln as a Western hick and snubbed him throughout the trial, refusing to dine with him or to consult the extensive brief he had prepared. And Stanton supposedly referred to him as "that giraffe," as that "creature from Illinois." Though he'd never been so mortified in his life, Lincoln remained in the courtroom as an observer and learned a great deal from the eminent Eastern attorneys who conducted the case, noting in particular how they had mastered all the technical data. Later, when he argued a patent case in the federal court in Springfield, Lincoln exhibited models of various reapers for the benefit of the jury. He explained in detail how power and motion traveled to different parts of the machine, to the sickle and the revolving rake and reel. During his presentation, Lincoln knelt down by one machine and pointed to various parts—and soon the jurors came over and got down on their knees as well. They were all amazed at his command of technical information.

Though Lincoln labored diligently in the federal and state circuit courts, he did his most influential legal work in the prestigious Supreme Court of Illinois. In all he participated in 243 cases before that tribunal and won most of them. It was here that he earned his reputation as a lawyer's lawyer, adept at meticulous preparation and cogent argument. It was here that he argued *Bailey* v. *Cromwell*. And it was here, in 1854, that he represented the Illinois Central Railroad in the famous McLean County tax case. In it, Lincoln persuaded the court to exempt railroads from county taxes because they were "public works" and already paid taxes to the state, a far-reaching decision that facilitated railroad growth all over Illinois. In yet another irony, however, Lincoln had to sue the Illinois Central for the bulk of his five-thousand-dollar fee, finally collecting it in 1857.

Thanks in part to the McLean County tax case, Lincoln became known in the 1850s as a railroad lawyer. And this was true to the extent that he and Herndon regularly defended the Illinois Central and various other railroad companies. After all, these were years of prodigious railroad construction all over the Midwest, and this in turn created a whole new area of law and legal practice in which Lincoln was eager to participate. Moreover, the coming of the iron horse marked the end of steamboating's golden age and precipitated a titanic struggle in the Midwest between rail and water interests for commercial supremacy. And that struggle offered lucrative rewards for attorneys like Lincoln who could command a mass of technical evidence.

Characteristically, Lincoln hadn't always defended the railroads against the river interests. On the contrary, he'd entered politics favoring steamboat over railroad transportation; and in 1840 and again in 1851, he'd represented the rivermen in legal actions against bridge and railroad enterprises. But he realized that railroad transportation was the way of the future—and Lincoln was never one to ignore the future—and so in the mid-1850s he swung over to the side of the railroads. He represented the Illinois Central alone in eleven separate appeal cases before the Illinois Supreme Court (although there is no evidence that he acted as a lobbyist for this or any other railroad, as charged by some of his political foes). At the same time, he and Herndon still accepted cases as they came to them, carrying on their remarkably varied and lucrative general practice.

Lincoln, of course, derived a lot more from the law than just a high income. For the law kept his mind open, affording him rare insight into the foibles and complexities of human affairs. It also kept his literary talents honed to a fine edge, schooled him in the structure of argument and the uses of interrogatories, and taught him the merits of accuracy, precision, and painstaking thoroughness. Yet, though he was known far and wide in the Illinois bar, Lincoln characteristically belittled his own achievements. "I am not an accomplished lawyer," he asserted in some notes he jotted down for a law lecture. "I find quite as much material for a lecture in those points wherein I have failed, as in those where I have been moderately successful." But he did offer some constructive advice to potential attorneys. "Resolve to be honest at all events," he beseeched them, "and if in your own judgment you cannot be an honest lawyer, resolve to be honest without being a lawyer."

If Lincoln was in exile from public office in the early 1850s, he still remained active in Whig affairs. He kept in touch with prominent Whigs around the state—his wide-ranging law practice helped him here—and served as the Illinois member of the Whig National Committee. Moreover, when Henry Clay died in June, 1852, Lincoln extolled him in a speech in Springfield, calling Clay "my beau ideal of a statesman," for whom Lincoln had worked "all my humble life." This overlooked the fact that Lincoln had opposed Clay in the Whig Presidential nomination of 1848. Nevertheless, Lincoln's address rang with praise for Clay's staunch nationalism and moderate antislavery views and for his role as an eloquent and effective compromiser. Above all, Lincoln applauded Clay's "devotion to the cause of human liberty," applauded his "sympathy with the oppressed every where" and his "ardent wish for their elevation." With Clay (as with Lincoln), "this was a primary and all controlling passion." Clay loved the United States "partly because it was his own country, but mostly because it was a free country; and he burned with a zeal for its advancement, prosperity and glory, because he saw in such, the advancement, prosperity and glory, of human liberty, human right and human nature. He desired the prosperity of his countrymen partly because they were his countrymen, but chiefly to show to the world that freemen could be prosperous."

Because of his concern for human liberty and advancement, Lincoln said,

Clay "ever was, on principle and in feeling, opposed to slavery." Unlike many contemporary Southerners, who regarded blacks as subhuman, Clay did not in the matter of human rights exclude Negroes from the human race. And yet Clay himself owned slaves. "Cast into life where slavery was already widely spread and deeply seated, he did not perceive, as I think no wise man has perceived, how it could at *once* be eradicated, without producing a greater evil, even to the cause of human liberty itself."

Thus Clay (like Lincoln) aligned himself against both extremes on the slavery issue. Those abolitionists who damned the Constitution as a Hell's agreement and burned copies of it at their rallies, those who clamored for disunion itself rather than allow slavery to continue another hour, were now getting "their just execration," Lincoln said, as Clay's feelings and opinions were powerfully arrayed against them.

Yet Lincoln also turned Clay's judgments against the opposite extreme— those Southerners who defended slavery as a glorious blessing for blacks and whites, insisting that it was condoned by the Bible and sanctioned by all the lessons of history. Lincoln had studied the proslavery arguments now thundering out of Dixie, had read in the newspapers how certain politicians and preachers, for the sake of perpetuating slavery, now assailed and ridiculed "the white man's charter of freedom—the declaration that 'all men are created free and equal.'" How strange, Lincoln cried out, that Southerners should mock this "self-evident truth" as a "self-evident lie." How strange to hear them demean and disparage the ideal upon which this Republic was founded. This, he grimly observed, was not heard in the early days of the Union. And then he compared proslavery Southerners to "that truly national man" Henry Clay, who as a leader of the American Colonization Society had championed gradual emancipation followed by colonization. And Lincoln, too, counted himself a friend of gradual abolition and colonization, because he considered this the only workable solution to slavery—and the problem of racial adjustment after the blacks were liberated. For he understood how much Southerners fretted over racial amalgamation—his term in Congress had shown him how Southerners equated emancipation with racial mongrelization and social chaos. All right, then, let Americans relieve the South of the troublesome presence of all those Negroes and return them to their native land. "Pharaoh's country was cursed with plagues," Lincoln warned, "and his hosts were drowned in the Red Sea for striving to retain a captive people who had already served them more than four hundred years. May like disasters never befall us! If as the friends of colonization hope, the present and coming generations of our countrymen shall by any means, succeed in freeing our land from the dangerous presence of slavery; and, at the same time, in restoring a captive people to their long-lost father-land, with bright prospects for the future; and this too, so gradually, that neither races nor individuals shall have suffered by the change, it will indeed be a glorious consummation."

Of course, gradual emancipation and colonization would depend entirely on the willingness of Southerners to cooperate. Yet Lincoln had no evidence that Southerners would ever voluntarily surrender their slaves, voluntarily give up their status symbols and transform their own cherished "Southern way of life" based on the peculiar institution. Virginians, for their part, had once considered colonization (in the aftermath of Nat Turner), but had rejected it as far too costly and complicated ever to carry out. And neither they nor their fellow Southerners were about to emancipate their blacks and leave them as free people in a white man's country. Consequently, they became adamantly determined that slavery should remain on a permanent basis, not just as a labor device but as a means of race control in a region brimming with Negroes. Yet Lincoln clung to the belief that slavery would eventually die out in the South and that Southerners were rational men who would gradually liberate their Negroes when the time came for them to do so. And he clung to the belief that somehow, when the time did come, the Republic would pay out all the millions of dollars necessary to compensate Southerners for their losses and ship more than three million blacks out of the country.

In 1853, Lincoln was riding circuit when reports came of new Congressional skirmishing over slavery in the territories. It appeared that Senator Stephen A. Douglas was trying to organize a Nebraska Territory out in the American heartland, but free-soil and proslavery forces were wrangling bitterly over the status of slavery there. Lincoln followed the course of Douglas's territorial bill as it was reported in the *Congressional Globe,* and he became melancholy again. Friends who saw him sitting alone in rural courthouses thought him more withdrawn than ever. Once when they went to bed in a rude hostelry, they left him sitting in front of the fireplace staring intently at the flames. The next morning he was still there, studying the ashes and charred logs. . . .

In May, 1854, while Lincoln was attending court in Urbana, news flashed over the telegraphs that a momentous new Kansas-Nebraska bill had emerged from Congress. When Lincoln read the provisions of the bill in the newspapers, he was "thunderstruck and stunned," he was aroused "as he had never been aroused before." In a single blow, the bill had obliterated the Missouri Compromise line and in Lincoln's view had profoundly altered the entire course of the Republic so far as slavery was concerned.

PART·FOUR

REVOLT AGAINST THE FATHERS

Lincoln and other Northern free-soilers had always considered the Missouri Compromise line an inviolable barrier of freedom, for it excluded slavery from the northern section of the old Louisiana Purchase territory. With the line overturned, slavery could now spread into a vast northern frontier once reserved as a free-soil domain. And the man most responsible for this infamous development was Stephen A. Douglas, though Lincoln had little idea what motivated the Little Giant beyond what seemed a monstrous moral indifference to slavery—and a calloused effort to court Southerners in Douglas's unrelenting political ambitions, including a desire for the Presidency.

In truth, though, Douglas had eradicated the Missouri Compromise line because he now thought it a nuisance which kept the slavery question ablaze in Congress. It seemed that every time he introduced territorial legislation —such as his initial Nebraska bill—Congress plunged into forensic convulsions about the Missouri line and the whole issue of slavery in the Western territories. So away with the line altogether. What Douglas wanted was a panacea that would silence the infernal slave controversy once and for all, so that he could get on with organizing the West and building the American empire. And that panacea, he decided, was popular sovereignty, which he now elevated to the lofty heights of "a sacred principle." By applying the cherished concept of local self-government and relegating the slavery problem to the frontier, Douglas convinced himself that Congressional agitation over slavery would stop and that everything would turn out all right.

So, with the help of several powerful Southern politicians and the pro-Southern administration of Franklin Pierce, Douglas rammed a new Kansas-Nebraska bill through a divided and tumultuous Congress. The final version of the enactment pronounced the Missouri Compromise line "inoperative," advancing an utterly fallacious argument that the Compromise of 1850 had "superseded" that of 1820 and had established popular sovereignty as the new formula for dealing with slavery in the territories. Accordingly, the bill created the new territories of Kansas and Nebraska, carved out of the northern section of the old Louisiana Purchase, and left the status of slavery up to the people who settled there: at some unspecified time, they could decide through their governments either to legalize the institution or throw it out. In the Senate, all but two Southerners supported Douglas's bill, not because they liked popular sovereignty, but because they wanted the "theoretical right" to extend slavery into all federal territories, including any new ones the United States acquired in the future. When Pierce signed the bill into law, Southerners could legally take slaves into most of the Western frontier from Mexico to Canada.

As it turned out, the Kansas-Nebraska Act was a monumental fiasco which inflamed the slave controversy worse than ever. A storm of free-soil protest broke across the North, as Whigs and Democrats alike raged against the repeal of the Missouri line and the treachery of their Southern colleagues. The Whig party, already demoralized and bitterly divided over slavery, began to disintegrate. The Democrats, too, were in obstreperous disarray, with free-soilers castigating Douglas as a proslavery "dough face" and a traitor to his section. Salmon Chase of Ohio circulated a ringing "Appeal of Independent Democrats," which branded Douglas's measure as part of a sinister "Slave Power" conspiracy to seize the frontier and strengthen Southern control of the federal government. In speeches widely quoted in the Northern press, Chase also amplified an argument he'd been expounding for several years: that the Founding Fathers had deplored slavery and tried to seal its doom by restricting its growth, that they regarded freedom and equality as the natural condition of men and envisioned a future America when the "solemn pledge" of the Declaration would apply to all. Yet the aggressive Slave Power had now taken over the government and dedicated itself to wrecking the noble dreams of the founders by extending slavery everywhere.

While free-soilers worried about the far Nebraska country, too, they fretted most about Kansas, for the territory lay due west of slaveholding Missouri and seemed wide open to an immediate proslavery invasion. Consequently, free-soilers summoned the North to rally against the Southern plotters and save Kansas from the grasp of the "Slave Power." "Come on, then, gentlemen of the slave states," cried William H. Seward of New York, "since there is no escaping your challenge, I accept it in behalf of the cause of freedom. We will engage in competition for the virgin soil of Kansas, and God give the victory to the side that is stronger in numbers as it is in right."

Through the spring and summer of 1854 a powerful "anti-Nebraska" movement swept the Northern states, as free-soilers mobilized to block slavery expansion and dismantle Southern power in Washington. Local anti-Nebraska groups held rallies in Wisconsin, Michigan, and the Midwestern states, called themselves Republicans one after another, and vowed to form a new political party dedicated to saving the West for free labor and free white men. Still, party lines were extremely blurred, with many free-soilers clinging tenaciously to their old affiliations—Whigs and Democrats—and simply calling themselves anti-Nebraska men. As a consequence, Northern free-soilers marched under various banners that year: some were Republicans, others were "fusion" men, still others were anti-Nebraska Whigs or anti-Nebraska Democrats. But all shared a common purpose—to overturn the hated Kansas-Nebraska Act and roll slavery and Negroes back into the South.

With the free-soil movement blazing across the North, Lincoln was back in Springfield, studying Chase's arguments and poring over free-soil editorials in the newspapers. He believed Chase and other free-soil exponents absolutely right about Douglas and his Southern allies: they had instituted a revolt against the goals and ideals of the Founding Fathers and altered the whole course of the United States regarding human bondage. It had become fundamental to Lincoln's thought that slavery would ultimately perish in America, that the Republic would one day cease to be a divided land, and that the liberating visions of the founders would at last be realized. But now he feared all that was changed. Now Douglas and his Southern-dominated party had reversed the policy of slave containment begun by the founders. Now under the deceitful guise of popular sovereignty Douglas and his cronies had opened the gates for slavery to grow and expand and continue indefinitely. Unless free-soilers stopped them, Southerners would drag manacled Negroes across the frontier, adapting slave labor to whatever conditions they found there, putting the blacks to work in mines and on farms, and making bondage powerful and permanent in America. Then the Republic would never remove the cancer which infected its political system, would never remove the wretched contradiction, as Lincoln put it, "of slavery in a nation originally dedicated to the inalienable rights of man."

By late summer Lincoln was in the thick of the anti-Nebraska fight, thundering against "the great moral wrong and injustice" of the Nebraska Act and the repeal of the Missouri line. He refrained from joining the new Republican movement, though, because he wasn't convinced that the Whig party was dead yet. In fact, some Whigs were talking about converting theirs into an all-Northern, free-soil party. Like many of his colleagues, Lincoln would gladly cooperate with all anti-Nebraska groups against a common enemy, but would remain in the Whig ranks so long as a revival of the old party seemed possible.

Lincoln's major target was the Little Giant himself, who hurried home to vindicate his Kansas-Nebraska doctrine and pull Illinois Democrats back together for the upcoming Congressional elections. Frankly Douglas was shocked at the free-soil backlash against him—shocked at the jeering crowds ("Judas," they screamed at him, "Traitor Arnold!") and blazing effigies he encountered on his trip back to Illinois. In Chicago a hostile crowd taunted him so mercilessly that he blew up, shouted and shook his fists, then tried to stare his hecklers into silence. And then he left, growling that the Chicago confrontation was part of a combination of political foes out to destroy him. All right, then, he would take his case to the people. He would stump the state. He would make Illinoisians listen to his arguments, convince them that popular sovereignty was not only a sacred principle but a free-soil device— that under its banners both freedom and Northern enterprises would triumph in the West. He would force all Democrats to embrace Kansas-Nebraska as a matter of party loyalty and ax anybody who refused. In September he set out across the state with a pack of anti-Nebraska men snarling after him.

Lincoln intercepted Douglas in Bloomington late in September, but Douglas refused to debate him there, insisted that this was his meeting and that people had come only to hear him. Undaunted, Lincoln addressed his own crowd and then followed Douglas to Springfield, where a state fair was in progress. Since heavy rains had transformed the fairgrounds into a quagmire, Douglas spoke in the statehouse, in the great hall of the house of representatives, with Lincoln pacing back and forth out in the lobby. In a voice grown hoarse and strained, the Little Giant rehearsed all his favorite arguments— that the Compromise of 1850 had made popular sovereignty the new formula for treating slavery in the territories; that popular sovereignty was "sacred" because it embodied the American principle of self-government; that slavery couldn't invade the West anyway because the soil and climate there would combine to keep it out. After Douglas finished, Lincoln stood on the stairway landing and shouted to the huge crowd that he would answer Douglas tomorrow, here in the same hall.

The next day, October 4, Lincoln stood by the table in the great hall, a long written speech before him—a speech that had grown out of all his thought and anxiety over the past few months, a speech that contained a grand summary of what other anti-Nebraska orators and editors were saying, but with Lincoln's own style, logic, and insights. He'd toiled hard on his composition, for he intended to publish it in the papers and reach audiences across Illinois. In fact, the *Illinois State Journal* would print an abstract of it the next day. But the complete manuscript would not be published until Lincoln gave the same speech at Peoria almost two weeks later. Thus his address became known as the "Peoria Speech."

It was stifling hot in the hall, and as Lincoln looked out over a mass of faces, his hair disheveled and his forehead bathed in sweat, he spoke with an urgent, searching eloquence. For he felt a powerful calling now—a sense of mission to save the Republic's noblest ideals, turn back the tide of slavery expansion, and restore America to the ground of her Founding Fathers. Since the great question before him was domestic slavery, he sang out in his shrill voice, "I wish to MAKE and to KEEP the distinction between the EXISTING institution, and the EXTENSION of it, so broad, and so clear, that no honest man can misunderstand me." He conceded that Southerners had a constitutional right to their property, so that the federal government could not molest the peculiar institution in those states where it already existed. This was the position of the founders, too, who treated slavery where they found it as a necessity, hiding "the thing" in the Constitution without direct reference to it, "just as an afflicted man hides away a wen or a cancer, which he dares not cut out at once, lest he bleed to death."

Still, these "old time men" expected slavery to die out someday—another reason why the words *slave* and *slavery* didn't appear in the Constitution—and they themselves sought to hedge and hem it into "the narrowest limits of necessity." Among other things, they prohibited bondage in the old Northwestern territories and outlawed the international slave trade. Then after the country acquired the Louisiana Territory, the Missouri Compromise of 1820—that "sacred compact"—saved the huge northern area for freedom. But the Missouri line, however sacred, applied only to the Louisiana Territory; it was not intended as a universal principle to be extended to the Pacific. Thus when the Republic obtained the Southwest from Mexico, Lincoln and other "Wilmot men" voted against expanding the Missouri line, because they were bent on saving all the new territory for freedom, in the spirit of the Founding Fathers. Yet the Wilmot men desired to retain the Missouri line, then thirty years old, and saw keeping the line where it already applied and stretching it to the Pacific as two entirely separate issues.

Then came Douglas's Nebraska Act and the repeal of the Missouri line, which "took us by surprise—astounded us," Lincoln cried out, and "we reeled and fell in utter confusion. But we rose each fighting, grasping whatever he could first reach—a scythe—a pitchfork—a chopping axe, or a butcher's cleaver." Because the repudiation of the Missouri Compact was an egregious wrong. It was wrong in letting slavery into Kansas and Nebraska, where it would take root by degrees, so that by the time people formed a territorial government slavery would be on the ground and would become one of the territory's "domestic institutions." And so it would go, across one territory after another, as popular sovereignty allowed slavery to spread into "every other part of the wide world" where men were inclined to take it.

Yet Lincoln had no prejudices against the Southern people themselves.

"They are just what we would be in their situation." They were not responsible for the origin of slavery, and he understood when they said it was hard to get rid of. "I surely will not blame them for not doing what I should not know how to do myself. If all earthly power were given me, I should not know what to do, as to the existing institution." His first impulse was to colonize the blacks in Liberia, but he realized that this couldn't be implemented at once—that it was a long-range solution. While it seemed to him that some gradual emancipation scheme could be adopted, he wouldn't condemn Southerners for their tardiness in this.

So when they reminded the North of their Constitutional rights, Lincoln acknowledged them "fully, and fairly." He would grant them legislation to return their fugitive slaves, since the Constitution required that people "held to Service or Labour in one State" must be sent back if they ran away. He would grant Southerners the disproportionate political power they enjoyed from the three-fifths clause in the Constitution, even though this degraded Northerners—was "manifestly unfair" to them. But he would not allow Southerners to augment such power by seizing the Western territories. He would not let them overthrow the Missouri line and send slavery flooding across the frontier under the popular sovereignty banners of "Judge Douglas."*

In the course of his argument, Lincoln flayed away at popular sovereignty itself as a fraudulent, insidious, and potentially devastating device. For one thing, it reduced a momentous national issue—the existence and growth of human bondage—to a purely local matter, equal in importance to the cranberry trade. For another, it had no historical precedent or legal justification whatever, since the Compromise of 1850 applied popular sovereignty only to New Mexico and Utah and did not establish it as a new universal formula (which was the case). At no time, Lincoln said, did the public ever demand that the Missouri line be removed and replaced with popular sovereignty or "the Nebraska doctrine." Nor was it certain—to touch on one of Douglas's "inferior" arguments—that the soil and climate of the West would exclude slavery *"In any event."* "This is a *palliation,*" Lincoln warned, "a *lullaby.*" In truth, in back of popular sovereignty was "a covert real zeal" for the spread of slavery, which Lincoln could not but hate. He hated it because of "the monstrous injustice of slavery itself," which deprived "our republican example of its just influence in the world."

Of all Douglas's contentions, though, Lincoln found most repugnant the claim that popular sovereignty was sacrosanct, because it applied to the

*Lincoln called him "Judge" because Douglas had served on the Illinois Supreme Court in the early 1840s.

slavery problem "the sacred right of self-government." Well, Lincoln also believed in self-government as an absolute and eternal right, but it had no just application as here applied. For if the Negro was a man, was it not a total destruction of self-government to deprive the Negro of the right to govern himself? "When the white man governs himself that is self-government; but when he governs himself, and also governs *another* man, that is *more* than self-government—that is despotism. If the negro is a *man,* why then my ancient faith teaches me that 'all men are created equal;' and that there can be no moral right in connection with one man's making a slave of another." Slavery, in sum, was a "total violation" of the sacred right of a man to govern himself, which was guaranteed by the Declaration and was "the sheet anchor of American republicanism." And that was why Lincoln detested popular sovereignty so—because under the pretext of "MORAL argument" it extended and compounded the evil of slavery and the violation of self-government. But he repeated that where bondage already existed, "we must of necessity, manage as we best can."

What was the solution? As a thoroughly national man, Lincoln contended that Americans must restore the Missouri Compromise line—must restore the national faith, the national confidence, and the national feeling of brotherhood. But if they failed to do this, if they allowed popular sovereignty to reign in the territories, there would be disaster on the Kansas prairies where proslavery and free-soil partisans were already gathering. Because what was involved out there, Lincoln said, was the struggle for human liberty. In his view, slavery embodied "the selfishness of man's nature" and opposition to it embodied man's love for justice. These two principles were "eternally antagonistic," and when brought together—as in the struggle for Kansas— they would lead inevitably to "shocks, throes, and convulsions." And once blood was shed there, would this not "be the real knell of the Union?"

Yes, the Nebraska doctrine was run with evils, but the one Lincoln "particularly objected" to was the new position it gave slavery in the body politic. In the beginning of the Republic, the founders were hostile to the principle of slavery and tolerated the institution strictly as a necessity. "But NOW it is to be transformed into a 'sacred right.' " Now in the Nebraska doctrine Americans were giving up the old faith for the new and were placing slavery "on the high road to extension and perpetuity." "Near eighty years ago we began by declaring that all men are created equal; but now from that beginning we have run down to the other declaration, that for SOME men to enslave OTHERS is a 'sacred right of self-government.' " But let nobody be deceived. The spirit of Nebraska and the spirit of '76 were "utter antagonisms." And yet the spirit of Nebraska was rapidly taking over. "Fellow Countrymen—Americans south, as well as north, shall we make no effort to arrest this? Already the liberal party throughout the world, express the apprehension 'that the one retrograde institution in America, is undermining the principles of progress, and fatally violating the noblest political system the world ever saw.' "

"Our republican robe is soiled, and trailed in the dust," Lincoln said. "Let us turn and wash it white, in the spirit, if not the blood, of the Revolution. Let us turn slavery from its claims of 'moral right,' back upon its existing legal rights, and its arguments of 'necessity.' Let us return it to the position our fathers gave it; and there let it rest in peace. Let us re-adopt the Declaration of Independence, and with it, the practices, and policy, which harmonize with it. Let north and south—let all Americans—let all lovers of liberty everywhere—join in the great and good work. If we do this, we shall not only have saved the Union, but we shall have so saved it, as to make, and to keep it, forever worthy of saving. . . ."

The audience in the great hall interrupted Lincoln many times with waves of applause. Though his best oratory had always been charged with a certain moral eloquence, his colleagues had never heard him give such an inspired address as this, never heard him declaim his antislavery convictions so fiercely and openly, never heard him defend the ideals of the Republic with such moral urgency as he had this day.

Then Douglas rose again and delivered a two-fisted rebuttal which brought his own supporters to their feet, clapping and shouting. Nevertheless, Douglas's intransigence on popular sovereignty had caused an irreparable split in his own party, which was underscored the next day when Lyman Trumbull denounced the Little Giant for repealing the Missouri line. An experienced and formidable Democrat who re-entered politics because of Kansas-Nebraska, Trumbull planned to run for Congress not as a Douglas man, but as an anti-Nebraska man.

On the same day Trumbull spoke, a group of zealous free-soilers—men like Owen Lovejoy and Ichabod Codding—held a Republican convention in Springfield, in an attempt to form a statewide Republican party. Contrary to what many writers have claimed, these pioneer Republicans were not "abolitionists" and "radicals" out to uproot slavery in the South itself and revolutionize the American system, but were adamant and outspoken antiextensionists who hoped to draw all free-soil groups into a single organization. As it happened, the Republicans were extremely impressed with Lincoln's address —for it expressed most of their own sentiments—and they strove to enlist him in their cause. They even placed his name on their state central committee, but did so without his approval. When Lincoln found out about it, he wrote Codding a sharp rebuke and demanded that his name be withdrawn. He didn't want to climb on the Republican train because he was still a Whig.

Yet as a Whig, an American, and a national man, he offered to stand with them and anybody else who "stood right."

Because most free-soil Whigs and Democrats refused to abandon their old parties, Illinois Republicans failed to establish a state party in 1854. Even so, the broad anti-Nebraska forces continued to swell, as more and more free-soil Democrats refused to let Douglas bully them into line, refused to endorse popular sovereignty as he demanded, and moved into the loose anti-Nebraska ranks. In all the free-soil uproar, Lincoln announced his candidacy for the state legislature as an anti-Nebraska Whig, but he had his eye on a higher office than that—he wanted a seat in the United States Senate. It so happened that James Shields was up for re-election to the Senate, running with the full support of the Little Giant and the regular Democratic machine. But with all the popular unrest against Douglas, it was entirely possible that anti-Nebraska forces would gain control of the legislature that November. If so, they would reject Shields and choose an anti-Nebraska man for the Senate (in those days, state legislatures selected U.S. senators). And Lincoln wanted desperately to be that man, wanted desperately to serve in the national Senate where Henry Clay had served, and there carry on the great battle against Kansas-Nebraska in full view of the American Republic—Senator Abraham Lincoln, fighting with all his parliamentary eloquence to block slavery expansion and save American liberty and global liberalism.

In the state elections, Lincoln and Trumbull both won their respective races, riding an anti-Nebraska sweep to power. In all, anti-Nebraska candidates captured nine Congressional seats and gained a commanding majority in the legislature, a triumph which reflected a popular repudiation of Douglas and the Kansas-Nebraska Act. Alas, though, Lincoln discovered that he'd made a bizarre oversight: a member of the legislature could not run for the U.S. Senate. So he resigned his seat and set about openly campaigning for the job. At the same time, his Whig friends from the Eighth Judicial Circuit —Davis, Logan, Swett, Lamon, and others—came to Springfield and went to work lining up support for him. Since the legislature contained more anti-Nebraska men than regular Democrats, Lincoln thought he stood a chance of winning the senatorship.

When the balloting began on February 8, 1855, Mary was in the galleries and Lincoln was over in his law office, following developments in a state of high tension. On the first ballot he emerged with 45 votes—only 6 shy of victory—while Shields collected 41 votes and Lyman Trumbull 5. Yet Trumbull's supporters—all anti-Nebraska Democrats, led by Norman Judd of Chicago—declared that they would never vote for a Whig and refused to endorse Lincoln. When in subsequent ballots the vote remained about the same, Democratic managers dropped Shields and pushed Joel Matteson, a rich and popular Democrat whom many Whigs even liked. At that, Lincoln's support eroded so dramatically that he wound up on bottom while Matteson rose close to victory. To block Matteson, Lincoln now threw his support to Trumbull, who on the tenth ballot finally won the senatorship. Stephen Logan, Lincoln's floor manager, broke into tears when he heard the announcement. And Judge Davis growled that it was all wretchedly unfair, that "Mr. Lincoln ought to have been elected."

Lincoln, too, was bitter. But if he'd held on to his dwindling support, he explained, Matteson would have triumphed and by implication so would popular sovereignty and Judge Douglas. So Lincoln sacrificed his own ambition for the sake of an anti-Nebraska victory. Ruefully he wrote a colleague that at least Matteson's defeat gave him pleasure and eased the pain of his own loss. With the regular Democrats wailing that Trumbull's election was a calamity for them, "it is a great consolation," Lincoln said, "to see them worse whipped than I am."

The night after the election, Lincoln attended a reception at the Edwards mansion and forced himself to appear in good spirits. He cheerfully congratulated Trumbull, a dry, well-educated fellow, Connecticut-born, clean-shaven, with round gold-rimmed spectacles and sandy hair. Mary, though, was hostile to both Trumbull and his wife Julia Jayne, one of Mary's old girlfriends. She thought that Lincoln's defeat was terribly unjust and that it was all because of Trumbull's "sordid, selfish" nature. She snubbed Julia Jayne and terminated their friendship. And she never forgave Norman Judd either.

Privately, Lincoln was extremely sad about his loss—his vision of himself orating in the Senate crushed now. When Henry C. Whitney called at his law office, Lincoln had surrendered himself to a severe attack of the hypo. Whitney had never seen him so melancholy.

After a while, though, Lincoln felt a little better. For one thing, Trumbull proved very solicitous, saying he would be glad to have Lincoln's views on how to combat slavery expansion and popular sovereignty. In truth, the main question now was how best to marshal anti-Nebraska forces and present a united front against Douglas's regular Democrats. Could the Whigs ever become the new free-soil party, given the recalcitrance of anti-Nebraska Democrats like Norman Judd? Or should there be a fusion with the Republicans after all? For their part, Orville Browning, David Davis, Simeon Francis, and other Whigs still shrank from enlisting with the Republicans, and Lincoln, too, was uncertain. As he wrote Republican Owen Lovejoy, the political atmosphere was so stormy he was afraid to do anything for fear of making a mistake. What if he joined the Republicans and then they disintegrated? Then where would he be politically? In 1855, moreover, it was not the Republicans but the Americans or Know-Nothings who seemed the new major party of the future. Begun as a nativist movement to halt immigration, suppress Catholics, and save America from the menace of "Popery," the Know-Nothings had broadened their appeal in many Northern states, drawing in anti-Nebraska men and abolitionists alike. Before anti-Nebraska forces could organize a successful new party in Illinois, Lincoln contended, they must win over the antiextensionists in the American ranks. Yet that seemed impossible so long as the Know-Nothings looked like a winning party. "I have no objection to 'fuse' with any body," Lincoln told Lovejoy, "provided I can fuse on ground which I think is right," and he believed many other anti-Nebraska men would do likewise if not for the infernal Know-Nothings, whose anti-Catholic bigotry Lincoln detested.

Meanwhile, conditions in Kansas had become increasingly combustible. By the summer of 1855, over nine thousand people were in the territory, most of them free-soilers from the Midwestern states. Enraged by this free-state "invasion," proslavery Missourians vowed to exterminate every "God-damned abolitionist" in Kansas, whereupon they stormed into the territory, terrorized free-state settlements there, and voted illegally in territorial elections. As a result, Kansas now had a "bogus" proslavery legislature and a spate of "bogus" laws, including slave codes and sedition measures which prohibited criticism of slavery. In reaction, free-soilers organized minutemen companies, laid plans for their own territorial government, and pleaded for the free states to send help for the cause of freedom in Kansas. With rival armed camps springing up on the Kansas prairies, violent confrontations seemed inevitable there—just as Lincoln prophesied.

As he studied his newspapers and received letters from a Whig associate in Kansas, Lincoln despaired of ever extinguishing slavery by peaceful means. The failure of Clay and other good men to implement gradual emancipation in Kentucky, "together with a thousand other signs," had shattered his hopes. Never had things seemed so desperate, so out of control. As he wrote a Whig friend, "Our political problem now is, 'Can we, as a nation, continue together *permanently—forever*—half slave, and half free?'" Lincoln did not know. The problem was "too mighty" for him. All he could do was hope that a merciful God would work out a solution.

And then came a letter from Joshua Speed, like an anguished cry from the dark of night. Speed, farming away in Kentucky, was painfully certain that his and Lincoln's views differed now, and he set forth his feelings about slavery. It may be wrong in the abstract, Speed said, but as a Southerner he would rather dissolve the Union than give up the right to own Negroes.

Who is asking you to give up your right to own slaves? Lincoln retorted. "Very certainly *I* am not," because slavery is your own matter and enjoys constitutional guarantees I have to acknowledge. But "I confess I hate to see the poor creatures hunted down, and caught, and carried back to their stripes, and unrewarded toils; but I bite my lip and keep quiet." Speed, he said, you ought to appreciate how I and the great body of Northern people crucify our feelings in standing loyally by the Constitution and the Union when it comes to slavery in the South. But "I plainly see you and I would differ about the Nebraska-law. I look upon that enactment not as a *law*, but as *violence* from the beginning. It was conceived in violence, passed in violence, is maintained in violence, and is being executed in violence." You ask where I now stand politically, Lincoln went on. That is a disputed point. I think I'm a Whig, but other men say there are no Whigs and Lincoln is an abolitionist. In Congress, I voted for the Wilmot Proviso "as good as forty times" and nobody tried to "unwhig me for that." "I now do no more than oppose the *extension* of slavery."

But one thing he knew for certain: "I'm not a Know-Nothing." "How could I be? How can any one who abhors the oppression of negroes, be in favor of degrading classes of white people? Our progress in degeneracy appears to me to be pretty rapid. As a nation, we began by declaring that *'all men are created equal.'* We now practically read it 'all men are created equal, *except negroes.'* When the Know-Nothings get control, it will read 'all men are created equal, except negroes, *and foreigners, and catholics.'* When it comes to this I should prefer emigrating to some country where they make no pretence of loving liberty—to Russia, for instance, where despotism can be taken pure, and without the base alloy of hypocracy."

The winter of 1855 and 1856 came on. By now a save-Kansas movement was flaming across the North, as numerous Kansas-aid societies solicited guns, money, and men for the free-state cause in the territory. In Springfield, Herndon and several other free-soilers formed a save-Kansas group and pronounced themselves "almost ready for revolution" to preserve Kansas for freedom. But Lincoln cooled them down. "Physical rebellions & bloody resistances," he insisted, would only hurt the free-state cause and the cause of freedom in general. The way to right a wrong was through political action and moral influence: let them mobilize public opinion against Kansas-Nebraska and make the national government stand once again for liberty and justice. But any attempt "to resist the laws of Kansas by force would be criminal and wicked," Lincoln said. Nevertheless, he subscribed twenty-five dollars for the Kansas-aid movement, stipulating that the money be used strictly to help free-state Kansans defend themselves against outside invaders.

Meanwhile old John Brown had arrived on the Kansas prairie with a wagonload of guns, knives, and artillery broadswords; and a column of drunken Missourians had threatened the free-state headquarters of Lawrence in the so-called Wakarusa War.

That winter Lincoln thought a good deal about the Republicans and Whigs, the latter so splintered over the slave controversy that the old party seemed doomed to extinction. By February, 1856, he'd concluded that old party labels—including his own Whig label —severely impeded the mobilization of anti-Nebraska forces and that a new free-soil party was imperative. In addition, news came that the American party had also split over slavery, that party bosses had refused to adopt a platform against slavery expansion, thus driving many antiextensionists out of the Know-Nothing and into the Republican ranks. As a consequence, the Republicans, now organizing on the national as well as the state level, loomed as the new major party of the future. And Lincoln was ready to join them and help shape the party in Illinois according to the principles set forth in his Peoria Speech. The party must stand with the abolitionists and against the South in blocking the spread of slavery, but it must stand against the abolitionists in upholding the Constitution and the fugitive slave law. In both cases, it would avoid "dangerous extremes" and occupy a national ground, holding the ship "level and steady."

On Washington's birthday, 1856, Lincoln joined with other anti-Nebraska leaders in calling for a convention to gather at Bloomington on May 29 and launch a new free-soil party in Illinois. Though referred to for now as simply an "anti-Nebraska" party, it would soon carry the official Republican name.

Lincoln not only helped make the arrangements for Bloomington, but vowed to "buckle on his armor" in the Presidential contest of 1856, with the Republicans intending to run an entire national ticket against the Democrats.

Lincoln led the way for various wavering Whigs, as Herndon, Browning, and others followed him into the Republican camp. Herndon even became a member of the Republican State Central Committee. Meanwhile Trumbull and Judd abandoned the Democrats and supported the Bloomington call as well, bringing in a procession of other anti-Nebraska Democrats.

In the last week of May, in the old tobacco-spattered courthouse in Danville, Lincoln and Judge Davis spoke intently about the situation in Kansas, where Missouri intruders and their proslavery Kansas allies had now murdered six free-state men in cold blood. And there were rumors that a storm was gathering on the Kansas border, that Missouri "border ruffians" were massing for still another invasion. Later Lincoln complained that "a man couldn't think, dream or breathe of a free-state there, but what he was kicked, cuffed, shot down and hung."

Deeply troubled about Kansas, Lincoln caught a train for Bloomington, riding in a nearly empty car as the Illinois countryside—windy grass, open prairie, occasional stands of timber—receded in the windows of the coach. He was apprehensive, too, lest old-line Whigs and important Democrats boycott the Republican convention. But his fears proved groundless, for he found Bloomington crowded with all manner of anti-Nebraska men. There were old Whigs like Browning and Dubois, Democrats like Judd, and liberal Republican pioneers like Lovejoy and Codding. There were also anti-German nativists and anti-nativist Germans, plus a motley assortment of other politicians who just hated Douglas, hated the "Slave Power," and hated Democratic President Franklin Pierce, a Northern "dough face." Like the national Republicans, the Illinois Republican party of 1856 was a coalition of disparate groups held together by a common objective: to keep slavery out of the territories and end Southern domination of national life.

Not long after Lincoln arrived there, Bloomington papers rolled off the presses with lurid headlines. In Washington, in the chamber of the U.S. Senate, a South Carolina congressman had attacked and severely beaten Senator Charles Sumner of Massachusetts, a Republican who'd recently given an anti-Southern speech on "The Crime Against Kansas." At almost the same time, another Missouri mob had sacked Lawrence, Kansas, destroying free-state printing presses, blowing up the free-state hotel, and robbing and burning homes there. In retaliation, John Brown and his little free-state company had massacred five proslavery men down on Pottawatomie Creek, dragging them out of their cabins in the dead of night and assassinating them with broadswords. A reign of terror had broken out in eastern Kansas, as

Missouri militia and their Kansas allies burned and pillaged free-state communities from Lawrence to Osawatomie. As Lincoln had predicted would
happen under popular sovereignty, civil war now raged on the Kansas prairie
—proof indeed that slavery was too volatile ever to be solved as a purely local
matter.

News of civil war in Kansas inflamed the Bloomington delegates, gave
them an even stronger sense of mission to organize their new party and drive
the Democrats out of office. As the convention got under way in Major's
Hall, with some two thousand delegates milling about inside, Lincoln played
a prominent role in the proceedings, helped draft a party platform and draw
up a slate of Republican state candidates. When business was done, the
convention called on him to give the keynote address, and he mounted the
platform "amid deafening applause." Aroused by the atrocities in Kansas and
the beating of Sumner, Lincoln spoke of the "pressing reasons" for the
Republican movement, of the urgent need to rally all free-soil men against
the westward march of the "Slave Power." Lamenting Southern threats of
disunion, he cried out that the nation must be preserved and that the Republicans must preserve it.

Lincoln evidently had no prepared script and spoke extemporaneously for
an hour and a half. Though usually very poor at impromptu oratory, he was
so inspired this day that he kept the entire assembly spellbound. Even reporters became entranced and laid their pencils aside or forgot to use them.
Herndon claimed it was "the grand effort of Lincoln's life," but because no
script or record of it was ever made, it became known as the "Lost Speech."

When Lincoln at last sat down, the crowd jumped up and cheered him
again and again. On Lincoln's keynote address, the Republican party was
born in Illinois and the Whig party, in which Lincoln had worked all his
political life, was dead. But on the way back to Springfield, he had little time
to bemoan the death of Whiggery in Illinois. For the Republicans must now
engage the Democrats and Know-Nothings in a critical Presidential election
that could well determine the future of the Union.

W henever he could in his busy round
of political and legal work, Lincoln investigated Southern proslavery arguments, scrutinizing Southern journals like the *Charleston Mercury,* to which
Herndon subscribed, and going over some of George Fitzhugh's writings.

A cranky Southern idealogue who produced his compositions in a ramshackle, bat-ridden Virginia plantation, Fitzhugh wanted Southerners to destroy capitalism or "free society," revive the halcyon days of feudalism, and enslave all workers—white as well as black. From these and other sources, Lincoln compiled what he considered the major points of the proslavery defense. First (as the racial theorists claimed), that Negroes were inferior subhumans whose natural lot was servitude. Second, that "all men are created equal" was a "self-evident lie." Third, that "master and slave is a relation in society as necessary as that of parent and child." Fourth, that because "free society" in the form of Yankee capitalism was "unnatural, immoral, unchristian, it must fail and give way to a slave society—a system as old as the world." For "two opposite and conflicting forms of society cannot, among civilized men, co-exist and endure. The one must give way and cease to exist—the other become universal." "Free society!" shrieked one Alabama paper. "We sicken of the name! What is it but a conglomeration of greasy mechanics, filthy operatives, small-fisted farmers, and moon-struck theorists? . . . This is your free society which the Northern hordes are endeavoring to extend to Kansas."

Pronouncements like these made Lincoln grimace. And sometime in this period he wrote out in notes and fragments a counterargument to guide him in his speechmaking. Volume on volume, he observed, had been written to defend slavery as "a very good thing." Yet Lincoln had never heard of a man who wished to take advantage of this good thing *"by being a slave himself."*

And the matter of slave versus free labor? Why, even the dumbest animal, even an insect—even an ant—knows it is entitled to the fruit of its own toil. Even "the most dumb and stupid slave that ever toiled for a master" understands that he is wronged, that he deserves the rewards of his labor.

And now the illogic of slavery. "If A. can prove, however conclusively, that he may, of right, enslave B.—why may not B. snatch the argument, and prove equally, that he may enslave A?—

"You say A. is white, and B. is black. It is *color,* then; the lighter, having the right to enslave the darker? Take care. By this rule, you are to be a slave to the first man you meet, with a fairer skin than your own.

"You do not mean *color* exactly?—You mean the whites are *intellectually* the superiors of the blacks, and, therefore have the right to enslave them? Take care again. By this rule, you are to be slave to the first man you meet, with an intellect superior to your own.

"But, say you, it is a question of *interest;* and, if you can make it your *interest,* you have the right to enslave another. Very well. And if he can make it his interest, he has the right to enslave you."

And then Lincoln distilled all his points down to one: slavery was *wrong* and opposition to it was *right*. And right must never submit to wrong . . . must never submit to the double wrong of extending slavery into the West. Would surrendering to wrong better our Constitution? our Union? our liberty? No, "there is no other way. . . . In the whole range of *possibility*, there is no other way."

As Lincoln made notations and prepared his campaign speeches, the Democrats held their Presidential convention down in Cincinnati, where they rejected Douglas in favor of James Buchanan, a Pennsylvanian with pro-Southern sympathies, as their nominee. Though the turmoil over Kansas-Nebraska had probably cost Douglas the nomination, the Democratic platform nevertheless endorsed popular sovereignty as the only solution to slavery in the territories . . . but whose interpretation of popular sovereignty? As it happened, the doctrine contained a fatal ambiguity which left the Democrats in a sectional uproar. At what point could the people of a territory decide to retain or prohibit slavery? If Southerners insisted that this could only be done at the time of statehood, Douglas vaguely argued that the institution could be dealt with at any territorial stage. Precisely because the doctrine was ambiguous, because it was open to the kind of interpretation Douglas gave it, many Southerners had never accepted popular sovereignty. And now, as Lincoln observed, Southerners increasingly contended that neither territorial governments nor the national government could prevent them from taking slaves into the West.

For Lincoln, all the confusion over popular sovereignty was a portentous sign for the Republicans. He wrote Trumbull that if they nominated former Whig John McLean, Lincoln's choice, they had a powerful chance of whipping Buchanan and Millard Fillmore, the latter a former Whig President now running as the American party candidate. But the national Republicans had other plans. Congregating in Philadelphia in mid-June, they chose John Charles Frémont, the celebrated Western explorer, as the Republican nominee and adopted a platform that not only blamed the Kansas civil war on the Democrats, but branded slavery as a relic of barbarism and demanded that Congress prohibit it in all territories by "positive legislation." The Republicans planned an intense and lively campaign, with "free soil, free speech, and Frémont" as their slogan.

Lincoln was at Urbana when he learned of Frémont's nomination—and learned, too, that the Illinois delegation had nominated him as Vice-President and that he'd picked up 110 votes before the convention went on to select William Dayton of Ohio. Lincoln was pleased, of course, but he seemed unassuming about it in public. No, he was not a candidate for public office, he told inquirers. No, he did not intend to run for Congress, out of his "clear conviction" that he would hurt the Republican cause if he did so. Which for Lincoln-watchers meant that he still had his eye on the national Senate.

And so, against the backdrop of civil war in Kansas, the campaign of 1856 began, with Republicans, Democrats, and Know-Nothings all pummeling one another with riotous oratory. By now the national Whig party was defunct, with anti-Nebraska men having moved over to the Republicans and some old-line conservatives throwing their support to Fillmore. The Republicans, for their part, made "Bleeding Kansas" a shibboleth and a rallying cry, contending that the war there was the inevitable result of Douglas's Kansas-Nebraska Act. Republican journals like Greeley's *New York Tribune* dispatched war correspondents to cover the Kansas conflict, and throughout the summer of 1856 Republican papers blazed with embellished accounts of battles and bushwhackings on the Kansas prairies. At the same time, Republicans waved copies of Fitzhugh's tracts, insisting that he spoke for all Democrats in advocating the enslavement of whites as well as Negroes. See? cried Republican campaigners. This is what will happen if Buchanan and his cronies win the election—we'll all end up in chains. How come? Because Buchanan, a fusty and enfeebled old bachelor, would do exactly what Southern bosses told him to do.

The Democrats, of course, countered with their own grisly prophecies. They contended that Frémont was not only a bastard, but a fanatical "black abolitionist" whose triumph would destroy the Union. As the leader of an all-Northern "abolition party," warned Southern Democrats, Frémont would emancipate the slaves and mongrelize the races. Southern papers claimed that slaves were rattling their chains across the South, because they knew that Frémont was their friend, knew that he wanted to liberate them and turn black men loose on Southern white girls. Think of the hideous scenes! Southern orators cried. Think of the consequences! If Frémont and his "abolitionist" henchmen won the White House, then the South would have no choice but to secede from the Union.

With Southerners beating the drums of disunion, Lincoln took to the hustings as "a Frémont man" and spoke to audiences as though Southerners were in the crowd. "All this talk about the dissolution of the Union is humbug—nothing but folly. *We* WON'T dissolve the Union, and *you* SHAN'T," because *"we* won't let *you."* As the campaign progressed, Lincoln stumped from central Illinois to Kalamazoo, Michigan, in an exhaustive effort to woo old-line Whigs and Fillmore men into the Republican ranks. Everywhere his argument was the same: a vote for "Fillmoreism" was a vote for "Buchanan and his gang." Vote Republican and you vote for free soil, free labor, and free men. Yet Lincoln refrained from agitating the Kansas question. Let Republicans avoid demagoguery, he insisted, and pitch their appeal to the man of reason. Let Republicans focus exclusively on the slavery issue, reminding men of reason that human bondage was "not only the greatest question, but very nearly the sole question of the day," and that Americans must never allow slavery to triumph in their government.

His speeches brought a lot of racist abuse from the opposition press, which belittled his "windy" declarations about "freedom" and "niggers." And some of his crowds were extremely hostile. In Petersburg, where most New Salem people had resettled, a mob hissed and hooted when he stepped off the stage, and local leaders denounced him as a "black Republican," a "high priest of abolitionism." As one witness said, Lincoln might just as well have tried to speak in Charleston, South Carolina, as in Petersburg, Illinois.

In November, having given some fifty speeches for Frémont, Lincoln returned home to Springfield, home to await the outcome, home to his wife and sons. But this year the Lincolns were politically divided, for Mary was no supporter of Frémont. On the contrary, she went for Fillmore, no matter what Lincoln thought of the Know-Nothings. "My weak woman's heart was too Southern in feeling," she wrote her sister Emilie, "to sympathise with any but Fillmore." After all, he'd "made so good a President & is so just a man." Besides, Fillmore saw the need for keeping foreigners "within bounds." If you Kentuckians had to contend with the "wild Irish" as we Illinois housewives, Mary informed Emilie, you too would vote for Fillmore.

And so the Lincolns cheered for different sides on election day, but both their candidates lost. Buchanan not only won the White House, but carried Illinois by a substantial margin. Nevertheless, the Republicans polled more popular votes than the American party and finished a respectable second behind the Democrats—a remarkable showing for a new political organization. And in Illinois the Republicans captured four out of nine Congressional seats and elected their entire state ticket, which gave them the governorship. For Lincoln the lesson was emphatically clear. A coalition of Fillmore and Frémont men could have defeated Buchanan. If in the next four years the Republicans could gather in Fillmore's supporters, they could sweep to power in 1860 and overcome the "black demon" of slavery.

Over the winter, Lincoln conceded that there were other reasons why the Republicans lost the Presidential contest. For one thing, thousands who agreed with Republican principles had turned against the party, and why?

Because, he noted, "we were constantly charged with seeking an amalgamation of the white and black races." For another, Republicans were assailed as being "the enemies of the Union," but of course neither charge was true. As he told a Republican banquet in Chicago, the election of 1856 was really a struggle by the Republicans to preserve the central idea of the Republic —that all men were created equal. For the Democrats, the election was an attempt to discard that idea and replace it with the principle that slavery was right and ought to be extended "to all countries and colors." Well, we Republicans have dedicated ourselves to blocking the advance of that idea, Lincoln said, so let us not despair. Let us resolve that free society is not and will never be a failure. Let us forget our own differences and animosities, unite, and go forward together in the cause of Republicanism—the hope of free men everywhere.

Which were grand words for the Republican cause, but some wondered about Lincoln's own career. Where was he going as a politician? What were his plans? During the winter, Lincoln pondered the career of Stephen A. Douglas, pondered his rocketing climb to national fame, and wrote on a scrap of paper how Lincoln had first met the Little Giant in Vandalia twenty-two years ago. "We were both young then; he a trifle younger than I. Even then we were both ambitious; I, perhaps, quite as much so as he. With *me,* the race of ambition has been a failure—a flat failure; with *him* it has been one of splendid success. His name fills the nation; and is not unknown, even, in foreign lands." But Lincoln had "no contempt for the high eminence he has reached," since Lincoln believed powerfully in the right of all men to elevate themselves. In truth, he hungered for the kind of self-made prominence Douglas enjoyed, would rather "stand on that eminence, than wear the richest crown that ever pressed a monarch's brow." Accordingly, Lincoln had every intention of challenging Douglas for his Senate seat in 1858. If Lincoln could whip the Little Giant, he could rectify the contrast in their reputations and win a spectacular victory for the Republican cause.

Since he'd lost most of the working part of last year in politics, Lincoln had to forgo campaigning in 1857 and tend his neglected law business. Nevertheless, he kept a high visibility on the social front, as he and Mary attended a party almost every night in early 1857. They even threw a "perfectly extravagant" affair of their own, to which Mary invited five hundred people, though because of "an unlucky rain" only three hundred showed up. After that came several more "grand" festivities, including "a very large party" given by Republican Governor William Bissell. The people were all beautifully dressed, Mary said, and the dancing was marvelous as she and Lincoln glided about the floor to the music of fiddles and violins.

At such gatherings, Lincoln and the other men congregated wherever they could and discussed the latest developments on the political front. Over the winter the situation had improved in Kansas, where a new territorial governor negotiated a truce between warring proslavery and free-soil forces and set about administering even-handed justice. In Washington, Democratic factions pressed Buchanan as to whose interpretation of popular sovereignty he accepted—Douglas's or the South's. On March 4, Buchanan announced in his inaugural address that the U.S. Supreme Court, consisting of seven Democrats and two Republicans, was about to solve the slave controversy once and for all. Two days later the court handed down the Dred Scott decision, which shocked Republicans everywhere.

Each judge wrote a separate opinion in the controversial case, but Chief Justice Roger Brooke Taney of Maryland, a gaunt Democrat and former slave owner, authored the one most often cited. The other six Democratic judges concurred with Taney, but the two Republicans vigorously dissented. First, Taney declared that free Negroes were not and never had been U.S. citizens, that the Constitution and the language of the Declaration of Independence did not embrace them as part of the American "people." At the time these documents were framed, Taney argued, Negroes were regarded as "beings of an inferior order, and altogether unfit to associate with the white race, either in social or political relations; and so far inferior, that they had no rights which the white man was bound to respect." As a consequence, the Declaration itself was a white man's document which had never applied to blacks.

As for slavery in the territories, Taney rendered a straight pro-Southern verdict: Congress had no authority to prohibit slavery in the national lands, because that would violate the property rights clause of the Fifth Amendment. As a result, the Missouri Compromise had never been constitutional. What was more, if Congress could not exclude slavery from the territories, then it could not authorize any territorial government to reject slavery either. All Congress could do in the matter was to protect the rights of property owners.

The Dred Scott decision sent the Republicans reeling. In one blow, Taney's Court had knocked the props out from under their antiextensionist platform and seemingly jeopardized the party's entire future. Fighting for their political lives, Republicans lashed back at Dred Scott as "cruel, inhumane, and unfair" and set about mobilizing Northern opinion against it. In Springfield, Billy Herndon inveighed against the decision, contending that North and South were clearly locked in mortal combat over slavery. By now "the fiery logic of sweeping events," as Herndon's biographer phrased it, had carried Billy out to the liberal flank of the Republican party, where he manned the barricades with men like Charles Sumner and Salmon Chase. Lincoln, too, was distressed about Dred Scott, observing that if neither Congress nor a territorial government could keep slavery from advancing west, then wasn't the net effect to legalize slavery in all the territories, as many Southerners demanded? Moreover, what did the decision do to Douglas's "much vaunted" doctrine of popular sovereignty, which allowed the people through their territorial governments to vote slavery in or out? Hadn't Dred Scott negated popular sovereignty as well?

A lot of people asked the same question, and in June Douglas addressed an assembly in Springfield about this and other pressing issues of the day. With Lincoln sitting in the audience, Douglas first discussed the Mormon uprising in Utah, declaring that the President ought to send U.S. troops there and that Congress ought to rescind Utah's territorial status. Then he turned to the Dred Scott decision. Initially he was troubled about it, Douglas admitted, but after much cogitation he concluded that it didn't contradict popular sovereignty after all. As he'd contended several times before, the people of each territory did not have to vote explicitly on whether to have slavery or not. No, they could exclude it through "unfriendly legislation"—that is, by refusing to enact police measures and slave codes necessary to sustain the institution. Thus in Douglas's view Dred Scott actually "upheld" popular sovereignty and the whole governing principle behind the Kansas-Nebraska Act.

As for Taney's opinion about Negroes and the Declaration of Independence, Douglas emphatically agreed. When the Founding Fathers spoke of equality, Douglas said, they meant equality among white people; they meant that British subjects in the colonies were equal to British subjects in England; and Negroes had nothing to do with it. Why? Because they were racially inferior and innately incapable of self-government. Still, Douglas wasn't arguing that Negroes had no rights at all. On the contrary, they were entitled to certain rights and privileges consistent with the welfare of the white communities in which they resided. And each community had the authority to decide what those privileges should be.

By contrast, the Republicans were a bunch of radical abolitionists who not only "resisted" the Court's decision (and thus defied law and order), but would have Americans believe that Negroes were included in the Declaration and were equal with white men. Yes, that was the Republican doctrine, Douglas said, tossing his head in defiance. And if it should triumph in America, then whites could look forward to racial amalgamation here—to the intermarrying of whites and blacks in direct violation of the "great law of nature" which held them apart. And the consequence would be the destruction of the superior white race, as the Republican program of race mixing lowered whites to the subhuman level of the blacks. . . .

For Lincoln it was the same old Democratic litany, the kind of race-baiting he thought had cost the Republicans thousands of votes in 1856. Well, Douglas was not going to get away with it now. Lincoln would make a public rebuttal, one that would convince white voters that Republicans were not racial amalgamationists . . . and that would also clarify and elucidate the Republican position in light of Dred Scott. It would have to be a persuasive speech; otherwise he and his young party would be in for a rough time of it in Illinois political races next year.

He spent two weeks preparing his argument. In the Illinois Supreme Court library he read all nine opinions in the Dred Scott decision and examined Republican editorials and speeches in reaction to it; then he assembled his reply, striving to refute the Little Giant not through fire-and-brimstone rhetoric, but through the sheer force of logic. And so Lincoln wrote, sometimes late into the June nights, this calm and judicious hymn:

"Fellow citizens, I am here to-night, partly by the invitation of some of you, and partly by my own inclination," to answer what Judge Douglas said about Utah, popular sovereignty, and Dred Scott. Let us start with Douglas's inconsistencies. In calling for Congress to interfere in Utah, the senator violated his own doctrines of local self-government and Congressional nonintervention in the territories. This, Lincoln wrote, proved what Republicans had said all along—that popular sovereignty was "a mere deceitful pretense for the benefit of slavery."

Now for the Republican opposition to Dred Scott. Republicans were not promoting resistance or disobedience to the decision, for that would be rebellion. Rather they regarded the idea in Dred Scott as erroneous, and through the power of criticism and persuasion they would work to have the court overrule itself, as it had often done in the past.

And Taney's opinion on Negro rights? Lincoln thought it "based on assumed historical facts which were not really true." Citing the dissenting opinion of Justice Benjamin Curtis of Massachusetts, Lincoln pointed out that in five of the original thirteen states—in Massachusetts, New Hampshire, New York, New Jersey, and North Carolina—free Negroes had the right to vote, undoubtedly played a part in the ratification of the Constitution, and certainly were included in the preamble, "We the people." What evidence existed that Negroes were not considered citizens at that time? There was no such evidence.

What of Negroes and the Declaration of Independence? For several years Lincoln had studied the question of equality and what he thought the Founding Fathers had meant by it. And now he offered a reply to both the proslavery theorists and to the Little Giant, whose own opinions had made "a mangled ruin" of the Declaration. Those who signed that document, Lincoln contended, did not intend "to declare all men equal *in all respects.* They did not mean to say all were equal in color, size, intellect, moral development, or social capacity." What they meant was that all men, black as well as white, were equal in their inalienable right to life, liberty, and the pursuit of happiness. "This they said," Lincoln wrote, "and this they meant." When they drafted the Declaration, of course, they realized that not all men then enjoyed such equality. No, they were simply declaring that the right existed, that it was to follow when circumstances permitted, serving as the standard maxim of free society in the future.

And now the matter of racial amalgamation. Judge Douglas declared that Republicans, in claiming that the Declaration applied to Negroes, desired therefore "to vote, and eat, and sleep, and marry with Negroes." Lincoln wrote: "Now I protest against that counterfeit logic which concludes that, because I do not want a black woman for a *slave* I must necessarily want her for a *wife.* I need not have her for either, I can just leave her alone. In some respects she certainly is not my equal; but in her natural right to eat the bread she earns with her own hands . . . she is my equal, and the equal of all others."

In his speech, Judge Douglas professed himself horrified at the thought of whites marrying with blacks. "Agreed for once," Lincoln said, "a thousand times agreed." There was, he conceded, "a natural disgust" among nearly all whites at the idea of racial amalgamation, and Lincoln and the Republican party opposed it too. Yet where was it that amalgamation occurred on the widest scale? Lincoln noted that of the 405,751 mulattoes now in the United States, 348,874 of them resided in the South. Thus slavery itself, which brought white masters and black slaves together, was the greatest source of race mixing in all the country. By demanding that slavery be excluded from the territories, that the two races be kept apart where they were not already together, Republicans were trying to *curtail* amalgamation, not bring it about.

But if they were against mixing the races, the Republicans still viewed the Negro as a human being, which brought Lincoln to one of the fundamental differences between the two parties. "The Republicans inculcate . . . that the negro is a man; that his bondage is cruelly wrong, and that the field of his oppression ought not to be enlarged. The Democrats deny his manhood;

deny, or dwarf to insignificance, the wrong of his bondage; so far as possible, crush all sympathy for him, and cultivate and excite hatred and disgust against him; compliment themselves as Union-savers for doing so; and call the indefinite outspreading of his bondage 'a sacred right of self-government.' "

On the evening of June 26, with an armload of legal volumes, Lincoln marched into the Illinois statehouse and read his speech before an applauding Republican crowd. Some Republicans, though, were disappointed in his logical, even-tempered style—what they wanted was not a scholarly lecture, but a rousing call to arms. Still, Lincoln was unperturbed: he'd written his speech to be read and contemplated, and it was. The *Illinois State Journal* not only published the entire address, but offered copies for sale; and Republican newspapers throughout Illinois either reprinted or excerpted from the speech. And so in this, his only political address in 1857, Lincoln struck the first blow in his long uphill fight for Douglas's coveted Senate seat.

Later that summer Lincoln took Mary with him on a business trip to New York City. On the way they enjoyed a belated honeymoon and visited Niagara Falls and Canada. But as their train swept on to New York City, signs of economic disaster were visible in one town after another. For a calamitous panic had stricken the country and banks and businesses alike had shuttered their windows and closed their doors. Republicans, of course, held the Democrats responsible, pointing out that Southern leaders had revised the tariff steadily downward and insisting that this had caused the depression. Playing it up for all it was worth, Republicans advocated a protective tariff to guard against future panics—an argument that would appeal to voters in the manufacturing states.

Lincoln desired a protective tariff, too, because he thought it would eliminate idleness and promote useful labor. But in truth his mind was on other things—on slavery, Kansas, and Senator Douglas—as he completed his business and set out with Mary to tour New York City. As they strolled along the wharves, with their ocean smells and cawing gulls, Mary talked about how she would love to visit Europe someday. But, she sighed, she supposed that "poverty was my position" and that she would never get to go. Then she teased her reticent husband, warning him that "I am determined my next husband *shall be rich.*"

Not that Lincoln wasn't trying to make all the money he could. Back in

Springfield, he went diligently to work in the state and federal courts, participating in all manner of litigation there. After taking the Illinois Central Railroad to court, he finally collected the $4,800 the company owed Lincoln & Herndon for the McLean County tax case of 1854. Nevertheless, Lincoln continued to represent the Illinois Central and other railroad firms as well. As a result of his association with railroad interests, he became involved that September in one of the most significant cases of the decade, that of *Hurd* v. *the Rock Island Bridge Co.*, which brought river and railroad forces together in a dramatic legal showdown.

The controversy began back in 1854, when Illinois authorized the Rock Island Bridge Company to build a railroad bridge across the Mississippi, one that would connect a railway in Illinois with another over in Iowa. The river interests filed a suit to block construction, but lost in the U.S. circuit court. In 1856 the bridge was completed—the first ever to span the Mississippi—and trains were soon chugging across heading west. That same year a steamer called the *Effie Afton* crashed into the bridge and exploded in flames, a collision that brought howls of protest from river interests along the Ohio and the Mississippi. When the boat's proprietor, with their support, sued the Rock Island Bridge Company for damages, various railroad men arrayed themselves behind the bridge owners. The case was now a cause célèbre, as both sides hired an imposing battery of attorneys and vowed to fight it out to the bitter end. At issue was whether this and perhaps any other railroad bridge could legally stand across the Mississippi.

Lincoln joined the case when a group of lawyers, railroad men, and bridge owners engaged him as co-counsel for the defense. With an eye to mastering all the facts, he paid a visit to the bridge site on the Mississippi, where he examined the river currents and interviewed the bridgemaster and engineer. He even sat down on the edge of the bridge, his long legs dangling over the side, and chatted about the river with the engineer's teen-age son. Then he hurried up to Chicago, where the case was scheduled for the federal district court, with Supreme Court Justice John McLean on the bench. The trial, which began on September 8, 1857, was protracted and complex, with attorneys for the steamboat arguing that the bridge was a hazard to navigation, while Lincoln and his associates contended that the case was a deliberate and sinister effort on the part of the river interests to have the bridge dismantled.

At last, on September 22 and 23, Lincoln stood before the jury and gave the summation for the defense. In graphic detail he described America's westward growth, especially in the booming states around the Great Lakes, and recited a host of facts and figures to demonstrate that more traffic passed over the Rock Island Bridge each year than went up and down the Mississippi. The opposition, of course, objected that this was not admissible evidence, but Lincoln convinced Judge McLean that his statistics illustrated the demands of a changing economy. Then, in restrained and dignified tones, Lincoln turned to the specific question at hand—the crash of the *Effie Afton* —and denied that the bridge was responsible. He discussed the draw of the bridge in detail and offered evidence that there were no crosscurrents about its piers. In fact, the currents were so mild (only five miles an hour) and the drawbridge so wide that another boat had preceded the *Effie Afton* through the draw and had done so without accident. Clearly the pilot of the first boat had exercised reasonable skill and care in navigating his vessel. But not so with the *Effie Afton*'s pilot. Although the boat was at anchor for a day, he'd never bothered to visit the bridge, to inspect its piers and currents. No, he'd dashed on "heedless of this structure which has been legally put there." As he addressed the jury, Lincoln produced a model of the *Effie Afton* and went on to explain how the accident had actually happened. Citing the testimony of witnesses, he observed that the starboard wheel had stopped working— had malfunctioned. And so it was mechanical failure, not the currents around the bridge, that had caused the boat to turn sideways and crash in flames. "Was it not her own fault that she entered wrong?" Lincoln asked. "So wrong that she never got right? Is the defense to blame for that?"

It was a virtuoso performance. And though the opposing attorneys took seven hours to make their closing arguments, the jury ended in a hopeless deadlock and the court dismissed the case. Yet because of the issues at stake, litigation continued for several years, until the United States Supreme Court finally decided in favor of the bridge company. Meanwhile, in the fall of 1857, Lincoln received at least four hundred dollars for his labors—and probably a good deal more money than that, given the importance of the *Effie Afton* case.

That winter another shock wave came from Kansas. Proslavery forces there had met in Lecompton, drafted a proslavery constitution in outright defiance of the territory's free-state majority, and applied for statehood on the basis of that document. In Washington, Buchanan and his Southern-dominated Cabinet gave Lecompton their blessings and demanded that Congress admit Kansas as the sixteenth slave state whether free-soilers liked it or not. In the rancorous debates that followed,

Douglas broke with the administration over the Lecompton "swindle" and sided with the Republicans in battling against it. For the Republicans it was an intriguing spectacle—the Democratic Little Giant standing on the floor of the Senate and raging against a Democratic administration ("Never have I seen a slave insurrection before," chuckled Republican Ben Wade of Ohio). As it turned out, though, Douglas did not oppose Lecompton because it was proslavery—for he cared not, he exclaimed, whether the proslavery clause in the constitution was voted up or down. No, he objected because the Lecompton convention had met illegally and had never fully or legally submitted the constitution to the people of Kansas for their ratification.

Yet there was more to Douglas's anti-Lecompton stand than that. He was fighting for his political future. His trips back to Illinois and his tangles with Lincoln had shown him how powerful free-soil sentiment was in Illinois and the rest of the Midwest. And the Little Giant, partly out of principle and partly out of ambition and partisan politics, was out to win back the support he'd lost there, wrench control of the Democratic party away from Southerners, make the Midwest the party's new power base, and then capture the Presidency in 1860 as a Northern man who opposed slavery and championed popular sovereignty. Still, his "bold stand" against Lecompton cost him dearly among Southerners (as he figured it would). On Capitol Hill, Southern leaders mounted a vicious attack against him: they and the pro-Southern Buchanan administration, which ran the regular Democratic machine, not only stripped him of his chairmanship of the Senate Committee on Territories, but vowed to drum him out of the party.

As he struggled savagely against Southern Democrats and an irate and vengeful Buchanan, Douglas seemed trapped in a spider web of contradictions. If he wanted to unite his party and restore its national character, his own ambiguous doctrines divided the Democrats and plunged them into confusion and bitter disputes. If he professed indifference as to whether slavery were voted up or down in Kansas, he kept insisting that popular sovereignty was a free-soil device that would bring about a Northern triumph in the West. If he regarded Negroes as repugnant and inferior, enjoyed taking racist swipes at Lincoln and the "Black Republicans," and argued that it wasn't his purpose to pass moral judgment on slavery, he also announced that bondage was "a curse beyond computation" on blacks and whites alike. At the same time, the senator from Illinois owned a large Mississippi plantation and some 140 Negroes. Furthermore, the profits from his plantation helped sustain him in politics so that he could trumpet popular sovereignty and the sacred right of self-government. And now this vague, blusterous, inconsistent man had declared war on the proslavery Lecompton Constitution—the unhappy fruit of his own territorial program.

Despite the opposition of Douglas and the Republicans, the Senate finally approved Buchanan's bill to make Kansas a slave state. But the House rejected the measure and substituted the so-called English bill, which summoned Kansans to vote on Lecompton in an open election and offered the territory a "bribe" in the form of a land grant if the constitution were ratified.

Meanwhile many Eastern Republicans were ecstatic, bedazzled, that Douglas had sided with them in the Lecompton struggle. In the *New York Tribune,* Horace Greeley not only applauded Douglas's stand, but asserted that Republicans ought to endorse him for re-election in 1858. If some Republican bigwigs balked at such a move, Greeley, Seward, and other Easterners all thought an alliance with Douglas would definitely strengthen the Republican cause, and Greeley urged Illinois Republicans to give the Little Giant their support. "His course has not been merely right," Greeley proclaimed, "it has been conspicuously, courageously, eminently so."

Greeley's course offended Lincoln deeply. From Springfield, where he followed the "rumpus" over Lecompton, Lincoln fired off a letter to Lyman Trumbull, who was now serving in the Senate. "Dear Sir: What does the New York Tribune mean by it's constant eulogizing, and admiring, and magnifying [of] Douglas?" Was this the sentiment of all Republicans in Washington? Were Republicans here in Illinois to be sacrificed?

Lincoln was beside himself with anxiety. By some monstrous trick of fate, was his party going to reject him and line up behind the champion of popular sovereignty, thus destroying all Lincoln had worked for in the free-soil cause since 1854? He wrote Trumbull that Republicans must stay out of the dogfight between Douglas and Buchanan. Lincoln thought them both wrong —although Buchanan was a little more wrong than the Little Giant. In January, 1858, Trumbull wrote back that "some of our friends here act like fools in running after and flattering Douglas," but Lincoln must be patient. Trumbull and the Illinois delegation had no intention of abandoning Lincoln for Douglas.

But for once Lincoln was all out of patience. For the signs all pointed to something ugly and sinister loose in the land. He'd read an editorial in the *Washington Union* of December 17, 1857, that slavery had a constitutional right not just to expand into the territories, but to spread into the free states as well. So what if Robert Toombs of Georgia had replied that no Southerner wanted to extend slavery into the North? Who could believe Toombs when George Fitzhugh advocated the liquidation of free labor in America and the enslavement of white workers as well as blacks? Who could believe Toombs when in Lincoln's opinion "the sentiment in favor of white slavery" now prevailed in the newspapers of Virginia and the Deep South? Who could believe Toombs when Senator James H. Hammond of South Carolina stood up in Congress and publicly extolled the virtues of slavery—see how happy and content the Negroes are in the South—and castigated Northern workers as "the mudsills of society"? And Lincoln was not the only Republican who saw a definite pattern, a "design," in Southern defenses of slavery and recent political events. In Washington, Republican senators like Seward himself, Zachariah Chandler, and William Pitt Fessenden expatiated on what they believed was a Southern conspiracy to nationalize slavery. In their view, the inexorable procession of events—Douglas's repeal of the Missouri line, Taney's Dred Scott decision, and Buchanan's proslavery program—all seemed part of a Southern plan to graft slavery on the face of the Republic and undermine civil liberties and even free government itself.

If Douglas was involved in a plot to nationalize slavery, how could Eastern Republicans shake his hand and pat his back and talk of supporting him? Did they not understand that he was the same old Douglas? That there remained profound and irreconcilable differences between him and Republicans? We must not hook on to Douglas's kite, Lincoln warned Republican leaders. We must maintain our own Republican identity. Douglas is not your man for the Senate. *I am your man. I, a pure Republican.*

In February Lincoln hurried up to Chicago for a conference with Norman Judd, chairman of the Republican State Central Committee. To his immense relief, Lincoln found Judd in an uproar about the Douglas flirtation. If the party supported him, Judd declared, he would resign. And Governor Bissell and Trumbull, who hadn't forgotten Lincoln's "sacrifice" back in 1855, also vowed to stick by him, as did virtually every other state leader. "The Republicans of Illinois," asserted Dr. Charles Ray, senior editor of the *Chicago Tribune,* "are unanimous for Lincoln and will not swerve from that purpose."

At that, party leaders called for a state convention to meet in Springfield on June 16, ostensibly to select candidates for administrative posts, but really to nominate Lincoln for the Senate. Then, in an unprecedented display of Republican unity, brought on by all the "outside meddling," ninety-five Republican county conventions named Lincoln as their "first, last, and only choice."

In April, Illinois Democrats held a state convention in Springfield with the Douglas forces in firm control. Lincoln, Judd, and Ray were on hand as observers and were jubilant when the convention endorsed popular sovereignty, as Douglas ordered. For that doctrine was anathema to all Republicans—and was proof, too, that Douglas had no intention of compromising his beliefs and forming an alliance with Eastern Republicans, which ended their flirtation. At the same time, a small faction of Buchanan or "Buchaneer" Democrats held their own convention in Springfield and planned to fight Douglas's bid for re-election. In high spirits about all this, Lincoln declared that "if we do not win, it will be our own fault."

With his own candicacy now assured, Lincoln set about preparing a speech for the upcoming Republican convention. As ideas came to him, he wrote them down on envelopes and scraps of paper, which he stored in his stovepipe hat. His speech would deal explicitly with the Slave Power conspiracy and with the disturbing question—one he'd contemplated for some time now —of whether a house divided against itself could stand.

In early May, 1858, Lincoln took time out from his hectic political activities to help an old friend. She was Hannah Armstrong, wife of Lincoln's old New Salem chum Jack Armstrong. Some time ago, Hannah had called on Lincoln in Springfield and said her boy "Duff" was in serious trouble with the law. According to witnesses, Duff Armstrong and two other fellows—James Norris and James Metzker—had attended a drunken camp meeting at Virgin's Grove on an August night back in 1857. A violent altercation broke out at a makeshift bar near the revival site, whereupon Armstrong and Norris allegedly mauled Metzker with fists and weapons. When he died three days later, authorities arrested Armstrong and Norris on murder charges and went on to convict the latter in a hurried trial that put him in prison for eight years. Armstrong, for his part, wound up in jail at Beardstown, on the bank of the Illinois River, where he awaited trial.

When Lincoln heard the story from Hannah, he agreed to serve as co-counsel in Duff's defense. The trial began in early May, 1858, in Beardstown's two-story brick courthouse. As Lincoln looked on, the prosecution's star witness, Charles Allen, testified that he was present when the fracas took place and that he saw clearly what happened because a full moon was shining directly overhead. According to Allen, Armstrong struck the victim in front of the head with a slingshot while Norris clubbed the back of his head with a wagon yoke. Obviously one or both blows killed him.

Lincoln rose to cross-examine. With his fingers hooked under his gallus straps, he asked the witness to repeat his story, especially the part where he couldn't have been mistaken about who did the killing. Then, with a sure instinct for the jugular, Lincoln produced an 1857 almanac and noted that at the very hour the brawl occurred, the moon was not directly overhead, but was low in the sky about an hour away from setting. "The Almanac," several jurors observed, utterly "floored the witness." After that Lincoln summoned his own witnesses, one of whom testified that the slingshot Armstrong allegedly used was in fact *his* weapon and that it was in his possession at the time of the fight. In his summation, Lincoln reminded the jury that if Allen was in error about the moon, he could be wrong—and most likely was wrong—about the rest of his testimony, too. The jury took an hour to find Armstrong not guilty.

The Armstrong trial became legendary—in part because a rumor circulated afterward that Lincoln had used a spurious almanac to trip up the prosecution's witness. Although the charge persisted through the years, there was not a shred of truth to it.

B̲y early June, 1858, Lincoln had completed his convention speech, "A House Divided," whose major themes had been percolating in his mind for several years. Still, he was a little insecure about the address, so on the eve of the convention he read the manuscript to a group of ten or twenty Republican friends in Springfield, including Billy Herndon. When Lincoln finished, several of his associates shook their heads, protesting that his phrasemaking was "dangerous" and would give the party a "radical" tinge. One even contended that "the whole spirit was too far in advance of the times." Lincoln turned to Herndon. And you, Billy? What do you think? Herndon jumped up. "By—God, deliver it just as it reads," he exclaimed. "The speech is true—wise & politic; and will succeed—now as in the future." Lincoln nodded and said he was determined to give the speech as written.

On convention day, June 16, Republican trains whistled into Springfield from all over Illinois, their coaches draped with Lincoln banners. The convention met in the great hall of the statehouse and unanimously nominated Lincoln for senator. It was the first time a convention had ever chosen a senatorial candidate, to be voted on in the legislature after the fall elections, and everybody remarked about how unprecedented the affair was. That night the convention reconvened with thunderous applause, and Lincoln stood once again by the speaker's table, reading from his manuscript.

'Mr. President and Gentlemen of the Convention.

"If we could first know *where* we are, and *whither* we are tending, we could then better judge *what* to do, and *how* to do it.

Thomas Lincoln, a traditional portrait. "His chief earthly pleasure was to crack and tell stories to a group of chums."

Sarah ("Sally") Lincoln in later years.

Mary Todd's imperious sister and brother-in-law who tried to prevent the Lincoln marriage.

Elizabeth Edwards

Ninian Edwards

Abraham Lincoln in 1846 when he was Congressman-elect.

Mary Lincoln in 1846. "They are very precious to me," Mary said of photographs above and left, "taken when we were very young and so desperately in love."

Seventeen-year-old Robert Todd Lincoln. An aloof, chilly introvert, Robert was "his mother's 'baby' all through."

William Wallace ("Willie") Lincoln, an affectionate, precocious boy who liked to read books and memorize railroad timetables.

Thomas ("Tad") Lincoln, who loved pranks and spoke with a lisp, whom Mary called "her little *sunshine*."

By the age of forty-nine, Lincoln was a highly successful lawyer and rising politician who loathed the nickname "Abe."

Mary Lincoln, aged forty-one. Though abnormally sensitive and neurotic about money, Mary was nevertheless a charming hostess and always fashionably dressed.

William H. Herndon, Lincoln's law partner, an intellectual gadfly who could talk nonstop about everything from philosophy to phrenology.

Judge David Davis, who rode the circuit with Lincoln and became his 1860 campaign manager.

Orville Browning, a suave, conservative Whig who later followed Lincoln into the Republican party.

Lincoln during the Great Debates of 1858. "The Republican party hold that this government was instituted to secure the blessings of freedom, and that slavery is an unqualified evil to the negro, to the white man, to the soil, and to the State."

Stephen A. Douglas: "I am opposed to negro citizenship in any and every form. I believe this government was made on the white basis. I believe it was made by white men, for the benefit of white men and their posterity forever."

A. Lincoln: "There is no reason in the world why the negro is not entitled to all the natural rights enumerated in the Declaration of Independence, the right to life, liberty, and the pursuit of happiness. I hold that he is as much entitled to these as the white man."

Presidential candidate, August, 1860. "Their thinking [slavery] right, and our thinking it wrong, is the precise fact upon which depends the whole controversy."

"We are now far into the *fifth* year, since a policy was initiated, with the *avowed* object, and *confident* promise, of putting an end to slavery agitation.

"Under the operation of that policy, that agitation has not only, *not ceased,* but has *constantly augmented.*

"In *my* opinion, it *will* not cease, until a *crisis* shall have been reached, and passed.

" 'A house divided against itself cannot stand.'

"I believe this government cannot endure, permanently half *slave* and half *free.*

"I do not expect the Union to be *dissolved*—I do not expect the house to *fall*—but I *do* expect it will cease to be divided.

"Either the *opponents* of slavery, will arrest the further spread of it, and place it where the public mind shall rest in the belief that it is in course of ultimate extinction; or its *advocates* will push it forward, till it shall become alike lawful in *all* the States, *old* as well as *new*—*North* as well as *South.*"

With that ringing prelude, Lincoln scanned distant horizons and spoke of the future of the Republic. We Republicans must place slavery back on the course of ultimate extinction, for that is how the American house will cease to be divided. If we do not do this, then the Democrats will nationalize slavery. "Have we no tendency to the latter condition?" he warned. Was it not so, that an "almost complete legal combination—piece of *machinery* so to speak—compounded of the Nebraska doctrine and the Dred Scott decision" was being worked out? Then in dramatic detail he traced "the design" and "its chief bosses" from the beginning. First came Douglas's Kansas-Nebraska Act, which opened the national territories to slavery behind the mask of popular sovereignty, of the sacred right of self-government. Second, as that act found its way through Congress, the Dred Scott case passed through the U.S. circuit court in Missouri, and both came to a decision in May, 1854, which was evidence that something sinister was afoot. And then came Buchanan's inaugural address, in which he announced that the Supreme Court was about to decide on slavery in the territories. Whereupon the Court handed down the Dred Scott decision, a straight proslavery verdict which Douglas ended up extolling here in Springfield. Then came the Douglas-Buchanan split over the Lecompton Constitution, over the mere fact of whether it represented the will of the people. Worse still, Douglas propounded his *"don't care"* policy about slavery. "He *cares* not," Douglas says, "whether slavery be voted *down* or voted *up."* And this policy of moral evasion was the next step in the design: it was calculated to make Northerners apathetic, so that they too won't care whether slavery advanced west or not. True, Lincoln admitted, we can't be absolutely certain that all this came from preconcert. But when we see "a lot of framed timber" gathered at different times by different workmen—by "Stephen, Franklin, Roger and James," in reference to Douglas, Pierce, Taney, and Buchanan—and when these timbers are put together and make the frame of a house, with all the pieces fitting, then it looks as though all the workmen operated "from a common plan." And the final piece, the last plank to complete the house, Lincoln believed, will be "another supreme court decision," one declaring that the U.S. Constitution forbids a *state* from prohibiting slavery. And Douglas's "care not" policy is clearing the way for such a decision by lulling Northerners to sleep. Then when we wake, we will find that the Court has made Illinois a slave state, that it has nationalized human bondage and made the entire Republic a slave house.

There are those, Lincoln went on, who whisper that Douglas is the man to destroy the conspiracy, that he is "a very great man" while the largest of us in Illinois are very small. Maybe so, Lincoln granted, but "a *living dog* is better than a *dead lion.*" And if Douglas isn't a *dead* lion, he's "at least a *caged* and *toothless* one" for the Republican cause of halting the spread of slavery and preventing its nationalization. "How can he oppose the advance of slavery? He don't *care* anything about it. His avowed *mission is impressing* the 'public heart' to *care* nothing about it." No, our Republican cause must be conducted entirely by "its own undoubted friends—those whose hands are free, whose hearts are in the work—who *do care* for the result." And so Republicans must stay together, united against a common danger. "Of *strange, discordant,* and even, *hostile* elements, we gathered from the four winds, and *formed* and fought the battle through, under the constant hot fire of a disciplined, proud, and pampered enemy." Are we to falter now when the enemy is "dissevered and belligerent?" No, "we shall not fail—if we stand firm, we shall not fail."

After all the cheers and ovations were over, some Republicans worried about the speech, fearing that it might be misconstrued as a call for war. But many others agreed with Herndon that the address was a "compact—nervous —eloquent" expression of Republican purposes. Across Illinois Republican journals gave "A House Divided" front-page exposure, and some country editors even brought it out in a pocket edition.

Herndon, of course, was enormously pleased. "I want to see 'old gentleman' Greeley's notice of our Republican convention," Herndon wrote Trumbull. "I itch—I burn, to see what he says." As it happened, Greeley also published Lincoln's speech, under the headline "REPUBLICAN PRINCIPLES." But he wrote Joseph Medill of the *Chicago Tribune* that in nominating Lincoln and rejecting Douglas, Illinois Republicans had ignored the near unanimous advice of Republicans in other states. "Now go ahead and fight it through," Greeley huffed.

When Douglas learned of Lincoln's nomination, he confided in a fellow Democrat that "I shall have my hands full. He is the strong man of the party —full of wit, facts, dates—and the best stump speaker, with his droll ways and dry jokes, in the West. He is as honest as he is shrewd, and if I beat him my victory will be hardly won."

PART·FIVE
YEARS OF METEORS

Why was he running against Douglas? Lincoln contemplated the notepaper in front of him and then wrote in his small, neat style: "I have never professed an indifference to the honors of official station; and were I to do so now, I should only make myself ridiculous." Yet the triumph of the Republican cause was more important to him than personal success (*"we* have to fight this battle upon principle, and principle alone," he would say later). But he realized that "the higher object of this contest"—the ultimate elimination of slavery—"may not be completely attained within the term of my natural life. But I can not doubt either that it will come in due time." And he was proud, "in my passing speck of time," to contribute what he could to that consummation.

On July 8 he turned up in Chicago, where Douglas opened his campaign with frenzied demonstrations. On the following night, the Little Giant stood on the balcony of the Tremont House and faced a huge Democratic rally, illuminated by torches and exploding fireworks. At Douglas's invitation, Lincoln joined him on the balcony and watched the back of Douglas's prodigious head, with his mane of black hair, as the Little Giant harangued the jubilant throng below. Douglas portrayed himself as a lonely defender of self-government battling an "unholy combination" of Republicans and Buchanan Democrats. Congress, he declared, had no power to force slavery on the people of a territory, or to take slavery away from them, but "must allow the people to decide for themselves whether it is good or evil." It was out of this conviction that Douglas had resisted the Lecompton Constitution— and virtually single-handedly, too. Then he attacked Lincoln's "House Divided" speech with both fists, flaying it as a radical cry for "a war of sections," to be fought relentlessly "until the one or the other shall be subdued and all the states shall either become free or become slave." Lincoln thus demanded a uniformity in local institutions which the Founding Fathers had never wanted and which, if achieved, would destroy state sovereignty and personal liberty. Then Douglas turned again to Negroes and the Declaration, repeating that this government "was made by the white man, for the

benefit of the white man, to be administered by white men." Since Lincoln
went for Negro equality, a sectional war, and uniformity of domestic institutions, the issues between him and Douglas were "direct, unequivocal, and
irreconcilable."

The next night, from the same balcony, Lincoln addressed a Republican
crowd almost as large as Douglas's "and five times as enthusiastic." He
insisted that the Republicans deserved the credit for whipping the Lecompton Constitution, since they provided most of the votes that killed it in the
House. "That's it!" people yelled below. "Good! Good!" And furthermore,
in "A House Divided" Lincoln was not advocating a sectional war, but was
only making a prediction—expressing an expectation—that a moral crisis
would one day have to come over slavery. He went on: "I have always hated
slavery, I think as much as any Abolitionist." But he'd been quiet about it
until this new era began—this new program to nationalize slavery and turn
the Republic away from the visionary plans of the Founding Fathers. For it
was the founders who had originally restricted the growth of slavery and
placed it on the road to its ultimate doom. And the Republicans agreed with
those old-time men, wanted like them to keep slavery from expanding,
wanted like them to put that institution on the high road to extinction. Yet
this didn't mean that Republicans desired uniformity in domestic institutions,
for all agreed that Illinois had no right to interfere with the oyster laws of
Virginia or the liquor laws of Maine. Slavery, though, was not "an exceedingly little thing" on a par with such laws. It was not, as Douglas maintained,
"something having no moral question in it," something equal to "the question of whether a man shall pasture his land with cattle, or plant it with
tobacco." No, slavery was "a vast moral evil" which concerned all Americans, an evil that must be confined to the South so that it could die a natural
death. Then Lincoln sought to turn the campaign and the Republic itself away
from racism. "Let us discard all this quibbling about this man and the other
man—this race and that race and the other race being inferior." Let us unite
as one people and "once more stand up declaring that all men are created
equal."

And so at Chicago the two combatants drew their first blood of the campaign. Their speeches not only filled up the front pages of Illinois papers, but
attracted considerable attention in the East as well. Said the *New York Times:*
"Illinois is from this time forward, until the Senatorial question shall be
decided, the most interesting political battle-ground in the Union."

Lincoln remained in Chicago for several days, conferring with the Republican State Central Committee. As they pored over maps and charts, they
agreed that northern Illinois—whose settlers had come mainly from the
Northeast—was solidly in the Republican column. They also conceded that

southern Illinois—Little Egypt—was a Douglas stronghold where men took Democratic politics as straight as they took their whiskey. The crucial area —the area that would decide the outcome—was Whiggish central Illinois, whose inhabitants had both Southern and Northern backgrounds. So Lincoln and his surrogate speakers would concentrate their energies here, in Sangamon, Coles, Macon, and contiguous counties.

At the same time, Republican managers cooperated some with the Buchanan Democrats, a small faction of castoffs and misfits, under orders from the President himself to use the patronage ruthlessly against Douglas and run rival candidates against Douglas's people. If the Buchaneers did that all over the state, Lincoln said, then "this thing is settled—the battle is fought." Yet Lincoln hotly denied that the Republicans had formed an alliance with the Buchaneers, denied that his party had compromised its principles, partitioned offices, or swapped votes with Buchanan men. In fact, Lincoln had warned his colleagues against "strange and new combinations," warned that "we must never sell old friends to buy old enemies." Yet Herndon wrote Trumbull that Lincoln "does not know the details of how we get along." The fact was that Republicans often swelled up Buchaneer rallies; and Herndon himself had frequent discussions with his father and brother—both Buchanan men—who made "no bones of telling me what they do."

On the advice of the central committee, Lincoln trailed the Little Giant across central Illinois, speaking at the same towns as he did. They even rode the same train from Bloomington to Springfield, with Lincoln traveling alone in a public coach while Douglas relaxed in a private car, bedecked with banners that blazed, STEPHEN A. DOUGLAS, CHAMPION OF POPULAR SOVE-REIGNTY. Attended by his lovely second wife, Douglas sipped brandy and smoked cigars with a bevy of advisers and reporters. An imposing brass cannon, mounted on a flat car and manned by Douglas supporters, boomed away as the train approached the prairie towns.

At Springfield, Douglas threw Lincoln on the defensive with stabbing accusations about his alleged racial views. And Democrats mocked and belittled him, saying he followed after Douglas because he couldn't get up his own crowds. Still, Lincoln's advisers urged him to stay with the Little Giant and hit him hard, be more aggressive, put *him* on the defensive. In fact, they rather insisted that Lincoln confront Douglas in open debate on the same platform. And Lincoln agreed that this would enhance his campaign. Accordingly, after talking with the central committee in Chicago, he challenged Douglas to fifty official debates around the state. Though Douglas had nothing to gain, he accepted the challenge because he was too competitive to back away. But he stipulated that they meet only seven times—in the convention towns of each Congressional district—and that he have four openings and rebuttals to Lincoln's three. While the terms favored Douglas, as the Republican press acidly observed, Lincoln agreed to the plan and the Great Debates were on.

For the first week of August, Lincoln stayed in Springfield, polishing his speeches. By now several Republicans urged him to nail Douglas on the contradiction between popular sovereignty and Dred Scott. If Douglas fell back on the argument about unfriendly legislation, he would trap himself since the South would not accept this "mere barren right." Lincoln thought it a good point, but doubted that Douglas cared about the South since "he was dead there already." If you could bring Douglas to an answer on this subject, Lincoln said, the Little Giant would repeat what he'd said before: that slavery couldn't survive in a territory without protective measures. "If this offends the South he will let it offend them; as at all events he means to hold on to his chances in Illinois."

At last it was time to go, time to pack his bag and tell Mary and the boys goodbye. Though Mary couldn't be with Lincoln, she assured him that he would defeat Douglas, assured him that he would "be a Senator and President of the United States, too." On the campaign trail, Lincoln would recall her remark about the Presidency with a warm chuckle. Still, as he told a young reporter, he would be satisfied just to be a senator.

On August 21, at Ottawa in northern Illinois, the two candidates met in their first debate, stood together on a lumber platform and faced twelve thousand cheering people in Ottawa's public square. For Americans in those days, stump political oratory was an exciting form of entertainment, a kind of open-air theater. And so people came from miles around Ottawa to witness what was billed as a mighty forensic duel between two political gladiators.

At last Douglas rose to speak, the sun blazing down in a dusty haze. He wore what Republicans dubbed his "plantation outfit"—a wide-brimmed white felt hat, a ruffled shirt, a dark blue coat with shining buttons, light trousers, and shined shoes. Behind him, Lincoln sat with his knees rising up at sharp angles, his eyes half closed, a carpetbag filled with notes and speeches at his side. He listened intently, though, for he was up against a veteran campaigner in the Little Giant, whose booming voice, menacing gestures, and hawklike scowls had delighted audiences and intimidated rivals for years. As he opened up on Lincoln now, Douglas shouted and clenched his fists

... and revealed the kind of pugilistic, race-baiting strategy he would employ throughout the debates. He hit Lincoln with seven interrogatories and demanded answers, quite as though he were a great trial lawyer putting on a show in an Illinois courthouse. He insisted that *he* was on the side of the Founding Fathers, that they had made America a divided land—half slave and half free—and that the Republic could remain that way. He accused Lincoln and Trumbull, who'd also taken to the stump against Douglas, of conspiring to destroy the Whigs, dissolve the Democrats, and clear the way for a new abolition party disguised as the Republicans, whose real purpose was to foment civil war, was to emancipate Negroes and make them the social and political equals of white people. Are you for this? Douglas asked the crowd. "No! No! Never!" someone cried. Are you for letting Negroes flood into Illinois "and cover your prairies with black settlements?" Do you want Illinois to become a free Negro colony? So that when Missouri abolishes slavery, 100,000 Negroes will run into Illinois to become citizens and vote on an equal basis with you? "No! No!" several whites cried. If you're for all this, then vote for Lincoln and the "Black Republicans." But I for one am against Negro citizenship. I want citizenship for whites only. "I do not question Mr. Lincoln's conscientious belief that the negro was made his equal, and hence his brother"—great laughter at that—"but for my own part, I do not regard the negro as my equal, and positively deny that he is my brother or any kin to me whatever."

When Lincoln rose to reply, the applause was so tumultuous—this was Republican northern Illinois—that he had to wait a few minutes to begin. As he addressed the crowd, he swung his arms awkwardly and his voice rose to "a shrill treble" when he became excited. Yet several who witnessed the Great Debates recalled the tremendous power of Lincoln's voice, which could be heard out on the fringes of a giant open-air assembly. To emphasize a point, Lincoln would bend his knees, crouch, and then spring up vehemently to his toes. He had brushed off Douglas's charge that he and Trumbull had conspired to destroy the Whig and Democratic parties. He was now talking about abolition and Negro equality.

He hated to spend time on these subjects, but Douglas kept raking them up and distorting Lincoln's position so wildly that he had to explain himself or risk being misunderstood in Illinois. Once again, he was not for molesting slavery in the South. Once again, he was not for Negro equality. There was "a physical difference" between the black and white races, he conceded, that would "probably" always prevent them from living together in perfect equality. And Lincoln wanted the white race to have the superior position so long as there must be a difference. Therefore, any attempt to twist his views into a call for perfect social and political equality with Negroes was "but a specious and fantastic arrangement of words, by which a man can prove a horse chestnut to be a chestnut horse."

Frankly, Lincoln went on, Negroes were not his equal or the equal of Douglas in moral and intellectual endowment. But they *were* equal to Lincoln, Douglas, and "every living man" in their right to life, liberty, and the pursuit of happiness, which included the right to the fruits of their labor.

Then he came to what he considered the central issue in this contest—the moral question of whether slavery would ultimately perish or ultimately triumph in the American Union. Douglas, however, kept evading this issue, and why? Because he was guilty of complicity in the proslavery conspiracy, the next step of which would be a Supreme Court decision legalizing human bondage in the free states as well as the territories. And Douglas, through his doctrine of moral evasion and "don't care," was working to clear the way for such a decision. . . . *"A lie,"* Douglas roared in his rebuttal. And he castigated Lincoln as "an ignorant man."

With their rival strategies clear now, both men took off in separate directions, to stump towns and villages individually until they met again at Freeport. By now the campaign had taken on a carnival air, as thousands of people clogged the towns where Lincoln and Douglas spoke. Everywhere Lincoln went, dust hung in the air and got in his eyes and mouth. Flags and banners draped buildings and swung across dirt streets. Bands played, glee clubs sang, and armed cadets marched smartly to the rattle of drums. In all the commotion, white "Republican" horses conveyed Lincoln to the speaker's stand in a fancy "Republican" carriage or a decorated wagon. As he addressed one audience after another, demonstrators waved signs and banners overhead. Those for the Little Giant included a sketch of a Republican domestic scene, which showed a white man with his "nigger" wife and "nigger" children. By contrast, a favorite Republican drawing depicted Douglas perilously astride two galloping horses, one labeled "Popular Sovereignty" and the other "Dred Scott." On another Republican sign, a train named "Freedom" roared down on Douglas's oxcart, while his Negro driver exclaimed, "Fore God, Massa, I b'lives we're in danger." After his speech and attendant festivities were over, Lincoln would hurry back to some noisy hotel, to catch a little sleep in an undersized bed and then revise and repair his speeches, writing every word of them himself.

Then it was on to the depot and another rocking railroad trip to the next town, and the next, with reporters scrambling after him just as they trailed after Douglas, and writing up accounts of the campaign that were fanatically partisan as always. Republican journalists, for their part, ridiculed Douglas's swagger, his dwarfish height, mammoth head, and ducklike walk, and described how he "foamed at the mouth" and dribbled "the saliva of incipient madness." On the other side, Democratic journalists taunted Lincoln's gangly arms, clownish legs, and apelike gestures. When Lincoln deprecated his "poor, lean, lank face," when he told how humble and little known he was in comparison to the "great" Douglas and insisted that he was the Republican standard-bearer only because *somebody* had to be, Democratic papers jeered at his "sniveling humility" and said he talked like Uriah Heep. What an "abject wretch" Lincoln was. And he was a coward, too, so frightened of Douglas that he sat on the speaker's stand with knees locked together and teeth chattering. One paper even called Lincoln a "Modern Titus Oates," in reference to the seventeenth-century English impostor, and noted that both Lincoln and Oates had low foreheads like baboons.

And so it went as the candidates crisscrossed Illinois and delivered scores of separate speeches in addition to their formal debates, campaigning in weather that ranged from boiling heat to blustery cold. As Lincoln pounded away at the Little Giant, so did a small army of other Republican speakers —Trumbull, Herndon, Swett, Lamon, and Davis—all stumping Illinois in Lincoln's behalf. Even Frederick Douglass, the eminent Negro editor and political abolitionist, came out from New York State and joined in the get-Douglas fight.

Getting the Little Giant, though, proved difficult indeed. After the first debate at Ottawa, Lincoln was uncertain how to proceed. At Ottawa, Douglas had posed several interrogatories which Lincoln could not ignore. Though he had his own ideas about how to answer them, he solicited advice from the Republican State Central Committee and other party managers. At Freeport, in far northern Illinois, Joseph Medill of the *Chicago Tribune* evidently met with Lincoln and gave him a summary of what he, Judd, and other Republicans thought should be done. The most important thing was to discredit Douglas in free-soil Illinois. Therefore Lincoln should "put a few ugly questions" to the Little Giant, such as "What becomes of your vaunted popular sovereignty in [the] Territories since the Dred Scott decision?" As Medill advised, Lincoln should pitch into Douglas, should be "saucy" and "dogmaticall" and "give him hell."

Acting on Medill's suggestions, Lincoln devised an aggressive strategy for Freeport and subsequent debates. Through his own interrogatories, arguments, and rebuttals, he would try to put Douglas on the defensive by exposing all his vague and contradictory posturings about slavery in the territories. At the same time, Lincoln would pound away on what he'd been saying all along—that popular sovereignty was sheer political fiction, a lullaby to delude Northerners and help clear the way for the spread and nationalization of slavery.

At Freeport, on the damp and chilly day of August 27, Lincoln stood in boots that glistened from the dampness and addressed a pro-Republican assembly of some fifteen thousand people. In answer to Douglas's interrogatories, Lincoln declared that yes, he was against the admission of additional slave states. No, he was not for repealing the fugitive slave law (even though he detested slave-catching). And no, he was not "pledged" to abolishing slavery in Washington, D.C. Then Lincoln struck Douglas with his own interrogatories. The most significant of them: Could the people of a territory "in any lawful way" exclude slavery before statehood? Moreover, if the Supreme Court ruled that a state could not prohibit slavery, would Douglas abide by the decision?

In his reply, Douglas iterated what became known as the Freeport doctrine. Of course the people of a territory could exclude slavery before statehood—through the principle of "unfriendly legislation," they could refuse to enact the necessary protective measures. Douglas pointed out that he'd argued this doctrine all over Illinois in past years, so there was "no excuse" for Lincoln's pretending not to know Douglas's position now. As to the question about the Supreme Court, Douglas was certain it would never hand down any such decision as Lincoln described, for that would be "an act of moral treason" to which no justice would descend.

Then Douglas turned to his favorite tactic against "you Black Republicans" ("White! white!" the crowd yelled). He believed that some people around Freeport, a Republican town near the Wisconsin border, thought that the Negro Frederick Douglass was a good man. "The last time I came here to make a speech, while talking from the stand to you, people of Freeport, as I am doing to-day, I saw a carriage and a magnificent one it was, drive up and take a position on the outside of the crowd; a beautiful young lady was sitting on the box seat, whilst Fred. Douglass and her mother reclined inside, and the owner of the carriage acted as driver." There was confusion in the audience, cries of *"Right!"* and "What of it?" Douglas bellowed back, "if you, Black Republicans, think that the negro ought to be on a social equality with your wives and daughters, and ride in a carriage with your wife, whilst you drive the team, you have a perfect right to do so." And of course you will "vote for Mr. Lincoln."

And so to Lincoln's dismay the race-baiting continued. From Freeport the two candidates headed south for Little Egypt, with Lincoln complaining that Douglas didn't care how much he distorted Republican racial policies if it won him votes.

On the night of September 14, Lincoln reached Jonesboro near the Kentucky border. He sat in front of his hotel for an hour, watching Donati's comet as it arced across the heavens with its tail of fire. Lincoln told himself that he wasn't afraid of the people in Little Egypt. He knew them better than

Douglas did. He was raised a little east of here, in Indiana. "I am part of these people," he said. But in the debate the next day, only twelve hundred showed up—and many of them Buchanan Democrats. Predictably, Douglas was unrelenting in his attacks against Lincoln's "House Divided" speech and "Black Republican" views. And Lincoln, in response, carefully left out what he'd said in northern Illinois about Negroes and the Declaration of Independence. He treated slavery as a national problem and appealed to the hostile gathering as "a national man." Then he narrated a scenario about Douglas's inconsistencies on slavery in the territories. First Douglas tells us that the people of a territory can vote to keep slavery out. But when asked whether they can do this before statehood, the judge tells us this is a matter for the Supreme Court to decide. Then comes the Court's Dred Scott decision, decreeing that a territory cannot prohibit slavery at all, thus making a shambles out of popular sovereignty. At that, Douglas "shifts his ground" again and tells us that the people of a territory can still ban slavery through "unfriendly legislation." But that argument was chimerical, Lincoln contended, as unreal and deceptive as all of Douglas's other doctrines. For slavery had plenty of vigor to plant itself in the territories without police measures. "It takes not only law but the *enforcement* of law to keep it out," Lincoln said. Then he posed another question designed to hurt Douglas in Illinois. If slave owners in a territory demanded Congressional legislation to protect slave property, would Douglas vote for or against such a measure?

Douglas made an evasive reply, vaguely reaffirming his belief in Congressional nonintervention in the territories. Then he accused Lincoln of ignoring his questions and misconstruing his arguments. No matter what Lincoln said, the Freeport doctrine and Dred Scott were not at all contradictory. If the Supreme Court recognized the right of a man to take slaves into the territories, it remained "a barren, worthless, useless right" unless the governments there enacted local protective legislation. If they refused, slavery would be excluded just as effectively "as if there was a constitutional provision against it."

Thus the candidates punched and counterpunched, each trying to discredit the other. From Little Egypt Lincoln made his way up to Charleston, in politically uncertain central Illinois, where a boisterous cavalcade escorted him into town in a moving cloud of dust. One flower-splashed wagon carried thirty-two young women, whose banners represented every state in the Union. Another young woman trailed behind on horseback, carrying a banner that read, "Kansas—I will be free." It was an ironic scene, given recent political developments in Kansas. On August 2, with federal troops guarding the polls and an anxious nation looking on, Kansas voters had overwhelmingly rejected the Lecompton Constitution, ensuring an eventual free-state triumph in that embattled territory. Yet, because of powerful Southern opposition in Washington, Kansas would not be admitted as a state until 1861.

In the Charleston debate, Lincoln started right off on the Negro question —the first time he'd done so. As he'd repeatedly argued, Negro equality was not the issue between him and Douglas. The moral question of slavery in a country with the Declaration of Independence—*that* was the central question in this contest. Yet Douglas kept avoiding that issue and dragging the debates back around to Negro equality and intermarriage, thus playing on the racial fears of white audiences. At Charleston, Lincoln wanted to make his position about Negroes absolutely clear, so that the issue would be settled once and for all (and whites in crucial central Illinois would know where he really stood). He was not and never had been in favor of Negro social and political equality with whites. He was not and never had been in favor "of making voters or jurors of negroes, nor of qualifying them to hold office, nor to intermarry with white people." Because he did not want a Negro woman for a slave did not mean he desired her for a wife. As he'd said before, "I can just leave her alone." Yet since Douglas fretted so about intermarriage, fearful that his own friends might marry Negroes if there was no law to prevent them, Lincoln would stand "to the last" behind the Illinois law against miscegenation.

Douglas, in reply, accused Lincoln of being a racial hypocrite, a chameleon. In northern Illinois he waved the Declaration of Independence and insisted that it applied to Negroes. But down here Lincoln said he was not for political and social equality with them. These Republicans! Douglas cried. "Their principles in the North are jet black, (laughter,) in the centre they are in color a decent mulatto, (renewed laughter,) and in lower Egypt they are almost white. (Shouts of laughter)." Well, Lincoln was right. "A house divided against itself cannot stand."

In his rebuttal, Lincoln denied that he favored Negro citizenship, as Douglas charged. Lincoln thought a state could make Negroes citizens—even though the Supreme Court said it couldn't—but he would not want Illinois to confer citizenship on blacks. Moreover, he insisted that he was entirely consistent on the race issue, that he said the same thing in southern and central Illinois as he did in the north.

Back on the train again, heading northwest to Galesburg, Lincoln argued with Douglas on slips of paper. Suppose it's true, he wrote on one fragment, that the Negro is inferior to the whites "in the gifts of nature." Then is it not unjust that whites should take from the Negro "any part of the little which has been given him?" The Christian rule of charity is to *give* to the needy. But to *take* from the needy is the rule of slavery.

And on another fragment: the larger issue before us is the subversion of the Declaration of Independence, the denigration of its equality doctrine as "a self-evident lie." Both the *Richmond Enquirer* and the *New York Day-Book* proclaim this and both are Buchanan papers. And Douglas too argues against the equality doctrine. So the "common object" is to destroy in the public mind and in public administration "our old and only standard of free government, that 'all men are created equal.' " In place of that we're now to have a different standard, one that advocates the natural, moral, and religious right of one class to enslave another. Yes, that was the grim proslavery conspiracy Lincoln saw at work in the United States. And once the conspirators achieved their ends, once they nationalized slavery and overthrew the Declaration of Independence, America would become a despotism based on class rule and human servitude. The Northern free-labor system would be expunged, the blessings of the Revolution obliterated, America's noble experiment in popular government vanquished from this earth.

Somehow Lincoln must make Illinois voters see what was going on, must make them understand that their own charter of freedom—the Declaration of Independence—was being undermined, must make them realize Douglas's own complicity in all this. And so at Galesburg on October 7, addressing twenty thousand people in a fierce and icy wind, Lincoln illustrated once again how Douglas's "don't care" attitude about slavery was helping to nationalize the institution. In the debate at Quincy, Lincoln reached for a metaphor to describe the "humbuggery" of popular sovereignty as any sort of antiextensionist device. Thanks to Dred Scott and Douglas's own inconsistent interpretations, "has it not got down as thin as the homeopathic soup that was made by boiling the shadow of a pigeon that had starved to death?"

On October 15 the candidates met for the last debate at Alton, in southwestern Illinois. Mary and Robert joined Lincoln for the final encounter, with Robert marching in the ranks of the Springfield Cadets. So hoarse that it hurt him to speak, Douglas insisted that he'd been totally consistent in his views and recited his familiar arguments about Negroes, the Declaration, and slavery, contending that "this government can endure forever, divided into free and slave States as our fathers made it,—each State having the right to prohibit, abolish or sustain slavery just as it pleases." By contrast, Lincoln declared that slavery was the only issue that had ever menaced the Union. But a threat was not removed by extending it and making it larger. "You may have a . . . cancer upon your person and not be able to cut it out lest you bleed to death," Lincoln said; "but surely it is no way to cure it, to engraft it and spread it over your whole body." He also asserted that he desired free territories not just for native-born Americans, but for "Hans and Baptiste and Patrick, and all other men from all the world" who wanted to find new homes, a place to "better their conditions in life."

After that it was back to Springfield, where Lincoln attended a giant Republican rally and gave the final speech of the most exhaustive political battle he'd ever waged. Then all he could do was wait and see how many Republicans and Democrats would be elected to the legislature.

Election day, November 7, was cold and rainy. Lincoln voted early, then spent the evening with friends at the telegraph office. When all the votes were in, Lincoln's candidates polled 190,000 popular votes and Douglas's 176,000. But with the Democratic holdovers in the legislature, Democrats outnumbered Republicans there 54 to 46. Lincoln knew that was how the balloting for senator would go, since no Democrat—not even a Buchaneer—would ever vote for a Republican.

So he'd lost again. After four months of arduous campaigning, after giving the best speeches he was capable of writing, after all the combined efforts of Judd, Trumbull, Davis, Herndon, and dozens of local Republican committees, Lincoln had lost to Douglas and his "don't care" doctrines; and he didn't think he could bear the pain. Still, nobody blamed him for the defeat. Friends and colleagues assured him that he'd "made a noble canvass," that "no man could have done more." And even if he held no future political office, said the *Chicago Press & Tribune,* the Republican party owed Lincoln a great deal, for he'd steadily upheld the principles of Republicanism and contributed speeches that would become "land marks in our political history."* No, nobody blamed him for the loss. Rather, many Illinois Republicans faulted Greeley and Seward, insisting that their courting of Douglas had given him free-soil respectability in Illinois. In their bitterness, various Illinois Republicans vowed never to support Seward for the Republican Presidential nomination.

Lincoln, too, was concerned that some Eastern Republicans might not have endorsed him fully, but he harbored no grudges, hated no one. "I am glad I made the late race," he wrote his old friend Dr. Anson Henry. "It gave me a hearing on the great and durable question of the age, which I could have had in no other way; and though I now sink out of view, and shall be forgotten, I believe I have made some marks which will tell for the cause of liberty long after I am gone."

In February, 1859, after the legislature had officially re-elected Douglas as senator, friends called on Lincoln at his Springfield law office. He was steeped in gloom. "I feel like the boy who stumped his toe," Lincoln said. "I am too big to cry and too badly hurt to laugh."

---

*In July, 1858, the *Chicago Press* merged with the *Chicago Tribune.*

Once again he resumed his law prac-
tice, complaining that "I have been on expences so long without earning any
thing that I am absolutely without money now for even household purposes."
But by practicing his own motto that "work, work, work, is the main thing,"
he soon restored his finances—and his political confidence. "The fight must
go on," he wrote his Illinois colleagues. "I have an abiding faith that we shall
beat them in the long run." He added, "I write merely to let you know that
I am neither dead nor dying."

Meanwhile a few Republican papers in Illinois were mentioning Lincoln
for the Presidency, remarking that his debates with Douglas had made him
the top Republican in Illinois—even more influential than Trumbull—and
had brought him to the attention of national party leaders. And Jesse Fell,
Lincoln's friend and a leading Illinois Republican, thought so too. One late
afternoon, after court adjourned in Bloomington, Fell took Lincoln by the
arm and walked him through the shaded streets. Fell had just visited in the
East and everywhere he went Republicans asked, "Who is this Lincoln we
read about in the papers, who ran Douglas such a fine race?" If Lincoln's
personal history and his speeches on the slavery question were brought
before the public, Fell said, Lincoln could become a serious Presidential
contender. True, Seward and Chase were the front runners, but both had
reputations as "radicals," which hurt them. Seward, in fact, had recently
caused a national sensation, announcing in a speech at Rochester, New York,
that North and South were locked in an "irrepressible conflict." Lincoln, on
the other hand, had no such "radical" image among national Republicans,
who might turn to him for fear that none of the front runners could win.

"Fell," Lincoln said, "I admit the force of much of what you say, and admit
that I am ambitious, and would like to be President." But Lincoln had to be
realistic. He didn't stand a chance against nationally known men like Seward
and Chase. And anyway, "there is nothing in my early history that would
interest you or anybody else; and as Judge Davis says, 'it won't pay.' " He
told Fell good night and walked on alone in the twilight.

The Presidency? Regardless of what he told Fell, Lincoln had ambivalent
feelings about it. Of course it was the highest and most prestigious office in
the Republic, a symbol of individual eminence that would guarantee one a
place in history. Yet he had no administrative experience, none at all, and

had little love for the mundane chores that attended an executive job, such as administering the hated patronage. In all candor, he wrote an Illinois editor who wanted to announce Lincoln's candidacy, "I do not think myself fit for the Presidency." And even if he was fit and did stand a chance (which he doubted), it was much too early to declare his availability. One must be cautious about these things, must not appear too eager for the job. And in any case his first preference was for the national Senate. In fact, by early spring of 1859 he'd made up his mind to challenge Douglas again, to fight for his Senate seat in 1864. For all of Lincoln's talents seemed to point him to the Senate—his literary gifts, his love of rational debate, of delivering logical and eloquent speeches from a prepared script.

Still, as a politician and a loyal party man, he would leave all options open, ready to take any responsible national office that came his way, any office that would simultaneously help the Republican cause and give him personal fulfillment. "Claiming no greater exemption from selfishness than is common," he told a Republican colleague, "I still feel that my whole aspiration should be, and therefore must be, to be placed anywhere, or nowhere, as may appear most likely to advance our cause." But whether he ran against Douglas in 1864 or sought some other national office, Lincoln must unite and expand the Republican party here in Illinois, using that as his base of power. And he must sell himself beyond the state, so that national Republicans would never again disparage him as an unknown.

And so in 1859 he embarked on an assiduous campaign to advertise himself and his party. He collected his 1858 speeches in a scrapbook and eventually arranged to have them published and distributed by an Ohio firm. He composed a statement for a group of Boston Republicans, to be read aloud at a Jefferson birthday celebration and given extensive press coverage. How ironic it was, Lincoln wrote, that the Democrats had abandoned their Jeffersonian heritage and that the Republicans—supposedly the descendants of the old Federalists—now defended Jeffersonian ideals. Today the Democrats argued that property rights were superior to human rights, whereas the Republicans were for the man and the dollar—but in cases of conflict placed "the man *before* the dollar." What was more, the Republicans were out "to save the principles of Jefferson from total overthrow in this nation." For the Democrats now condemned those principles as "self evident lies," as "glittering generalities," as applying only to "superior races." Such expressions, Lincoln said, were calculated to subvert the principles of free government in order to restore "classification, caste, and legitimacy." Yes, "they are the vanguard—the miners, and sappers—of returning despotism." And "we must repulse them or they will subjugate us." He concluded with a grave warning to the Democrats. "This is a world of compensations; and he who would *be* no slave, must consent to *have* no slave. Those who deny freedom to others, deserve it not for themselves."

Meanwhile, Lincoln wrote Republican leaders beyond Illinois and offered sober advice about party strategy. He emphatically opposed an anti-alien amendment adopted in Republican Massachusetts, asserting that the American mission was to elevate men, not degrade them. And he counseled an Indiana party leader that Republicans must keep the whole political theater in mind, that local and state conventions must do nothing that would damage the party somewhere else. For example, Republicans in Ohio and New Hampshire must not openly defy the fugitive slave law, because that would alienate Illinois conservatives, who would say that Republicans were against the Constitution. Moreover, Massachusetts Republicans must look beyond their own bailiwick in tilting against foreigners, for that would ruin the party in the old Northwest, with its heavy German population. We must all pull together, Lincoln said, and unite against the expansion and nationalization of slavery—the issue that called our party into existence—and concentrate on beating Douglas here in the North.

For the Little Giant, Lincoln wrote another Republican, was still "the most dangerous enemy of liberty, because the most insidious one." In Springfield, Lincoln followed Douglas's course as reported in the newspapers, observing that the Little Giant was now driving for the Democratic Presidential nomination. But the irony was that he couldn't get it without Southern support. And the Freeport doctrine—an overt appeal to Northern free-soil sentiment —had now completely alienated the South. True, he'd propounded the idea of unfriendly legislation several times before. But reiterating it in all the heat and excitement of the 1858 debates had created a national sensation, enraging Southern leaders even more than his stand against Lecompton. Thus when Douglas returned to the Senate in February, 1859, Southern Democrats greeted him with cold and adamant hostility. What was more, they now demanded a federal slave code for the territories, regarding this as the only weapon that could combat Douglas's Freeport doctrine. And since the champion of Congressional nonintervention could never endorse such a code, Douglas and Southern Democrats were locked in their own irrepressible conflict which Lincoln predicted would disrupt their party.

By September, 1859, the Little Giant was trapped in a terrible paradox: he was trying to hold the nation and his party together with an ambiguous principle that was helping to tear them both apart. What would he do now? Lincoln thought Douglas himself didn't know. "Like a skillful gambler he will play for all his chances" to win the Presidential nomination. If Southern bosses forced a slave code test on him, Douglas would oppose it as he'd fought against Lecompton—and then would present himself to the North as a great hero. Should that happen, Lincoln said in his correspondence, Eastern Republicans must stand clear of him and avoid their "errant folly" of 1858.

Then Douglas dropped a bombshell. In the September issue of *Harper's Magazine,* the Little Giant published an article designed to placate Southerners in his bid for the Presidency. Citing historical "facts" out of context and belaboring the "lessons" of history, Douglas argued that the Founding Fathers had established popular sovereignty—or Congressional nonintervention—as the standard formula for dealing with slavery in the territories. Then he altogether abandoned the Freeport doctrine—or the principle of unfriendly legislation—and now insisted that the people of a territory could merely "control" slavery as other property.

The *Harper's* article, of course, brought a barrage of denunciations from Republican ranks. And Lincoln, too, was irate about Douglas's newest position, which demonstrated once again that the Little Giant "never lets the logic of principle, displace the logic of success." For Lincoln, the doctrine of "controlling legislation" was an overt and infamous concession to proslavery expansionists. And he meant to let the public know about it.

He would do so in Ohio, where Republicans had invited him to speak and help their candidates in upcoming state elections. As it happened, the Little Giant stumped Ohio just before Lincoln did, repeating much of what he said in the *Harper's* article. In mid-September, Lincoln and Mary caught a train for Columbus, with Lincoln studying Republican editorials against Douglas and compiling his own arguments. As Lincoln worked on his speeches, Mary no doubt talked about Robert, who'd recently gone off to Phillips Exeter Academy in New Hampshire. She missed him desperately. Sometimes, she said, "it almost appears, as if light & mirth, had departed with him." Nervous and upset, she traveled a lot with Lincoln these days.

At Columbus, Lincoln addressed an impressive crowd from the steps of the statehouse, with many prominent Ohio Republicans in attendance. Eager to demonstrate that he had no equals in standing up to Douglas, Lincoln flailed popular sovereignty as "the miner and sapper" for the movement to nationalize slavery and the most pressing danger the Republicans faced. He dismissed Douglas's latest arguments about the founders and popular sovereignty as historical nonsense. In contending that the Fathers embraced Congressional nonintervention, Douglas ignored the true facts of the case. He ignored the Northwest Ordinance of 1787, a Congressional measure in which the founders barred slavery from the Northwest Territories. He also ignored the Missouri Compromise which prohibited slavery from half of the Louisiana Territory. In truth, Douglas's entire course in the matter of popular sovereignty was a succession of muddled and inconsistent stands, proving it a delusion, a sham. Though Douglas hailed the Dred Scott decision as "right," he espoused the Freeport doctrine—the idea of unfriendly legislation—which contradicted the Court's ruling. When one cleared away all the trash and words and chaff, Lincoln maintained, one found that Douglas was propounding "a bare absurdity—*no less that a thing may be lawfully driven away from where it has a lawful right to be.*"

Well, Lincoln said, that argument got him in trouble in the South. So Douglas switched his ground again and promulgated his latest doctrine—that the Dred Scott decision "does not carry slavery into the Territories beyond the power of the people of the Territories *to control it as other property.*" He no longer contended that slavery could be driven out. Now it could only be controlled as other property. What does this mean, Lincoln asked free-soil Ohio, except that the Negro is to be taken into the territories and used and abused as property? Please, what do proslavery men "want more than this?"

Then it was on to Cincinnati, where Douglas himself had spoken only a few days before. Since the Little Giant had branded Lincoln and Seward both as warmongering radicals, Lincoln aimed his remarks at Kentuckians across the Ohio River, as though they were in his audience. "I think Slavery is wrong, morally, and politically." I want to block its spread and won't object if it is gradually terminated in the whole Union. But you Kentuckians "differ radically with me upon this proposition." You think slavery is a good thing and ought to be expanded and perpetuated. Therefore you should nominate Douglas for President, for he is more wisely for you than you are for yourselves. But whether you run Douglas or not, we Republicans are going to "stand by our guns" and beat you in a fair election. Yet we won't hurt you. We will treat you as Washington, Jefferson, and Madison treated you, and will leave slavery alone where it already exists among you. We will remember that "you are as good as we are" and that there are no differences between us except those of circumstance. But a warning about your repeated threats to split up the Union if we win the Presidency. How will disunion help you? If you secede, you will no longer enjoy the protection of the Constitution; we will no longer be obliged to return your fugitive slaves. What will you do, build a wall between us? Make war on us? You are brave and gallant, but man for man you are no braver than we are, and we outnumber you. You can't master us, and since you can't, secession and war would be the worst of follies. . . .

These were shrewd Republican speeches, calculated to present Republicans as peaceful, liberty-loving nationalists and to injure a befuddled Douglas in the North. The addresses brought Lincoln considerable praise from Republicans in and out of Ohio—and more invitations to speak in areas with forthcoming elections. In October, the party carried the Ohio state elections; and Lincoln received part of the credit for the victory.

After the Ohio tour, Lincoln spoke at numerous Republican rallies—and even journeyed up to Milwaukee to address the Wisconsin Agricultural Society. In that speech and in other addresses and writings of this period, he gave a graphic defense of the Northern free-labor system, one he'd been working on for several years now. His purpose was to refute proslavery defenders like Hammond and Fitzhugh who proclaimed that Southern slave labor was superior to the "wage slavery" of the North. Characteristically, Lincoln's arguments were a mixture of points borrowed from other Republicans and his own trenchant insights.

Southerners, he observed, maintained that their slaves were better off than the hired worker in Northern society. Apparently they thought that Northern laborers were "fatally fixed in that position for life." But "how little they *know!*" For that was the basic error in Southern thought from which flowed all their other mistaken attitudes. In fact, it was the genius of the free-labor system that there was no permanent class of hired workers. Northerners were free to move up, progress, enjoy social and economic mobility. "I was a hired laborer," Lincoln recalled, working for twelve dollars a month. Yet the system gave him the opportunity to improve himself, to rise above his humble origins and become a self-made professional man. So in the free-labor system "the hired laborer of yesterday labors on his own account to-day; and will hire others to labor for him to-morrow." Lincoln described in story form how the system operated. "The prudent, penniless beginner in the world" leaves home with his capital—two strong hands and a willingness to work—and chooses his employer and mode of labor. His employer pays him "a fair day's wages for a fair day's work," he saves frugally for a couple of years, buys land "on his own hook" or goes into business for himself, marries, has sons and daughters, and in time has enough capital to hire another beginner. And this free-labor system, this "progress by which the poor, honest, industrious, and resolute man raises himself"—as Lincoln had raised himself—is "the great principle" Republicans intend to take into the territories, "which belong to us, which are God-given for that purpose." For the free-labor system "opens the way for all—gives hope to all, and energy, and progress, and improvement of condition to all." And those who remain as workers—and he conceded that many did—cannot fault the system for their condition. They remain as workers either because they have dependent natures, or because of improvidence, folly, or misfortune. For the doors are always open for them to better themselves. The hope for self-improvement is always there.

This was why Lincoln and his colleagues hated slavery so, "for pure slavery has no hope." In the South, it was a fixed condition which prevented the Negro from "eating the bread which his own hand earns." Thus in Lincoln's view, slavery not only besmirched the ideals of the Declaration, but violated the principles of self-help, social mobility, and economic independence—all of which lay at the center of Republican ideology. All of which gave the Republicans a vision of a future America—a better America than now existed —an America of thriving farms and bustling villages and towns, an America of self-made agrarians, merchants, and shopkeepers who set examples and provided jobs for self-improving workers . . . an America, though, that would never be should slavery, caste, and despotism triumph on these shores.

In October Lincoln was back in Illinois, working in the Urbana circuit court, when the papers blazed with reports from Harpers Ferry in northern Virginia. According to the *Chicago Press & Tribune,* a band of Northern abolitionists—most of them young, five of them black—had tried to capture the remote mountain town, seize the federal arsenal there, and ignite a full-scale slave rebellion. For two days the raiders shot it out with local militia and threw Harpers Ferry into bedlam, until a column of U.S. marines under Colonel Robert E. Lee arrived and captured them. The leader of the attack was old John Brown, late of Kansas fame, who warned Southerners that God had appointed him to liberate their slaves "by some violent and decisive move."

Though no slave uprising had occurred and Brown had been jailed, Harpers Ferry produced "a profound sensation" in the Southern states, where people reacted with even greater hysteria than had followed Nat Turner's rebellion back in 1831. For thousands of Southerners, from poor whites in Virginia to rich planters in South Carolina and Mississippi, Harpers Ferry was hardly the work of a handful of independent Yankee revolutionaries. On the contrary, Southerners thought it the vanguard of a Northern abolitionist-Republican juggernaut that would plunge the South into a racial blood bath. When Democratic papers erroneously linked several national Republicans with Brown's venture, distraught Southerners considered this dramatic, conclusive proof that slave insurrection was what the Black Republicans had desired all along. As a consequence, the Republican party and Brown-style revolutionary violence were forged like a ring of steel in the Southern mind.

In Washington, Southern leaders kept up a steady roll of accusations against the Republicans. They cried out that Harpers Ferry was the inevitable results of Seward's irrepressible-conflict doctrine, that Seward, Sumner,

Chase, Thaddeus Stevens, and many other Republicans had all masterminded Brown's invasion of Southern territory. In righteous indignation, Southern leaders launched a Senate investigation to root out Republican suspects. And Northern Democrats, hoping to capitalize on Harpers Ferry in forthcoming state elections, emphatically agreed with their Southern colleagues. Douglas announced that Harpers Ferry was indeed the "natural" and "logical" consequence of the teachings and doctrines of the Northern Republican party. An Ohio Democratic paper, reflecting conservative Northern opinion, blamed the raid not on Brown, "for he is mad," but on Black Republicans who "induced him thus to resort to arms to carry out their political schemes." Such men, the paper declared, "must answer to the country and the world for this fearfully significant outbreak."

The Republicans backpedaled for all they were worth, trying desperately to disassociate themselves from Brown's attack. Republican leaders belittled Harpers Ferry as the work of a solitary fanatic and insisted that the party had neither advocated nor sanctioned invasions of the South, but had always maintained a hands-off policy regarding slavery there. But with local elections approaching, many Republicans were apprehensive. "We are damnably exercised here about the effect of Brown's wretched fiasco in Virginia upon the moral health of the Republican party," a Chicago editor wrote Lincoln. "The old idiot—the quicker they hang him and get him out of the way the better."

Lincoln, too, was distressed about Harpers Ferry—distressed at how Southerners wrongly blamed it on the Republicans, distressed at how it exacerbated sectional hostilities worse than ever. Maybe old Brown was "mad," Lincoln conceded. But the more he read about Brown's life in the newspapers, the more he thought Brown a man of "great courage" and "rare unselfishness." And Lincoln thoroughly sympathized with his hatred of slavery and contended that the institution itself fostered outbreaks like Harpers Ferry. Nevertheless, Lincoln would never approve of Brown's raid, because no reasonable man could ever condone violence and crime. And, alas, where had Brown begun his career in violence? In the Kansas civil war, in all the strife and bloodshed that had marked that hapless territory for five convulsive years. And the cause of all the turbulence there, of course, was Douglas's pernicious Nebraska doctrine.

With Brown much on his mind, Lincoln stumped Kansas in November and December, 1859, speaking at prairie settlements in lashing wintry winds. The free-soilers now ruled the territory, having elected a free-state legislature and framed a free-state constitution. They had also adopted a Negro-exclusion policy which kept out free Negroes from the North as well as slaves from the South. Like most other Northern whites, free-state Kansans were dedicated white supremacists who wanted to make the American West not just a free country, but a white man's country.

Lincoln was glad that Kansas now had a free government, but insisted that this would have transpired without civil war had the principle of the Northwest Ordinance been applied there in the first place. He also talked about John Brown. On December 3, the day after Brown died on the gallows in Virginia, Lincoln told a crowd at Leavenworth that hanging the old man was just, "even though he agreed with us in thinking slavery wrong. That cannot excuse violence, bloodshed, and treason. It could avail him nothing that he might think himself right." But Lincoln warned Southerners that "if constitutionally we elect a President, and therefore you undertake to destroy the Union, it will be our duty to deal with you as old John Brown has been dealt with."

Then it was home to Springfield for the winter, home to reconsider his options after this long and momentous year. With Congress degenerating into a circus of threats and accusations over Harpers Ferry, Seward's Presidential prospects seemed seriously jeopardized. For the Democrats wouldn't let up on him and his irrepressible-conflict speech as the major cause of the Harpers Ferry attack. Since most Northerners had condemned Brown's act and approved of his hanging, would the Northern electorate go for a man incessantly accused of fomenting slave rebellions? Lincoln, for his part, still thought Seward would get the Republican nomination. But Lincoln's friends were not so sure. Davis, Swett, Lamon, and Judd all believed his chances were on the rise. They pointed out that many Illinois papers were discussing him as a favorite son and that his speeches in Ohio, Wisconsin, and Kansas had made a favorable impression on party bigwigs. Also, he came from a crucial state and had no national enemies and no national ideological label, all of which made him an attractive possibility in this uncertain winter.

Norman Judd, chairman of the Republican State Central Committee and chief Lincoln promoter in the preconvention campaign, pointed to another significant plus for Lincoln. The key to the victory in 1860 depended on which candidate could carry the populous states of the lower North—New Jersey, Pennsylvania, Ohio, Indiana, and Illinois, whose southern sections were similar in sentiment to the border South. It was in the lower states of New Jersey, Pennsylvania, Indiana, and Illinois that Frémont had lost and that Seward and Chase were weakest. Lincoln, on the other hand, could have a strong appeal in these "doubtful states," thanks to his Kentucky background and Illinois moorings.

Lincoln insisted that he would rather go to the Senate than the White House. But leaving all possibilities open, he finally wrote a short autobiographical sketch for Jesse Fell, secretary of the Republican State Central

Committee. Though Lincoln remarked that "there is not much of it, for the reason, I suppose, that there is not much of me," he did stress his Pennsylvania and Quaker ancestry. Fell sent the sketch to a newspaper in Pennsylvania, one of the critical states a Republican candidate must have in order to win, and the journal published an article about Lincoln widely copied in the Republican press.

Meanwhile Judd attended a meeting of the Republican National Committee, held in New York, and casually persuaded his colleagues to hold the national convention in Chicago. It was an excellent neutral site, Judd said, since Illinois had no clear Presidential contender. Back in Illinois, Lincoln men rejoiced, for the choice of Chicago boded well indeed for Lincoln's incipient candidacy.

In January, 1860, Judd and Lincoln's Eighth Circuit chums met with him in a secret Springfield caucus and officially launched a Lincoln for President movement. Though he was still dubious about his chances and still looked to the Senate contest of 1864, Lincoln gave the movement his blessing. But whatever national office he ultimately sought—whether the senatorship, the Presidency, or even a Cabinet post—he must control the state party organization, uniting Illinois Republicans firmly behind his candidacy.

The trouble was that he had no such unified support. In truth, some Illinois Republicans considered him too "liberal" or too inexperienced for the Presidency. His friend Orville Browning opposed his candidacy and favored Edward Bates of Missouri. Senator Trumbull frankly went for Judge McLean. And Herndon—loyal Billy, who usually stood by Lincoln come what may— scoffed at his chances and dropped out of his inner circle. In the state at large, there was sympathy for Bates in southern Illinois, some for Chase and Seward in the north.

Lincoln worried about his Republican opposition in Illinois. Unless the Illinois convention delegates gave him a first-ballot endorsement, he feared that his chances for any national office would be seriously impaired. In early February he wrote Judd that he wasn't in a position where it would hurt him to lose the Presidential nomination, but he *was* in a position where it would injure him not to secure the Illinois delegation (for that would cast doubt on his state party leadership and erode his appeal). He asked Judd: "Can you not help me a little in your end of the vineyard?"

As it happened, Judd had close political ties with the *Chicago Press & Tribune,* which had been leaning toward Chase. But a week after Lincoln wrote Judd, the *Press & Tribune* came out for Lincoln instead, declaring him the first choice of all Illinois Republicans and exhorting them to form Lincoln for President clubs.

Meanwhile Lincoln himself struggled to hold the state party together, reconciling feuds and internecine rivalries—and thus enhancing his reputation as the Illinois party leader. With Lincoln men booming his favorite-son candidacy throughout Illinois, he now put the finishing touches on a speech he would give in New York City. The invitation had come the previous fall, and he'd spent his spare time poring over Republican speeches in the *Congressional Globe,* Greeley's editorials in the files of the *New York Tribune,* and the six-volume *Debates on the Federal Constitution.* In late February, with considerable apprehension about addressing a cultured New York audience, Lincoln caught a train for the East.

When he arrived in New York, Lincoln found that he was to speak not at Henry Ward Beecher's Brooklyn church, as originally scheduled, but at the Cooper Institute in New York City. He also learned that another committee had taken charge of his appearance and that it enjoyed the support of Greeley and other anti-Seward Republicans, who hoped to block the senator's nomination. Thus Lincoln's New York visit took on a new significance in terms of his own Presidential possibilities, as anti-Seward forces treated him like a celebrity and lavishly entertained him at the Astor House. On February 27, the day of his address, he called at Mathew Brady's studio and had a photograph taken that could be reprinted and circulated in the Northeast. The photograph revealed a beardless fifty-one-year-old Lincoln, with a receding hairline, a mole on his right cheek, and a firm and steady gaze in his eyes. He wore a new black broadcloth suit, a vest, a stiff white shirt, and a black tie. With his left hand resting on a stack of books, he looked like a learned statesman, tall, straight, and sure of himself.

That night, as a snowstorm raged outside, fifteen hundred shivering people gathered in the Cooper Institute to hear "the Westerner." Presently Lincoln entered the hall and mounted the platform with his escort, acknowledging the polite applause and eying the skeptical newsmen in the front row. After a distinguished New Yorker had introduced him, Lincoln rose with nervous hands and read his manuscript in a high voice. He wanted, he said, to demonstrate once and for all that Republicans and not Judge Douglas were right about the Founding Fathers and slavery. Drawing from the writings of Chase and Greeley and from his own research, Lincoln illustrated how the founders voted again and again to let Congress regulate slavery in the territories. Of the thirty-nine men who signed the Constitution, two voted for the 1787 ordinance banning slavery in the old Northwest and sixteen approved (and Washington signed into law) a Congressional act which implemented the ordinance. Two of the original founders voted for Congressional control of slavery in the Louisiana Territory; and two subsequently voted to prohibit slavery in the Missouri Territory. In sum, twenty-one of the thirty-nine original founders voted at one time or another to exclude slavery from the national lands. They never believed that the Fifth Amendment forbade Congress to touch slavery there. Thus Douglas was wrong in arguing that the framers of the government endorsed the principle of Congressional nonintervention in the territories. Douglas said, "Our fathers, who framed the government under which we live, understood the question just as well, and even better, than we do now." Well, Lincoln asserted, let those who believe this speak as they spoke and do as they did. That was all the Republicans asked —that slavery be branded as an evil not to be extended, as the Founding Fathers so branded it.

Then he spoke to Southerners, though he doubted they would listen. You are a just people, he said, but you unjustly accuse us of being outlaws and "black" Republicans. You say we are sectional because we don't exist in the South. But the truth is that *you* won't let us run candidates there; *you* have restricted us to the North. Moreover, you and not we have discarded the old policy of the founders regarding slavery; you have instituted a new policy for the old; and "we resisted, and still resist, your innovation." More insidiously, "You charge that we stir up insurrection among your slaves. We deny it; and what is your proof? Harper's Ferry! John Brown!! John Brown was no Republican; and you have failed to implicate a single Republican in his Harper's Ferry enterprise." Your accusation "is simply malicious slander," since we accompany all Republican declarations with assurances that we have no Constitutional right to interfere with slavery in the Southern states.

And now a word to the Republicans. We must strive to keep harmony in the Republic. Though much provoked, *"let us do nothing through passion and ill temper."* If the Southern people will not listen to us, let us calmly consider their demands and see what will satisfy them. First, they insist that we leave them alone. This we do, but they won't believe us. What then do they want? They want us to stop calling slavery wrong and join them, in acts as well as words, in calling slavery *right.* Only when the whole atmosphere is "disinfected from all taint of opposition to slavery" will they quit blaming us for their troubles. And what is their ultimate demand? While they don't say so yet, logic dictates that they will next call for the nationalization of bondage. If they now hold that slavery is socially elevating and morally right, they cannot stop demanding that it be nationally recognized as a social blessing and a legal right. If slavery is right, then all laws and all state constitutions against it must be swept away. "All they ask, we could readily grant, if we thought slavery right; all we ask, they could as readily grant, if they thought it wrong. Their thinking it right, and our thinking it wrong, is the precise fact upon which depends the whole controversy. Thinking it right, as they do, they are not to blame for desiring its full recognition, as being right; but, thinking it wrong, as we do, can we yield to them? Can we cast our votes with their view, and against our own? In view of our moral, social, and political responsibilities can we do this?"

"If our sense of duty forbids this, then let us stand by our duty, fearlessly and effectively. . . . Neither let us be slandered from our duty by false accusations against us, nor frightened from it by menaces of destruction to the Government nor of dungeons to ourselves. LET US HAVE FAITH THAT RIGHT MAKES MIGHT, AND IN THAT FAITH, LET US, TO THE END, DARE TO DO OUR DUTY AS WE UNDERSTAND IT."

As Lincoln spoke, "the vast assemblage frequently rang with cheers and shouts of applause," said Noah Brooks of the *New York Tribune*. When he finished, people gave him a standing ovation, waved hats and handkerchiefs overhead, rushed up to congratulate him. Brooks claimed that nobody had ever made such a favorable first impression on a New York audience, and Lincoln himself was surprised and flattered. The next morning, four New York papers published the entire speech; and the Young Men's Republican Union, which sponsored the address, brought it out later as a pamphlet.

On February 28 Lincoln left for New England, partly to visit Robert and partly to do more campaigning. Impressed with his oratory, party leaders invited him to speak in Connecticut, Rhode Island, and New Hampshire, where local elections were coming up. He rode across Connecticut in an overheated train, passing frozen farms and snow-covered villages, and stopped at Providence, Rhode Island, to address a crowd that included the Republican governor. Then on to New Hampshire, where "Bob" was attending Phillips Exeter Academy. Last year, Robert had applied for admission to Harvard College, only to fail fifteen of sixteen subjects on the entrance exams. Determined to redeem himself, Robert entered Exeter in order to prepare for Harvard's rigorous tests.

After a brief reunion, Robert accompanied his father to Concord and Manchester, where Lincoln was introduced as the next President of the United States. Then on March 3 they were back in Exeter again. By now Lincoln's reputation as an orator had spread widely among Northeastern Republicans, and invitations to speak came in from such cities as Pittsburgh, Philadelphia, and Newark, New Jersey. But Lincoln declined them all. By the time he finished in New England, he wrote a New Jersey Republican,

"I shall be so far worn down, and also will be carried so far beyond my allotted time, that an immediate return home will be a necessity with me."

The next day was Sunday, and Lincoln devoted it entirely to Robert; it was the closest they had ever been, largely because Willie and Tad weren't around to demand Lincoln's attention. Lincoln was proud of Robert, proud of his determination and cultivated manners, proud of how easily he mixed with the sons of distinguished Eastern families. Once he entered Harvard, he would enjoy a formal education Lincoln had never had (he liked to joke, somewhat wistfully, that he'd never even seen the inside of a college). On Robert's "orders," Lincoln attended church with him. They had a leisurely dinner together and afterward gathered in the boardinghouse with Robert's friends, one of whom entertained Lincoln with a banjo.

Lincoln wrote Mary about the visit and asked about his "dear little fellows," Willie and Tad, both of whom had been ill. He confessed that he was unhappy with his tour. "I have been unable to escape this toil," he complained. "If I had foreseen it I think I would not have come East at all. The speech at New-York, being within my calculation before I started, went off passably well, and gave me no trouble whatever. The difficulty was to make nine others, before reading audiences, who have already seen all my ideas in print." Anyway, he would be home by the trains as soon as possible. "Kiss the dear boys for Father. Affectionately A. Lincoln."

The next day he traveled down to Hartford, where he spoke at City Hall and was escorted to his hotel by a contingent of Young Republican "Wide Awakes," so named because they marched in formation by torchlight, all dressed in uniforms and flowing capes. At a corner bookstore in Hartford, Lincoln chatted with bewhiskered Gideon Welles, founder of the *Hartford Evening Press.* A powerful Connecticut Republican and a staunch foe of Seward, Welles expatiated on the evils of Seward's political machine and thought Lincoln might give the New York senator a good run for the nomination.

By the time he returned to New York, Lincoln was beginning to think so too. As it turned out, his tour earned him a substantial reputation among Eastern party leaders, who considered his Cooper Institute speech a major Republican address. In fact, the *New York Tribune,* the Washington Republican Club, and the *Illinois State Journal* all made pamphlet copies available for mass circulation. Moreover, Lincoln's triumphal tour came at a strategic time for him, with the national convention just a month and a half away and a stop-Seward movement gathering momentum across the Northeast.

Back in Illinois, Lincoln found that his success at the Cooper Institute had had a tremendous impact on Illinois Republicans, convincing many of his intra-party opponents that he was presidential timber after all. In his Cooper

Institute address, he'd demonstrated once again that he was unsurpassed when it came to elucidating the moral principles and goals of Republicanism. Also, Illinois Republicans found him the only candidate they could unite behind, since Seward and Chase were anathema in southern Illinois and Bates of Missouri was unacceptable in the north. Because of Lincoln's availability and his growing national stature, Illinois Republicans rallied behind him for the Presidency. By April, 1860, the state Republican machine, deftly run by Norman Judd, was leaving no stone unturned in Lincoln's behalf; and Joseph Medill of the *Press & Tribune* was now in Washington, promoting Lincoln there as the only Republican who could carry the crucial lower North.

Trumbull wrote from Washington and asked if Lincoln was now a serious contender. "The taste *is* in my mouth a little," Lincoln replied, and he conceded that his name was in the field. As he informed Ohio Republicans, his strategy was to "give no offense" to delegations already pledged and "leave them in a mood to come to us if they shall be compelled to give up their first choice." Meanwhile he and his friends mailed out adroit letters in an attempt to line up delegates in uncommitted states; and Lincoln casually disparaged Seward and Bates, pointing out that neither man could carry Illinois if Douglas became the Democratic nominee. And he bluntly advised Trumbull to "write no letters which can possibly be distorted into opposition, or quasi opposition to me," because that would cost Trumbull the support of Lincoln's own "peculiar friends." Up for re-election as senator that year, Trumbull took the hint and stopped promoting Judge McLean. But frankly he didn't think Lincoln could defeat Seward.

Well, maybe he couldn't. By April's end the Seward boom was roaring across the West, where the New Yorker picked up whole delegations from Minnesota, Wisconsin, and California. Though Indiana would go to Chicago unpledged, Lincoln hadn't acquired a single delegate beyond Illinois, and he grew doubtful again. The only good news came from Charleston, South Carolina, where Douglas had failed to win the Democratic nomination. Steadfastly refusing to endorse the Little Giant and popular sovereignty, cotton-state delegates stormed out of the hall and the convention disintegrated. Unable to nominate Douglas, Northern Democrats left Charleston with plans to meet again at Baltimore. The long-expected Democratic explosion had happened at last.

May came. Still anxious about getting the Illinois delegation, Lincoln was on hand when the state convention met in wind-swept Decatur, assembling in a ramshackle tent structure like that being erected in Chicago for the national convention. His worries were groundless, though, as the Republican state organization crushed a feeble move for Seward and locked up the Illinois delegation for Lincoln. On the first day of the convention, May 9,

Lincoln's cheering colleagues lifted him overhead and passed him hand by hand down to the platform, where he looked like the "worst plagued" man one witness had ever seen. But more highjinks followed. Lincoln's cousin John Hanks and another fellow marched down the aisle carrying a banner tied between two rotted fence rails. "Abraham Lincoln, the Rail Candidate for President," the banner read. "Two rails from a Lot of 3,000 Made in 1830 by Tnos. Hanks and Abe Lincoln—Whose Father was the First Pioneer of Macon County."* At that the delegates broke into a thunderous demonstration, stomping and shoving so hard that part of the roof awning collapsed on top of them. When the crowd called for a speech, Lincoln pointed at the banner and said, "I suppose I am expected to reply to that." As much as he detested "Abe" and disliked hickish symbols, he let it all go, remarking that he didn't know whether he'd split those two particular rails or not, but he'd mauled better ones since becoming a man. Again the delegates shouted and whooped and flung their hats in the air. And so the "rail splitter" image was born, the symbol of Lincoln as humble "Abe" of the common people, a homespun hero brimming with prairie wit and folk wisdom—a symbol Lincoln's backers hoped would give him an electric popular appeal.

By now suave and shrewd David Davis had emerged as Lincoln's top manager, as the head of a Lincoln for President cadre that included Swett, Dubois, Lamon, Logan, and the ubiquitous Judd. After the state convention moved to place him in nomination in Chicago, Lincoln conferred with his managers about the odds he faced. With Thurlow Weed of New York orchestrating the Seward boom, the senator now had 150 convention delegates—183 shy of victory. For their part, Chase, Bates, and Cameron had around 50 delegates apiece and Lincoln only 22. Nevertheless, Lincoln was a serious dark horse whose major strength lay in the weakness of his opponents. Though Chase had been a U.S. senator and governor of Ohio, he had a reputation as a Republican "radical" and did not have the support of the entire Ohio delegation. Bates of Missouri was a lackluster individual, a one-time slave owner and former Know-Nothing who would alienate Germans in Illinois and Wisconsin. And Simon Cameron of Pennsylvania was a renegade Democrat who'd amassed a personal fortune

*The banner should have read John Hanks.

in public office, some said by unsavory means. Moreover, his principles were so vague and his loyalties so suspect that he had opposition in his own state and little appeal anywhere else.

No, the man to beat still was Seward. Could it be done? Or was Seward as unassailable as Trumbull contended? As they analyzed Seward's record and political profile, Lincoln's managers spotted certain cracks in his armor they hoped to exploit. True, he was the national party leader, a position he and "boss" Weed, his long-time associate in New York politics, had assiduously cultivated for several years. But Seward had also made a lot of enemies. His Whiggish arrogance enraged former Democrats, and his ties with Weed alienated many Eastern Republicans, who accused the Seward-Weed machine of unscrupulous politics. But perhaps the most damaging of all was Seward's public image as a dangerous "radical." In the early days of the party, he'd hobnobbed with Republican liberals—mainly to gain their support in his drive for party leadership—and had matched their rhetoric adjective for adjective. It was his ringing oratory about "the higher law" and "the irrepressible conflict"—that and the fact that Democrats blamed him for Harpers Ferry—which gave him a reputation as an extremist. Yet in back of his public bluster, Seward wasn't a radical at all. On the contrary, he was a practical, calculating man of the political arena who disdained extremism and never let abstract moral principles get in the way of compromise and adjustment. And now, with Weed coaching him from the wings, Seward was trying hard to erase his radical image. In February, 1860, he gave a speech in the Senate that was so mild and so conciliatory that no Southerner could take offense. Now the real Seward was on stage, a Seward who expected to win the Presidency and administer a balanced and eminently practical Republican government.

But try as he might, Seward could not scrub away the radical tinges in his political profile. As a consequence, many Republican analysts were convinced that he could not carry the doubtful states of the lower North. And without them, a Seward ticket was doomed to fail.

At Chicago, Davis and his lieutenants would harp away on this theme, insisting over and over that Seward simply couldn't win. Meanwhile Davis would import thousands of Illinois Republicans into Chicago, to demonstrate for "honest Abe" and create a Lincoln atmosphere around the convention. At the same time, operating out of Lincoln headquarters in the Tremont House, he would send his lieutenants to parley with specific delegations, try and get them to abandon their first choices after the initial ballot and go for Lincoln on the second and third. Above all, Lincoln's managers would sell him as the only candidate who had all the ingredients necessary for victory in November—who had no derogatory national image, had offended no

Republican groups or factions, and had the strongest appeal in the crucial lower North. At the same time, they could argue that Lincoln was a deeply principled and dedicated Republican: in a tough political contest, he could be counted on to defend the Republican cause without letting the standard down.

In mid May, Davis and company hurried up to Chicago while Lincoln remained in Springfield, waiting anxiously for news. He was almost certain he wouldn't win. He told friends that if Seward failed to get the nomination on the first ballot, the convention would probably turn to Chase or Bates. But his managers were a good deal more optimistic than he. From Chicago, where Davis was working for Lincoln like a man possessed, going almost entirely without sleep, came a procession of messages and telegrams. From Davis and Dubois: "We are quiet but moving heaven and earth. Nothing will beat us but old fogy politicians." From Davis again: "Am very hopeful—dont be Excited—nearly dead with fatigue—telegraph or write very little." From delegate Knapp of the Springfield district: "Things are working; keep a good nerve—be not surprised at any result—but I tell you that your chances are not the worst. We have got Seward in the attitude of the representative Republican of the East—*you* at the West. We are laboring to make you the second choice of all the delegations we can where we cannot make you first choice. We are dealing tenderly with delegates, taking them in detail, and making no fuss. Be not too expectant, but rely upon our discretion. Again I say brace your nerves for any result."

Then came a message from Dr. Ray of the *Press & Tribune:* "A pledge or two may be necessary when the pinch comes." Suddenly apprehensive, Lincoln sent urgent instructions to Davis: *"Make no contracts that will bind me."* And Davis did as he was told.

The balloting began in Chicago at ten on Friday morning, May 18. Down in Springfield, Lincoln had steeled himself for another failure, another rejection. He talked with James Conkling in the latter's law office, but was too restless to stay. So he headed for his own office and passed time with a couple of law students there. Around noon, the coeditor of the *Illinois State Journal* burst in with the first ballot results: Seward 173 1/2, Lincoln 102, with the other candidates far back in the counting! Lincoln was astonished. Unable to remain in his office, he went to the *Journal* office and sank into a chair, visibly nervous. Presently another telegram came with the second ballot tallies. Seward had picked up 11 delegates, but Lincoln had gained 79—including Pennsylvania's 48 delegates. At this point, Lincoln had all of the three critical states and heavy support from the entire lower North. Everyone in the office was keyed up now. How would Ohio go on the next ballot? Could Seward, whose support came chiefly from the upper North, crack Lincoln's hold on the lower states? The minutes dragged unbearably by. At last a messenger arrived with another telegram. "Lincoln opened it," a witness said, "and a sudden pallor came over his features. He gazed upon it intently nearly three minutes."

TO LINCOLN YOU ARE NOMINATED. . . . Vote just announced—whole no 466—necessary to choice 234—Lincoln 354 votes—on motion of Mr. Evarts of NY the nomination made unanimous amid intense excitement.

"Well," Lincoln said, "we've got it." At that everybody in the *Journal* office broke into a cheer. Yes, he'd actually won. The strategy of his managers had worked. The Chicago delegates had made a hard decision that the leading Republican was a loser and must give way to a man who could win. And so they nominated the more "available" Lincoln, the favorite son of Illinois, known for his eloquent speeches.

More telegrams arrived from Chicago: "God bless you we are happy & may you ever be." "We did it glory to God." After learning that Hannibal Hamlin of Maine was the Vice-Presidential nominee, Lincoln shook hands all around the room and said, "There's a little woman down at our house would like to hear this. I'll go down and tell her." Outside, church bells began to toll and cannon to boom. Republicans were dancing and singing in the streets, hurrahing deliriously when they saw him pass.

For the rest of the day, Lincoln greeted friends and neighbors who came by to congratulate him and Mary, who was "the very creature of excitement." That night Springfield roared with Republican parades, exploding fireworks, and blaring bands. At last a jubilant procession came winding through the streets and surrounded the Lincoln home at Eighth and Jackson. Lincoln stood outside in the glare of torches and pronounced this not so much a personal victory as a triumph for the Republican cause. He said he would invite the entire crowd into his house if it were big enough to hold them. "We will give you a larger house on the fourth of next March," someone shouted.

In all the commotion of the next few days, Lincoln managed to seclude himself long enough to study the 1860 platform. It was less strident about slavery than the 1856 platform had been. It no longer branded slavery as a relic of barbarism and no longer demanded a Congressional law against slavery expansion. That should be enacted only when it was "necessary." Not that the party was softening toward slavery, though. As Lincoln noted, the platform gave territorial governments a solemn choice about slavery: either they outlawed human bondage or Congress would do it for them. One way or the other, slavery must be contained in the South, as Republicans had always insisted.

Other planks opposed any attempt to reopen the international slave trade, castigated disunion threats as "contemplated treason," and derided popular sovereignty as "deception and fraud," as Lincoln had derided it. If keeping slavery out of the territories remained a primary Republican objective, it was no longer the only goal. Other planks called for homestead legislation, a transcontinental railroad, and some degree of tariff protection, all designed to attract specific economic interests and implement the Republican vision of the good society. As for the tariff, Lincoln understood that Southerners and their Northern allies had revised it downward so that it now stood at the level of 1816, the year the first protective measure was enacted. Lincoln, too, desired to reverse the trend and provide a moderate, carefully adjusted system of protection. But he didn't want the tariff to become "a perpetual subject of political strife." And so he vowed never to force a tariff on Congress, but never to veto a reasonable one either.

To contend with the crowds and handle his mounting correspondence, Lincoln set up office in the spacious governor's room in the statehouse and employed young John Nicolay as his private secretary. A native of Bavaria and a gifted journalist, Nicolay had been working as custodian of records for the Illinois secretary of state. Lincoln, an "assiduous student of election tables," had often come to his office, admired Nicolay's punctual and fastidious ways, and thought his writing talents made him an ideal personal secretary.

And so in the governor's room Lincoln endured the long and arduous wait for election day. As was customary for Presidential nominees in that time, he did no campaigning himself, relying on Republican workers to disseminate his printed speeches. Here in the statehouse people paid him court by the hundreds; they packed the halls to see the nominee and gave him so many gifts—axes, wedges, log chains—that his office resembled a museum. At the same time, artists came to paint his portrait, and he sat for hours in sultry summer heat as the painters stroked and dabbed at their canvases. Mary always passed final judgment on the portraits while the boys ran loose like deer. "Tad was everywhere at once," said one observer, "being repeatedly recaptured by his mother." And then there were the newspapermen. Since the "dark horse candidate" was largely unknown to the national public, journalists flocked into town for interviews. They looked at his house and lot, jotting down notes for their articles, and bombarded the nominee with questions about himself and his policies: Who would you prefer for the Democratic nominee? Who will be in your Cabinet? What will you do if you win and the South secedes? Will there be war? For their part, visitors to the statehouse had mixed impressions of Lincoln. One found him "slouchy, ungraceful, round shouldered, leans forward (very much in his walk) is lean and ugly every way." But a newsman from Utica, New York, described him as "a high-toned, unassuming, chivalrous minded gentleman."

Newspaper reports hardly satiated public curiosity about the Republican nominee. Letters fell on his desk wanting to know about his early career, his youth, his family background. Because he lacked national renown, Republican papers throughout the North needed biographical information to promote their candidate, to sell the electorate on his talents and Presidential appeal. As Horace Greeley said, thousands of Americans did not know Abraham Lincoln, so Greeley thought some inexpensive biography ought to be published and widely distributed by Republican campaigners. There was already one profile in circulation, a scissors-and-paste job which called him "Abram" Lincoln—an error repeated so often that Lincoln himself was irked. Then in early June John Locke Scripps of the *Chicago Press & Tribune* came to Lincoln with a proposal to publish a short campaign biography. Grudgingly Lincoln consented to the project, so long as the book was scrupulously accurate and contained only information Lincoln authorized. Accordingly he furnished Scripps with a factual autobiography in which he surrendered some information about his early life and traced his political career down to the 1856 Presidential campaign. Throughout the autobiography Lincoln referred to himself in the third person—as "A." when he was a boy and as "Mr. L." when he became a man. On the basis of Lincoln's sketch and additional notes supplied by Nicolay, Scripps fashioned a thirty-two-page pamphlet biography. Jointly published by the *Press & Tribune* and the *New York Tribune,* it sold more than one million copies.

Several other "lives" appeared that summer, but Lincoln did not endorse them. He refused to authorize or even read the proof sheets of a biography published by Follet, Foster & Co. of Columbus. Given the fiery nature of the campaign, he'd been advised not to speak or write a word for the public that could be used against him. How, then, could he dare "send forth, by my authority, a volume of hundreds of pages, for adversaries to make points upon without end. Were I to do so, the convention would have a right to reassemble, and substitute another name for mine." "I *authorize nothing,*" he said of the Follet, Foster biography, "will be *responsible* for *nothing.*"

And to what Republican faction would he be responsible? Which of the vague party groups would he favor? Was he a "radical" like Charles Sumner

or a "conservative" like his friend Orville Browning? Since Lincoln had no
ideological image within the party, several people wanted to know where he
stood. But Lincoln refused to pin an ideological label on himself—and rightly
so. For in 1860 the differences between "radicals," "moderates," and "conservatives" were largely rhetorical, more of degree than of kind. True, there
was a group of Republicans loosely known as "radicals"—men like Chase,
Sumner, Giddings, Wade of Ohio, Greeley, Stevens, and Lovejoy of Illinois.
But "radical" was and is a misnomer for them—an inaccurate and emotion-
charged label that has persisted through the years. While they hated slavery
and wanted it to die out, they sought no fundamental transformation of the
American system, desired no radical break from basic American ideals and
traditions, shared no program for radical change and reform. If anything,
they were generally progressive, nineteenth-century "liberals" who often
took up this or that social reform, spoke out forthrightly against the wrong
and anachronism of slavery (which more than anything else got them labeled
as "radicals"), wanted to repeal the fugitive slave law, and refused to surrender another inch of territory to Southern "man stealers." Yet they were not
abolitionists. They did not agree with the Garrisonians or any other abolitionist group which ignored Constitutional and legal restraints and demanded that slavery be immediately abolished in the South itself. Nor did
they agree with Gerrit Smith, one of John Brown's secret backers and head
of the Radical Abolition party, who called on the federal courts to uproot
slavery in Dixie. No, Republican liberals were confirmed party men who
conceded that under the Constitution both Congress and the courts lacked
the power to remove human bondage in the Southern states. Unlike the
abolitionists, they were content to fight slavery strictly in modes provided by
the Constitution—that is, to outlaw the institution in all areas where the
federal government enjoyed exclusive jurisdiction, such as the national lands
and the national capital. But as one liberal argued, only a Southern state could
legally eradicate slavery within its borders. Moreover, liberal Republicans
were often at odds with one another on such issues as currency and the tariff,
so that they had no consensus on anything except blocking slavery extension
and perhaps repealing the national fugitive slave law. By comparison, the
so-called conservatives appeared more flexible than the liberals, more willing
to compromise on tangential issues, more willing to tolerate the fugitive slave
law. And arrayed in between was an amorphous cluster of "moderates." Yet
men moved around so much on specific matters that few remained clearly and
consistently within any of these vague, artificial categories. In truth, when all
is said, Republicans of 1860 probably agreed with one another more than
they differed. They all stood together on the Republican platform. They all
opposed the Dred Scott decision, popular sovereignty, and the sinister de-

signs of the Slave Power. They all championed the Northern free-labor system and the principles of personal liberty, self-help, and social mobility. Although most of them (liberal and conservative alike) were white supremacists, they all hoped that human bondage would one day terminate in America. And because they condemned slavery as an evil and desired its ultimate extinction, they all looked like "radicals" to Democratic politicians and voters, in North as well as South.

Lincoln himself resisted being categorized as anything but a Republican. If he often called the Republicans a "conservative" party, he meant that in a historical sense—that they held the same ground as the Founding Fathers when it came to slavery and the inalienable rights of all men to life, liberty, and the pursuit of happiness. Like the liberals, though, Lincoln was against compromising Republican principles, against surrendering a single Republican ideal for the sake of expediency or adjustment with slaveholders, against lowering the Republican platform "a hair's breadth" to let the likes of Douglas on it. Let the Republicans do their duty, he'd proclaimed at the Cooper Institute, and resist slavery as a moral wrong regardless of the trials ahead. He stressed the moral wrong of slavery as much as any liberal did. And though he was not identified as a "radical" or "liberal" within the party (and would not so identify himself), such Republicans could find little in Lincoln's speeches—save perhaps his acquiescence in the fugitive slave law —to disappoint them. On the contrary, Republican liberals extolled his addresses, applauded his nomination, and campaigned indefatigably in his behalf.

As the party standard-bearer, Lincoln shunned alliances with any faction, clique, group, or man. His purpose, he said, was to unite the party, preserve its identity, defend its principles, and deal fairly with all. So he told Thurlow Weed when the latter visited him in Springfield shortly after the Chicago convention. Lincoln found the tall and robust New Yorker deeply anguished over Seward's defeat. But in spite of all the stories he'd heard about Boss Weed, Lincoln detected "no signs whatever of the intriguer." Weed assured him that New York was safe without condition, that all Seward's men wanted was *"fairness,* and fairness only." And Lincoln promised they would have it. Then he felt as though Weed was looking him over, "keeping up a show of talk while he was at it." Lincoln thought he went away satisfied.

Afterward Lincoln instructed Davis to write the Seward forces and repeat what he'd told Weed—that Lincoln would make alliances with nobody, favor nobody, make deals with nobody. To win the election, he said, he needed the help of the entire party—Easterners and Westerners, liberals and conservatives, former Whigs and former Democrats.

And so he told others who visited him, who dined at his home and shared in the pomp and ceremony, parties and parades, that continued to punctuate his days. In his office at the statehouse, he surveyed political developments with a careful eye, sending out advice to party leaders and asking for the latest news. A new party was in the field now—the Constitutional Union party, consisting of Know-Nothings and die-hard Whigs. With John Bell of Tennessee as their candidate, they ignored the slavery issue and spurned a party platform, vowing simply to stand by the Constitution, the Union, and the laws in this dark and dangerous time. Because of their evasive course, the Republicans dismissed them as "Do-Nothings" and "Bell Ringers."

And then came portentous news from Baltimore: on June 18 the Northern Democrats nominated Douglas on a platform vaguely endorsing popular sovereignty. Southern Democrats now broke entirely with the Northern wing and nominated John C. Breckinridge of Kentucky. Their platform endorsed a federal slave code for the territories. In a desperate effort to restore his ruptured party and salvage his own candidacy, Douglas set out to stump the Union—the only candidate who did so.

Rejoicing over the Democratic split, Lincoln wrote Simeon Francis: "I hate to say it, but it really appears now, as if the success of the Republican ticket is inevitable." And in a news interview in the statehouse, he poked fun at Douglas. If the Little Giant had "great hardihood" and "magnetic power," Lincoln said, he also had "the most audacity in maintaining an untenable position." And now he was at it again: in the North, South, and West, Douglas continued to champion popular sovereignty with all its ambiguities and contradictions, insisting that it was still the best solution to slavery in his time. His erratic behavior, moreover, only added to the storm of confusion that surrounded the man. In the South, he announced that he was no longer running for the Presidency, but was trying only to prevent secession and save the Union. Yet he refused to withdraw in favor of a candidate more acceptable to the South, refused to participate in any fusion movement against the Republicans, because he was the only man, he declared, who could whip Lincoln—even though the senator regarded a Lincoln victory as just about inevitable.

As Lincoln followed the movements of his Democratic rivals, he complained of a sore throat and a headache. Willie had come down with a severe case of scarlet fever; and Lincoln thought he had an inferior form of the same disease. And Mary wasn't feeling well either. Though she enjoyed all the social functions that attended the campaign, she missed Robert terribly. That summer Harvard finally admitted him, proving what Lincoln told one of Robert's friends: "that you *can* not fail, if you resolutely determine, that you *will* not." Mary was proud of "Bobby," yet she felt an emptiness inside with him gone; sometimes she was *"wild* to see him." And she worried about

Willie, too, and worried about the campaign and all the ugly things being said about her husband, especially in the South. She dreaded another failure. "I scarcely know, how I would bear up, under defeat," she wrote a friend, speaking for Lincoln as well. "I trust we will not have the trial."

B̲y midsummer the campaign was in full swing, a confused and raucous contest among four candidates—five if you counted Gerrit Smith of the Radical Abolition party. Nearly all Republican leaders were now stumping for Lincoln, with Chase addressing Ohio rallies and Seward—loyal Republican that he was—barnstorming New England and the Midwest. In all corners of the North, party workhorses drummed up support for the national and state tickets. There were barbecues, rallies, and parades, with young Republican Wide Awakes—wearing glazed hats and flowing capes, flaming torches in their hands—marching through Northern streets in zigzag lines that resembled fence rails. On a thousand platforms, all draped with banners and Union flags, Republican orators heralded their party as the spirit of the modern age, an age of free labor, free soil, and free men. They appealed to local and special interests, stressing their railroad and tariff planks and their promise of free homesteads on the frontier. They portrayed Lincoln as a man of the people, a celebrated Western lawyer who personified the "distinctive genius of our country and its people." In Illinois, the prairie towns boomed and blazed with demonstrations, bonfires, and parades. Lovely Republican ladies flourished banners that proclaimed: "Westward the star of Empire takes its way, We link-on to Lincoln, as our mothers did to Clay." And Republican crowds sang a boisterous campaign song:

> Ain't I glad I joined the Republicans,
> Joined the Republicans, joined the Republicans,
> Ain't I glad I joined the Republicans,
>   Down in Illinois.

In Springfield, hammer-and-nail stands sold fence rails and other Lincoln curios, until the city's streets, said an Indiana newsman, "resemble a Hindoo bazaar." On August 8 there was a mammoth Republican rally on the fairgrounds, with cannon roaring and bands, Wide Awakes, and Young Americans for Lincoln all marching about in noisy exultation. "A Political Earthquake! THE PRAIRIES ON FIRE FOR LINCOLN," headlined the *Illinois State Journal,* which carried the emblem of an elephant bearing a banner in its trunk, "We Are Coming!"—probably the first use of the elephant as a Republican symbol. When Lincoln arrived on the fairgrounds, the crowd lifted him bodily out of his carriage and bore him overhead to a speaker's stand. Straightening his suit and stovepipe hat, he faced the cheering multitude and said he was gratified at such expression of feeling for him. He knew that in 1864 they would all fight hard—as hard as they were fighting now—for another man who stood for truth, "though I be dead and gone" by then.

Yes, his friends said, the Republicans were doing all they could for him, singing his praises and demonstrating for him in boundless enthusiasm. On the other side, meanwhile, the Douglas Democrats were doing everything they could to smear and belittle the Republican "rail splitter." There were malicious whispers that Lincoln was a bastard, that his real father was Abraham Enloe, or Henry Clay, or even John C. Calhoun. At their rallies, Democratic orators mocked Lincoln's "traitorous" Mexican War stand and led their crowds in a chant: "Mr. Speaker! Where's the spot? Is it in Spain or is it not? Mr. Speaker! Spot! Spot! Spot!" At the same time, Douglas newspapers dismissed Lincoln as an inept party hack and said he had no record as a statesman—as some Seward papers had said of him just after the nomination, though they supported him now. The Democratic journals insisted that Lincoln had done nothing in any legislative or executive capacity —he'd never even had an executive capacity—that entitled him to the Presidency. Nothing in all the history of his life indicated that he possessed the intelligence or the attributes for the highest job in the land. On the contrary, Lincoln was illiterate, coarse, and vulgar, a pettifogging abolitionist who lusted for the emancipation and equality of Negroes like the rest of his party. . . . With Horace Greeley roaring back, "That is not so!" And it *wasn't* so, Lincoln would snap, and perhaps he would wonder again how long there would be knaves to peddle and fools to gulp such demagoguery. No, he was not an abolitionist, as abolitionists themselves never tired of pointing out. Garrison, for his part, thumbed the Republicans as "a cowardly party," and Wendell Phillips berated Lincoln himself as "a huckster in politics." As Lincoln expected, the political abolitionists rallied behind Gerrit Smith and railed at Lincoln for ignoring "all the principles of humanity in the colored race, both slave and free." They did not read his speeches, he might complain. Did not understand the nuances and complexities of his position, did not realize that slavery could not be removed in the South without plunging the country into a conflagration. Nevertheless, some unaffiliated abolitionists and younger Garrisonians supported him, because they thought a Lincoln Presidency would inaugurate "a new and better era." And many blacks felt the same way. In the few Yankee states where they could vote (in New York, Massachusetts, and other New England states), Negroes said they would vote for Lincoln and even formed Black Republican clubs in some places, for Lincoln's election would at least be a step forward . . . a single stroke of the clock toward a future time when all people (as Lincoln would put it), the old and the young, the rich and the poor, the grave and the gay, of all sexes and tongues, and colors and conditions, would be guaranteed liberty and equality of opportunity before the law. But not all Negroes were for him. A black leader here in Illinois suggested that Lincoln was as odious to the antislavery cause as John C. Calhoun. And Frederick Douglass, the most eminent black man in the North, contended that there was little difference between Lincoln's party and the Democrats. Still, Douglass admired Lincoln personally, said he was "one of the most frank, honest men in political life."

Which was radically different from what they said about him in Dixie. Ah, Lincoln could scarcely believe the vehemence and tumult his candidacy had caused in the South, where his name would not even appear on the ballot in ten states and where he was burned in effigy in windows and public squares. Yes, Lincoln, you are a scourge in Southern eyes. You are another John Brown, a mobocrat, a Southern-hater, a lunatic, a chimpanzee. Read what Southern papers say about you. Here comes Billy Herndon with the latest issues of the *Charleston Mercury* and other Southern journals. You are "a horrid looking" wretch, "a blood-thirsty tyrant," who has sent abolitionist spies into Southern communities to circulate "Lincolnisms" and goad the slaves to rebellion. You have caused mobs to form and vigilantes to organize in Dixie; they burn your propaganda, beat up on suspected Yankee agents, arrest Negroes accused of cheering for you in some slave quarters, a shack, a shed. Yes, a Great Fear was loose in the South, where Democrats called Lincoln the greatest "ass" in the United States, a "sooty" and "scoundrelly" abolitionist whose running mate was a mulatto and whose victory would toll the bells of doom for the white man's South. The loudest invective and grimmest prophecies came from the fire-eaters—Robert Barnwell Rhett, William Yancey, and their ilk—who'd been agitating for Southern independence for years now. And Southern crowds were listening intently to them, because the long chain of disquieting events—from the rise of Garrison and Nat Turner down to Brown's raid and the gathering Republican juggernaut —had united them all, the ruling planters, the yeoman farmers, and the poor whites, against Lincoln and his "Black Republican, free love, free Nigger" party. Their orators preached that Lincoln would free more than four million Southern slaves and wipe out untold millions of dollars in slave investments, that he would give Negroes all the federal jobs in the South and urge them to copulate and marry with white women. Yes, that was how they saw Lincoln in Dixie. Having lived for twenty years in a closed and suspicious society, dedicated to suppressing dissent and defending racial slavery as the corner-stone of their white men's world, they would not believe anything Lincoln or his party said about their slaves. Were Republicans not against slavery in the territories? Then they must be against it in the South as well. Had Lincoln not proclaimed that a house divided against itself could not stand, that this nation could not endure half slave and half free? Had Seward not declared that North and South were locked in an irrepressible conflict? Never mind their disavowals. Never mind their talk about Constitutional guarantees. Never mind their twitterings about "ultimate" extinction. Extinction was extinction. Once the Republicans took power, then pledges of "ultimate" extinction would disappear like any other campaign promise, and the next thing Southern whites would know, Republican troops would be invading their farms and plantations and liberating their slaves at gunpoint. "I shudder to contemplate it!" cried an Alabama white man. "What social monstrosities, what desolated fields, what civil broils, what robberies, rapes, and murders of the poorer whites by the emancipated blacks would then disfigure the whole fair face of this prosperous, smiling, and happy Southern land."

If there were voices of reason in the Southern wind, if (as Lincoln pointed out) there were Southern Unionists who begged their neighbors to stay calm, considered him less of a threat than Seward, called for a "fusion" movement to derail the hurtling Republican express, the voices of doom seemed to prevail in Dixie. If Lincoln won, they warned, the Southern states would secede and establish an independent slave nation. Even if they had to drench the Union in blood and cover it with mangled bodies, shrieked an Atlanta paper, "the South, the loyal South, the Constitution South, would never submit to such humiliation and degradation as the inauguration of Abraham Lincoln."

But Lincoln did not take Southern threats seriously. In his Springfield office, he read pro-secession editorials and tossed them off. "The good people of the South," he remarked, "have too much good sense and good temper to attempt the ruin of the government." In fact, his advisers informed him that Southern Unionism was much too powerful for secession to triumph. Also, he'd heard it all before, all this talk about disunion if the Southern people didn't get their way, and so he called it an "empty threat" and announced that he would not give in to them, abandon his platform, or get out of the race. In one interview he gave during the campaign, a representa-tive of New England's commercial and manufacturing interests spoke to him in urgent tones. Because Southerners bought their manufactured goods from them and sold them cotton, Eastern capitalists wanted desperately to appease the South and forestall secession, for that would be economically devastating.

Therefore they implored Lincoln to be conciliatory, to compromise with Southerners, to offer them some "conservative" promise. But Lincoln replied with a fiery no. He would not barter away the moral principle involved in this contest "for the commercial gain of a new submission to the South."

And anyway, with the Republicans on the verge of victory, why should Lincoln surrender anything? In the early state elections, the party carried Vermont and Maine, Pennsylvania, Ohio, and Illinois. The signs were truly auspicious, the Republican time was undoubtedly here. With four major candidates in the field, Lincoln would almost surely win most of the populous Northern states with their heavy electoral votes. And so a quick note to Seward: "It now really looks as if the Government is about to fall in our hands."

On election day, November 6, telegraph offices across the nation prepared to flash the earliest returns. All that day Lincoln was in an amiable mood, scarcely alluding to the election or his candidacy. Around nine that evening, he and Jesse Dubois and several other colleagues went over to the Springfield telegraph office to catch the early returns. Lincoln sprawled across an old sofa as the superintendent read off the initial tallies from Illinois. They were all good, very good. By now Trumbull had arrived and was studying the state races through his spectacles, enough to know that he would be re-elected U.S. senator. The telegraph was clicking again: returns indicated that Lincoln had carried New England and the Northwest and hinted at a mighty Republican sweep of the upper and lower North. Now private messages began to tick in—one from Cameron said that Pennsylvania was certain to go Republican. "If we get New York," said Trumbull, "that settles it." Through it all Lincoln registered no emotion; only when he carried Springfield did he show excitement, uttering something between a crow and a cheer. At last came the news they had all been waiting for: Lincoln had won New York and thus virtually assured himself of victory. Dubois grabbed the telegram and ran over to the statehouse, screaming, "Spatch! spatch!" A huge Republican crowd was gathered there, and when Dubois burst in with the news, men threw their hats, yelled "like demons," and rolled over and over on the carpet. Outside, Republicans paraded arm in arm in the streets; others climbed to their rooftops and sang in ecstasy, "Ain't you glad you joined the Republicans, down in Illinois!" Mercy Conkling reported that the streets scenes were "perfectly *wild;* the republicans were . . . *singing, yelling! shouting!!* Old men, young, middle aged, clergymen and *all!''*

Back at the telegraph office, returns from the South were starting to come in. "Now," Lincoln said, "we shall get a few licks back." After a while he went with friends to a banquet which Mary and other Republican women had prepared. When Lincoln entered the room, the ladies crowded around him and beamed. "How do you do, Mr. President." They escorted him to a seat not far from Mary, who was radiant tonight and terribly proud of him and their shared triumph. The Republican women, wrote a correspondent, made over Lincoln with "solicitous attention," serving him coffee, fetching him sandwiches, and singing him "vigorous Republican choruses."

Still, he couldn't get his mind off the returns. He suggested that Mary go home without him—he would join her later. Then he went back to the telegraph office and stayed there until he was certain of his election. He would come in some distance ahead of Douglas in popular votes, with Breckinridge third and Bell fourth. And if he would not have a majority of the popular votes, Lincoln would defeat his combined opponents in the electoral college, so that nobody could label him an "accidental" President who owed his victory to the split of his foes.*

Around 1:30 A.M. he went home through the tumultuous streets—home to Mary and his boys. Later he tried to get some sleep, but he was too restless to sleep—too concerned about his awesome new responsibilities. He was really President of the United States. How successful would he be? How would he measure up in the trials that lay ahead? Yes, he was President. He who came from Thomas and Nancy Lincoln, who used to dream of Washington and Jefferson as he plowed an Indiana wheat field, was now the leader of the Republic as they had been. After all the bitter political defeats of the last six years, he had finally won. His party had won. The cause of liberty and free government had won. And if he might have preferred to serve that cause in the national Senate, fate had decreed otherwise: in the inscrutable rush of time, Lincoln had been carried to a different destination at the other end of Pennsylvania Avenue.

Beyond his home, the celebrations went happily on, as Republicans yelled in the streets and sang in the statehouse throughout the night. At 4 A.M. an impassioned crowd rolled out a huge cannon and made it "thunder rejoic-

---

*In the final tallies, Lincoln collected 1,866,452 popular votes and his combined opponents 2,815,617 (Douglas, 1,376,957; Breckinridge, 849,781; and Bell, 588,879). But in the electoral college, Lincoln had 180 votes; Breckinridge, 72; Bell, 39; and Douglas, 12. Lincoln carried California and Oregon and every Northern free state except New Jersey, which gave him 4 electoral votes and Douglas 3.

ings," the concussions of the big gun rattling Lincoln's own windows some distance away.

Next day the President-elect rose early and headed for the governor's office to work on a list of possible Cabinet members. In the Deep South that day, telegraph dispatches screamed with news about Lincoln's election, and people thronged the streets of Southern cities with talk of secession everywhere.

PART·SIX

MY TROUBLES
HAVE JUST BEGUN

"Well, boys," he told newsmen the day after his election, "your troubles are over now, mine have just begun." Somehow that hectic day, with friends, politicians, and reporters streaming into his office, he completed a list of men he wanted most in his Cabinet, men who were honest and incorruptible and who would satisfy all Republican factions—geographical, political, and ideological. Since he lacked administrative experience and frankly admitted it, he also wanted the best talent the party had to offer. So at the head of the list went William H. Seward, leading Republican and powerful United States senator, who would serve as Lincoln's liaison man in Washington during the winter. Other musts were Chase of Ohio and Bates of Missouri. Lincoln added Montgomery Blair of Maryland, Nathaniel Banks of Massachusetts, Welles of Connecticut, Dayton of Ohio, and Caleb Smith of Indiana. By sundown he'd finished his list; from these names he hoped to fashion the Cabinet of the Lincoln administration.

In the following week came the deluge: an army of office seekers swarmed into his office in the statehouse, insisting that they had got him elected and that he owed them Cabinet positions, postmasterships, customs posts, and commissioners jobs, not to mention Lincoln's debts to their brothers and uncles and cousins who were also loyal Republicans. . . . Damnable patronage. How he disliked office seekers—"vultures," he called them—and the headache of distributing federal positions. He wanted to put the best men in office and yet be fair to everyone, placate all Republican groups and factions, and reward all his friends. Honesty and loyalty warred in him as he dealt with office seekers. Yes, I'll look into it. Yes, I remember you and will do what I can. At the same time there came gaggles of newspapermen, special delegations, interviewers, and party bosses (what will Pennsylvania get? New England? Ohio? former Whigs? former Democrats?). His visitors averaged some 150 people a day, all squeezing his swollen hand and wanting this or wanting that.

Then there was the hate mail from the South. No Chief Executive had ever received such malignant correspondence, most of it demanding Lincoln's immediate execution by the dagger, the gun, the gibbet, the hangman's noose. One letter contained a sketch of the Devil stabbing Lincoln with a three-pronged fork and pitching him into the fires of Hell. Such missives left Mary deeply apprehensive. And though Lincoln said that no "decent man" could write such things, he was troubled, too. For what about the other signs? The broken panes of glass by his front door? The warnings from General Winfield Scott in Washington that he really was in danger? The confidential letters from political friends that an assassination plot was being concocted in New Orleans? What did they augur? What was he to make of the signs?

One day, unable to bear the mob at the statehouse, he came home to rest, lying down on a sofa in his chamber. He glanced across the room at a looking glass on the bureau and saw himself reflected at almost full length. But his face had two separate and distinct images. Startled, he got up and approached the glass, but the illusion vanished. He lay back down, and the double image reappeared, clearer than before. Now one face was flushed with life, the other deathly pale. A chill passed through him. Later he told Mary about it and she became very upset. She interpreted the vision to mean that he would live through his first administration, but would die in his second. Lincoln tried to put it out of his mind, but "the thing would come up once in a while and give me a little pang, as though something uncomfortable had happened."

It was now the third week after his election, and reporters and politicians who called on him were worried about the slave states. Clearly a dangerous secession movement was sweeping the Deep South with its heavy concentrations of slaves: five states had already called for secession conventions or submitted the matter to their legislatures. From all directions came letters that Southerners were in earnest this time, were really going to destroy the Union in order to save their slave regime from Lincoln's grasp. From all directions came appeals for him to clarify his position, make a statement, announce a pacification program, publish his Cabinet choices, make a grand gesture, *do* something.

Do something? What could he do he hadn't already done? Lincoln sat in the governor's office with his legs crossed, resting an elbow on his knee and stroking his chin where a stubble of a beard was growing. A little girl in Westfield, New York, had written and said his face was so thin he really should grow a beard. And on a whim he'd started one. No, he told his interviewers (and repeated in his correspondence), he still didn't take secession seriously, still thought it all humbug, still thought Southern Unionism too strong to let secession take place. And most Republican strategists agreed with him. Yes, he was aware of Southern complaints. If they wanted to know his views on slavery, they could read the "conservative" Republican platform or look at his published speeches. You say Republican papers don't circulate in the South? That is the South's fault, not ours. No, I am not hostile to the Southern people and neither is Seward or any other Republican. No, I will not make a public statement to allay Southern fears. Southern papers and politicians would misrepresent whatever I say anyway. In fact, politicians in the South have stirred up this whole crisis just to scare us into making concessions. And we're not going to concede anything.

Nevertheless, one New York reporter found him reading a history of the South Carolina nullification crisis of 1832, studying how President Jackson had handled that dilemma. Not only had Jackson issued a proclamation to the people of South Carolina, but he'd threatened to hang Calhoun and hurl an army into that belligerent state to enforce federal laws. Lincoln said nothing one way or the other about Jackson. But he did remark that "self respect demands of me and of the party that has elected me that when threatened I shall be silent."

But the pressures on him to say something were tremendous, so on November 20 he ran up a trial balloon just to see what would happen. In a speech Trumbull gave in Springfield, Lincoln inserted a couple of conciliatory passages, reiterating once again that the Republicans would not lay a hand on Southern slavery. Predictably, the opposition press held the speech up as "an open declaration of war" against the slave states. "This is just as I expected," Lincoln said, "and just what would happen with any declaration I could make." He doubted that the South would understand him anyway, since the South "has eyes but does not see, and ears but does not hear." He resolved again to say and do nothing about the secession threat until his inauguration.

In the meantime, he was concerned less about the South than about the Cabinet. In late November he conferred with Hamlin up in Chicago, inviting suggestions about who should go into the Cabinet from New England. It was the first time Lincoln had met Hamlin, who was tall, slump-shouldered, and olive-skinned. He, too, was being maligned and indicted in the South, where a couple of men even offered to buy "the mulatto" for a modest price. Hamlin carried no gun, insisting that his fists were all he needed to guard against Southern assassins.

After their conference, Lincoln scratched Banks from his Cabinet list and designated that either Welles or Charles Francis Adams would represent New England. At some point on his trip, perhaps on the train ride back to Springfield, he chatted with newsman Don Piatt about Southerners and secession. "They won't give up their offices," Lincoln argued. But Piatt thought they were deadly serious and predicted that in ninety days the sections would be at war. "Well," Lincoln said, "we won't jump that ditch until we come to it." He added: "I must run the machine as I find it."

Back in Springfield, Lincoln told a Philadelphia journalist that he was actually optimistic about the South. "I think, from all I can learn, that things have reached their worst point in the South, and they are likely to mend in the future." He kept telling himself that secession simply wasn't going to happen. When all was said, the majority of Southerners loved the Union too much to wreck it.

On December 8 Lincoln made his first Cabinet appointment, inviting Seward to become Secretary of State. In a private letter, Lincoln denied rumors that it was a complimentary offer he expected Seward to reject. On the contrary, Lincoln had wanted Seward in the Cabinet since the Chicago nomination and now anxiously awaited his reply. Several days later came a note that Seward was thinking it over. After Christmas he officially accepted.

Meanwhile, on December 15, Edward Bates visited Lincoln in Springfield, and Lincoln offered him the post of Attorney General. In his late sixties, a short Missourian with a scrubby white beard and reddish hair, Bates had never joined the Republicans and still called himself a Whig. He was un-happy with the Republican platform—it was too "radical" for him—and considered Lincoln an "unexceptional" man who was as committed to "ex-tremist" antislavery doctrines as Seward was. So why did Lincoln choose Bates for the Cabinet? Because he was a Whig, a conservative, a border Southerner, and a fine lawyer, thus filling all of Lincoln's requirements for one kind of Cabinet member. Bates growled that if not for the present troubles, he would reject the offer. He accepted only "as a matter of duty." As his first duty, Lincoln had him prepare an opinion on the constitutionality of secession.

By now Congress had convened, Bu-chanan had delivered a flatulent message saying that the Southern states had no right to secede but he had no right to stop them either, and House and Senate committees had been established to work out compromises that might keep the slave South in the Union. In the Senate, where Seward orchestrated Republican moves, John J. Crittenden of Kentucky proposed what became known as the Crittenden Plan, which recommended several Constitutional amendments. Let the Missouri Compromise line be revived, Crittenden sug-gested, and let slavery be abolished in the territories north of 36°30′ and let it be guaranteed and protected south of the line. Then at the time of state-hood, let the citizens of the territories from either side of the line decide whether to keep or eradicate slavery (which was the old Southern interpreta-tion of popular sovereignty). In addition, let there be an amendment which would forever prevent Congress from enacting any law, or framing any future Constitutional amendments, that would interfere with slavery in the Southern states, thus safeguarding that institution for as long as Southerners desired it.

In the ensuing debates, with secession flames crackling in the border as well as the Deep South, some Republicans began to waver. Maybe Thurlow Weed was right, maybe slavery extension was now a dead letter and Republi-cans ought to accept the Crittenden Plan. Maybe they should even endorse Douglas's popular-sovereignty program, because "anything," as one man wrote Trumbull, "is better than Civil War."

Lincoln kept himself informed about the Congressional deliberations, re-coiled at some of the conciliatory trends he spotted in Washington, and fired off orders to Trumbull, Elihu Washburne, and other Republicans there. "Let there be no compromise on the question of *extending* slavery." "There is no possible compromise upon it, but which puts us under again, and leaves all our work to do over again." "The dangerous ground—that into which some of our friends have a hankering to run—is Pop. Sov. Have none of it." We are not, he said, going to let the Republican party "become 'a mere sucked egg, all shell and no meat,—the principle all sucked out.' "

Anyway he did not think that conciliation would lead to peace. As he understood it, all these compromise measures were really designed to bring about the spread of slavery and put the country once again on the high road to a slave empire. Whether it was popular sovereignty or the revival of the Missouri line, "it is all the same. Let either be done, & immediately filibuster-ing and extending slavery recommences." Therefore he was "utterly op-posed" to the Crittenden Plan and to any concessions whatever on slavery in the territories. "On that point hold firm, as with a chain of steel," he instructed Republicans in Washington. "The tug has to come, & better now, than any time hereafter."

So it was that Lincoln rallied Congressional Republicans to his side. And his course won him accolades from the Republican ranks. Lincoln was all right, asserted a Massachusetts Republican, because he kept the party firm and steady in its purpose. As a consequence, Senate Republicans rejected the Crittenden Plan and fought hard to keep it off the floor, though most senators from the Deep South didn't care about it anyway. As they warned their own people, any compromise over slavery was now out of the question.

Meanwhile Lincoln summoned Thurlow Weed to Springfield. On December 20, with a secession convention under way down in Charleston, South Carolina, they secluded themselves in Lincoln's parlor and spent the day talking about the Cabinet. Lincoln conceded that he was inexperienced in Cabinet-making and said he needed Weed's help, particularly in selecting former Democrats. They discussed several men, including Simon Cameron of Pennsylvania. Republicans there had persuaded Lincoln that they deserved a Cabinet appointment and one faction was booming Cameron as the man. When Weed warned that a lot of other Pennsylvanians despised him, Lincoln wanted to know what other choice they had? They turned to Monty Blair of Maryland. Weed opposed him emphatically and urged Lincoln to take Henry Winter Davis instead. But Lincoln couldn't make up his mind between the two. As they went back over Lincoln's tentative Cabinet, Weed, an ex-Whig, complained that former Democrats outnumbered former Whigs by four to three. "You seem to forget," Lincoln said, "that I expect to be there."

A messenger brought ominous news from Charleston. The convention there had unanimously voted to take South Carolina out of the Union. Cannon boomed all over the city, militia paraded about, and men and women danced in the streets. South Carolina was free at last, free of abolitionist threats, free of Yankee invasions, free to control her own destiny, free at last.

The news upset Weed. Unless you want a war on your hands, he warned Lincoln, you must compromise on slavery in the territories. But Lincoln disagreed. Your views, he declared, are not those of most Republicans and are certainly not mine. "While there are some loud threats and much muttering in the cotton states," Lincoln said, he remained unalterably opposed to concessions that "would lose us everything we gained by the election." The best way to avoid disaster, he insisted, was through wisdom and forbearance.

By now secession movements were sweeping like fire storms across the other cotton states, with Unionist minorities fighting desperately to quell the winds. On December 22, away from the crowds for a few moments, Lincoln penned a letter to Alexander H. Stephens of Georgia. Lincoln had served in Congress with him and had long admired his eloquence and old-time Unionism. Back in November, Stephens had stood in the Georgia legislature and given an impassioned speech against secession, imploring Southerners to do nothing until the Republicans committed an overt act against them. Lincoln read the speech in the newspapers and wrote Stephens for a revised copy; Stephens replied that he hadn't made any revisions—and went on to warn Lincoln of the terrible responsibilities on him in the present crisis. Now, three days before Christmas, Lincoln sent him a letter intended *"for your own eye only."* "Do the people of the South really entertain fears that a Republican administration would, *directly,* or *indirectly,* interfere with their slaves?" If so,

"I wish to assure you, as once a friend, and still, I hope, not an enemy, that there is no cause for such fears." In fact, the South was in no more danger in this respect than in the days of Washington. "I suppose, however, this does not meet the case. You think slavery is *right* and ought to be extended; while we think it is *wrong* and ought to be restricted. That I suppose is the rub. It certainly is the only substantial difference between us."

On the same day that Lincoln wrote Stephens, rumors flew about that President Buchanan was about to surrender the three Union forts in Charleston harbor. "If that is true," Lincoln snapped, "they ought to hang him." Still, Lincoln could hardly believe that Buchanan would hand the forts over to South Carolina secessionists. What an insult to national authority that would be. If it happened, Lincoln would make a public announcement that he would retake the forts in March. "There can be no doubt," he told Herndon, "that in *any* event that is good ground to live and to die by."

As it happened, Buchanan didn't "surrender" anything. Two of the forts were not even garrisoned and South Carolinians seized them and ringed the harbor with guns. But a small Union force dug in at Fort Sumter, whose Union flag was the last symbol of federal authority in South Carolina. Though Buchanan attempted without success to provision Sumter, Republicans remained extremely suspicious of him. Certain that the administration was brimming with secessionist sympathizers, Seward established contacts there and kept a close surveillance of what was going on.

Lincoln was in despair about Simon Cameron and the fight that now swirled around him. A wealthy businessman whose empire included bank, insurance, and railroad interests, Cameron made no secret of the fact that he wanted to be Lincoln's Secretary of the Treasury. His enemies, though, cried out in protest, insisting that to make him Treasury Secretary was equal to appointing a bank robber as president of a bank. One Cameron foe contended that he was so evil and unscrupulous, with such a stench of corruption about him, that he would contaminate Lincoln's entire administration. Yet Cameron's friends all argued that he was honest, sincere, and capable. Whatever Lincoln did about him, he would get into another hateful family feud. Finally he listed Cameron's pros and cons, determined to let the facts decide his case. First, the cons. Back in the 1830s, as commissioner for the Winnebago Indians of Wisconsin, Cameron had allegedly swindled the Indians out of their lands and pocketed the proceeds, thus earning him his infamous nickname, "the Great Winnebago Chief." In 1849 he supposedly tried to bribe an entire state convention in Pennsylvania. In 1857, when the legislature there chose him as United States senator, his enemies accused him of bribing three legislators as well as a would-be investigator. Yet none of these charges had ever been proven, which was a sort of plus. Another plus was the fact that most Pennsylvania Republicans supported him. In any case, as a former Democrat and an incumbent United States senator, he was the only Pennsylvanian suitable for a Cabinet post.

Late in December, after a good deal of soul-searching, Lincoln invited Cameron to Springfield and found him tall and silver-haired, with a jutting nose, pinched mouth, receding chin, and sagging neck. He was a solicitous fellow, but rather insistent about becoming Treasury Secretary. Lincoln handed him a letter, dated December 31, which stated that "at the proper time" Cameron would be appointed to head either the Treasury or the War Department. But after he'd gone, Lincoln had nagging doubts about his own decision. For one thing, a Pennsylvania Republican called and recounted Cameron's misdeeds in vivid detail. When Lincoln demanded that the man document his accusations, he promised to do so "with fearful fidelity." Also, Trumbull, Hamlin, Blair, and Chase all objected to Cameron's appointment and demanded a Cabinet member with clear Democratic antecedents and uncompromising antislavery views. Cameron was such a chameleon, his critics said, that nobody knew where he stood on anything but making money —usually by nefarious means.

"Under great anxiety," Lincoln wrote Cameron on January 3, 1861, and rather curtly withdrew his Cabinet offer, insisting that developments in and out of Pennsylvania now made it impossible for him to join Lincoln's official family. But as luck would have it, Cameron had already shown Lincoln's offer to his friends—and it had somehow leaked to the papers. Lincoln was scandalized. Now everybody involved would be furious at him.

On January 4, Chase came to Springfield at Lincoln's request, and the two had a long talk about the Cabinet troubles. Chase was tall and clean-shaven, a three-time widower who'd endured his share of suffering. A meticulous man who insisted on neat and orderly ways, he spoke with a slight lisp and regarded people with an air of imperious solemnity, like a priest looking down on his errant flock. His sanctimonious ways had earned him a lot of dedicated enemies over the years. Still, he was a capable senator and a brilliant Cincinnati lawyer, once known as the attorney general for runaway slaves because he'd defended so many in court. Because he was a liberal and a former Democrat with pronounced antislavery convictions, Lincoln had pretty much decided to appoint Chase as Treasury Secretary. Also, his honesty and unequivocal rectitude might deflect criticism should Lincoln keep Cameron after all. Though Lincoln couldn't make an official appointment at this time (he was trying to avoid further Cabinet controversy), he asked Chase if he would accept the Treasury post. But Chase was grave and reluctant. Frankly, he would rather be in the Senate. Too, he was unhappy that Lincoln had named Seward to the Cabinet, inasmuch as Chase questioned Seward's integrity and his commitment to the Republican cause. Nevertheless, Chase was impressed with Lincoln, and he left Springfield contending that here was one man you could depend on. Lincoln may make mistakes as all men do, Chase told some friends, but you could rely on him to defend Republican principles. On January 7 Lincoln wrote Trumbull that Chase would go to the Treasury.

He turned back to Cameron. Thus far his detractors had failed to substantiate a single corruption charge. And his feelings, Lincoln learned, had been deeply hurt when the President-elect had dropped him from Cabinet consideration. On January 13, Lincoln wrote him again and sincerely apologized if he'd seemed offensive, noting that he'd withdrawn the offer under mounting anxiety and perhaps hadn't been as guarded as he should have been. He therefore enclosed a new and kinder letter to take the place of the offensive one. And he promised Cameron that if he appointed a Pennsylvanian to the Cabinet before he reached Washington, he wouldn't do so without consulting Cameron and weighing his views and wishes. With that Lincoln put the matter aside for now. But he remarked that Pennsylvania had given him "more trouble than the balance of the Union, not excepting secession."

Lincoln was depressed about his Cabinet. Here were the leaders of the party—and some of his own closest friends—fighting like alley cats for the spoils of his office. As soon as he mentioned a name—Judd, Smith, Blair, or Henry Winter Davis—here came "an army of patriotic individuals" to fume and fuss about it. He simply couldn't take any more. He would not invite anybody else into the Cabinet until he reached Washington, lest he be "teased to insanity" to make changes.

Reports from the Deep South were most discouraging. Between January 9 and 11, while Lincoln brooded over the Cameron problem, Mississippi, Florida, and Alabama all seceded from the Union with joyous celebrations in their capitals. On January 19 Georgia also went out, with Stephens himself finally coming around once secession was a *fait accompli.* By February 4, conventions in Louisiana and Texas had voted to leave the Union, too, and delegates from six of the seceded states had gathered in Montgomery to establish a Southern Confederacy. As the editor of the *Montgomery Mail* explained: "In this struggle for maintaining the ascendancy of our race in the South—our home—we see no chance for victory but in withdrawing from the Union. To remain in the Union is to lose all that white men hold dear in government. We vote to get out."

For Lincoln the secession of the Deep South was hard to comprehend and hard to accept. Yet he made no public statement about it, acknowledged no miscalculation on his part as to the powerful appeal secession had for all white classes in the Deep South. He seemed dazed—or resigned—as he read about the developments in Dixie.

The question now was how to hold the border slave states in the Union. Seward and his Republican Congressional colleagues had a plan, which they sent out to Lincoln by a special messenger. To forestall further secession movements, Seward in the Senate and Thomas Corwin and Charles Francis Adams in the House all wanted to admit the New Mexico Territory—which contained twenty-two slaves already—as a slave state. They also desired a Constitutional amendment that would guarantee bondage where it already existed—basically the same proposal that Crittenden had offered. Lincoln read over the so-called Corwin or Adams plan with a wary eye, complaining that he would rather die than yield the integrity of the government, than appear to be buying the right to take office.

He wrote Seward that he was still inflexible on slavery expansion, still opposed to all compromises that would allow slaves onto the soil of the federal government and put the institution back on the road to extension and permanence. Yet he would not object if the South were given guarantees about fugitive slaves, the internal slave trade, the existence of slavery in Washington, D.C., "and whatever springs of necessity from the fact that the institution is amongst us," which seemed to suggest that Seward go ahead with the proposed thirteenth amendment. And New Mexico? Since the territory had been organized on the basis of popular sovereignty and since a few slaves were already there, Lincoln didn't care much about New Mexico, so long as "further extension were hedged against."

But that was absolutely as far as he would go. And perhaps he anticipated what would happen in Congress anyway, for compromise between Republicans and Democrats was really impossible, even with the departure of Deep Southern representatives, who'd resigned and gone home. Northern and border-state Democrats were not going to accept the Republican platform on which Lincoln had been elected. And Republicans, with Lincoln at the controls in Springfield, were not going to support the Crittenden Plan—pushed by the Democrats—with its provision for slave expansion. In fact, a majority of Republicans even voted against the New Mexico bill—which was Seward's sop to the border—and killed the move to make New Mexico a slave state. In the end the Republicans voted down the Crittenden Plan as well, thus standing firmly on their party platform as Lincoln exhorted them to do. Only the proposed slave amendment made it through Congress over strong Republican opposition. Not that the amendment or the failure of Congressional compromise impressed the seceded states one way or the other. They were out, gone from the Union, and building a new slave-based nation in Dixie, and no amount of promises, amendments, and guarantees was going to bring them back.

Still, Seward and border-state Unionists would not give up. Thanks to their manipulations, a Peace Convention had assembled in Washington on February 4, consisting of delegates from twenty Union states (the seceded states ignored the affair), and spent several weeks debating various plans to restore the Union. In part, Seward's objective was to stall for time, keep the border states occupied with the Peace Convention and compromise proposals so they wouldn't secede. And it worked, too, as the border slave states decided against secession for now and waited to see what Lincoln would do after his inauguration, to see whether he would commit "an overt act" against the South.

Lincoln, for his part, refused to endorse the Washington Peace Convention. He didn't even want Illinois to send delegates. Once again he said he was not going to buy his right to be peacefully inaugurated. Once again he said he would rather die first.

B y now Washington crawled with rumors that Lincoln was going to be assassinated. Reports came to General Scott that Southern rebels planned to seize Washington, block Lincoln's election, shoot him dead. And there was always the danger that somebody might gun him down on his trip east. The worst danger seemed to be in pro-secessionist Baltimore, where gangs of "pluguglies" who hated Lincoln were crying for his head. Or so the army said. Seward, too, was worried about Lincoln's safety, and he and Scott took great care in mapping out his train route across their stricken land.

All finally agreed that he should leave on February 11—one day before his fifty-second birthday. Meanwhile, with letters piling up about Southern plots, Lincoln tried to concentrate on his inaugural address. Each day he secluded himself in "a dingy, dusty, and neglected back room" above a store in Springfield's square, and here wrote out a slow and painful draft. The nights were lonely for him, since Mary was away on a shopping trip in New York. Determined to show Eastern society that she was no uncouth lout from the frontier, she sailed from one New York shop to another, outfitting herself with the latest fashions as advertised in *Godey's Lady's Book* . . . and shocking people with her indiscreet and impertinent political remarks, speaking out on policy and personalities with a cheerful disregard for consequences. Back in Springfield, Lincoln could hardly wait for her to return with Robert. On three successive January nights he drove down to the depot and waited vainly for them in the snow and the cold. He was "delighted," said a reporter, when they finally arrived on January 25, Mary refreshed from "the winter gayeties of New York" and Robert dressed in conspicuous elegance.

On January 30, with his inaugural speech weighing on his mind, Lincoln left for Coles County to visit his stepmother Sally. He rode part of the way in the caboose of a smoky freight, reminiscing with old lawyer friends about their days on the circuit. The next day he took a buggy out to Sally's home-stead, a cold wind burning his face. Companions recalled that he spoke affectionately of Sally Lincoln, said she was the best friend he'd ever had. He remembered the "sad condition" of his family when his father married her and what a cheerful and positive change she'd brought about.

When he reached her home in the village of Farmington, he stepped down and hugged her with tears stinging his eyes. She was an old and wrinkled woman now—perhaps he regretted having seen so little of her in the last thirty years. They spent the day together, holding hands sometimes and recalling the shared events of their lives. He visited with the Johnstons and the Hankses, too, and then sometime that day went to see his father's grave, the prairie wind moaning around him. Lincoln ordered a stone marker for Thomas's grave. At least the old man should have a marker.

Sally rode with Lincoln to nearby Charleston, where they enjoyed an evening reception with acquaintances. Then at last it was time to leave, time to say farewell to his stepmother, farewell to this old and fragile woman— the only one of his family for whom he felt any love. It was an emotional goodbye. Embracing him, she worried about his safety, worried that she might not see him again. He stroked her face, waved at her one last time, and caught an afternoon train back to Springfield.

In those final days in Springfield, Lincoln pleaded for "the utmost pri-vacy," so that he could arrange his personal affairs, rent his house, say goodbye to old friends. Orville Browning saw him in the Chenery House, where the Lincolns stayed until their departure, and found Lincoln firmer than he expected. Lincoln doubted that anything "but evil" would come of the Washington Peace Convention. What would satisfy the South? Only the "surrender of everything worth preserving," he said.

On Sunday the tenth, his last full day in Springfield, Lincoln strolled down to his law office to complete some business and see "Billy." It had rained all night and through the morning, but the clouds had cleared away now and it was a warm and muddy afternoon. In their office, Lincoln chatted pleasantly with Herndon—no hint now of their estrangement this past year—and tidied up his lawbooks. Then he lay down on the sofa and stared at the ceiling for a long time. "Billy, how long have we been together?" Over sixteen years, Herndon replied. "We've never had a cross word during all that time, have we?" No, Herndon said, we haven't. Lincoln reminisced about their practice, asked Herndon a blunt question about his drinking, which Billy answered, and then rose to go. By the way, Lincoln said, the Lincoln & Herndon sign outside should remain where it is. Just because he was President didn't mean that their partnership was over. "If I live," he said, "I'm coming back some time, and then we'll go right on practising law as if nothing had ever happened."

He took a last look around the office, then went down the stairs and out into the muggy afternoon, complaining to Herndon about the unhappy features of his "new job." "I am sick of office holders already," he grumbled, and dreaded what lay ahead. He said he felt a deep sorrow about leaving his "old associations," a deeper sorrow than most could ever imagine, and it was all the stronger because he had a strange feeling he might never return.

February 11, departure day, came in with low-hanging clouds and drizzling rain. Over in the Chenery House the Lincolns were already up, making final traveling arrangements. Earlier they had disagreed about their journey east. Because of the danger he faced, with various authorities warning him of plots to capture Washington and assassinate him en route, Lincoln didn't think Mary and the boys should accompany him on the same train. Too, his trip would take him across much of the North—all Seward's idea, to expose him to the public and rally Union morale—and promised to be a grueling ordeal. Many of Mary's friends also said she should go to Washington later. But Mary wouldn't listen to them. "The plucky wife of the President," said a reporter, announced that "she would see Mr. Lincoln on to Washington, danger or no danger." They finally agreed that Lincoln and Robert would leave today, with Mary and the younger boys catching a later train and joining them in Indianapolis.

By eight o'clock that morning, some one thousand people had gathered

at the depot of the Great Western Railroad to see Lincoln off. He shook
hands in an overheated waiting room and then went outside, where the
crowd called for a speech. He really hadn't planned to say anything, but the
sight of his friends and neighbors—the men and women, the lawyers, judges,
and politicians he'd known all these years—moved him to make a few impromptu remarks from the train's rear platform. "My friends—No one, not
in my situation, can appreciate my feeling of sadness at this parting. To this
place, and the kindness of these people, I owe every thing. Here I have lived
a quarter of a century, and have passed from a young to an old man. Here
my children have been born, and one is buried. I now leave, not knowing
when, or whether ever, I may return." And with a task before him greater
than even Washington had to bear, Lincoln hoped and trusted in God that
all would be well.

In addition to Robert, his traveling party consisted of Nicolay and young
John Hay, who was going along to help with secretarial chores; Browning
and Governor Richard Yates, who would go as far as Indianapolis; and Ward
Lamon, who was loaded with guns as Lincoln's self-appointed bodyguard.
The train crawled out of the station with clanging bells and moved off into
a rain-streaked sky, heading east for Indiana.

From Indianapolis, the Presidential
train puffed across open country at thirty miles an hour, with flags and
streamers snapping in the wind. According to the itinerary Seward had
worked out, Lincoln would travel across Ohio, western Pennsylvania, back
to Ohio again, across New York State and then down through New Jersey,
Pennsylvania again, and on to Washington, D.C. In the Presidential coach,
outfitted with a rich carpet, tassels, and heavy dark furniture, Mary chaper-
oned Willie and Tad while Lincoln chatted with his military escort and
members of his entourage. Among them was an effervescent young fellow
named Elmer Ephraim Ellsworth. Not long ago Ellsworth had organized
sixty Chicago boys into a crack Zouave company and had taken them on a
tour of the East, where they won many trophies. After that he'd worked in
Lincoln's law office, and Lincoln and Mary had become so fond of Ellsworth
that they invited him along on the trip. While Ellsworth entertained the
younger Lincoln boys, Robert renewed his acquaintance with gregarious
John Hay, who'd attended school with Robert back in Illinois. In another

part of the coach, Lamon picked out ballads on his banjo as the flat brushlands of Indiana passed in the windows. Finally there was William Johnson, "a colored boy" who worked as Lincoln's hired servant.

For twelve exhausting days, the train chugged through endless towns and villages, with crowds waving Union flags as the coach of the President-elect swept by. In the larger cities, Lincoln was obliged to attend receptions and give speeches—now from the platform of his coach, now from a hotel balcony, now from the steps of a statehouse. And wherever he spoke, people wanted to know about secession and what he would do about the crisis. Trying to avoid any utterance that might ignite Virginia and the combustible border, Lincoln was extremely guarded in what he said now, assuring his audiences that he would make an explicit statement of policy in his inaugural address. Still, since people were apprehensive about the crisis, he did comment on it from time to time. But he was such a poor extemporaneous speaker—he needed a written script to be eloquent—that his remarks were often dismally trite. Moreover, he was tired and unsure of himself, confused about the Southern dilemma, unable to comprehend the extent and finality of what had happened in the Deep South.

At Columbus, Ohio: "There is nothing going wrong. . . . We entertain different views upon political questions, but nobody is suffering anything." All we need to overcome the present difficulty is "time, patience, and a reliance on that God who has never forsaken this people."

At Pittsburgh, Pennsylvania, addressing a crowd of umbrellas in a heavy downpour: "There is really no crisis except an *'artificial one'*" whipped up by "designing politicians" in the South, a crisis which could never be justified from the Southern view. If people on both sides would only keep their "self-possession," the crisis would clear up on its own.

At Cleveland, Ohio, where people stood in snow, rain, and deep mud to hear him: Why, he asked, are Southerners so incensed? "Have they not all their rights now as they ever had? Do they not have their fugitive slaves returned now as ever? Have they not the same Constitution that they have lived under for seventy odd years? . . . What then is the matter with them? Why all this excitement? Why all these complaints?"

And so it went from one stop to the next. There was no real crisis. If the trouble in the South were left alone, it would take care of itself. Lincoln did not have to save the country; the country "will save itself." It was up to the people, not the President, to preserve the Union. Yet, if necessary, I will "put my foot down firmly." And then there was an occasional attempt at humor to ease his embarrassment. At one depot he brought out Mary, short, plump, and smiling, to stand beside him on the platform. This, he announced, "is the long and the short of it."

On February 16 the Presidential train wound along the shores of Lake Erie and stopped at Westfield, New York, long enough for Lincoln to kiss Grace Bedell, the little girl who'd convinced him to grow his beard. As the train sped eastward toward Albany, Lincoln received word that down in Montgomery, Alabama, Jefferson Davis had been sworn in as President of the Confederacy, with none other than Alexander H. Stephens as his Vice-President. Delirious crowds and booming cannon had greeted Davis's inaugural, and an actress named Maggie Smith had danced on a United States flag. At Albany, Lincoln admitted in public that he was exhausted. "I have neither the voice nor the strength to address you at any greater length," he murmured at the statehouse, with soldiers and police holding back the crowd.

Then on to New York City, where opposition newspapers mocked Lincoln's awkward gestures and Western gaucheries. What a provincial hick the President-elect was. Hostile editors and politicians alike ridiculed him for saying "inaugeration" and for wearing black kid gloves to the opera—yes! —and hanging his big ugly hands over the rail of his box. How, they wailed, could this "baboon" be our President?

From New York, the Presidential train rolled south across New Jersey, passing through neat little villages with their two-story frame homes and spired churches, the engineer clanging the engine bell so that children and chickens would get out of the way. Lincoln spoke at Trenton and then rode on to Philadelphia, where he checked his family into the Continental Hotel and surrendered himself to another packed and noisy reception. Around eleven that evening, Norman Judd called Lincoln to his rooms for "a secret meeting" with Allan Pinkerton, a Chicago private detective whose agency now worked for the Philadelphia, Wilmington, and Baltimore Railroad. A short, bewhiskered man who spoke with a slight Scottish burr, Pinkerton informed Lincoln that his detectives had uncovered a well-organized plot in Baltimore, now a rabid secessionist city crawling with pro-Confederate pluguglies. Lincoln was scheduled to change trains there, and the plotters intended to kill him as he took a carriage from one station to the other. There could be no doubt of the conspiracy, Pinkerton and Judd both insisted. Lincoln must catch a train for Washington tonight.

This Lincoln firmly refused to do, pointing out that tomorrow was Washington's birthday and that he'd promised to speak at Independence Hall in the morning and Harrisburg in the afternoon. Still, if they really thought he was in danger, he would try to get away from Harrisburg in the evening. Then they could tell him what to do. But "whatever his fate might be," Lincoln said, he would not "forgo his engagements for the next day."

Lincoln retired to his room and tried to rest. There was a knock on the door. It was Seward's son Frederick, with a letter from the Secretary of State and General Scott. The letter warned that there was definitely a plot to kill Lincoln in Baltimore and that he must avoid the city at all costs. After young Seward had gone, Lincoln supposed that he would have to believe it now, though he hated to. He hated to admit that the country was so torn with violent passions that a fairly elected President was threatened with assassination by his own people.

The next day, Washington's birthday, Lincoln addressed an audience in historic Independence Hall, where the Founding Fathers had signed the Declaration of Independence. Lincoln spoke with great feeling about the Declaration, observing that all his political ideals stemmed from that document. He'd long believed, he said, that the founders had fought for more than just American independence from Britain. They had also struggled for the great principle in the Declaration "giving liberty, not alone to the people of this country, but hope to the world for all future time." This was why he wanted to save the Union now—to preserve that noble promise for all succeeding generations. "But, if this country cannot be saved without giving up that principle. . . ." He paused. "I was about to say I would rather be assassinated on this spot than to surrender it."

On the train to Harrisburg that afternoon, Judd took Lincoln aside and rehearsed a clandestine getaway plan Judd had worked out with railroad officials and trusted army officers. At dusk, a special train would convey Lincoln back to Philadelphia, where he would be ushered in disguise aboard a sleeping coach. A night train would pull it to Baltimore and another would take it to Washington in secret. Tomorrow the regular Presidential train would go on to Baltimore as scheduled, with Judd and the military escort on board to protect Lincoln's family and traveling companions.

Lincoln didn't like the idea of sneaking into Washington, but decided to "run no risk where no risk was required" and consented to Judd's scheme. Yet Lincoln insisted that Mary be told about it, "as otherwise she would be very much excited at his absence." Mary learned of the plan later that day in Harrisburg, and she was understandably upset. After all, she considered it her duty and her right to travel with Lincoln and stand by his side in case of danger. But now he was to be taken from her and hurried through a hostile city where men were plotting to murder him. What if he were murdered? What if she lost her husband whom she loved more than life itself and depended on "for everything"? "Fear threw her into an unreasoning panic," wrote her biographer, and evidently she made a scene. But there was nothing she could do; the plan had been arranged; Lincoln was going ahead without her. In tearful resignation, she demanded that Ward Hill Lamon accompany her husband, for she could count on Lamon to protect Lincoln with his life.

That night, disguised in "a brown Kossuth hat" and an overcoat, Lincoln waited in a carriage somewhere in West Philadelphia. Lamon was with him, armed with two revolvers, two derringers, and two large knives. In the darkness someone approached. . . . It was Pinkerton. Everything was going as planned. They escorted the President-elect to the depot and sneaked him into the last sleeping car of the Baltimore train, where he climbed into a berth reserved for the "invalid brother" of a Pinkerton detective. The berth was so short that Lincoln had to double up his legs—and then he lay there, unable to sleep, as the train jerked forward and droned down the track. Maybe Judd and Pinkerton knew best about Baltimore, but Lincoln wasn't proud of himself for what he was doing. What if something should happen to Mary and his boys? The image of a mob of pluguglies beating on the Presidential coach and screaming his name . . . a fight with the soldiers . . . gunshots. He tried to rearrange his legs and sleep. But sleep wouldn't come. At three-fifteen in the morning, the train passed through the empty streets of Baltimore and left Lincoln's car at Camden Station, to be picked up by the night train. As Lincoln lay there, he could hear a drunk singing "Dixie" on the platform outside. ". . . *I wish I was in Dixie. Hooray! Hooray! In Dixie land I'll take my stand, To live or die in Dixie.*" At last the car jumped forward, hissed, and ground out of Camden Station, hooked to the Washington-bound train.

Lincoln reached the capital at dawn, February 23, and rode to Willard's Hotel with Congressman Washburne from Illinois. It had been a bad night; Lincoln had not slept at all. Down on the bank of the Potomac, a white shaft of marble rose against the horizon; it was an unfinished monument to Washington. And over there was the Capitol with its uncompleted dome. Scaffolds enfolded the cupola, and cranes stretched over the dome, obscuring the sky. There were stacks of building material all around the Capitol and the unfinished Treasury Building. Washington was a dirtier, ranker city than Lincoln remembered, with a plethora of livery stables and rancid saloons. Pigs rooted in the dirt streets slanting off from Pennsylvania Avenue, and sewage marshes lay at the foot of the President's park south of the old mansion. At the northern edge of the garbage-strewn Mall ran an open drainage ditch, "floating with dead cats and all kinds of putridity," said an observer, "and reeking with pestilential odors." Even now, in the early morning, a stench hung over the city worse than any Lincoln could recall. And there, he noticed, was evidence of Washington's large black population—squalid shanties clinging to the flanks of fine Southern-style mansions. Well, Washington always was a Southern city. Some blacks were already up and at work—slaves most likely. But at least the slave pens and auction blocks were gone now.

They came to Willard's Hotel, a fortresslike structure which boasted of running tap water in every room. Here in Suite 6 the Lincolns would stay until the inauguration. Inside, Lincoln found a letter waiting for him: "If you don't Resign we are going to put a spider in your dumpling and play the Devil with you." There followed nine lines of obscene abuse, ending with "you are nothing but a goddamn Black nigger."

That afternoon Mary arrived with the boys. Lincoln was much relieved to find them all right, though Mary was extremely shaken and had a severe headache from the ordeal in Baltimore that day. No violence had broken out there, but frenzied crowds had greeted the Presidential train and shouted for Lincoln. It was a terrible experience for Mary—the menacing faces, the yells. She would never forget it.

So the Republican President had finally arrived, the opposition press sarcastically noted. When the details of his secret night ride got out, papers of all persuasions mocked Lincoln without mercy and published scathing cartoons about "the flight of Abraham." One story claimed that he'd stolen into Washington disguised in a Scotch plaid cap and a long military cloak, and the Northern press published the story as true. *Harper's Weekly* even carried a grotesque cartoon, "The Mac Lincoln Harrisburg Highland Fling," which depicted a scarecrowish Lincoln dancing in Scottish kilts.

It was the beginning of a relentless smear campaign against "this backwoods President" and his "boorish" wife, particularly on the part of Democratic papers. Their taunts about his crudities and illiterate manner wounded Lincoln to the core, but he never replied to such journalistic abuse, tried to accept it as one of the hazards of his job. Mary, however, was mortified and permanently hurt, for she was proud of her manners, her upbringing, her social grace, and she was proud of her gifted and intelligent husband. She hated editors who called him a hick. And her hurts and furies were intensified when the leading ladies of Washington, most of them from the border South, proceeded to snub her, quite as though she were a country hussy who smelled of the barnyard. Well, Mary would show them. She was determined to prove herself the best First Lady Washington had ever seen. She would dress better, furnish the White House better, and entertain better than any of those snobs could ever do.

Inaugural week was a nightmare. For one thing, a mob of rabid and persistent office seekers would not leave Lincoln alone. "It was bad enough in Springfield," he told a reporter, "but it was child's play compared with this tussle here. I hardly have a chance to eat or sleep. I am fair game for everybody of that hungry lot." Then there were the endless delegations—Buchanan and his Cabinet, senators, congressmen, and others who pestered him about his Cabinet choices and Southern policy.

Stephen A. Douglas also came to Suite 6 and pleaded with Lincoln to endorse the Washington Peace Convention. They both had children, Douglas said, and exhorted Lincoln "in God's name, to act the patriot, and to save our children a country to live in." Lincoln thanked Douglas for his visit, but still refused to approve of the convention, which soon adopted a compromise program very like the Crittenden Plan.

There were delegations from Virginia and other border states as well. One group came from the Virginia secession convention, which, in waiting to see what Lincoln would do, would vote neither to secede nor to adjourn. The Virginians urged Lincoln to give them "a message of peace" to take home to Virginia, but he would only say that Southerners would be protected in all their legal rights. Another delegation, consisting of border state Unionists (three of them Virginians), came from the Peace Convention itself and told Lincoln what he must do. He must avoid coercion at all costs. He must evacuate Fort Sumter, whose Union flag aggravated the Confederates. And he must offer "satisfactory guarantees" to the eight slave states still in the Union. Seward had assured border Unionists that the crisis would disappear within sixty days after Lincoln's inauguration. Now the delegation wanted assurances from Lincoln himself.

For once Lincoln was blunt and specific. He would, he said, support the proposed slave amendment to the Constitution. But he would never guarantee slavery in the territories. Moreover, why shouldn't he collect revenues in the Deep South and retake all the forts the rebels had captured? Please, how was it coercion to uphold the integrity of the government? But evidently he did make the three Virginians an offer: if they could persuade the Virginia secession convention to disband, he would give up Sumter.

But the Virginians refused his proposal. All they could promise was devotion to the Union. And so the delegates from the Peace Convention left with nothing agreed on and nothing solved. In late February, the convention sent Congress a plan which among other things proposed that the Missouri line be revived and extended westward, with slavery to be protected south of the line. The Republicans, of course, repudiated the proposal. But so did Virginia leaders and many other border Southerners. As a consequence, the Senate rejected the convention plan, and the last effort at compromise came to an unceremonious end.

Somehow, in all the commotion of inaugural week, Lincoln managed to complete his Cabinet, with rival factions hassling him down to the last appointment. One of his more persistent visitors was old Frank Blair, Sr., a powerful Maryland Republican who tenaciously promoted his son Monty. Here he came again, his face aged and desiccated, buttonholing Lincoln in Suite 6 and elaborating on Monty's virtues. Though Lincoln made up his own mind about Cabinet choices, he liked old man Blair and ended up making him a kind of ex-officio Cabinet member, an "elder statesman" of the Lincoln administration whose political clout the President could use. A former Democrat who resided on a plush country estate in Silver Spring, Maryland, Blair had edited the *Congressional Globe* for fifteen brilliant years and had gone on to become one of the founders of the national Republican party. What was more, his son Frank Blair, Jr., was an influential Republican out in border-state Missouri, another man whose support and counsel Lincoln needed.

As it turned out, Lincoln's final Cabinet appointments were not much different from the original list he'd drawn up the day after his election. Seward and Bates were officially in already. He now sent Chase the portfolio as Secretary of the Treasury. And happily for Lincoln, the Cameron problem had resolved itself, with rival Pennsylvania factions, afraid that the state might lose a Cabinet post altogether, rallying behind Cameron as their man. So Cameron would go to the War Office. Caleb Smith would head the Interior Department, Welles the Navy Office. And despite Seward's nagging insistence that Henry Winter Davis was the better man, Monty Blair would serve as Postmaster General. In truth, Lincoln thought Blair would be more amenable to the border South because the Blair clan had Kentucky ancestry. This overlooked the fact that Blair sneered at Seward's attempts to pacify the border and woo back the seceded states. "Violence," Blair snorted, "is not to be met with peace." Regardless, Blair would run the Post Office.

Blair's appointment triggered Lincoln's first Cabinet crisis. Because Lincoln had rejected his advice, Seward tendered his resignation on the very eve of the inauguration. For Lincoln it was pretty unsettling. He realized that Seward, with his overweening confidence, his certainty that he was a more talented statesman than the President-elect, hoped to control the Cabinet and the administration itself. So this resignation thing was doubtless a power move. Seward seemed to reason that Lincoln, unable to do without Seward's superior abilities, would offer to kick Blair out of the Cabinet and accept Seward's choice if only he would stay. Lincoln told Nicolay he couldn't "afford to let Seward take the first trick" in their struggle for administrative leadership. Therefore Lincoln asked him to "countermand the withdrawal" and later had a confidential chat with Seward, hinting that he might appoint Dayton of Ohio as Secretary of State. At that Seward withdrew his resignation. Lincoln had won the first trick.

And so it was done. Lincoln's Cabinet was a collection not so much of incompatibles as of opinionated, strong-willed, ambitious individuals, each of whom desired a prominent voice in decision making. Taken together, they made up a balanced Cabinet, representing all the discordant elements in the party—former Whigs and former Democrats, liberals and conservatives, Easterners, Westerners, and border Southerners. Lincoln thought them all capable men whose talents he needed to make his a successful administration.

As Inauguration Day approached, Lincoln worked on a final draft of his speech. A lot was at stake in what he said, and Lincoln revised and polished his paragraphs with meticulous care. On March 3 he asked Seward to criticize the speech; and Seward returned it with the advice that Lincoln offer more concessions to the South. He insisted that Lincoln remove one offensive sentence—that he would recapture all federal forts and arsenals the rebels had taken. Since last December Lincoln had argued that this must be done, but Seward now convinced him to drop the statement lest he alienate Southern Unionists, on whom any pacification program must depend.

But Lincoln would not concede much else. For him, the future of freedom and self-government depended on him to stand firm. After all, he had been freely and fairly elected. He would not violate the Republican platform and his "pledges" to the people who had voted for him and his party. He would preserve the Union and the principle of self-government on which the Union was based: the right of a free people to choose their leaders and to expect the losers to acquiesce in that decision. If Southerners did not like him, they could vote him out in 1864. But they had no right whatever to separate from the Union, and he was not going to let them go, for that would set a catastrophic precedent that any unhappy state could leave the Union at any time. No, he regarded the philosophy of secession as "an ingenious sophism" which had no logical, historical, or legal defense. The Constitution specifically stated that the Constitution itself and the national laws made under its authority were the supreme law of the land. Therefore the states could not be supreme as the secessionists claimed; the Union was supreme, perpetual, and permanent, and could not be legally wrecked by a disaffected minority. The principle of secession was disintegration. And no government—not this one or any other in history—had ever been established which allowed for its own destruction.

And there was more. Some men seemed to think he should apologize for his victory, and it grated on him. He had worked hard for the Presidency, once he'd decided to forgo a career in the Senate and pursue the White House. Why should he have to whine and wheedle for a job he'd rightfully won? No Democratic President had ever had to bargain for his right to assume office. Well, Lincoln wasn't going to either. He would be restrained and reassuring in his speech, but he would not surrender the government or the Republican principles on which he'd been elected.

At dawn, March 4, heavy storm clouds hung low over Washington, threatening rain. Cavalry and artillery clattered through the early morning streets, and crowds already milled about the city, eager for festivities to begin. There were assassination rumors in the wind, and if many people feared for Lincoln's safety, others hoped for trouble.

In Suite 6 that morning, Lincoln read the inaugural address to his family, then asked to be left alone for a while. And alone now, he listened to the sound of marching troops outside and then uttered something like a prayer. When at last the mantel clock stroked twelve noon, Lincoln was dressed in a new black suit, shined black boots, and a stovepipe hat which went well with his clipped Quaker's beard. He carried a gold-headed cane. Mary very much approved of his appearance: he looked distinguished, almost handsome.

Buchanan, white-haired and muttering, called to escort him on the traditional carriage ride to Capitol Hill. The clouds had cleared off now and sunshine bathed the streets and buildings in liquid light. The two Presidents said little to one another as the carriage bumped over the cobblestones of Pennsylvania Avenue, part of a gala parade that featured horse-drawn floats and strutting military bands. Double files of cavalry rode along the flanks of the carriage and infantry marched behind. Hundreds of people lined the sidewalks, and Lincoln stared at their faces (some frowning, some cheering) and nodded from time to time. Up above, people were gathered at the windows of buildings, saying things he couldn't make out in all the din. And troops were everywhere, deployed by General Scott to guard against assassination. Cavalry on skittish horses cordoned off intersections. Infantry mingled with the sidewalk crowds, and sharpshooters peered over rooftops on both sides of the avenue. It was as though the country were already at war.

The carriage pulled up at the Capitol, whose marble front was specked with tobacco juice. Thousands of people moiled about the East Plaza, where an enormous inaugural platform extended from the Capitol's east wing. Plainclothes detectives stood around eying the people, and soldiers watched from the windows of the Capitol and the roofs of adjacent buildings. On a nearby hill, artillerymen manned a line of howitzers, prepared to rake the streets at the first sign of assassins.

After Hamlin took his oath of office inside, Lincoln filed out onto the giant platform with some three hundred other dignitaries. With a brisk March wind blowing, he stood at the podium now, looking out over the ocean of people at sunlit buildings in the distance. He unrolled his manuscript, put on a pair of steel-rimmed spectacles, and read in a nervous treble voice.

First, he reassured Southerners that his policy toward them was one of forbearance, not coercion. Southerners seemed afraid that his Republican government would endanger their property, their peace, and their personal security, but that was not so. As he'd said all along, he would not menace the institution of slavery. He had no right to menace it. He quoted from the Republican party platform about the power of states to control their own domestic institutions.

What was more, he now publicly endorsed the projected Thirteenth Amendment, which Congress had passed and Buchanan had just signed, an amendment forbidding Congress ever to interfere with slavery in the states. Lincoln gave the amendment his blessing because it was consistent with Republican ideology.

He spoke at some length about the supremacy of the national government, directing his remarks mainly at Southern Unionists. If he convinced them that he and his party would stand by the flag, maybe they would overthrow secession—as Seward kept arguing—and end the crisis. Thus Lincoln contended that the Union was perpetual and could not be destroyed, that secession was constitutionally illegal, and that violent resistance to federal authority was "insurrectionary" and "revolutionary." He vowed to enforce federal laws in all the states—as the Constitution enjoined him to do—and to defend and maintain the Union. But he promised that the government would shed no blood—unless it was forced to do so. Moreover, he would "hold, occupy, and possess" those forts still in Union possession—which included Fort Pickens in Florida's Pensacola Bay and Fort Sumter in Charleston harbor. Yet beyond what might be necessary to hold these places, there would be no invasion, no use of force. He did not, however, specifically rule out the use of force to keep Pickens and Sumter.

Having strongly defended the principle of federal authority, he now offered as many concessions as he could without violating his oath of office. He would defer patronage appointments in the South for now, would collect tariff duties off shore, and would deliver mail in the South only if Southerners let him.

And so to his conclusion. "In *your* hands, my dissatisfied fellow countrymen, and not in *mine,* is the momentous issue of civil war. The government will not assail *you.* You can have no conflict, without being yourselves the aggressors. *You* have no oath registered in Heaven to destroy the government, while *I* shall have the most solemn one to 'preserve, protect and defend' it.

"I am loth to close. We are not enemies, but friends. We must not be enemies. Though passion may have strained, it must not break our bonds of affection. The mystic chords of memory, stretching from every battle-field, and patriot grave, to every living heart and hearthstone, all over this broad land, will yet swell the chorus of the Union, when again touched, as surely they will be, by the better angels of our nature."

Lincoln turned and faced Chief Justice Taney, architect of the infamous Dred Scott decision, so old and shriveled that his face resembled that "of a galvanized corpse," and Lincoln took his oath as the sixteenth President of the United States. Up on the hill, cannon belched smoke into the wind and concussions exploded like thunderclaps over the East Plaza. With people cheering him along the way, Lincoln and Mary rode with Buchanan up to the White House. Here Buchanan took his leave, shaking Lincoln's hand and wishing him luck. In the executive office, old Winfield Scott was relieved that none of the threats of violence had even been attempted. "Thank God," the general sighed, "we now have a government."

On March 5, with Southerners damning Lincoln's address as a declaration of hostilities, the War Department sent him a stunning report from Major Robert Anderson, commander of Fort Sumter. The garrison was now surrounded by rebel batteries, and Anderson had grave doubts about his ability to hold Sumter at all. His supplies were running out, would be exhausted in six weeks. Moreover, any attempt to relieve the garrison, so that he could hold on, would require a force of "seventy thousand good and well disciplined men."

Lincoln was dismayed. Since Anderson was a Kentuckian, Lincoln bluntly asked the War Department if he could be trusted. Assured of Anderson's loyalty, Lincoln did not know what to do. His first day as President and already there was a military crisis. He was no military man, had never commanded anything beyond a company of ill-disciplined volunteers, and he felt inadequate to assess Anderson's situation. So he prevailed on General Scott for an opinion. Scott was General in Chief of the army, a hero of the Mexican War, a professional soldier. True, he was seventy-five years old, suffered from dropsy and vertigo, so enfeebled that he couldn't mount a horse or climb the White House steps without help. Nevertheless, he was a towering military fixture in Washington, a legendary soldier whom Lincoln respected to the point of diffidence.

Scott replied that night. Frankly he thought it was too late to save Fort Sumter, as "we cannot send the third of the men in several months, necessary to give them relief." "I now see no alternative but a surrender, in some weeks," the general said. "Evacuation seems almost inevitable, & in this view our distinguished Chief Engineer concurs—if, indeed, the worn out garrison be not assaulted & carried in the present week."

So what was Lincoln supposed to do—let the fort go? Pull Anderson out? Scott's reply gave him the hypo. In public and in private, he'd said over and over that he would hold Sumter and people had cheered him lustily. What would the North say if he abandoned the very place he'd promised only yesterday to hold, occupy, and possess? How would his party react if he seemed to grovel at the feet of Southern insurrectionaries?

In "great anxiety" about what to do, Lincoln consulted for several days with his Cabinet Secretaries and with high-ranking officers of the army and navy. They would gather around a big oak table in his executive office and argue and deliberate. There was Seward with his Havana cigar, witty and sure of himself. There was Welles, with his long-haired wig and snow-white beard, his mouth turned down in what seemed a permanent frown, suspicious and distrustful of everyone. There was Blair, tall and slim, with his acerbic tongue and narrow eyes. There was Chase, handsome and forthright, speaking with his mild lisp. There was Cameron, with his thin mouth and deep-set eyes; General Scott, with his imperial presence and trembling hands; and several other army and naval officers, imposing in their glittering uniforms. And there was Lincoln himself, hair somewhat disheveled, bags under his eyes, listening intently to the arguments of his advisers.

They all agreed that Fort Pickens should be reinforced and held. Since Pensacola Bay was not so combustible as Charleston harbor, a show of Union strength at the Florida fort was less likely to precipitate conflict. Accordingly Lincoln ordered that a troopship anchored near the fort should reinforce Pickens without delay.

But their discussions about Sumter were disconcertingly chaotic. When Lincoln asked about throwing in reinforcements, the navy regarded the danger as slight, but the army thought it would be a disaster. All right, what if Lincoln forgot about reinforcements and sent only a provisioning flotilla? At first Welles supported the idea, then changed his mind and favored letting Sumter go. Seward, for his part, wanted to evacuate emphatically. We can defend Fort Pickens, he contended, and make that a symbol of federal authority with minimal risk. But Sumter must be given up. The situation in Charleston, with rebel batteries dotting the harbor and troops and politicians crowding the city, was so inflammable that an aggressive move there would almost certainly obliterate Unionism in South Carolina and detonate a civil war. No, the best policy was to surrender Sumter and gain time so that Unionists in South Carolina and the rest of the Deep South could consolidate their strength, take over the seceded states, and return them to the Union. This, he insisted, was the only way to end the crisis and avert civil war. But Seward said nothing in Cabinet about his secret negotiations with Confederate emissaries. On his own initiative, Seward assured the Southerners that Sumter would be evacuated, and they relayed the message on to Jefferson Davis, assuming that Seward spoke for Lincoln as well. Though Lincoln had not authorized Seward's statements, Seward took it for granted that the President would follow his advice, that this "little Illinois lawyer" would make Seward a kind of prime minister and let him run the administration and pacify the South.

Blair sharply dissented from Seward's evacuation plan. In truth, he saw something fishy in Seward's pacifism and Scott's, too. Were they cowards? traitors? Both he and old man Blair beseeched Lincoln to hold Sumter come what may. Be like Andrew Jackson, they exhorted the President. Issue a proclamation. Threaten to send an army into South Carolina to deal with those traitors.

With Blair counseling one thing and Seward another, Lincoln directed his Secretaries to submit written opinions, and his wording hinted that he was now leaning toward provisioning Sumter without reinforcements. Assuming that supplying the fort was possible, Lincoln stated, did the Cabinet think it wise? Only Blair and Chase said yes. The other Secretaries sided with Seward and voted to evacuate.

In all the confusion of voices, Lincoln could not reach a decision. If he followed Seward's advice and stood by the flag at Fort Pickens, would the country accept the surrender of Sumter as "a military necessity" and not accuse him of backing down? Now he leaned toward Seward's argument, now away from it. Finally he jotted down the points for and against evacuation, so that the facts themselves might guide him. To hold Fort Sumter would require a large force (Scott estimated at least 25,000 men) and would almost surely start a bloody conflict. And the thought of war made Lincoln wince. Look, the fort was of inconsequential military value anyway. If he gave the place up, it might remove a source of irritation to the South and deprive secession of considerable popular support. Surrender might also demonstrate how conciliatory he could be and silence those in North and South alike who raged on about Republican coercion.

On the other hand, Sumter had become such a powerful Union symbol that withdrawal might well demoralize the Republican party and show that his administration had a "want of pluck," regardless of what he did at remote Fort Pickens. Worse, evacuation might convince the secessionists that they had won a victory for their side.

So what to do? The pressure on him seemed unbearable. The fate of the Union, of his party, of his own career, could well hinge on his decision. Never had an inexperienced President been caught in such a dilemma. Later he confided in Orville Browning that "all the troubles and anxieties of my life" could not equal those that attended the Sumter nightmare. They were so great, he said, that he did not think it possible to survive them.

Finally he postponed any decision for now. He still had four weeks before the Sumter garrison ran out of provisions—enough time to send some trusted men down to South Carolina on fact-gathering missions, so that he could make up his mind on the basis of solid evidence. Among those he sent to Charleston was Stephen A. Hurlbut, an old Illinois friend who had been born and educated in South Carolina. Hurlbut would visit Sumter, talk with rebel leaders in Charleston, and report back to Washington toward the end of March.

Then all Lincoln could do was wait. That and tend to the day-to-day grind of his office, making dozens of diplomatic appointments and distributing other "loaves and fishes" to office seekers who mobbed the White House. On top of everything else, he also held public receptions on prescribed days, standing in a "flood of light" with bands blaring and endless people filing past. By now his nerves were gone. As his personal secretaries observed, the Sumter crisis, the diplomatic correspondence, and the demands of patronage subjected him to such mental strain that he felt like a prisoner in his own office, and the "audible and unending tramp of the applicants outside impressed him like an army of jailers." He felt, he said, "like a man letting lodgings at one end of his house, while the other end was on fire."

By the end of March, a lot of Republicans were distressed at Lincoln's indecisiveness. "WANTED," headlined a *New York Times* editorial, "A POLICY." So far, the paper declared, the Lincoln administration seemed timid and dangerously adrift. It had no policy adequate to meet the secession crisis and to deal with an "active, resolute, and determined enemy." Lincoln must do something or the Union would lose everything, "even honor." Letters were coming in, too, some for letting the South go in peace and be rid of her. But others demanded that Lincoln get tough with the rebels and protect federal property—a position most Republicans seemed to hold.

On March 27, back in Washington, Hurlbut handed Lincoln a detailed report of what he'd found in South Carolina. Contrary to what Seward claimed, South Carolinians had "no attachment to the Union," and some even wanted a clash with Washington to unite the Confederacy. Moreover, Unionism seemed about as dead everywhere else in the South, convincing Hurlbut that the seceded states were "irrevocably gone." What would pacify the rebels? Nothing but "unqualified recognition of absolute independence." Lincoln had been trying to decide whether to reinforce or merely provision Sumter. Hurlbut now informed him that the rebels would open fire on any ships Lincoln sent to Charleston harbor, including a strictly provisioning flotilla.

Should he evacuate then? Hurlbut was certain that withdrawal wouldn't solve anything. If Lincoln surrendered Sumter, the rebels would then demand that he give up Fort Pickens as well. "Nor do I believe that any policy which may be adopted by the Government will prevent the possibility of armed collision."

So there it was. A rude awakening for a President who'd once placed such faith in the potency of Southern Unionism, who'd thought that Southerners at bottom loved the Union as much as he. No matter what he did about Sumter, a violent showdown with the rebels appeared unavoidable. If he evacuated the fort, it would just postpone the shooting.

The next day came another shock—a memorandum from General Scott advising that Lincoln surrender both Sumter *and* Pickens so as to appease the loyal slave states. Lincoln could scarcely believe this. He considered Scott a great military man, but this memorandum was clearly political advice from a native Virginian that strayed far beyond military considerations. Lincoln was bitterly disillusioned. After a Cabinet dinner that evening, he summoned his Secretaries into another room and with much "emotion" and "excitement" told them what Scott had said. He instructed them to turn in another opinion about Sumter tomorrow morning. But after they had gone he realized that he couldn't depend on generals or Cabinet Secretaries to make up his mind for him. He had to decide the matter for himself. He returned to his office and kept the lamps burning all night.

By dawn, March 29, he'd reached a decision. He informed his Cabinet that he would dispatch a supply fleet to Fort Sumter and leave it up to the rebels whether to start a civil war or not (as he'd said in his inaugural address). That same day he instructed the War and Navy departments to start outfitting a relief expedition for Charleston harbor. Three days later, at Seward's urging, he ordered an expedition to sail for Pickens as well.

All the Cabinet except Seward and Caleb Smith now supported Lincoln's Sumter policy. This left Seward himself in a desperate dilemma, since he'd promised the Confederate government that Sumter would be abandoned. And he'd said the same thing to Virginia Unionists. Yet here Lincoln was, announcing that he was going to hold the fort. Here was this prairie lawyer, this upstart who'd taken the nomination and the Presidency away from Seward, rejecting his council and threatening to demolish the delicate negotiations Seward had privately undertaken to save the Union. What an incompetent and myopic man Lincoln was. He had "no system," Seward complained, "no relative ideas, no conception of the situation." Unless Seward did something drastic, the government would drift into ruin. Consequently, on April 1 he sent Lincoln a memorandum entitled "Some thoughts for the President's consideration."

"We are at the end of a month's administration," the document read, "and yet without a policy either domestic or foreign." Since without a policy the Union faced catastrophe, Seward now offered one for Lincoln to adopt. First, the country must be told that slavery was not the issue in the present exigency; the real issue was union or disunion, and slavery must be left out of it. Next, the government must get out of Sumter, but hold the Florida garrison and establish a blockade. Then the government must arouse "a vigorous continental *spirit of independence* on this continent against European intervention." With Spain, France, and England all making designs on Mexico, the United States should demand an immediate "explanation" and if it were unsatisfactory must declare war.

But whatever policy was chosen, it must be pursued energetically either by the President or by "some member of his Cabinet"—in other words, Seward himself. Then all debate must stop. Everybody in the administration must agree with the policy and execute it dutifully.

Obviously Lincoln was appalled, for the Secretary of State was not only upbraiding his own boss, but offering to take over the administration. Lincoln wrote Seward a blunt reply, but evidently chose not to rebuke him in writing and never sent it. Instead he apparently confronted Seward in private, rejected his advice, and disagreed that the administration had no policy. On the contrary, Lincoln said, he was doing what he vowed to do in his inaugural address—to hold Sumter and Pickens. Moreover, Seward had read the address and given it his distinct approval. On the matter of who should carry out "whatever policy we adopt," Lincoln declared that *"I must do it."* Because Lincoln was President and nobody else. Finally, he had no intention of squelching Cabinet debate. "I wish," he asserted, "and suppose I am entitled to have the advice of all the cabinet."

As Seward found out, Lincoln could be tough when pushed too far. "Executive force and vigor are rare qualities," Seward wrote his wife. "The President is the best of us."

Still, Seward had one more trump card left to save his Sumter policy. On April 4 he made arrangements for John B. Baldwin, a Virginia Unionist, to talk with Lincoln about the Virginia secession convention and Fort Sumter. Lincoln was anxious for the convention to adjourn, but Baldwin insisted that Lincoln must pull out of Sumter first, warning that if he didn't and if a shot were fired, then "as sure as there is a God in heaven the thing is gone" and Virginia would secede in forty-eight hours. If Lincoln had once offered to give up the fort if the Virginia secession convention would disband, he made no such offer now. "Sir," he said of Baldwin's demand, "that is impossible."

On that same day, Lincoln directed that Assistant Navy Secretary Gustavus Fox command the Sumter expedition, to consist of three warships, a gunboat, and a steamer containing two hundred soldiers and a year's provisions. At the same time, the President sent a special messenger to inform Major Anderson that a relief fleet was on its way.

Irritating news arrived from Pensacola Bay. Thanks to a tangle of contradictory orders, the troopship anchored near Fort Pickens had not reinforced the garrison as Lincoln had instructed. Which meant that Pickens and Sumter might both be lost unless the administration moved swiftly and got relief flotillas under way to the two forts as soon as possible. On April 6, with tension building in the White House, Lincoln dispatched a message to the governor of South Carolina that a supply fleet was coming down to provision Sumter. The Union ships would not open fire, Lincoln said, unless South Carolina resisted them or bombarded the fort.

There was another debacle. Through a mix-up of orders, the flagship of Fox's Sumter expedition was transferred to the Pickens fleet and sailed off for Florida. Transported with rage, Fox accused Seward of switching the warship in a deliberate move to sabotage the Sumter relief attempt. But, alas, the fault lay with Lincoln himself. In all the confusion that surrounded the countdown to Sumter, he had signed orders that gave the vessel to both expeditions. He tried to retrieve the errant ship, but to no avail. On April 9, without its flagship, Fox's little fleet set out on its fateful mission to Charleston harbor.

That same day Lincoln stood at his office windows, looking out at a dull and leaden sky. For the last couple of days it had rained so much that water swirled in muddy eddies through Washington's streets. Yes, Fox was on his way at last. The decision was made; Lincoln had put his foot down. The Union would defend its flag snapping in the sea wind over Sumter. He had refused to knuckle in to Southern insurrectionaries. He had stood by his oath of office and done what he thought most Union men wanted him to do. If the rebels opened fire, the momentous issue of civil war was indeed in their hands.

On April 13 telegraph messages about Sumter came in rapid-fire sequence. The Confederates, feeling betrayed by Seward's false promises, had opened up on Sumter the day before and were pounding the garrison with their harbor batteries. Then came the news that the rebels had allowed Fox to evacuate Anderson and his men. Fort Sumter had fallen. On April 14, with an air of solemn resignation, Lincoln announced to his Cabinet that the rebels had fired the first shot, forcing on him the decision of "immediate dissolution, or blood." Therefore he would mobilize 75,000 militia to suppress the rebellion and call for Congress to convene in special session on Independence Day. Officially, his administration interpreted the present conflict not as a war between the states, but as a domestic insurrection against the national government. Since secession was constitutionally illegal, Lincoln refused to concede that the Deep Southern states had ever left the Union. Rather he contended that rebellious citizens had taken over in Dixie and established a "pretended" Confederate government which Washington would never recognize. Lincoln's objective now was to suppress Southern rebels as rapidly as possible and restore the national authority in the territory they had seized.

Later that afternoon, Stephen A. Douglas swept into Lincoln's office and offered his support. The Little Giant "cordially" agreed that Union troops should be mobilized to suppress the rebels, except that "I would make it 200,000," Douglas said. "You do not know the dishonest purposes of those men as well as I do." The two old adversaries "spoke of the present and future, without reference to the past," and studied a map on Lincoln's wall, discussing strategic points which should be strengthened. Later Douglas released a press statement that he stood resolutely behind the President; then the Little Giant hurried home to rally Illinois Democrats to the Union cause.

On April 15 Lincoln's proclamation for 75,000 militiamen went out to the states, and it forced vacillating men in both North and South to choose their sides. In the border South, secession conventions sprang into action, for the specter of federal troops invading the Deep South, shooting at rebel Southerners and liberating slaves, was more than even Southern Unionists could bear. On April 17 the Virginia convention adopted a secession ordinance which the voters eventually ratified. Virginia joined the Confederacy, and the rebels moved their national capital to Richmond. Within the next two months, Arkansas, North Carolina, and Tennessee also seceded and became Confederate states, with Maryland, Missouri, and Kentucky threatening to go out as well.

Lincoln was depressed at the reaction of the border states, especially Virginia. What really embittered him was the lightning speed with which many of Virginia's Union men had joined the secessionists and voted for disunion. Many times he'd heard them proclaim their lasting devotion to the national flag. Yet when Lincoln chose to defend the flag against insurrectionists, these *"professed* Union men" almost instantly became insurrectionists themselves.

At that, Virginians not only seized the federal arsenal at Harpers Ferry (as John Brown had done) and captured the federal Navy Yard near Norfolk, but imported large bodies of troops from the "so-called" Confederate states, sent delegates to the rebel Congress, and allowed the "insurrectionary government" to transfer its capital to Richmond. Yes, Lincoln was angry with the people of Virginia. They had allowed "this giant insurrection" to make its nest within their borders, within sight of his office windows where he could see the chimneys and church steeples of Alexandria. If he had to, he would use force against those people over there. He had no choice but to deal with the rebellion where he found it.

PART·SEVEN

STORM CENTER

If Lincoln was going to suppress "this giant insurrection," he must have capable generals as well as a trained army, but there lay the problem. At the time Sumter fell, the small regular army was a shambles, depleted by Southern resignations and run by senior officers like Scott who were too old and infirm to lead a field command. Only Scott and muttering old John Wool had ever led armies, the largest of which had been the fourteen-thousand-man expeditionary force Scott had taken into Mexico back in 1847. None of the junior officers had even directed the evolution of a brigade.

Scott himself admitted that a younger officer should take field command of the army. When Lincoln's call for troops went out, Scott recommended Colonel Robert E. Lee of the First Cavalry, whose family mansion could be seen up on Arlington Heights south of the Potomac. Lee had graduated from West Point, fought in the Mexican War, served on the Texas border, and captured John Brown at Harpers Ferry a year and a half ago. The colonel was married to Martha Custis—daughter of George Washington's adopted son—and had an awesome reverence for the first President, whom Lee strove to emulate. Lincoln liked Scott's suggestion and authorized Frank Blair, Sr., to invite the colonel into Washington for a talk. On April 18, the day after the Virginia convention voted for secession, Blair met with Lee in his Washington home and unofficially offered Lee field command of the Union army, insisting that he spoke for the President. But Lee declined the offer. He conceded that he opposed secession, considered it "anarchy, nothing but revolution," and that he abhorred civil war as well. Though he didn't say so now, Lee also regarded slavery as a "moral and political evil"—yet he enjoyed the grace and status of plantation life and showed no willingness to emancipate the two hundred or so slaves he managed on his Virginia estate. No matter how he felt about the slave system, he detested Northern abolitionists, often vented his temper on "those fanatics," and blamed them for the present crisis. And who knew what revolutionary and abolitionist turns a Republican government might take as the conflagration progressed? So, no, he told Blair in his courteous Virginia way, he could not fight against Virginia or the South, could not "raise my hand against my relatives, my children, my home." After that Lee had a talk with old Scott, also a Virginian, who lectured the colonel about his duty and patriotism. Then Lee rode back to Arlington.

The next Lincoln heard of him, Lee had taken command of the Virginia state troops and then enlisted in the Confederate Army, to fight for the preservation of slavery—the very thing he professed to oppose. In Lincoln's view, Lee was a strange and inexplicable man. Yet he was only one of many supposedly loyal Southern officers who violated their oaths of allegiance and went over to the rebels. Another Virginian, Captain John Bankhead Magruder of the artillery, came to see Lincoln, stood right here in his office and "repeated over and over again" his "protestations of loyalty," only to resign his commission and head for the South. It gave Lincoln the hypo. He referred to Lee, Magruder, and all like them as traitors.

So Lincoln must rely on Scott and his inexperienced junior officers. And for fighting men he must depend on the three-months militia he'd ordered up from the states. But where were they? Reports had come that some Pennsylvania units and the Sixth Massachusetts were on the march, but nobody knew when they would arrive. Only a skeleton force and a few ragtag volunteers now defended Washington. And rumors multiplied of impending rebel attack, of midnight assassinations and abduction plots which left Washington a vortex of fear.

On April 19 came news of a riot in Baltimore. The Sixth Massachusetts had been fired on while marching through the city, whereupon soldiers and a mob of mutinous civilians had waded into one another. Approximately four infantrymen and nine civilians had been killed, and many others had been injured. At last the regiment got away to Washington, but the mayor of Baltimore and governor of Maryland both wired Lincoln that the city could not be controlled, that no more troops must pass through that secessionist town unless they were prepared to fight for every foot.

As night closed over Washington, rumors flew about that howling Maryland mobs were on their way to burn the capital. Too weary to do much until tomorrow, Lincoln retired to his chamber. But at eight o'clock the next morning a Baltimore delegation intercepted him as he hurried out to see General Scott, and insisted that no more Union soldiers march through Baltimore. "If I grant you this concession," Lincoln said, "you will be back here to-morrow, demanding that none shall be marched around it." Still, he realized how inflammable Baltimore was, so he and General Scott finally worked out a compromise with authorities there. If they would keep their "rowdies" in Baltimore and control their riotous people, the army would shuttle troops around the city by alternate routes.

But some Maryland leaders objected even to that. In fact, a second Baltimore delegation accosted Lincoln and announced that Union troops could not "pollute" the soil of Maryland itself. "Our men are not moles," Lincoln exclaimed, "and can't dig under the earth; they are not birds, and can't fly through the air. There is no way but to march across, and that they must do." When the delegation implored him to make peace with the South "on any terms," Lincoln bristled. "You express great horror of bloodshed, and yet would not lay a straw in the way of those who are organizing in Virginia and elsewhere to capture this city. The rebels attack Fort Sumter, and your citizens attack troops sent to the defense of the Government, and the lives and property in Washington, and yet you would have me break my oath and surrender the Government without a blow. There is no Washington in that —no Jackson in that—no manhood nor honor in that."

Meanwhile Washington, D.C., was in a state of panic. So far only the Sixth Massachusetts had come to the defense of the government. Where were all the other troops? Where was the other Massachusetts regiment, the Rhode Island regiment, the Seventh New York, all of which were allegedly sent, allegedly on the march? Lincoln mounted "the battlements of the executive mansion," as Hay put it, and scanned the Potomac through a telescope, looking for a sign of troop transports. Then he focused on Alexandria, where he could see rebel flags flying over buildings. Reports claimed that rebel troops were massing there, preparing to strike at Washington, and Lincoln could see enemy tents on the verdant hills nearby. At night rebel campfires flickered against the Virginia sky. It seemed that the White House was in a nest of rattlesnakes. Right here in Washington Southern sympathizers wore secession badges in the streets, and Seward claimed that the executive departments were crawling with disloyal men. "We were not only surrounded by the enemy," Nicolay recorded, "but in the midst of traitors," and the possibility was growing every hour that Washington might fall. If so, what would happen to Lincoln? To Mary and his boys?

On Sunday, April 21, in an atmosphere of intense foreboding, Lincoln called his Cabinet Secretaries to an emergency meeting. With Washington trapped between a secessionist Virginia and a hostile Maryland, they unanimously agreed that Lincoln must assume broad emergency powers or let the government fall. Accordingly he directed that Welles empower several private individuals—including Welles's own brother-in-law—to forward troops and supplies to embattled Washington. He allowed Cameron to authorize one Alexander Cummings and the governor of New York to transport troops and acquire supplies for the public defense. Because he thought the government was alive with traitors, Lincoln himself selected private citizens known for "their ability, loyalty, and patriotism" to spend public money without security, but without compensation either. He told Chase to advance two million dollars to three New Yorkers for the purpose of buying arms and making military preparations. Lincoln conceded that these emergency actions were "without authority of law," but argued that they were absolutely indispensable to save the government. And his Cabinet emphatically agreed.

Yet maybe even these measures were too late now. Two days passed and still no troops arrived. A White House staffer saw Lincoln pacing in his office, pausing from time to time to glance out the window. Why didn't the troops come? The governors of Indiana, Pennsylvania, Ohio, Illinois, New York, and Rhode Island had all promised to send thousands of men. Where were they? On April 24, a day "of gloom and doubt," Lincoln spoke with men of the Sixth Massachusetts in the White House. "I don't believe there is any North," Lincoln said. *"You* are the only Northern realities."

At last they came. At noon, April 25, the Seventh New York filed off a train in Washington and marched up Pennsylvania Avenue with snapping flags and a blaring band. The arrival of those troops gladdened Lincoln's heart, Hay said. In the next few days soldiers reached Washington by the hundreds, scrambling off steamers and trains from all over the North, until by April 27 some ten thousand were dug in along the Potomac and thousands more were on the way.

As though by a miracle, Washington was now transformed into an armed camp, with cavalry and caissons rumbling through the streets, barracks and hospitals springing up around the city, and tents dotting the landscape south of the capital, where men were renovating forts and throwing up earthworks. From the White House Lincoln could see soldiers everywhere on the sidewalks, could hear the staccato cry of bugles and the crack of musketry from army firing ranges.

As volunteers continued to stream in, here came none other than Elmer Ellsworth, leading a smart-looking Zouave outfit he'd raised in New York. Young Ellsworth was dressed in a red cap and red shirt and was armed with a sword, a huge revolver, and an awesome bowie knife that could halve a man's head like an apple. The Lincolns gave him a warm reception in the White House, where he'd lived before going off to New York. In truth, he was so much a part of the family that he'd once caught the measles from

Willie and Tad. For Lincoln, his boundless enthusiasm was a welcome relief from the dark days that had just passed.

As May came, a kind of heady optimism settled over Washington. Men talked about how the war would be over in ninety days. Military bands entertained Lincoln on sunny White House lawns; and couples rode out to picnic at the Great Falls. "You couldn't discover from anything but the everywhereness of uniforms and muskets," Nicolay said, "that we are in the midst of revolution and civil war."

In the springtime lull, with a fragrance of lilacs filling the White House, Lincoln and his Cabinet adopted a series of additional emergency measures designed to bring the rebellion to a speedy end. In a procession of orders and proclamations, the President declared a blockade of the Southern coast, added 22,000 men to the regular army and 18,000 to the navy, called for 42,000 three-year volunteers, and put national armories into full production. At the same time, he made certain that the Union war effort remained thoroughly bipartisan, handing out military commands to loyal Democrats as well as Republicans and summoning Union men of all political persuasions to help him save the government in this "great trouble."

Meanwhile, Lincoln dealt harshly with "the enemy in the rear"—with what he called "a most efficient corps of spies, informers, suppliers, and aiders and abettors" of the rebellion who took advantage of "Liberty of speech, Liberty of the press and *Habeas corpus*" to disrupt the Union war effort. Consequently he suspended the writ of habeas corpus and authorized army commanders to declare martial law in various areas behind the lines and to try civilians in military courts. Lincoln steadfastly defended such an invasion of civil liberties, contending that strict measures were imperative if the laws of the Union—and liberty itself—were to survive this "clear, flagrant, and gigantic case of Rebellion."

At the outset, responsibility for suppressing disloyal activities was divided among the State, War, and Navy departments, with Seward's State Department playing the largest role. Convinced that treason lurked everywhere, in every bureau, post office, customs house, regiment, and ship of war, Seward took extraordinary steps to root out subversives. He not only censored the telegraphs and the mails, but utilized government agents, United States marshals, Pinkerton's detectives, city police, and private informers to maintain surveillance of "suspicious" persons and to help arrest them. In May, Lincoln became apprehensive about such activities. "Unless the *necessity* for these arbitrary arrests is *manifest,* and *urgent,"* he wrote in an executive memorandum, "I prefer they should cease."

So did Chief Justice Taney. While on circuit duty, he rebuked Lincoln for usurping power in suspending the writ of habeas corpus. Only Congress could legally do that, Taney argued, and he admonished the President not to violate the very laws he had sworn to uphold. "Are all the laws, *but one?* to go unexecuted," Lincoln replied later, in reference to habeas corpus, "and the government itself go to pieces, lest that one be violated?" Besides, the Constitution did not specify which branch of the government could suspend the writ, so that Lincoln didn't think he had broken any laws or violated his oath of office. Therefore the government would continue to imprison people who were known disloyalists.

Meanwhile Lincoln was pulling all the wires he could to keep the rest of the border slave states loyal to the Union. If Kentucky, Maryland, and Missouri went out, then the rebellion would not be expunged in any ninety days, but would consume Washington itself and cost Lincoln "the whole game." Missouri remained in the Union thanks to a couple of adamant administration men—Frank Blair, Jr., and General Nathaniel Lyon, the latter a petulant little New Englander whom Lincoln had placed in command of the federal arsenal in Saint Louis. Cheered on by Blair, who kept Washington apprised of Missouri developments, Lyon smashed secessionists there with lightning blows: he dispersed the pro-secessionist state militia, put the pro-Confederate governor and legislature to flight, then turned the "vacant" state government over to Union men who looked to Blair as their leader and who cooperated fully with the Lincoln administration.

So Missouri stayed. But for a while it was touch and go with Maryland and Kentucky. Maryland, for one, absolutely had to be kept out of rebel hands; otherwise the United States capital would be in the center of an insurrectionary state—a potential calamity that Lincoln shuddered to contemplate. At first he resorted to diplomacy rather than force to keep Maryland, wooing the loyalist governor, keeping the lid on explosive Baltimore, and even allowing the state legislature to convene though it might vote for secession. Only if Marylanders actually took up arms against the government would Lincoln order their arrest, let the army bombard their cities. Though General Benjamin Butler occupied Baltimore without Lincoln's approval, the President's strategy ultimately paid off: Baltimore quieted down and Maryland remained in the Union. Less than a month after the Baltimore riot, the state sent a thousand volunteers to Lincoln's army.

Later, though, Seward became convinced that the mayor of Baltimore and the secessionist members of the legislature were actively helping the rebellion. Therefore, with Lincoln's approval, the Secretary of State had them all arrested and imprisoned. When Marylanders cried out in protest, Lincoln replied that while "public safety" obliged him to withhold the details, he assured everyone that he had "tangible and unmistakable evidence" that those imprisoned were involved in traitorous activities. After the state elections, however, Lincoln released most of the political prisoners once they had taken an oath of allegiance. Through a mixture of gentle diplomacy and strong-arm arrests, Lincoln "redeemed" troublesome Maryland.

Even more critical than Maryland was Lincoln's home state of Kentucky. If he lost her, he couldn't control the strategic Ohio River and couldn't hold Missouri either, which meant that Kansas would be cut off from her Unionist moorings and that important wagon roads to Western gold and silver mines would be lost. Worse, a rebel Kentucky would give the Confederates a potential base from which to launch attacks against Illinois, Indiana, and Ohio. And if Kentucky seceded, it would likely inflame Maryland again and all of Lincoln's work there would be destroyed.

So he sought with all his power to pacify Kentucky, to soothe her worried leaders and divided people. He knew that secession had split up families and turned entire communities into hostile camps, one for the Union and the other for the rebels. And though the state government pronounced Kentucky "neutral" in the conflict, for neither North nor South, men enlisted in both armies. In Louisville, Union volunteers marched along one side of the street and passed rebel recruits on the other. Brothers dressed in opposing uniforms rode the same train to the fronts. Secession had cut the heart of Kentucky in two, yet Lincoln worked like the deftest of surgeons to save her life for the Union. He ignored the secessionist governor and treated with the pro-Union legislature. He promised Kentucky he would not touch her slaves, invade her territory, burn her cities, shoot her troubled people. At the same time, though, he carried on clandestine operations to mobilize Unionists inside the state: he sent out loyal Kentucky officers to recruit volunteers there, and he relied on Unionist friends to distribute federal guns among loyal Kentucky outfits. When the governor howled that Unionist forces had occupied his state, Lincoln retorted that federal recruiting camps consisted solely of native Kentuckians. In the end, it was a combination of Lincoln's own program and Confederate anxieties that kept Kentucky in the Union. Unable to abide Kentucky neutrality, the Confederacy threw an army into the state and so enraged the pro-Union legislature that it voted to fight under the Union flag.

As summer approached, Lincoln found himself in the middle of a huge administrative mess, with every agency from the White House to the army in utter disarray. Never before had an administration had to cope with a massive insurrection or raise, equip, and supply such large field armies. Here was Lincoln, both inexperienced and unsystematic, having to solve complex problems for which precedents and guidelines were virtually nonexistent. His Cabinet members, moreover, lacked clear lines of authority in managing a civil war and often worked at cross-purposes and even interfered in one another's tasks. Welles, for his part, complained that Seward and Cameron meddled in navy matters as though Welles's office did not exist. And Chase, in addition to his Treasury chores, was as busy raising and inducting regiments into the army as was Cameron himself. Administratively, Washington was a whirlwind of confusion and chaos.

The worst troubles were in Cameron's War Department, where a minuscule staff tried in vain to direct the military colossus now being assembled. In truth, there was such a swell of patriotism and surge of volunteering across the Union, with Republicans and Democrats alike swamping military centers, that the understaffed War Department could not possibly accommodate or even keep track of all the regiments arriving in Washington each day. And Simon Cameron, as confused as he was injudicious, sometimes refused to accept state regiments, which sent both governors and officers squalling at Lincoln. As if he didn't have enough minutiae to worry about already, in addition to fashioning national and international policies, now he had to tend to individual regiments as well. He hated to interfere with his Cabinet Secretaries in their duties, but in the case of rejected regiments he did. Time and again he sent over memoranda advising Cameron to take the latest outfits offered from Massachusetts, from Indiana and Michigan. When rival recruiters quarreled over whose units should be enlisted, Lincoln told Cameron to induct them all.

What a headache it was to build an army. Not only did Lincoln have to contend with unhappy recruiters; he also had to sign hundreds of officers' commissions in both the regular and the volunteer armies. And he had to allot so many generalships to each state and fill each position himself. The result was a plethora of "political generals," as Lincoln handed out commissions to both Republicans and Democrats—to men like James Shields of Illinois (Lincoln's old dueling opponent) and Nathaniel Banks and Benjamin Butler of Massachusetts. Why let politicians become officers? "To keep them from fighting against the war with their mouths," a Lincoln friend explained. Still, the dual army system, which piled layers of volunteers on top of the regular army nucleus, only added to the growing military jumble. To make matters worse, there were shortages of everything from shoes to guns. "The plain matter-of-fact is," a harried Lincoln wrote one officer, "our good people have rushed to the rescue of the Government, faster than the government can find arms to put into their hands." And finding military supplies led to still more bureaucratic tangles. Because his department was understaffed, Cameron turned to state governors and private citizens for help. The result was widespread disorder, as agents representing governors as well as the Ordnance, Quartermaster, and Commissary bureaus of the War Department vied with one another in spending federal funds for war matériel.

Then there was the Negro problem, haunting Lincoln as always. Caught up like whites in the rush of patriotism, black men across the North wanted to fight for the country as their forebears had done in the Revolution and the War of 1812. The trouble was that a federal law now barred Negroes from enrolling in state militias, and the all-white regular army was not about to accept any blacks. Even so, Negroes in various cities organized their own outfits and set about drilling until white authorities made them stop. When blacks urged the administration to let them serve, Lincoln told them no. As both the *Illinois State Journal* and the *New York Tribune* said, this was strictly a white man's war.

It was also strictly a war to save the Union and not to free the slaves, as Lincoln repeatedly asserted. Yet slavery was inextricably involved in the conflict—was the reason the South had seceded in the first place—and was bound to create a problem wherever Union troops touched rebel territory. For example: at the tip of Virginia's Yorktown peninsula, where General Butler now commanded Fort Monroe, fugitive slaves were flocking to his lines. And rebel whites were demanding their return, as though they were still protected by the laws of the very government from which they had seceded and which they were now resisting by force of arms. Butler refused to hand the blacks over and pronounced them "contraband of war," which seemed only a few steps shy of emancipating them.

Lincoln was concerned about Butler's action, for he did not want this misconstrued as emancipation. He feared that the slightest move in that direction would alienate Northern Democrats and send the critical border spiraling into the Confederacy. No, his policy was to remain consistent with Republican promises and to bring the South back with the peculiar institution still intact. Nevertheless, after talking the matter over with the Cabinet, he approved of Butler's move. The fugitive slave law no longer applied to those in rebellion against the government and their slaves need not be returned. After that, word spread across the slave grapevine from Virginia to Tennessee; and many a slave, after running through the woods all night, reached Union lines and said, "I'se contraband."

On May 24, 1861, Lincoln studied a map on his office wall and stuck blue pins on the Virginia side of the Potomac. During the night, Union troops had dashed across the Washington bridge under a full moon and seized Lee's plantation and Arlington Heights. Then they had occupied Alexandria itself, tearing down those rebel flags that had long been an insulting spectacle to Lincoln. It was the first offensive undertaken by the unwieldy Grand Army of the United States, and Lincoln and official Washington were in high spirits about it. Too, Lincoln had a personal interest in the operation since Colonel Ellsworth and his New York Fire Zouaves were part of the attacking force.

Lincoln was in his library when a young captain brought him the news that Ellsworth had been killed. Ellsworth? Dead? But how? In Alexandria that morning, the captain explained. He'd taken a rebel flag down from a hotel, and a secessionist had riddled him with a double-barrel shotgun. The captain left with his regrets and Lincoln stood at the window, looking out over the Potomac. He was crying. He turned abruptly, for here came Senator Henry Wilson of Massachusetts and a reporter. "I cannot talk," Lincoln said. Ellsworth is dead and it's "unnerved me." He left, muttering "poor fellow" to himself.

Mary and the boys were crushed. She accompanied Lincoln down to the Navy Yard to see Ellsworth's body; and Lincoln had him brought to the White House, where he lay in state and received a full military funeral. Meanwhile somebody handed Mary the blood-spattered rebel flag Ellsworth had torn down in Alexandria. But neither Lincoln nor Mary could bear to look at it, and she put it away. And so Ellsworth was gone, one of the first Union casualties in the war—cheerful and affectionate Ellsworth, who had played with Tad and Willie like a boy himself and who had almost made Lincoln feel young again.

A week or so later came news of another death. Out in Illinois, where he'd

damned secessionists with his customary gusto and rallied loyal Democrats
to the Union cause, Stephen A. Douglas had died of "acute rheumatism."
He was only forty-eight years old. Lincoln had the White House decorated
in black drapes in memory of his long-time adversary. But his grief was still
for Ellsworth.

The talk along embassy row was that
the President was a duffer in foreign affairs. Lord Richard Lyons, the British
minister to the United States, considered Lincoln "a rough farmer" and
sniffed to Lord John Russell, British Foreign Minister, that Lincoln had an
"ignorance of everything but Illinois village politics." Most ambassadors
thought that Seward would mastermind Union diplomacy.

Seward expected this, too. After all, the President was untutored in foreign
matters and relied on Seward a great deal, especially since Seward had served
on the Senate Foreign Relations Committee and had some expertise in international affairs. In truth, Seward was determined to rule his own fiefdom and
boasted that because his chief had an "utter absence of any acquaintance with
the subject," Lincoln would let him shape and control foreign policy.

And his policy turned out to be stridently pugnacious so far as Britain and
France were concerned. On May 21 Seward came to Lincoln's office with a
"manifesto" addressed to Charles Francis Adams, the United States minister
in London. Seward had written the communiqué because he was certain that
Great Britain was drifting toward Confederate recognition. While the Queen
had affirmed British neutrality in the American conflict, she had also acknowledged the Confederacy as a belligerent power—that is, as a responsible
government conducting war. Seward, for one, was furious that Britain had
rejected Washington's interpretation of the war as a domestic insurrection
and had conferred respectable international status on the pretended Southern
government. In Seward's view, Britain was now traveling the high road
toward full recognition of the Confederacy as a nation. Should that happen,
France and the other European countries were sure to do the same, thus
admitting the Confederacy into "the family of nations." Then the Confederacy would seek military alliances and even armed intervention to guarantee
her sovereignty. All of which meant disaster for the Union and left Seward
in a rage. "God damn 'em," he said of England and France. "I'll give 'em
hell."

Thus his "bold remonstrance" to Adams. Lincoln read over the document with growing uneasiness, for Seward demanded that the British not only accept the Union blockade, but eschew Confederate recognition and refrain from any further intercourse with rebel commissioners abroad. If Britain refused, it meant war with the United States. According to Seward's instructions, Adams was to read this to Lord Russell's face.

Lincoln said he wanted to study the letter, and after Seward departed the President summoned Senator Charles Sumner of Massachusetts, a liberal Republican with considerable knowledge of British affairs. Sumner had recently become chairman of the Senate Foreign Relations Committee and had traveled extensively in England and maintained many friendships there. An arch, sophisticated bachelor with B.A. and law degrees from Harvard, Sumner even looked English, with his tailored coats, checkered trousers, and English gaiters. He was so conscious of manners, he admitted, "that he never allowed himself, even in the privacy of his own chamber, to fall into a position which he would not take in his chair in the Senate. 'Habit,' he said, 'is everything.' " A humorless, high-minded man, he hated slavery and spoke out with great courage against racial injustice to black people. Back in 1856, he'd almost been beaten to death by Congressman Preston Brooks of South Carolina and had gone off to Europe to convalesce. He had rich brown hair streaked with gray, a massive forehead, blue eyes, and a rather sad smile. Mary was terribly impressed with him. And so was Lincoln.

In his office, Lincoln showed Sumner the manifesto Seward had written. The senator was appalled. Was Seward insane? This would cause war with Britain, would be a national calamity. The best way to keep Britain out of American affairs was to placate her, pacify her. You must not allow Seward's letter to go off, Sumner warned. "You must watch him and overrule him."

What Lincoln did was to compromise. He softened the offensive language in Seward's manifesto, deleting a remark about the United States and Great Britain being enemies, and marked the document "Confidential," to be read only by Adams and shown to no one else. But Lincoln let Seward's basic policy stand: if Britain recognized the Confederacy as a nation or tried to intercede in her behalf, it would mean hostilities with the Union.

Lincoln liked using Sumner as a check against Seward and decided to make it a standard practice. Therefore he designated Sumner as his chief adviser on foreign policy, authorized him to go through all foreign correspondence since the inauguration, and henceforth solicited Sumner's opinions on a variety of international questions, particularly the blockade. As "a responsible member of the Lincoln administration," states his biographer, Sumner now possessed "a virtual veto over foreign policy," and he became a regular guest at diplomatic dinners. In return for such power, he gave tacit approval of Lincoln's war policies and became a valuable Lincoln man on Capitol Hill.

Not that Seward was now squeezed out of foreign policy decisions. On the contrary, Lincoln prevailed on Seward and Sumner both for advice, realizing that he needed more than one view in an area where he knew he was ignorant. He still treasured Seward's talents, still gave him considerable freedom in running the State Department, and still sought his opinions on a multitude of other problems. Only when Seward became too tempestuous and too independent did Lincoln reign him in.

Moreover, now that Seward had given up trying to run the administration, Lincoln liked him as a man and thoroughly enjoyed his company. Sixty years old and slightly stooped, Seward resembled a jocular bird chewing on a Havana cigar. His nose was hooked in a beak, his ears stuck out, his voice was husky, his eyebrows thick and grizzly, and his silver hair always disheveled. He was a celebrated raconteur, loved to pun and banter, often braying so hard at his own wit that it left him hoarse. A chain talker, he entertained guests at his house on Lafayette Square with "a regular Niagara flood" of chatter, gossip, and uninhibited profanity. And how he could entertain, throwing lavish dinner parties that lasted four hours and went through eleven courses, complete with imported wines and brandy. Yet he was a man of many moods—now an effusive storyteller, now a cynic, now a show-off, now a tough and serious administrator. In all, he was a man of immeasurable self-esteem, so certain of his own greatness that he tipped his hat to any stranger who appeared to recognize him.

Well, never mind his conceit. Lincoln was attracted to Seward because of his warmth and buoyant humor. Soon the President took to dropping by Seward's home in the evenings or taking him on an occasional Sunday outing. They found one another vastly amusing, engaged in preposterous dialogue and swapped nonsensical stories and off-color jokes that left them helpless with laughter. It was marvelous therapy for Lincoln's hypo. Moreover, the two men found a common bond in the fact that both had troubled wives. If Mary suffered from chronic headaches and fears of poverty, Frances Seward was plagued by periods of deep depression, convinced that her nerves were diseased and her vascular system deranged. Given his own melancholy, Lincoln could sympathize with Frances—and with Seward's worries about her.

Lincoln's intimacy with Seward caused resentment and jealousy among the other Cabinet Secretaries. They complained that he seldom held regular Cabinet meetings and depended much too much on Seward's advice. Chase, the Cabinet liberal and a busybody himself, viewed Seward as a conservative "opportunist" and deeply distrusted him. Though Lincoln included Chase on most major policy decisions at this time, Chase fretted that Seward still wanted to manipulate Lincoln and rule the administration. And Bates and Welles thoroughly agreed. Welles himself was certain that Seward lusted for power, that this was why he chummed around with the President, told him what to do, and treated him with a familiarity that in Welles's opinion bordered on disrespect. Moreover, Welles could not stand Seward's cocky ways, could not bear it when Seward summoned Welles and the others to Lincoln's infrequent meetings. Who did this egotist think he was anyway? At night, in the privacy of his home, Welles confided in his diary that Seward was a nefarious schemer who'd sold out to "the interests" and who used Lincoln and other men for his own sinister designs. Welles was suspicious of most men in government, for he saw only the darker side of humankind, contending that the "real motives" of people were almost always evil. Though he affected an air of reticent calm, of cautious deliberation, he was boiling with emotional insecurities and a pathological mistrust of his colleagues, especially Seward. From behind his tranquil mask, Welles watched Seward's every move and smoldered at his flirtations with the President. Yet what could one expect? Welles believed Lincoln a naïve and passive man who could not make a single decision without first running to Seward.

For his part, Lincoln seemed unaware of Welles's misanthropic bitterness. All he perceived was that façade of unruffled calm. Out of Welles's hearing, he playfully called him "Father Neptune" because of his flowing white beard. As for Seward, Lincoln pretty much ignored what the other Secretaries thought of him. Lincoln now trusted Seward implicitly and judged him one of the most capable men in government.

Moreover, in the matter of diplomacy, maybe a little of Seward's truculence was necessary. Maybe getting tough with Great Britain would bolster Union confidence and "maintain the pride and dignity of our government," as Seward argued. In truth, by mid June Seward's belligerence seemed to be working. Great Britain announced that she would respect the Union blockade, which was a diplomatic victory for Washington. Nevertheless, the weight of English opinion probably did not sympathize with the Union. As various Englishmen observed, Lincoln's abstract Constitutional arguments in defense of the Union sounded like lame excuses for sheer bullying. Why shouldn't the South gain her independence, as the thirteen colonies had done? Anyway, with Lincoln and Seward insisting that emancipation was not a goal in the war, there seemed nothing morally right in the Union cause. As *Punch* said, "An afterthought only is 'Justice to niggers.' "

So English opinion tended to sympathize with the underdog rebels, who with only five or six million whites were fighting an established government with some twenty million and most of the railroads and factories. British policymakers, on the other hand, favored the Confederacy for entirely practical reasons: Confederate independence might cause more defections from the Union, which in turn would divide North America into three and maybe four impotent states, eradicate any potential U.S. commercial rivalry, and leave mighty industrial Britain in sole command of the world's markets.

By the summer of 1861, Lincoln's day had become a set routine. After a "sleep light and capricious," he rose at first light and was at work in his office by seven. His "shop," as he called it, was a large room on the second floor of the White House. It contained a marble fireplace, two hair-covered sofas, a big oak table for Cabinet meetings, and a desk with documents and correspondence stuffed in pigeonholes. On the wall were framed military maps and a faded oil painting of Andrew Jackson. Lincoln usually wrote at a table between two high windows, sitting in a large armchair with his legs crossed. This June he worked on his message to Congress, writing in his usual slow and laborious manner, often whispering phrases to get them right before he wrote them down. From time to time he would pause to stare out the windows, which afforded a sweeping view of the south lawn, the Smithsonian Institution, the unfinished Washington monument, and the distant Virginia hills, covered now with Union tents, wagon trains, and bawling cattle.

At nine he breakfasted with his family, downing an egg and a cup of coffee, and then returned to the shop with callers already assembling downstairs. For an hour or so he studied a digest of the day's news prepared by his personal secretaries, Nicolay and Hay, and discussed important correspondence. Nicolay (or "Nico") was twenty-nine now, an emaciated fellow with blue eyes and a slow smile, who wrote love letters to a girl back in Illinois. Hay, at twenty-three, was a boyish gadfly who loved wine and cheese and laughed compulsively at anything. Both young men idolized Lincoln, would do anything for him, and would one day write an encyclopedic history of his administration. For Mary Lincoln, though, they had nothing but adamant hostility. They did not understand what pressures and hurts lay behind her temper flares, and with youthful intolerance they disparaged her as a raging "Hellcat" whose shrieks made the White House tremble.

Nicolay and Hay did their best to guard Lincoln from intruders, cranks, and politicians who tried to push past them to see "the new man," but Lincoln was hard to protect. At first he placed no limits on visitors, and they accosted him incessantly from dawn to dusk, making it impossible for him to tend his other chores. So with Nicolay and Hay the President worked out a system: he would see people on Monday, Wednesday, and Friday from ten till two, on Tuesday and Thursday from ten till noon, when he was supposed to meet with the Cabinet. His personal secretaries, situated in an adjacent office, were to head off anybody who tried to interrupt him other than during office hours. With a system thus established, Lincoln proceeded to "break through every regulation as fast as it was made," his secretaries complained. If they told some caller to come back during regular hours, Lincoln would open his door and invite the person in anyway. It infuriated Nicolay and Hay. How could there be an office system if Lincoln ignored the system? The trouble was that Lincoln loathed rules and red tape and generally did as he wanted, as though the White House were his Springfield law office.

Officially, Lincoln would throw his door open at ten and let in a river of raucous humanity—interviewers, politicians, office seekers, businessmen, sobbing mothers who wanted their sons released from the army, and pretty young wives who flirted with him to promote their military husbands. At times Lincoln enjoyed all the attention he received—particularly from the young women. And he was courteous and attentive to most everyone—the jobless, the infirm, the promoters, the parvenus—who passed through his doors. He avoided false enthusiasms, never lied and exclaimed "I am delighted to see you" when he wasn't delighted. Usually he would greet people with "What can I do for you?" Then he would listen, stroking his beard, and would promise to do what he could if the request were reasonable. If he was in a hurry to get rid of someone, he would crack a joke and with both of them laughing would ease the caller out the door. Sensitive himself, he tried to give everybody something, if only "a quaint phrase" or a memorable poem. Once he recited an entire poem of Oliver Wendell Holmes to a group of enchanted ladies.

Still, some visitors thought him less a man of culture and polished wit than a shrewd "old codger" like Andrew Jackson. A New York lawyer was astonished at Lincoln's language, reporting that he said "thar" for *there,* "git" for *get,* "kin" for *can,* "one of 'em" for *one of them,* and "I hain't been caught lyin' yet, and I don't mean to be."

Lincoln claimed to like his "public opinion baths," deemed them an indispensable way of finding out what people were thinking. But sometimes the crowds and the endless demands were too much, his temper would blaze, and he would call somebody "a damned rascal" or explode at a persistent visitor: "Now go away! I can't attend to all these details. I could as easily bail out the Potomac with a teaspoon." In truth, the long visitations wore him out. He complained about "the numerous grist ground through here daily," from some senator desiring a war with France to some poor woman after a Treasury job, and remarked that each caller took away a special piece of his vitality. "When I get through with such a day's work," he sighed, "there is only one word which can express my condition, and that is—*flabbiness.*" Nicolay and Hay estimated that he spent three-fourths of his time meeting with people. The crowds were so great that sometimes even U.S. senators had to wait ten days to see him.

One class of visitors Lincoln always welcomed. These were inventors and gunsmiths who came to promote some newfangled weapon. Fascinated with the tools of war, Lincoln loved to examine and talk about them. Whenever he could, he would slip away from the morning crowds and witness the testing of some new gun out at the firing ranges. One day he took Hay down to the Navy Yard to watch ordnance experts test fire the great Dahlgren gun into the Potomac. The cannon exploded with a tremendous concussion and the eleven-inch shell ricocheted across the water like a thrown boulder, churning up a thirty-foot column of spray with every jump, until at last it rolled over in the waves. "The President," Hay said, "was delighted." On other mornings, Lincoln might go over to his private rifle range, in the Treasury Park beyond the south lawn, and fire some new rifle himself, blazing away at a target pinned to a woodpile there. The President not only tried out the new breach-loading rifle, which he thought superior to the muzzle-loaders used by the army, but helped in its eventual introduction into his armed forces. He was also instrumental in the development and introduction of a rudimentary machine gun.

So, with office work, public opinion baths, and occasional target practice, went the President's mornings. At noon on Tuesdays and Thursdays, Lincoln was scheduled to hold Cabinet meetings, but he seldom did so on a regular basis. Though Chase and other Cabinet Secretaries continued to fuss about this, there wasn't much they could do to change Lincoln's ways. As in his office routine, he remained an incorrigibly haphazard administrator who seemed oblivious to bureaucratic regulations. Rather than confer with the full Cabinet twice every week, he liked to consult with individual Secretaries about matters that concerned them. If there was a problem with the army, he simply saw Cameron about it. Similarly, he met with Welles about naval matters, with Chase about finances, with Seward and Sumner about diplomacy. Most of the time he called full Cabinet meetings only to discuss vital questions of national policy. Still, he wasn't consistent about individual consultation either and often sought out his favorites, Seward and Chase, for opinions about strictly military problems, which of course offended Welles and Cameron. At other times, Lincoln didn't bother to confer with anybody, but simply made decisions himself.

Sometime in the afternoon Lincoln would run "the gauntlet" down to the family rooms in the West Wing and have a biscuit and a glass of milk for lunch. Midafternoon found him back in his shop doing paper work—signing commissions, composing important letters, sending out memos to various department heads about requests received that morning, asking Chase to look into this, Cameron into that. "The lady—bearer of this—says she has two sons who want to work," Lincoln wrote on one memo. "Set them at it, if possible. Wanting to work is so rare a merit, that it should be encouraged."

At four he would get away from the White House and go for a carriage ride—his only regular source of relaxation. Sometimes Mary would accompany him. If she were out of town, then Seward or an old Illinois crony would go along. Often they would rattle across the Washington bridge to review troops in Virginia, visit officers in their quarters, chat with the men at army messes, with the odor of biscuits and pork in the wind. Then they would head through the woods, enjoying the fresh air, the cawing birds, the tangle of limbs against the June sky. To guard against assassination, a cavalry escort accompanied Lincoln on his country rides, although he protested all security measures as a nuisance. Why would anybody want to kill him? What could that possibly gain? And in any event, anybody who really wanted to shoot him could do it with or without a military patrol. When mounted guards and infantry were detailed to protect the White House, Lincoln dismissed them because they made him feel like an emperor. But he grudgingly consented to the cavalry escort for his carriage rides, even though they made such a racket with their jingling spurs and clanging sabers that he and his companion could scarcely hear one another. The story goes that Lincoln liked to prod his coachman and try to outrun the cavalry—the President's carriage careening down dusty roads, the flustered soldiers trying in vain to catch him.

Then it was back to the White House for dinner at six. Usually Lincoln dined with his family and a few friends or special guests. State dinners, wretchedly formal affairs, began at seven-thirty, and Lincoln ate lightly from one or two courses, took a sip of wine, and then joined the men in the sitting rooms, though he abstained from the cigars and brandy. Once or twice a week there were evening dress receptions, or levees, when he had to greet a flood of people in a capacious White House room. Herman Melville met him on one such occasion and said he shook hands "like a man sawing wood at so much per cord." The levees, too, were painful ordeals which left his right hand so swollen and sore that he could not use it for hours.

On rare evenings when there were no receptions and no pressing work to do, Lincoln would relax with Seward and a small circle of Illinois associates —men like Browning or Ward Hill Lamon, now United States marshal of Washington, D.C. To ease the exhaustions of the day, Lincoln would let his humor flow, until "he was once more the Lincoln of the Eighth Circuit," Hay said, "the cheeriest of talkers, the riskiest of story tellers." When he reached his punch line, observed one visitor, Lincoln "wrinkled up his nose" and "showing all his front teeth gave a very wheezy catching laugh and in his glee fell to scratching himself on the elbows."

His White House tales gave rise to a new item called "Lincoln stories," but Lincoln always denied their originality. "You speak of Lincoln's stories," he once told a correspondent. "I don't think that is a correct phrase. I don't make the stories mine by telling them. I am only a retail dealer." Yet he conceded that humor was his therapy. "Some of the stories are not so nice as they might be, but I tell you the truth when I say that a funny story, if it has the element of genuine wit, has the same effect on me that I suppose a good square drink of whiskey has on an old toper; it puts new life into me."

The opera and theater restored him, too. Whenever he could spare the time, he and Mary would dress up, climb into the Presidential carriage, and sally forth into the Washington night to attend one of several theaters. Mary loved the theater more than any other form of entertainment. And Lincoln considered it a "wonderful" way to escape the troubles of his office. He preferred Shakespearean productions—not the tragedies, which he liked to read, but the comedies with their risqué scenes and pungent dialogue. Lincoln enjoyed almost any humorous play, "no matter how absurd or grotesque." He was even delighted with John Brougham's "travesty" of *Pocahontas,* and cracked puns and chortled at its "delicious absurdity." If Mary were out of town, Lincoln would often slip out alone to see a play or go to the opera. Or he might take Hay to a concert, for he liked music of all kinds, from military bands to dancing choruses. At one concert in Ford's Theater, he and young Hay amused themselves in a private box, applauding the singers and carrying on "a hefty flirtation with the M—— Girls in the flies."

For Lincoln, though, a night out was rare. He spent most evenings in his White House office, working late into the night, toiling on his Congressional speech or some other composition. In the summer he had to raise the windows, which let in waves of insects to buzz around the lamps and flutter against the walls. And then there were the awful smells, wafting in from drainage ditches and the unreclaimed Potomac flats, a stench Hay likened to that of "twenty thousand drowned cats." After a June shower came squadrons of mosquitoes, which spread malarial fevers through the White House until most everyone there shook and cried from the ague.

To escape the mosquitoes and sweltering Washington heat, the Lincolns retreated up to the Retired Soldiers' Home, situated in the woods a couple of miles to the northwest. Here the elevation was higher and the nights cooler than down in the capital. Mary thought the Soldiers' Home "a very beautiful place" where "we can be as secluded, as we please," so the Lincolns moved into a cottage on the grounds and it became a summer White House. In the mornings Lincoln would ride alone into Washington to spend his whirlwind days, then would return at night to relax in cool seclusion with his family. Here, when Mary and the boys were asleep, he would do his late-night work by lamplight. Later, while waiting for sleep to come, he would read his worn copy of Shakespeare's tragedies, turning again and again to *Hamlet* and *Macbeth.* Or he might peruse a volume of poetry—of Burns, Whittier, or Holmes—and recite half aloud those sad and pathetic stanzas he knew by heart and loved so well. "Green be the graves where her martyrs are lying! Shroudless and tombless they sunk to their rest . . ." It was the beginning of a stanza from Holmes which moved Lincoln almost as much as "Mortality." Then at last, when all was quiet except for the groan of trees in the wind, Lincoln would sink into a restless sleep, often filled with dream sequences about war and distant voices and a phantom ship moving in the fog.

Outside Lincoln's White House windows, Washington was aswarm with politicians gathering for the special session of Congress, to begin on July 4. The congressmen were too excited to fuss much about the heat. They hurried across the Potomac to inspect the Grand Army, a motley force of thirty thousand under the field command of General Irwin McDowell, who'd installed his headquarters in the captured mansion of Robert E. Lee. In Washington, senators and congressmen reveled in all the military activity—the bands, troops, and wagon trains which clogged the city's streets. But one question pervaded conversation in Washington and much of the rest of the Union: When was the army going to fight? Over two months had elapsed since Lincoln's proclamation. What was the administration waiting for? Let the President send the army forward. Let McDowell smash the rebels in Virginia, march to Richmond, and end the rebellion at once. Yes, the cry of "On to Richmond" rang across the land. And the volunteers, too, were eager for a swift and decisive victory. They had joined up to defend the Union and prove their heroism, had marched off to war with young ladies throwing flowers in their paths and orators singing on about how the young manhood of the Union would crush the traitors of the South. They had posed before exploding cameras and sent daguerreotypes of themselves in uniform back to their families and sweethearts. They had gathered around glowing campfires and talked in hyperboles about the excitement of battle, even though few of them had been in combat before and seen a comrade die. For the volunteers, it was all a Picture Book War, a time of pomp and pageantry—of fierce drums and ringing bugles, of strutting drum majors and marching bands, of whipping banners and fluttering flags. It was a time when everybody from volunteers to civilians was swept up in the romance of war, in the thrill and glory of it all.

In the public outcry for battle, Lincoln called Generals Scott and McDowell to a Cabinet meeting on June 29, to discuss a summer offensive with McDowell in command. He was an unproven officer—the first in the history of the Republic to lead so large a field army. Though he loved to eat and could consume an entire watermelon for dessert, McDowell abstained from spirits, tobacco, and coffee. He was a huge man, forty-two years old, "square and powerfully built." Lincoln directed his attention to a map on the office wall and said he wanted McDowell to attack a rebel force concentrated at Manassas railroad junction, just across Bull Run about twenty-five miles southwest of Washington. But General Scott shook his imperial head. The Grand Army, he insisted, could not be ready for an offensive until the autumn. And in any event, an advance against Manassas was not wise, not wise. The old general suggested that the blockade be tightened and an expedition sent down the Mississippi, thus sealing off the Confederacy as though a huge anaconda snake were coiled around her borders. And then? And then wait for Unionist sentiment to rise in the South and overthrow the rebellion.

Lincoln didn't agree. Though he appreciated the strategic importance of the Mississippi, he was leery of Scott's judgment now, thanks largely to his advice about giving up Sumter and Pickens. Also, Lincoln had abandoned the old argument that Southern Unionism would somehow extinguish the rebellion. No, only military force could do that, an all-out attack, a battlefield victory. And so he turned back to his map. The rebel army was split, with some twenty thousand men at Manassas and the rest up in the Shenandoah Valley. While a Union force neutralized the rebels there, McDowell could crush those at Manassas Junction, sweep down to Richmond, and end the war.

Now it was McDowell's turn to object. He protested that his men were too green to fight at this time, that he must train and discipline them before throwing them against the rebels. "You are green, it is true," Lincoln said; "but they are green, also; you are green alike." So go and prepare for battle, he told McDowell. The public would not tolerate further delay. Unless the Grand Army fought a battle now, public morale might fall disastrously.

On July 4, Congress assembled with talk of an impending battle echoing across the capital; and Lincoln sent over an Independence Day message in which he set forth the central issue in this contest. And that was whether a constitutional republic, a democracy, could preserve itself. There were those in Europe who argued that rebellion and anarchy were inherent weaknesses of a republic and that an enlightened monarchy was the more stable form of government. "Must a government, of necessity, be too *strong* for the liberties of its own people, or too *weak* to maintain its own existence?" Lincoln asked. No, he believed that America's popular government could survive the present crisis. But to do so the government must meet force with force. It must teach dissidents "the folly of being the beginners of a war." It must show the world "that those who can fairly carry an election, can also suppress a rebellion," and that a constitutional republic was a workable system which offered hope for people everywhere, as he'd long contended. "This is essentially a People's contest," he declared. "On the side of the Union, it is a struggle for maintaining in the world, that form, and substance of government, whose leading object is, to elevate the condition of men—to lift artificial weights from all shoulders—to clear the paths of laudable pursuit for all—to afford all, an unfettered start, and a fair chance, in the race of life."

These were noble words indeed, but what about black people? What about the slaves? Did they not deserve "an unfettered start, and a fair chance, in the race of life," just like white people? How could Lincoln deliver these eloquent phrases and yet insist that the termination of human bondage in a "free" America was not a war objective? So went the arguments of abolitionist leaders like Garrison and Douglass, who exhorted Lincoln to issue an emancipation decree.

Nor were abolitionists the only people who objected to Lincoln's policy about slavery. So did liberal Republicans Sumner, Wade, and Chandler. Before and after his Congressional message, they secluded themselves with Lincoln in his White House office, and Sumner even accompanied the President on his carriage rides. Zachariah Chandler, senator from Michigan, was a Detroit businessman who'd amassed a fortune in real estate and dry goods. A restless, rawboned New Englander who'd migrated west to make money and history, he was smooth-shaven and wore an eternally grim expression, with his mouth turned down at the corners. Bluff Ben Wade, senator from Ohio, was short and thick-chested, with iron-gray hair, sunken black eyes, and a square beardless face. A pugnacious individual known for "a certain bulldog obduracy" and a readiness to duel with Southerners, Wade had little patience with indecisive, slow-moving Presidents. He hated slavery as Sumner and Chandler did. But Wade was also prejudiced against Negroes, complained about their "odor," and growled about all the "nigger" cooks in Washington, remarking that he'd eaten food "cooked by Niggers until I can smell and taste the Nigger all over." Like many other Republicans, he thought the best solution to America's race problem was to ship all the blacks back to Africa.

Now, in their conferences with Lincoln, the three senators wanted to make the annihilation of slavery a Union war objective. Before secession, of course, they had emphatically endorsed the party's hands-off policy as far as slavery in the South was concerned. But civil war had removed their constitutional scruples about that. Now they argued that either the President or Congress could remove the peculiar institution by the war powers, and they wanted the President to do it. If he emancipated the slaves, it would maim and cripple the Confederacy and hasten an end to the rebellion. Sumner flatly asserted that slavery and the rebellion were "wedded" and would stand or fall together.

Lincoln was sympathetic to their argument. Personally he detested human bondage as much as they did. He was also anxious to maintain a close working relationship with Republican liberals, who, though a minority, still controlled most of the powerful Congressional committees. Moreover, he had great respect for the liberals, thought them completely dedicated Republicans, and referred to men like Sumner as the conscience of the party.

Yet he could not free the slaves, would not free them. As President, he was responsible to the entire country, which obliged him to move with extreme caution and care; and right now most of the country seemed steadfastly opposed to emancipation even as a war measure. Should he now violate his pledge to leave slavery alone as an institution and issue an emancipation proclamation, the consequences, he feared, would be calamitous. Emancipation would almost surely drive the loyal border states out of the Union, alienate Northern Democrats and destroy the bipartisan war coalition that began after Fort Sumter, and maybe even ignite a racial powder keg in the North. Then the Union really would be lost. So, no, he would stick to his policy about fighting strictly to save the Union. He would crush the rebellion with the army and restore the national authority in the South with slavery still intact. At the same time, Lincoln and his party would implement their policy of slave containment, sealing it up in the Southern states and putting it on the road to ultimate doom.

Though Wade and Chandler disagreed with him, Sumner and other Congressional liberals grudgingly acquiesced in Lincoln's policy for now. As the President's personal friend and foreign policy adviser, Sumner told himself that at bottom Lincoln was "a deeply convinced and faithful anti-slavery man" and that the sheer pressure of war would eventually force him to strike at slavery.

As the Congressional session got under way, Republicans on Capitol Hill seemed fairly united about Union war aims. All of them wanted to preserve the Union and fight the war through to a swift and successful conclusion. All were eager to sustain the President in resisting the rebellion, and so they legalized most of the emergency measures he'd adopted last spring. In complete agreement with Lincoln's slave policy, Congressional Republicans also supported the so-called Crittenden-Johnson Resolutions, which declared that the sole purpose of the war was to restore the Union without overturning state institutions. While Sumner abstained from voting, most of his liberal colleagues reluctantly consented to the resolutions.

On the matter of slave contraband, though, the liberals mustered almost unanimous Republican support for a confiscation bill, which they pushed through Congress and sent to Lincoln for his signature. The most controversial provision was one which authorized the seizure of rebel slaves actively employed in the Confederate war effort. Predictably, border state Democrats like Crittenden of Kentucky howled that the provision was nothing less than a general emancipation act. And Midwestern Democrats warned that the bill would bring several million "woolly headed, thick-lipped" Southern Negroes into the North and "Africanize" the region. But most Republicans thought all this absurd. Since confiscated slaves would be carefully controlled, Southern blacks were not going to flood into the free states. In wartime, moreover, the government had every right to seize enemy property —including slave property—as legitimate contraband. Furthermore, since the bill affected only those slaves used for rebel war purposes, it was hardly a general emancipation bill which eradicated slavery as a state institution. No, it was an entirely legal war measure designed to weaken Confederate military forces and help terminate the rebellion.

Since Lincoln had already upheld Benjamin Butler's contraband policy, he approved of the confiscation bill and signed it into law. Later he said that he faithfully enforced the measure, including the controversial slave provision. But with racial antagonisms flaring up in the wake of the act, Lincoln still insisted that emancipation was not a goal in the war.

The telegraphs were humming with the news: at last, after nearly a month of preparations, the Grand Army was on the move. And most everyone—even old Winfield Scott—was predicting a glorious Union victory. With politicians and their wives riding out in carriages to watch the show, McDowell marched his troops southward across Bull Run and on the morning of July 21 hurled them against the left side of the rebel lines near Manassas Junction.

In Washington that hot and humid Sunday, Lincoln was extremely apprehensive about the battle. He even went to church that morning. Afterward he received word that the two armies were fighting in the woods along Bull Run, but reports were terribly confusing: the Grand Army was first carrying the field, then retreating, then carrying the field again. In "deep anxiety," Lincoln called at army headquarters, only to find General Scott sound asleep. The old man roused himself long enough to reassure Lincoln that McDowell was going to win. By five-thirty the battle did seem a Union victory, so Lincoln went on his usual carriage ride. But when he returned, his personal secretaries gave him a message from Seward that the battle was a Union disaster, that McDowell was falling back and crying for Scott to save Washington. Without changing his expression or uttering a word, Lincoln hurried over to army headquarters, where Scott and his staff confirmed the news. Yes, the Grand Army was in retreat, all was lost. But how did it happen? Who was to blame? Nobody seemed to know. Scott was dazed, his staff in a state of shock.

Lincoln returned to the White House and lay down on a sofa in his office. As night closed over Washington, heavy clouds swirled across the moon, "forming a most fantastic mass of shapes in the sky," said a newspaper correspondent. It began to rain. In his office Lincoln could hear crowds milling around outside, waiting for further reports from the front. Around midnight Chandler stormed into the White House. He was wet and furious. He'd been at Bull Run, he told Lincoln, and had seen what had happened. He and Wade had sat in a carriage like dozens of other congressmen, correspondents, and spectators, some of whom had brought picnic baskets and opera glasses. The Grand Army attacked with puffs of artillery smoke and delayed concussions, which delighted the civilians. For several hours the woods south of Bull Run rang with shouts, musket fire, and cannonades. Then suddenly the Union Army was in full retreat, falling back along the very road where the spectators were gathered. What followed was utter pandemonium, as soldiers tangled with civilians and then broke in panic, throwing away canteens, knapsacks, and muskets as they ran. Chandler and Wade pulled their carriage across the road and waved revolvers at the fleeing troops. *Stop, you cowards! Go back and fight!* But it was no use. The entire army disintegrated, and soldiers and spectators alike stampeded back to Washington. What had caused this? A premature advance? Raw volunteers? No, Chandler exclaimed, stupid generalship had brought about the Bull Run fiasco. But he admonished a grieving Lincoln to stand firm, admonished him to call up 400,000 more troops and show "the rebels that the government is not discouraged a whit, but is just beginning to get mad."

After Chandler came more witnesses to recount what had happened at Bull Run that day. Lincoln stayed up all night listening to their stories. Never had he felt so hurt, so shaken. At dawn, with rain falling in torrents outside, Lincoln stood at a White House window as remnants of McDowell's army staggered in from Virginia. He saw soldiers move like wounded phantoms through the fog and rain, until they collapsed on lawns and sidewalks along Pennsylvania Avenue. He saw nurses circulate among them, tending to their wounds and handing out cups of steaming coffee. Some men were barefoot, others were wrapped in blankets, their faces obscured in the fog.

All that day Washington waited in dazed panic, expecting a full rebel attack to come at any time. But the Confederates were as disorganized in victory as the federals were in defeat and could not mount a follow-up assault. By July 23 the Union army had regrouped and retrenched on Arlington Heights, and the danger to Washington passed.

For Lincoln, for the Union at large, Bull Run was a shocking and sobering defeat. The Picture Book War was over now, and so was the naïve optimism that had prevailed in Washington since the late spring. With more than 500 Union men killed in action and over 2,600 wounded and missing, the war was real now—profoundly real. And it wasn't going to end in any ninety days either, for the rebel army was better organized and more determined than Union commanders had ever expected. Though Bull Run was really a battle of errors on both sides, the rebels committed fewer blunders and had better leadership when it counted. With Southerners exulting in their "great victory" and vowing to fight harder than ever, Union men had no idea how long the conflict would last now.

There were recriminations, of course, as newspapers and politicians bemoaned Bull Run as a national disgrace and searched about for scapegoats. Republicans and Democrats blamed one another, the New York Herald blamed the abolitionists, the regular army blamed the volunteers, Chandler still blamed imbecilic officers, and just about everyone blamed the man at the White House windows. Yes, the final responsibility for Bull Run came to rest on the President himself. General Scott even hinted that Lincoln had made him fight against his better judgment, and said he should be removed as General in Chief for letting the army advance. On the other hand, Lyman Trumbull asserted that Lincoln was not nearly decisive enough, that in fact he was a weak and irresolute President.

Lincoln's confidence was badly shaken, but he was trying desperately to make up for Bull Run. In the painful days that followed that debacle, he did his best to formulate a coherent military strategy to subdue the rebels. First he wanted to tighten the blockade (as Scott had suggested earlier) and replace short-term with long-term volunteers. Then, with the whole military picture in mind, he proposed three coordinated movements to seize strategic points in both the Eastern and Western theaters: the Union army in Virginia would strike once again at Manassas, a second force would push down the Mississippi toward Memphis, and a third, acting in concert with the Mississippi expedition, would move out of Cincinnati into east Tennessee, an area brimming with Union sympathizers Lincoln hoped to liberate.

Now to get these operations in motion before autumn set in. The details of the Tennessee offensive still had to be worked out. But John Charles Frémont, whom Lincoln had put in charge of the Western Department with headquarters in Saint Louis, would raise an army there and drive against Memphis. The former Presidential candidate was a popular and powerful figure in the Republican party, with a lot of liberal friends in and out of Congress. Forty-eight now, with gray hair and a musical voice, Frémont was a tempestuous, independent man and an untested army commander. But Lincoln hoped he would prove himself and assured Frémont of the President's faith in him. In fact, Lincoln gave him carte blanche in running his department, told him to use his judgment and do the best he could in getting up the Mississippi River expedition.

To carry out the second offensive at Manassas, Lincoln demoted McDowell and ordered young George B. McClellan from Ohio to take field command of what was now called the Army of the Potomac. Back in May, McClellan had cleared the rebels out of western Virginia with a force of Ohio volunteers, and both Scott and Chase had brought him to Lincoln's attention. A bright, conceited officer with a fondness for chewing tobacco, McClellan seemed to personify the American success story. Born in Philadelphia and raised in material comfort, he attended West Point, where he graduated second in his class, and then served in the Mexican War as an exemplary engineer. In the 1850s he observed the Crimean War for the American army and developed such expertise in the art of war that he wrote a manual about it. Then he became disillusioned with army life and resigned to enter the railroad business. By the close of 1860 he was president of an Ohio railroad firm with a salary of ten thousand dollars a year and was married to Ellen Marcy, daughter of a career army officer. Politically, McClellan was a staunch Douglas Democrat who sympathized with Southern slavery. Then came the war, the flare of patriotism, the rush of volunteers. Swept up in all the military excitement, McClellan re-enlisted in the army and rose rapidly to major general in command of Ohio's volunteers. When Lincoln called him to Washington, McClellan was only thirty-five years old.

The general reached Washington on July 26 and the next day Lincoln summoned him to the White House to look him over. He was stocky and stiff, spoke in a "quick, clear, low voice," and stood about five feet seven, with red hair, a red mustache, and a rather boyish face. He was fiercely proud, fiercely intense about his role in the war. Already official Washington was swooning over him, for he seemed the very picture of what a general should be, with his military bearing, his impeccable military good looks.

Frankly, though, Lincoln had mixed feelings about his brash young general. Was he mature enough, experienced enough, to command a large field army? Lincoln liked McClellan's self-confidence, but there was something about the general—a cocky streak, a tendency to embellish things—that bothered Lincoln some. Still, he wanted to believe the best of McClellan— wanted him to win brilliant conquests. And there was no doubting his knowledge of the theory of war, as Scott pointed out.

Later that day, Lincoln asked McClellan to attend a Cabinet meeting, but neglected to invite Scott as well—an oversight that irritated the old general. In Cabinet Lincoln told McClellan: General, it is up to you to save the country. Though he didn't think much of Lincoln, McClellan beamed and basked in all the attention, assuring the President that he could be counted on. From a White House window, Lincoln watched him gallop away with his aides pounding after him, a superb horseman riding with confident ease toward the jungle of tents along the Potomac.

As the days passed, Lincoln made frequent visits to the army camp and was impressed with the infectious and spectacular zeal with which McClellan set about reorganizing his forces. When he assumed command, McClellan pointed out, there was no army at all, "Merely a collection of regiments cowering on the bank of the Potomac." But he was now whipping the Army of the Potomac into an efficient fighting machine that would surely reclaim the glory lost at Bull Run. As McClellan drilled his troops, little cavalcades of reporters and politicians followed Lincoln out to pay him court. "I find myself in a new and strange position here," McClellan wrote his wife. "Presidt., Cabinet, Genl. Scott and all deferring to me." By some strange miracle, he mused, "I seem to have become *the* power of the land. I almost think that were I to win some small success now, I could become Dictator or anything else that might please me—but nothing of that kind would please me,— *therefore I won't* be dictator. Admirable self denial!" Then more seriously: "It is an immense task that I have on my hands, but I believe I can accomplish it."

Back at the White House, Lincoln told himself that he'd made sound military decisions: while old Scott remained as General in Chief, McClellan in the East and Frémont in the West would lead their armies off to victories.

Yet Lincoln was sinking into a bad depression, worn down with worries about the war and with doubts about himself. Nicolay said it was impossible to describe "the strain of intellect and the anguish of soul he endured." So many people were berating him as unfit for his job, contending that the war and the Presidency were plainly too much for him. Horace Greeley wrote that "You are not considered a great man" and urged him to avoid any more catastrophes like Bull Run and sign an armistice with the rebels on their own terms. Greeley later changed his mind and enjoined the administration to fight on, but his vacillations were symptomatic of a confused and divided people whose certitudes had crumbled like sand. And in their woe, men lashed out at the President, the target in the center of the storm.

One evening Lincoln had a private talk with Orville Browning, who was filling out Douglas's term in the Senate. Lincoln confessed that he was despondent and "not at all hopeful" about his future. Browning tried to cheer him, insisting that he did himself a great injustice in feeling as he did. Lincoln had control of his fortunes and was in a position to make his one of the most memorable names in history. Lincoln must hold the reins of the government with a steady hand, "must be firm, earnest, and, if need be, even inexorable."

In August, everything Lincoln had tried to build seemed coming unhinged. He'd hoped that his armies in the East and West would fight in concert. But McClellan announced that the Army of the Potomac wasn't ready for battle, and the Western Department was in chaos. No sooner had Frémont arrived in Saint Louis than he began bombarding Washington with shrill complaints. He argued that there was no way he could attack Memphis. In fact, unless the government sent him reinforcements at once, he could not even hold Missouri. Not only were people there divided in their loyalties and his own army undermanned and mutinous, but a large rebel force had invaded the state. *I'm about to be overwhelmed,* he wired Washington. *Why do you not send me reinforcements?* To aggravate matters, Frémont and Frank Blair, Jr., were tangled in a hot and destructive feud for control of Missouri politics—something Lincoln evidently knew little about. All he knew was that Blair wrote a procession of letters—passed along by brother Monty—which accused Frémont of a myriad of evils. Frémont was hysterical and rash, his command rife with corruption, his hold over the troops rapidly disintegrating.

Blair thought him utterly unfit to run his department.

In late August came more alarming news. When the rebels whipped a Union force in southwestern Missouri and advanced northward, Frémont issued a proclamation to meet the crisis. Dated August 30, the edict placed Missouri under martial law and ordered that the slaves of rebels there be seized and "declared freemen."

Lincoln, of course, was stunned. He hadn't authorized Frémont's proclamation—and in any case it went far beyond the confiscation act, which applied only to slaves employed in the rebel war effort. Predictably, the proclamation brought dangerous rumblings from the loyal border, where Unionists interpreted it as an official emancipation act. Speed wrote Lincoln that Frémont's "foolish" edict would destroy the Union party in Kentucky and incite a slave rebellion. Unless the President revoked Frémont's proclamation, others warned, "Kentucky is gone over the mill dam."

On the other hand, abolitionists and liberal Republicans cheered the proclamation and hailed Frémont as a Union hero. Lincoln was concerned about liberal Republican opinion, but he was even more concerned about the border. So on September 2 he wrote Frémont "in a spirit of caution and not of censure," explained that his proclamation would "alarm our Southern Union friends" and perhaps ruin the Union cause in Kentucky, and asked that Frémont modify his proclamation so that it accorded with the confiscation act. At the same time, Lincoln dispatched Monty Blair and a succession of army officers to talk with Frémont and check on his troubled department.

As Blair traveled west, Frémont's wife, Jesse, raced east to Washington to defend the general's position and policy. At the White House, Lincoln ushered Jesse Frémont into the Red Parlor, but there was such hostility between them that he did not even offer her a seat. She handed him a letter from her husband, and Lincoln read it under the chandelier. Frémont declared that he would not "change or shade" anything in his proclamation. If the President desired to amend the section on emancipation, he must "openly direct me to make the correction." Lincoln folded the letter and turned back to Mrs. Frémont. "I have written to the General and he knows what I want done." Yes, she replied, but we think it best that I present his views since our enemies enjoy your personal confidence. Lincoln did not know what she was talking about. Enemies? What enemies? Though she did not name the Blairs, she was quite sure Lincoln knew whom she meant. She went on: My husband understands the English feeling for emancipation, and both of us think his proclamation will win us friends abroad.

"You are quite a female politician," Lincoln said sarcastically. He ignored her sneer, dismissed her argument, and told her bluntly that this was "a war for a great national idea, the Union, and that General Frémont should not have dragged the negro into it." Bristling, she warned Lincoln that it would be hard on him if he opposed her husband. If he did, she asserted, then Frémont would "set up for himself."

What a woman! She "taxed me so violently," Lincoln told Hay afterward, "that I had to exercise all the awkward tact I have to avoid quarrelling with her." But he hadn't heard the last of Mrs. Frémont. When she found out about Frank Blair's letters and Monty's trip west, she flew into a rage. She accused Lincoln of dispatching Monty Blair to spy on her husband and sent the President a note demanding that he show her Frank Blair's letters—the ones about corruption in Frémont's department. Lincoln could scarcely believe her impertinence. Of course he would not let her look at his correspondence. Moreover, he hadn't sent Monty Blair or anyone else out to spy on Frémont. Lincoln was trying to help the general, not undermine him.

What a nightmare this had become. Still, Frémont had to be controlled, his proclamation modified. On September 11 Lincoln wrote him again and conceded that the general was on the ground and perhaps understood the military situation better than Lincoln did. But political decisions must be made by the President and Congress, not by commanders in the field, and Frémont's proclamation was an unauthorized political act. Therefore, as Frémont requested, Lincoln "very cheerfully" ordered him to revise the slave provision so that it conformed to the confiscation act. And since Frémont had asked that this be an open order, Lincoln sent a copy to the press.

The reaction was swift and predictable. Border Unionists applauded Lincoln's order and contended that he'd saved their states for the Union a second time. But liberal Republicans were dismayed. Sumner moaned that Lincoln "is now a dictator," and Wade sneered that Lincoln's views on slavery "could only come of one, born of 'poor white trash' and educated in a slave State." If the liberals were disenchanted, the abolitionists were furious. They derided Lincoln's order as "pigheaded stupidity" and vowed to launch a public campaign to reverse his "proslavery" policies. In Boston and other cities, they established Emancipation Leagues and set about distributing literature to arouse public opinion and convert this into a war for slave liberation.

In all the uproar, Lincoln tried to vindicate himself. He argued that Frémont had no right to free all the slaves in his department. that such an act didn't come "within the range of *military* law, or necessity." Had Lincoln allowed the proclamation to stand unchanged, Kentucky, Maryland, and Missouri itself might have gone over to the rebels and the result would have been a Union disaster. Moreover, Lincoln s critics persisted in misunderstanding what he'd actually done. He hadn't *revoked* Frémont's order about the slaves, but had simply made it coincide with the Congressional confiscation act . . . with the very law his liberal critics championed. Frémont was still free to seize slaves who served the rebel army as foragers, laborers, and the like.

Still, the Frémont controversy left Lincoln more depressed than ever. He told Nicolay that things were dismal everywhere. Not only was Frémont "ready to rebel," but the Treasury was empty and the War Department a mess. In fact, Cameron himself was becoming an impossible problem, was "utterly ignorant and regardless of the course of things," was incapable of "organizing details or conceiving and advising general plans." On top of that, Cameron was "selfish" and "openly discourteous" to Lincoln himself. And then there was the military situation. The rebels had infested Kentucky and "virtually seized" Missouri. October was here and no Union army in either theater was prepared to advance. And the opinion of almost everyone —of Chase, Bates, the Blairs, Browning, and Adjutant General Lorenzo Thomas—was "that everything in the West, military & financial, is in hopeless confusion."

Yet Lincoln did little to rectify such conditions. He left Frémont in command in the West. He left Cameron in the War Department, even though there were whispers of corruption there as well. It was as though Lincoln rejected his own instincts, covered his eyes, and waited for inept men to reform themselves so that he would not have to go through the pain of sacking them. Meanwhile a Congressional subcommittee—followed later by a Commission on War Claims—investigated the Western Department and found that much of what Blair charged was true. Believing that Lincoln had authorized him to do anything he liked, Frémont had indeed squandered public funds. He'd surrounded himself with a retinue of California cronies who made extortionate profits by securing army contracts without competitive bidding, which was illegal. In addition, Frémont had given favorites "the most stupendous contracts" for railroad cars, horses, mules, tents, and other equipment, much of it inferior, some of it worthless.

At last, when Adjutant General Thomas filed a similar report, Lincoln had no choice but to relieve Frémont of command or risk a public scandal. He did so in late October, 1861. Yet Lincoln did not think Frémont personally dishonest. "His cardinal mistake," the President said, "is that he isolates himself, & allows nobody to see him; and by which he does not know what is going on in the very matter he is dealing with." Yet Lincoln seemed reluctant to admit that the waste, corruption, and extravagance in the West were characteristic of the entire War Department.

Contrary to what a lot of writers have said, Frémont's dismissal did not cause an irreparable split between Lincoln and Republican liberals. True, the liberals were bitter about the whole Missouri episode, but many of them admitted that the famous Pathfinder was a maladroit administrator. Also, they were pleased when Lincoln put General David Hunter—another liberal commander—in charge of the troubled Western Department. Lincoln did so in part to appease them.

On October 20, a few days before he relieved Frémont, Lincoln sat on the White House lawn with his old friend Edward Baker, now a colonel in one of McClellan's regiments. It was a golden autumn day and the leaves and foliage were ablaze with colors. Willie played in some leaves nearby, calling out from time to time for his father and Baker to watch. Lincoln leaned against a tree and Baker stretched out on the grass, both of them talking earnestly about the war. The conflict had now entered its sixth month and the Union had won no victories. McClellan had drilled the Potomac Army into a potent fighting force of some 75,000 men, and the pressure from the press and the country was tremendous for him to fight. ("Don't let them hurry me, is all I ask," the general had said at one point, and Lincoln promised him that "You shall have your own way.") Lincoln was glad at least that tomorrow McClellan would reconnoiter rebel entrenchments at Leesburg, some forty miles up the Potomac. Baker was to take part in the operation and was pretty excited about it. As they talked, their conversation drifted back to the old days, and they reminisced about Whig politics and Sangamon County. Presently they rose and Willie ran over. Baker lifted him up and kissed the boy, remarking that he'd known Willie since he was a baby. Mary came out and gave Baker some flowers, and the four of them chatted intimately on the White House lawn. Then Baker mounted and rode out of the White House gates, and Lincoln watched him until he disappeared from view.

The next day Lincoln was at McClellan's headquarters when telegraph dispatches came in about the Leesburg operation. An army lieutenant asked Lincoln into an inner office and broke the news to him. There had been a fierce skirmish at Ball's Bluff near the town. Baker had led an attack against a rebel position there, and enemy artillery had cut him down. Colonel Baker, the lieutenant said, was dead.

Lincoln left the office with his hands pressed hard against his chest, stumbled when he stepped out into the street, and returned to the White House without a word to anyone. Ellsworth's death had hurt, but Baker was an old, old friend and his loss, Lincoln said, "smote like a whirlwind." He told Mary and the boys about Baker and then retired to his office. That night, with rain pelting against his windows, Lincoln was still there, pacing back and forth and recalling his visit with Baker only yesterday. And in another part of the White House, Willie Lincoln wrote a poem on tear-stained pages, a poem about Baker that was mature and moving and deeply felt. The poem appeared later in the *Washington National Republican*, whose readers were astonished that a ten-year-old boy could pen such lines.

> *There was no patriot like Baker,*
> *So noble and so true:*
> *He fell as a soldier on the field,*
> *His face to the sky of blue. . . .*

> *No squeamish notions filled his breast,*
> *The* Union *was his theme,*
> "No surrender and no compromise,"
> *His day thought and night's dream.*

> *His country has her part to play,*
> *To'rds those he left behind,*
> *His widow and his children all—*
> *She must always keep in mind.*

Beyond the White House, a terrible outcry rose over Ball's Bluff, as newsmen and politicians alike called it another Union fiasco and bewailed the loss of patriots like Baker. Never mind that Baker had led a reckless charge in what was supposed to be a reconnaissance mission, as McClellan kept pointing out. No, administration critics saw Ball's Bluff as symptomatic of incompetent leadership from the White House to the battlefield, and several Republicans demanded that Baker's immediate commander be held account-able for his death.

A few days after Ball's Bluff, Wade, Chandler, and Trumbull came to see Lincoln, and they were angry and grim. The public demanded a battle and a victory, they informed the President, and would not tolerate any more debacles, any further drift and delay. For over two months McClellan had been drilling and organizing his army. Why didn't he attack the rebel force at Manassas? Why wouldn't he fight? Was it because he was a coward? Because he was a Democrat who sympathized with Southerners and con-doned slavery? Lincoln defended McClellan's "deliberateness," but the sena-tors were in no mood for his excuses. In truth they had their doubts about Lincoln, too. Away from the White House, Wade called Lincoln "a fool" and Chandler thought him as "weak" and "inefficient" as Trumbull did. And they weren't liberal voices howling in the wilderness either. By the fall, nearly all Congressional Republicans believed Lincoln too inexperienced for his job—and nearly all were becoming exasperated with George McClellan.

With Wade and his colleagues clamoring for a battle, Lincoln had a confidential discussion with McClellan and warned him that the senators reflected a public demand which neither of them could ignore. McClellan became quite excited. He had everything at stake in this. He was preparing the army for a mighty triumph, but his men were not yet ready for battle. Moreover, he was certain that the rebels had over 130,000 men dug in at Manassas. Before he could advance, he must have 273,000 men. Only then could he attack Manassas, take Richmond, and win the war. Only then could he carry this *"en grand."* Furthermore, he resented the meddling of politicians who knew nothing about warfare—about assault and supply—and who were bent on forcing him into a premature offensive. Did the President want another rout? another Bull Run? Well, of course, Lincoln said, "You must not fight until you are ready." Still, McClellan gave the impression that he *would* advance, maybe late in November. "I have a good notion," Lincoln said, "to go out with you and stand or fall with the battle."

Meanwhile another problem had broken out that vexed the President. Scott and McClellan had clashed almost from the outset, and the friction between them had developed into a bristling animosity. For his part, the old general thought McClellan grossly exaggerated rebel troop strength at Manassas and insisted that he had the superior force. McClellan, on the other hand, told Lincoln that Scott was his greatest "antagonist"—even worse than the politicians—and argued that their ideas were so widely divergent that cooperation between them was impossible. Besides, Scott was so old that he constantly fell asleep at his desk. How could a man like that know what was going on?

And Scott, carrying his body around with a huge and painful effort, complained to Lincoln about McClellan's impudence. How dare he go over Scott's head and deal directly with the President himself? How dare Lincoln invite this junior officer to Cabinet meetings and snub the General in Chief? But then Scott's anger would trail off into a sigh. He was really getting too old for this job—too old to cope with the responsibilities, the headaches. But when he offered to retire, Lincoln balked at letting him go. Scott was such an institution in Washington, had such a long and distinguished career, that Lincoln couldn't bring himself to accept Scott's offer.

McClellan, of course, was infuriated. "I am daily more disgusted with this administration—perfectly sick of it," he wrote his wife. "There are some of the greatest geese in the Cabinet I have ever seen—enough to task the patience of Job." It left the general sick and deeply troubled when he saw "the weakness and unfitness of the poor beings who control the destinies of this great country."

As it happened, most of the Cabinet believed that Scott could *"command no longer"* and evidently favored McClellan as his successor. Finally Lincoln came around to their view. On November 1 he officially accepted Scott's resignation and designated young McClellan as General in Chief, though without Scott's rank of lieutenant general. So, Lincoln told McClellan, you are now supreme commander. "Draw on me for all the sense I have, and all the information." But he cautioned McClellan that the supreme command "will entail a vast labor upon you."

"I can do it all," the general replied.

During the next several weeks, Lincoln and Seward held frequent strategy sessions with McClellan, either at army headquarters or at McClellan's house in Washington. At first the General in Chief simply tolerated Lincoln and his anecdotes ("Isn't he a rare bird?" McClellan said). But in time he came to resent Lincoln's intrusions with a passion. Who the hell did this amateur think he was, coming out and bothering the General in Chief in his preparations? McClellan began confiding in his friend Edwin Stanton—the same Stanton who, in the McCormick Reaper case years before, had denigrated Lincoln as that "giraffe" from Illinois. Now McClellan took to hiding out at Stanton's house to escape "browsing" Presidents and other meddlesome politicians. And one night in mid November he even gave Lincoln an incredible snub. Accompanied by Seward and Hay, Lincoln called at McClellan's house, only to find that he was attending a wedding. So they waited for him in the parlor. When the general returned, he refused to see the President and the Secretary of State, walked passed the very room where they were sitting, and went upstairs to bed. Hay, transported with rage, dubbed this "unparalleled insolence" and "a portent of evil" to come. But on the way back to the White House, Lincoln remarked that at times like these it was best not to worry about etiquette or personal dignity.

But deep down Lincoln was aggravated. After that he stopped calling on McClellan and summoned him to the White House instead. Hay and Nicolay were overjoyed.

As November passed, Lincoln and McClellan at last worked out a strategic plan for the Eastern and Western theaters, a plan that entailed the kind of concerted action Lincoln had suggested after the disaster at Bull Run. They divided the chaotic Western Department into two separate commands, appointing Henry Halleck to run the Department of the Missouri and Don Carlos Buell to head the Department of the Ohio with headquarters at Louisville. As Halleck pushed down the Mississippi, Buell was to drive into east Tennessee, liberate the Unionists there, and cooperate with McClellan once he began campaigning in Virginia.

Inevitably there were delays. Buell, a harsh, aloof soldier with a penchant for meticulous preparation, informed Washington that it was almost impossible for him to invade east Tennessee, since railroads there were inadequate and his logistics problems insuperable. This meant that Buell couldn't help McClellan in Virginia either, that the General in Chief was left with only the Army of the Potomac to fight all those hordes of Confederates he believed were entrenched at Manassas. Therefore McClellan postponed his offensive as well, announcing once again that the Potomac Army was not ready to advance.

Lincoln tried to rationalize McClellan's behavior, told himself that the general was a professional soldier who knew what he was doing. Moreover, McClellan sent word that he now had another plan in mind, a daring operation nobody in North or South had anticipated. But he did not reveal what the plan involved.

No, Lincoln said, he would not issue an emancipation edict. He must adhere to his border-state policy, because to alienate the border was to lose the war, the Union, everything. So he repeatedly told liberal Republicans in the fall and winter of 1861. Yet their arguments in favor of emancipation struck a responsive chord in him. With the war dragging on, they insisted that slavery must be attacked in order to weaken the Confederate ability to fight. And they pointed out that more and more Congressional Republicans were coming around to their view (in December, in fact, House Republicans would vote down the Crittenden resolution which demanded that state institutions be left alone in the conflict). Moreover, there was the inescapable fact that slavery had caused the war, was the reason why the rebel states had left the Union, and was now the cornerstone of the Confederacy. It was absurd, liberal Republicans contended, to fight a war without removing the thing that had brought it about. Should Lincoln restore the Union with slavery preserved, Southerners would just secede again and start another war to defend the peculiar institution, so that the present struggle would have accomplished nothing, nothing at all. In addition, as Sumner said, emancipation might well prevent Great Britain from recognizing the Confederacy as a nation and intervening in the American conflict. With England's strong antislavery tradition (England had abolished slavery in the British Empire back in the 1830s), Sumner insisted that she would never help the rebels if Lincoln made emancipation a Union war objective. As it was, though, Lincoln's hands-off policy convinced the British that slavery was not involved in the war and invited them to befriend and assist the rebels.

At first glance, Lincoln seemed doomed no matter what he did. Yet there might be a way out, a way to accommodate an emancipation scheme with his border policy and still be consistent with all Republicans had ever promised about slavery as an institution. What he had in mind was a gradual, compensated emancipation scheme to be financed by the federal government and implemented voluntarily in the border states. The more Lincoln thought about the idea, the more excited he became. In late November, he even drafted a model plan to be tested in Delaware, with its relatively small and harmless slave population. According to Lincoln's plan, Delaware would gradually free her blacks over the next thirty years, and the federal government, shelling out five hundred dollars for every slave held there in 1860, would compensate Delaware slave owners for their losses. Just before Congress convened in December, Lincoln confided in Browning that he was contemplating a similar plan for the other border slave states and that it would cost the government about one-third what it had thus far expended on the rebellion. After the border states liberated their blacks, Lincoln said, the federal government should try to colonize them in central America or Liberia. Without colonization, Lincoln understood only too well, most Northern whites would never accept emancipation, even if it were carried out by the states.

Yes, his plan seemed the best way out of an impossible dilemma. Should the loyal border adopt a gradual and compensated program, a Presidential decree would never be necessary. Once the rebellion was suppressed, the plan could be applied to the other Southern states and their liberated blacks could be colonized as well. Thus by a slow and salubrious process would the cause of the civil war be removed and the future of the Union guaranteed.

In his annual message to Congress, sent to Capitol Hill on December 3, Lincoln urged that federal funds be appropriated to colonize all slaves freed by the confiscation act or by the states. But he hinted that he would not issue an emancipation proclamation, because he did not want the war degenerating into "a violent and remorseless revolutionary struggle." In every case, he was striving "to keep the integrity of the Union prominent as the primary object of the contest." While the government must use all "indispensable means" to crush the rebellion, Lincoln added, "we should not be in haste to determine that radical and extreme measures, which may reach the loyal as well as the disloyal, are indispensable."

In saying this, Lincoln had something in mind besides emancipation. By now, Cameron and several Congressional Republicans were prodding him to recruit Negro soldiers. But out of deference to the loyal border, he still refused to do so. He would let the navy enlist free Negroes and fugitive slaves alike, employing them on warships as well as naval installations, because the navy was not a physical presence in the border. But the army was such a presence, and Lincoln feared that Southern loyalists would turn against the government if it resorted to armed Negro troops. In Cabinet, he made it explicitly clear that he would use blacks as foragers and laborers for the army, but not as soldiers. And most of the Cabinet agreed with him.

Cameron, however, cheerfully disregarded Lincoln's policy. In his annual report of 1861, released simultaneously with Lincoln's Congressional message, the Secretary of War publicly contradicted his chief and declared that the slaves should be emancipated and enlisted in a Negro army, observing that this would be an excellent way to punish "rebel traitors." When Lincoln found out about the report, he tried desperately to retrieve it and ordered Cameron to delete the unauthorized statements about Negroes. But both versions found their way into the press—some newspapers even published them side by side—so that the Lincoln administration seemed more confused than ever.

Lincoln was "greatly aggrieved" about the entire affair. Why had Cameron released an unauthorized report? What had he to gain? Maybe he sincerely believed in emancipation and the use of Negro fighting men and hoped to pressure Lincoln into embracing his views. On the other hand, there were reports of scandal in Cameron's office—Lincoln was aware of them—and maybe Cameron was maneuvering to protect his job. He'd been consorting with Republican liberals of late; perhaps his report was an attempt to get their support, so that Lincoln would be reluctant to fire him. Whatever the reasons, Lincoln conceded that Cameron's behavior placed a "severe strain" on their relationship. Yet Lincoln still kept him in office.

Meanwhile, abolitionists and Republican liberals openly condemned Lincoln's stand against federal emancipation and exerted all their powers of persuasion to change his mind. Chief among them was Charles Sumner, who visited Lincoln regularly and beseeched him to stop protecting the very institution that had caused the rebellion. One day, as Lincoln sat in the Senate galleries, Sumner gave an impassioned eulogy to Edward Baker, who for a time had served as senator from Oregon. Gesticulating dramatically, Sumner described how Baker had died at Ball's Bluff and then—looking straight at Lincoln now—Sumner cried that slavery was "the murderer of our dead Senator." A correspondent said that Lincoln started violently at Sumner's remark, quite as though he had been stabbed.

When Sumner learned that Lincoln was maturing a program of gradual and compensated emancipation, he hurried to the White House and implored Lincoln to introduce it. "I want you to make Congress a New Year's present of your plan," the senator said. But Lincoln's Delaware scheme, like his armies, had bogged down in delays. For one thing, his Kentucky friends disliked the whole idea and urged him to shelve the project. For another, nobody in Delaware seemed impressed with it either, even though Lincoln was promoting it there himself. In fact, no bill incorporating his proposals ever emerged from the Delaware legislature.

Though his state program was stalled, Lincoln assured Sumner of his firm antislavery convictions. And in a long and candid discussion about emancipation, he remarked that he really wasn't very far behind the senator himself. "Mr. Sumner," Lincoln said, "the only difference between you and me on this subject is a difference of a month or six weeks in time."

"Mr. President," Sumner replied, "if that is the only difference between us, I will not say another word to you about it till the longest time you name has passed by."

In mid December, a crisis was brewing in foreign affairs, as an incident in the Bahama channel, north of Cuba, threatened to escalate into war between Great Britain and the Union. Back in November, an American warship under the command of Charles Wilkes had intercepted a British vessel called the *Trent,* seized a couple of rebel commissioners on board, and hauled them to Boston as contraband. Overnight, Wilkes and his crew became national heroes, with newspapers ballyhooing the "Trent affair" and politicians roaring with anti-British delight. Initially Lincoln, too, was pleased, though he predicted that the captured "traitors" would turn out to be "white elephants," more difficult to hold than to set free.

The British, of course, were enraged, and indignation swelled through Parliament and the press against this insult to the British flag, this egregious violation of international law. "You may stand for this," Prime Minister Palmerston thundered in Cabinet, "but damned if I will!" Whereupon the British government ordered its warships to prepare for hostilities and dispatched eight thousand troops to Canada. Then the Cabinet sent Washington an official demand, which Lord Lyons handed to Seward on December 19:

the United States must release the prisoners forthwith and apologize for violating British neutrality. If the U.S. refused, Lyons had private instructions to sever diplomatic relations and return to London.

The British reaction caused great consternation in Washington. Sumner hurried to the White House with disturbing correspondence from his English friends, who wrote that a collision seemed imminent. What did Sumner think? Sumner thought a war with England would be a catastrophe and would all but ensure Confederate independence. And Seward for once agreed with him. Though Seward was famous for his anti-British posturings (the English thought him "an ogre fully resolved to eat all Englishmen raw"), Seward consulted with McClellan about the *Trent* affair, and McClellan warned him flatly that the Union could not fight Great Britain and the Confederacy at the same time. Seward admitted that "if the matter took that turn," the prisoners would have to be liberated.

Lincoln, too, wanted to avoid a violent showdown with England; he, too, realized that the United States could fight only one war at a time. The trouble was that too many Americans had cheered the seizure, and Lincoln had to be extremely careful in how he backed down. At all events, he must not sacrifice the national honor, must not appear to crumble under British pressure, for that would be a disastrous blow to an administration already accused of spineless incompetence. So how to resolve the dilemma without violence? Look, he told Sumner, let's cut through all this diplomatic protocol. You bring Lord Lyons here for a personal talk with me, and I'll show him I'm for working this thing out peacefully. But Sumner rejected this as improper. The best solution, he counseled, was to submit the question to a country like Prussia for arbitration.

Lincoln was ready to arbitrate if diplomacy failed. But as it happened, the British had cooled down and were equally anxious to settle the question without bloodshed. Accordingly, they dropped the demand for an apology, so long as the United States assured them that Wilkes had acted without authority. But they still insisted that the rebel prisoners be released. So on Christmas Day Lincoln called a full Cabinet meeting, with Sumner in attendance, to decide what to do about the captured rebels. As they belabored the need to save American honor, Bates noted that Lincoln and several Cabinet members seemed reluctant to admit the obvious truth that Wilkes was simply in the wrong. By international law, he had the right to search the *Trent* and confiscate rebel war supplies, but not to impress people from the decks of a neutral ship. In fact, in the years that led to the War of 1812, the United States had vehemently protested similar tactics on the part of the British navy.

In the end, though, everybody at the Christmas meeting agreed that Sumner was right, that the Union could not win "in a super added war with

England." So they yielded to necessity and elected to turn the prisoners loose. "It was a pretty bitter pill to swallow," Lincoln said later, but he hoped the U.S. could get even with Britain once the rebellion was over. Then "we could call her to account for all the embarrassments she had inflicted upon us."

So the *Trent* affair ended peacefully. In Congress, several patriots seethed at Lincoln's unmanly "capitulation" to Great Britain, but Sumner defended the administration with a masterful discourse on international law. Seward meanwhile drafted a long reply to the British, with clever touches intended for home consumption. He pointed out that Wilkes had indeed acted without authorization (therefore the U.S. need not apologize for the seizure). Furthermore, Seward mused, Wilkes's gravest error was in duplicating the very thing the British had done back before the War of 1812—boarding neutral vessels and seizing people there. Now, in demanding that neutral rights be honored on the high seas, Britain was endorsing "an old, honored, and cherished" American principle. Thus the U.S. "cheerfully" released the prisoners and sent them on their way.

However galling it might have been to give them up, Lincoln could be glad about one thing: Great Britain had passed up a chance to enter the war and make the Confederacy an independent nation. For once reason had prevailed . . . thanks to Fate perhaps, or God. Still, 1861 was drawing to a close and the war continued, with a storm of troubles blowing up both in and out of the White House. Over eight months had passed since he'd made his stand at Sumter, and now it was the Christmas season, the wind at his office windows carried a breath of winter, and none of his armies was fighting anywhere. In the midst of the holiday festivities—the Christmas gifts and laughing faces and organ music and White House serenades—a brooding Lincoln had little to be joyful about.

Hadn't Lincoln noticed? Mary wondered. Had he been so busy that he hadn't heard all the commotion and hammering in the White House? By Christmas time, Mary had renovated the entire mansion except for Lincoln's office. When they had first moved in, the White House was a shambles—the walls smudged, the furniture shoddy and broken down, the carpets stained with tobacco juice. Mary, of course, was not about to live and entertain in such a dilapidated place. So she obtained twenty thousand dollars from Congress to repair and refurbish the Executive Mansion and vowed to transform it into a showplace for the winter social season—and demonstrate to the socialites who'd snubbed her that she had exquisite taste.

Through the spring, summer, and fall, Mary traveled back and forth to New York and Philadelphia on shopping expeditions. She bought imported drapes, custom-made carpets, ornately carved furniture, glittering vases, and a seven-hundred-piece set of Bohemian cut glass, not to mention a $1,100 set of china emblazoned with the national emblem. She spent money with abandon, picking up anything for the White House that caught her fancy and acquiring boxes of gloves and assortments of gowns, cashmeres, and shawls, which she charged to her personal account. Back in Washington, she watched with a critical eye as workmen installed her purchases and servants set about scrubbing windows, painting walls, and putting new gaslights in the rooms. By the Christmas season, Mary had completed her renovations and Robert had come home for the holidays. She was "having *so much bliss,*" being with her family in a thoroughly modernized White House, a lovely mansion befitting the most lavish state entertainments.

As the rounds of receptions and state dinners began in earnest, White House visitors found the mansion truly stunning to behold. They wandered through the rooms admiring the Parisian upholstery, the rich and colorful carpets, the expensive furniture, the French drapes and gold tassels, and the Swiss lace curtains that hung about the windows. And there was the First Lady herself, fiercely fashionable, dressed in elaborately trimmed and long-trained gowns cut low enough to show off her ample bosom, tending to her guests with impeccable social grace. Ben: Perley Poore, the celebrated journalist and Washington observer, declared her the most charming White House hostess since Dolley Madison.

Away from the public functions, though, Mary felt lonely and unfulfilled. Lincoln spent so little time with her in the White House, with the war keeping him in his office late into the nights. Did he appreciate her remodeling of the White House? Yes, of course he did. Did he like the family rooms? She had redecorated them, too, so that they could have comfortable surroundings when he came in late at night, exhausted from his work. Yes, he liked the rooms. As always, she worried about his health, too, worried that he didn't get enough fresh air and sleep. She saw so little of him and came to miss him with an emptiness inside. Bored in the days, she formed a kind of salon in her Blue Room, where she entertained a coterie of male friends. Sumner became a frequent caller and often escorted her to the opera when Lincoln was unable to go. But some of Mary's men friends were rogues and sycophants who used her for their own insidious purposes, who saw that behind her façade of authority was an insecure woman who craved attention and succumbed to flattery. One such fellow was Henry Wikoff, who thrilled Mary with a flow of learned chatter. According to an acquaintance, Wikoff could talk about anything, could "gossip of courts and cabinets, of the *boudoir* and the *salon,* of commerce and the church," of Dickens and Thackeray, of Buchanan and Pierce, "of the North and the South, of the opera and the theater." In reality, though, this charmer was an agent of the *New York Herald* who preyed on Lincoln's vulnerable wife to obtain inside scoops from the White House. When Lincoln discovered the truth about him, he barred Wikoff from ever setting foot in the White House again, and he asked Mary to be more discreet in what she said. And Mary? She just laughed about Wikoff—oh, he was fun for a while—and surrendered herself to other flatterers, mistaking their overtures for genuine affection and companionship.

Her closest confidante, though, was her Negro seamstress, Elizabeth Keckley. A poised, handsome woman in her thirties, "Lizzie" had suffered a great deal in her life, and yet she wasn't bitter. On the contrary, she was one of the most dignified and compassionate women Mary had ever known. Born a slave in Virginia, Lizzie had spent more than twenty-five years in bondage and could speak with firsthand authority about life under the lash. She'd seen Negroes beaten and families broken up, had seen a boy sold away from his mother so that the master could pay for some hogs. By the time she was eighteen, she was living with her master and mistress in North Carolina. She was a pretty girl, with a spirit that aggravated many white people. To break her stubborn pride, the village schoolmaster, "a hard, cruel man," approached her with a whip and ordered her to take her dress down. When she refused, he tore the dress off and flogged her—and then for good measure struck her with a chair. Later another white man raped her and made her his mistress. "I spare the world his name," Lizzie said, but "he persecuted me for four years," and she bore his son. In time, a new master took her out to Missouri, where she married a Negro named James Keckley, who turned out to be "dissipated, and a burden." In 1855, with the help of several kindly whites, she bought her freedom and that of her son. In 1860 she left her husband and came to Washington to live and find employment. Ironically enough, she worked as a seamstress for Jefferson Davis and his wife until they left Washington during the secession crisis. After the war broke out and Lizzie came to the White House, her son was killed in a battle out in Missouri, so that she was alone now.

Lizzie had a strong influence on Mary, gave her rare insight into the cruelties of slavery from a black woman's view, and helped convert Mary to decidedly abolitionist views. Moreover, because Lizzie was a warm and sympathetic woman, Mary confided almost everything to her. In her private chamber, where Lizzie fitted her dresses, Mary talked freely about her money troubles and her terror of poverty (even though as President Lincoln now earned $25,000 a year), about her private debts to New York shops and her fears that Lincoln would find out. She talked about her jealousies, her peeves, her wounds, her need for lovely clothes, her tremendous pride in the way she'd restored the White House. . . .

But Mary's pride turned to panic when all the bills came in. She'd exceeded her appropriation by $6,700, and she was terrified. She knew Lincoln would never approve. In her misery, she called on Benjamin French, Commissioner of Public Buildings, and begged him to plead her case with Lincoln. Tell him, she said in tears, "that it is common to overrun appropriations —tell him how *much* it *costs* to refurnish."

When French interceded in Mary's behalf, Lincoln became furious and refused to cover the excess bills with government funds. "It can never have my approval," Lincoln stormed. "I'll pay it out of my pocket first—it would stink in the nostrils of the American people to have it said the President of the United States had approved a bill over-running an appropriation of $20,000 for *flub dubs* for this damned old house, when the soldiers cannot have blankets."

Congress finally settled Mary's dilemma by burying an extra appropriation in the White House budget for the ensuing year. But what a target Mary became now, as the very people she wanted most to impress—Washington socialites—gossiped about her extravagance. And though she had her public defenders, such as Ben: Perley Poore, she never got over the malicious slander aimed at her, from slurs about her manners and excesses to savage accusations that the First Lady was a rebel sympathizer.

Lincoln detested the way Washington society treated Mary and in their brief intervals together he did what he could to comfort her. And Mary, in turn, strove all the harder to be a loyal, understanding wife to him. Yet there were tensions between them now, brought on not just by the Wikoff episode and the budget fiasco, but by the crushing pressures of the war and the Presidency on them both. Because of the stress they were under, their marriage and their intimacy inevitably suffered.

As the war ground on, the strain on Mary began to take its toll. Increasingly distressed and unsure of herself, she became terribly jealous of Lincoln, so possessive that at public functions she hated to let him out of her sight. Although she could be a flirt around other men, she flew into a rage when other women paid attention to her husband. Of course some did so because they wanted a favor. But others found him genuinely attractive and liked to talk with him. More secure around women than in his younger years (credit Mary and their marriage in good part for this), Lincoln was as open and direct with women as with men. He refused to treat females as ornamental trinkets, as childlike creatures one engaged in mindless prattle. He listened attentively and respected their opinions, and women enjoyed being around him. But Mary could not bear to see him with someone else—especially if she were young and pretty. Mary fussed at him to stay away from other women; she didn't even want him speaking to them. When Lincoln protested that he had to talk with somebody, that he couldn't "stand around like a simpleton, and say nothing," Mary flared at him: "You know well enough, Mr. Lincoln, that I do not approve of your flirtations with silly women, just as if you were a beardless boy, fresh from school."

According to Keckley, Lincoln was perplexed about Mary's jealousies, but as always he avoided a quarrel and simply teased her about them. And the young women at public functions? Lincoln respected Mary's feelings, didn't want to hurt her. But he wasn't going to be a henpecked husband either. At receptions he continued to speak graciously with all the lady guests.

O ver the holiday season of 1861–62, Lincoln reached a painful decision about his troublesome Secretary of War. Not only was Cameron rude, incompetent, and insubordinate (no explanation from him about the Negro troops controversy), but there were Congressional revelations about Cameron's office that scandalized Lincoln and his entire administration.

Congressional suspicions about Cameron had begun as early as July, 1861, when complaints of certain irregularities in War Department spending flooded into Washington. Twice Congress had demanded that Cameron provide specific information on all government contracts awarded since the beginning of Lincoln's administration. But Cameron refused to send anything to Capitol Hill. As a consequence, the House established a special committee on contracts to investigate; and the committee, after hearing testimony and examining war expenditures, produced a 1,109 page indictment of maladministration in Cameron's office. The committee's principal objection was that the Secretary of War and his maze of agents—some unscrupulous, others inept—had ignored competitive bidding and bought exclusively from favorite middlemen and suppliers, many of them "unprincipled and dishonest." Thanks to a combination of inefficiency and fraud, the War Department had purchased huge quantities of rotten blankets, tainted pork, knapsacks that came unglued in the rain, uniforms that fell apart, discarded Austrian muskets, and hundreds of diseased and dying horses—all at exorbitant prices. In one instance, the War Office had sold a lot of condemned Hall carbines for a nominal sum, bought them back at $15 apiece, sold them again at $3.50 apiece, and bought them back again at $22 apiece.

The list of abuses seemed endless. One Boston agent, charging the government a percentage of the contracts he arranged, made $20,000 in one week. Another agent acquired two boats for the War Department at a price of $100,000 each—after the navy had rejected them as unsafe—and one of the vessels sank on its first voyage. Then there were the activities of one of Cameron's own political lieutenants, Alexander Cummings, whom the Secretary had appointed supervisor of army purchases in New York City. By Cummings's own authority, the government spent $21,000 for straw hats and linen pantaloons and bought such "army supplies" as Scotch ale, selected herring, and barreled pickles. In addition, Cummings had contracted for 75,000 pairs of overpriced shoes from a firm that occasionally loaned him money.

The House committee, of course, bemoaned such "prostitution of public confidence to purposes of individual aggrandizement" and castigated Cameron's department for treating Congressional law as "almost a dead letter," for awarding contracts "universally injurious to the government," and for promoting favoritism and "colossal graft." "Public good requires another man in his place," cried the loyal press. "Mr. Lincoln should rise to the dignity of his position, and remove him."

By early January, 1862, Lincoln had concluded that Cameron must go. There was no evidence that he'd enriched himself in the contracts scandals, but it was obvious that if the government intended to win the war, Lincoln had to find a better man to head the War Department and manage the Union's huge armies. Since Cameron had recently expressed a willingness to step down, Lincoln accordingly asked for his resignation and on January 13, 1862, appointed him United States Minister to Russia. On the same day the President named Edwin Stanton—Cameron's chief legal adviser—to run the tangled War Office.

On Capitol Hill, men were glad to get rid of Cameron, and Thaddeus Stevens cracked: "Send word to the Czar to bring in his things of nights." But Congress was not going to let Cameron go without a reprimand: just before he sailed for Russia, the House officially censured him for entrusting men like Cummings with public money and for adopting a policy which flagrantly damaged the public service.

Lincoln, for his part, could have let the matter drop, but he was too honest to conceal his own responsibility in the War Department troubles. In a subsequent message to Congress, the President insisted that he and all other department heads were equally accountable with Cameron "for whatever error, wrong, or fault was committed in the premises." When war broke out, Lincoln explained, Congress was not in session, the capital was threatened with occupation, the nation was about to disintegrate. Therefore Lincoln had met with his Cabinet Secretaries, and they had all decided that they must resort to emergency action or let the government fall. Lincoln described how he'd authorized private individuals—including Cummings himself—to buy war supplies and forward them to Washington. He recounted the other expedients he and the Cabinet had adopted that Sunday of April 21, contending once again that they were absolutely necessary to save the government in those critical days after Fort Sumter. He did not deny that misdeeds had occurred in War Department operations, nor did he claim that the House was wrong in censuring Cameron. But he was not willing, he said, to let that censure fall on Cameron alone.

Meanwhile Lincoln met with Stanton and warned that he expected a great deal of his new Secretary. Stanton had to clean up the War Department from top to bottom, eliminate the waste, and develop it into a centralized agency that would mobilize and administer the Union war machine, keep generals in line with administration policies, and maintain order and security on the home front. All this was essential, Lincoln said, if the rebellion were ever to be suppressed.

Stanton was confident he could do the job. But he was still astonished that Lincoln had appointed him Secretary of War. After all, Stanton had humiliated Lincoln back in the McCormick Reaper case. And in Washington this past year, Stanton had so vilified this "imbecilic" President, this "original gorilla," that even McClellan had winced at his language. But Lincoln made it clear that he bore Stanton no ill will. If the McCormick Reaper episode had been one of the most humiliating episodes of his life, Lincoln had put that aside now. He never carried a grudge, he said later, because it didn't pay. Stanton had exceptional skill and that alone was why Lincoln had promoted him.

Stanton had skill all right. On his first day at the War Department, he told his staff they were going to clean up this rat's nest and get it moving in a hurry. He worked at a killing pace, arriving at nine in the morning and seldom leaving until after ten at night. He was a gnomelike man, with a heavy chest, short legs, and a thick beard streaked with gray. He wore small, moonlike spectacles and fixed on people with "a perpetually irritable look in his stern little eyes," as one man said. Born in Ohio and raised by a widowed mother, he'd worked his way through Kenyon College and become a successful Steubenville lawyer. But in the 1840s he suffered a series of devastating tragedies. First a cherished little daughter died of some undiagnosed ailment, and Stanton was almost paralyzed with grief. Two and a half years later his wife also died. "This calamity has overwhelmed me," Stanton wept, and at night he would wander through his house with a lantern, calling out for his wife, his "bride," his Mary. Then his brother Darwin, sick with a fever, committed suicide by cutting his throat, and Stanton ran grieving into the night. The loss of his daughter, wife, and brother in so short a time left Stanton a hardened man, causing him "to turn a stern face to the world," as his biographers have said. In 1847 he moved to Pennsylvania, eventually married a woman sixteen years younger than he, and became a renowned Pittsburgh attorney. In 1856 the Stantons moved to Washington, where he served as a government lawyer and Buchanan's last Attorney General.

If most people found Stanton unpleasantly intense, he was nevertheless honest and incorruptible. Nobody could buy him off. Although he was a Democrat, he had a baffling way of getting along with everybody from conservative Democrats to liberal Republicans. But he was unswerving in his loyalty to the Union and wanted the government to fight this war through with everything it had. Socially, he loved the good life, and he and his second wife Ellen entertained in splendid style. Ellen was a frail, aristocratic woman, "white and cold as marble," Hay observed, "whose rare smiles seem to pain her." Mary could not abide Ellen Stanton, but Lincoln found something appealing in this cold and sickly woman. She was one of three people in all his life to whom he gave a copy of his treasured poem "Mortality."

Lincoln kept a close watch on Stanton and liked what he saw. With single-minded zeal, Stanton transformed the War Department into a superbly efficient agency, consisting of three chief assistants and a score of harried secretaries. He undertook an extensive audit of government contracts, canceling some and adjusting others, until he'd saved the government some seventeen million dollars. He streamlined army purchasing, reorganized the supply system, kept an accurate record of new units for the army, and assimilated a mass of complicated military data which Lincoln found indispensable. At Lincoln's orders, he also assumed sole jurisdiction over internal security matters and created a corps of civilian provost marshals to deal with them.

As an administrator, Stanton was a brusque, irascible martinet. He intimidated army contractors who called at his office. He banged on his desk, yelled at his aides, barked out orders, and drove his staff as hard as he pushed himself. Yet he sometimes broke under the strain. After one typically hectic day, an assistant found him alone in his office, sobbing, with his head on his desk.

Because of the war, Lincoln and Stanton spent a great deal of time together, absorbed with military problems and questions about the army. In time they developed a lasting respect for one another, shared a love and deep concern for the volunteer army, and trusted each other implicitly. Given the hostility between their wives, they couldn't socialize much, but they found ways to get together. When Mary was gone on a shopping trip, they took carriage rides in the afternoon, huddled close in conversation, or sequestered themselves in the White House or the War Department. Lincoln came to understand Stanton better than any man, realized that underneath that gruff exterior was an honest and thoroughly loyal human being who could be counted on to obey orders.

From the outset, Stanton shared the President's worries about McClellan and his endless delays when it came to attacking the enemy. McClellan might be his friend, but Stanton agreed with the general's critics. "This Army has got to fight," growled the Secretary of War. "The champagne and oysters on the Potomac must be stopped."

PART·EIGHT
THIS FIERY TRIAL

Ever since the *Trent* crisis, McClellan had become an almost impossible problem. In late December the general fell ill and lay in bed for three weeks, as immobilized as the Army of the Potomac. When Lincoln called at his house, the President was not allowed to see the afflicted general, who was down with typhoid and must have uninterrupted rest. Moreover, he refused to send any word to Lincoln and the Cabinet about his plan of operation, so that nobody in the government had any idea when he would ever take the field.

By the end of December, Congressional Republicans were in a tirade. On all fronts Union military operations were at a standstill. McClellan was in bed . . . faking illness, fumed some Republicans, so he wouldn't have to fight. "All is quiet on the Potomac" became a slogan of ridicule not only against McClellan but against the administration as well. On December 31 a new watchdog committee—the Joint Committee on the Conduct of the War, consisting of five Republicans and two Democrats—confronted Lincoln about the lack of action. Wade was chairman, and among the other members were Senator Chandler and Senator Andrew Johnson, the latter a pro-Union Democrat from Tennessee. A profane, self-educated man who hated Southern planters, Johnson had denounced secession in Tennessee while holding his audience at bay with a revolver. In their White House conference, Wade accused Lincoln of "murdering" the country with his inept military commanders and his want of "a distinct policy" about slavery. Lincoln didn't argue with Wade or the other worried patriots on the committee, whose task it was to ferret out traitors, thieves, and incompetents in the administration and the army. In truth, Lincoln thought Wade and his colleagues good, sensible men and wanted to keep the lines of communication open with them—and beyond them with Capitol Hill itself.

Still, he admitted that the military situation was calamitous. He conferred with Attorney General Bates about what to do, and Bates implored Lincoln to take charge of the army and be Commander in Chief, just as the Constitution designated. "The Nation requires it," Bates said, "and History will hold you responsible." But Bates feared he spoke in vain. The President was a wise and excellent man, Bates recorded in his diary, "but he lacks *will* and *purpose,*" and his entire administration was adrift, as Cabinet members worked without guidance or policy.

But Lincoln hadn't ignored Bates's advice completely. At the Library of Congress he checked out books on the science and strategy of war and put himself through a rigid course on how to command an army. And he sent off telegrams to Buell and Halleck in the West, urging them to mount a concerted operation into east Tennessee, where "our friends . . . are being hanged and driven to despair." But Halleck replied that he knew nothing about Buell's plans. And Buell reported that he hadn't the resources to invade east Tennessee. Lincoln did not understand the two generals. In studying the maps on his wall, the President saw no good reason why they couldn't strike at Knoxville, and in his telegrams he told them so.

On the night of January 6 the Joint Committee called again about McClellan. Wade was beside himself with rage. What, he demanded, did the President know about McClellan's plans? It was a fact that the Army of the Potomac outnumbered the rebels at Manassas, so why did McClellan not attack? Was it not true that he sympathized with the South? That his heart was not in the war? That he intended to save slavery? That he was conniving to become dictator of a reunited, slaveholding nation? Wade and several of his colleagues pronounced McClellan unfit for command, with Wade demanding "in the most undiplomatic plainness of speech" an immediate change in war policy. Seward and Chase were present, and they sided with Lincoln in defending McClellan, all three expressing their confidence in the general. But after the meeting, Lincoln sent a note to McClellan that he'd better see the Joint Committee as soon as his health permitted. "Today, if possible."

Meanwhile more telegrams out to Halleck and Buell. Please name as early a day as possible when you can move in concert against Tennessee, Lincoln said. "Delay is ruining us, and it is indispensable for me to have something definite." But Buell couldn't name the day he would be ready to advance. And Halleck said he didn't have enough men to help Buell, so that concerted action was out of the question. "It is exceedingly discouraging," Lincoln wrote on the back of Halleck's letter. "As everywhere else, nothing can be done."

He called on Quartermaster Montgomery Meigs, an able and candid man, and slumped in front of the fire in Meigs's office. A cold winter wind moaned outside. "General," Lincoln said, "what shall I do? The people are impatient; Chase has no money and tells me he can raise no more; the General of the Army has typhoid fever. The bottom is out of the tub." Meigs suggested that

Lincoln confer with McClellan's division commanders, but the President had just about made up his own mind. He left Meigs's office and held several conferences with Seward, Chase, and Generals McDowell and William Franklin. Lincoln told the generals he was greatly distressed about the state of affairs and announced that if McClellan didn't want to use the Army of the Potomac, Lincoln wanted to borrow it. He asked what the generals thought of starting active operations with the Army, and McDowell recommended an early advance through Manassas. Lincoln, too, favored an attack in that direction and later told Orville Browning, "I am thinking of taking the field myself."

On January 13, the same day he called Stanton to the War Department, Lincoln set down what he called "my general idea of the war" and how it should be fought, a strategic plan which derived from his military studies and his own notions about concerted action that had been percolating in his mind since Bull Run. "In the midst of my many cares," Lincoln sent Buell and Halleck a joint communiqué which outlined his strategic conceptions about the war. We have the superior forces, he said, but the rebels, by shifting troops across interior lines, have the greater ability to concentrate their manpower "upon points of collision." To defeat them, our armies must menace the enemy simultaneously at various points. If the rebels make no change in their forces, we can launch concerted attacks all across the front. Or, if they weaken one spot to reinforce another, we can hit the weakened point and smash through to victory. Now how did this apply to the West? Halleck could harry the rebels in western Kentucky and along the Mississippi. At the same time, Buell could worry them in eastern Kentucky and south of that in eastern Tennessee. At some point the rebels would have to weaken—and when they do, we can drive through their lines and rout them. Lincoln thought this so good a plan that neither general could object.

That same January 13, he held another conference with McDowell, Franklin, and members of the Cabinet. But this time McClellan got out of bed and showed up at the meeting, for he was convinced that a plot was afoot to remove him from command and hand the Army of the Potomac back to McDowell. As he faced Lincoln and his colleagues, the young general was in an ugly mood. He was rude to McDowell and belligerent toward everyone else. There were whispers and embarrassing silences. Finally Lincoln urged the general to reveal his plan of operations in detail, and so did Chase. Though he'd defended McClellan up to now, Chase was tired of his stalling and wanted to know when and where he intended to fight. But McClellan indignantly refused to discuss his strategy. He knew what he was doing. He didn't need the approval and judgment of "such an assembly" as this, some of whom "were incompetent to form a valuable opinion." Moreover, there were others here who could not keep a secret (he looked straight at Lincoln), so that if he told his plans they would end up in the Washington papers. At that, the general refused to give the group any further information unless ordered by Lincoln "in writing." But he did say that he would move Buell's army into battle very soon. Encouraged by even a hint of action, Lincoln said, "Well, on the assurance of the General that he will press the advance in Kentucky, I will be satisfied, and will adjourn this council."

Two days later McClellan went on the carpet in front of Wade's committee and they fussed with one another, and afterward Wade and Chandler griped to Lincoln about McClellan's "infernal, unmitigated cowardice." By now Stanton had added his voice to the clamor for McClellan to move. When two more weeks elapsed and nothing happened anywhere, Lincoln lost all patience with his generals. On January 27 and 31, he ordered the Army of the Potomac to march on Richmond by way of Manassas and instructed army and naval units in Kentucky, on the Mississippi, and in the Gulf of Mexico to begin simultaneous operations against the enemy. All movements were to commence on or before February 22, 1862. Lincoln had consulted nobody about these war orders, not even Stanton. In all the feuding around him, Lincoln just issued them.

They got action of a sort. For the first time, McClellan took the administration into his confidence, revealed his plan in detail, and explained why he preferred it to Lincoln's. It was unfeasible, the general argued, to attack the rebels at Manassas as the President wanted, because they were much too powerful there. So here was what McClellan intended to do: he would move the Army down the Potomac, land at Urbana on the Rappahannock, and strike at Richmond before the Confederate army at Manassas could pull back. Once McClellan captured Richmond, all Virginia would fall. The rebels would have to evacuate Tennessee and North Carolina. Union armies would then mount offensives in all theaters, crush the Deep South, and end the rebellion.

While Lincoln appreciated McClellan's candor, appreciated his confiding fully in the administration, the President had nagging doubts about the Urbana operation. Frankly Lincoln still believed the main attack should take place at Manassas. But McClellan was a professional military man, and Lincoln hated to override him. Reluctantly the President told McClellan to go ahead, and the general once again set about his preparations.

In the White House later, Mary brought up McClellan and said he was "a humbug." But why? Lincoln protested. "Because he talks so much and does so little," Mary replied. But Lincoln, who wanted to believe the best of everyone, remarked that McClellan was a capable young officer who had to contend against tremendous pressures. Mary cut him short.

McClellan can make plenty of excuses for himself, she said. He needs none from you.

How Lincoln cherished his younger boys, Willie and Tad. As the war continued, Lincoln found them increasingly an antidote to the depressions and anxieties of his job. He treasured their fleeting moments together, when he could forget about McClellan, forget about feuding generals and irate politicians, and could relax with his sons in their world, sharing in their fun and reading them stories with his spectacles on. He loved to recount the antics of his "two little codgers" and bragged about them to anyone who would listen.

Both were merry fellows who found the White House a place of boundless revelry, and the Lincolns let them frolic with few restraints. The boys ran and shouted in the corridors, broke into Lincoln's office in the middle of conferences, and chased one another around unamused politicians. Tad, who instigated most of their mischief, once aimed a toy cannon at a Cabinet meeting. In addition, he liked to stand at the front of the grand staircase and collect a nickel "entrance fee" from callers on their way to see the President. With Lincoln's help, the boys converted the White House and its grounds into a veritable zoo—they collected ponies, kittens, white rabbits, a turkey, a pet goat which slept in Tad's bed, and a dog named Jip which curled up in Lincoln's lap at meals. When the boys weren't chasing animals through the mansion, they were holding fairs and minstrel shows up in the attic. One day Tad discovered something there that offered irresistible deviltry: the White House bell system, with cords running down to various rooms below for Lincoln or the staff to pull when they wanted something. Tad set all the bells to clanging at once and threw the White House into bedlam. Was the place on fire? The bells bewitched? At last members of the staff climbed into the attic and found Tad yanking the bells and giggling helplessly. On another occasion, Tad stole into the White House kitchen and ate all the strawberries intended for a state dinner. The steward raged at the boy and pulled his hair until his mother rescued him. And then there were the war games, inspired by the martial atmosphere in Washington. There were bloody battles in the attic or out on the lawns. There were military parades through the White House corridors, with the boys and their friends marching in single file, blowing on dented horns and banging on tin drums. There were reconnaissance missions on the White House roof, the boys hiding behind the rails and searching for "rebs" through telescopes. And there were solemn court-martials, in which the boys tried a soldier doll named Jack, found him guilty of desertion, shot him to death, and buried him in the White House garden. One day, though, they darted around Hay and burst into Lincoln's office, explaining in a breathless rush that they had tried Jack again and had to shoot him for desertion and bury him like they always did, but the grouchy gardener said it dug up the roses so they wanted "Paw" to fix up a pardon. Lincoln said he reckoned he could do that and drew out a sheet of executive stationery. "The Doll Jack is pardoned by order of the President." And he signed it "A. Lincoln."

Because the boys loved the army, Lincoln often took them along when he visited McClellan's camps across the Potomac. They looked up to soldiers in wide-eyed reverence, gloried in the marching bands and drilling regiments. In reviews, they rode behind their father, tipping their hats at troops just as he did. When he left them back at the White House one raw and windy day, the boys proved themselves nothing if not resourceful. As Lincoln reviewed McClellan's army, riding a horse that was too small for his long legs, a rickety mule cart came bouncing over the field. A grinning Negro boy was driving and Willie and Tad were hanging on happily, their wooden swords held at salute.

Still, life for the Lincoln boys wasn't all play. Willie, who was eleven this last December, had a serious side and could behave quite like an adult sometimes. Blue-eyed and precocious, he sat in church in rapt attention, listening to the preacher while Tad played with a jackknife on the floor of his mother's pew. When he was tired of romping with his younger brother, Willie liked to seclude himself in Mary's room, where he would curl up in a chair and read a book of poetry or jot down verse on a writing pad, much as his father used to do back in Indiana. He had a sense of history, too, and kept a scrapbook of significant events, pasting in clippings about his father's inauguration, the war, and the deaths of important people like Edward Baker. Willie was a gentle boy, so affectionate that Mary came to depend on him desperately for family companionship. He would, she remarked, "be the hope and stay of her old age."

By contrast, eight-year-old Tad was nervous like his mother, a hyperactive child with a speech impediment who was so slow to learn that some claimed he still couldn't read. Mary did hire tutors for the boys, but Tad had "no opinion of discipline," Hay grumbled, and his tutors one after another resigned in frustration. Nevertheless, Lincoln refused to worry about Tad, insisting that he would learn his letters in time. So what if the boys were a little spoiled? He and Mary were determined to "let them have a good time" while they could. They would have to grow up soon enough as it was.

Could this be true? Lincoln read the dispatches with pleasant surprise. Buell had actually ventured forth and defeated the rebels in a battle in eastern Kentucky. And Halleck, too, had finally come alive and sent a column under Brigadier General Ulysses S. Grant to operate down the Tennessee River in western Kentucky and Tennessee. Buell and Halleck weren't cooperating as Lincoln had directed, but at least they were fighting. Then came even better news: in February Grant drove into northwestern Tennessee, captured Fort Henry on the Tennessee River, and then stormed Fort Donelson on the Cumberland, pounding the shell-torn garrison until it met his terms of unconditional surrender. Lincoln and Stanton congratulated one another as they read the reports, and Lincoln happily noted that many of Grant's men hailed from Illinois. Subsequent reports indicated that Grant's victories had cracked the Confederate line in Kentucky and forced the opposing rebel army, its flanks now exposed, to retreat down into Tennessee. Though Halleck, from his Saint Louis desk, claimed most of the credit, Lincoln himself nominated "Unconditional Surrender" Grant for promotion to major general. In this long and dismal winter, Grant alone had given the President something to cheer about.

While Grant was pounding rebel river garrisons in Tennessee, both Willie and Tad came down sick. There was nothing to worry about, a physician assured the Lincolns. It was just a fever. They should both recover all right. But Willie did not improve. He developed a chill. Suddenly his fever flamed out of control, and the White House secretaries could hear his cries in the night. Lincoln and Mary came and went, doing what they could to comfort him, and Mary was extremly upset. She'd planned a lavish party complete with dinner and martial music, and the invitations had already gone out. When the physician again encouraged them not to worry, the Lincolns decided to go through with the affair. But in the course of the evening, they took turns slipping away from the guests to be with their stricken boy.

He grew weaker as the days passed. More physicians came to the White House, but nothing could be done for him. All Lincoln and Mary could do was bathe his face with a wet cloth and look on helplessly as the boy's life ebbed away. Lincoln stayed up night after night, giving Tad his medicine and keeping a vigil at Willie's bedside. "The President," Bates wrote in his diary, "is nearly worn out, with grief and watching."

At last, on February 20, at five in the afternoon, Willie Lincoln died. Mary collapsed in convulsions of sobbing and Lizzie Keckley led her away. Lincoln, overwhelmed with grief, looked down at Willie. "My poor boy," he murmured, "he was too good for this earth . . . but then we loved him so." Then out into the corridor and down to Nicolay's office in numb despair. "Well, Nicolay, my boy is gone—he is actually gone." The tears came now—let them come—as Lincoln found his way to Tad's room, lay down with the sick boy, and tried to tell him that Willie would no longer play with him, that Willie was dead. Tad was incredulous for a time. Then he too began to cry.

The Brownings came at once to help. Elizabeth remained with Mary through the night, and Orville took care of the funeral arrangements. On February 24 a minister conducted services in the East Room while Willie lay in a metal coffin in the nearby Green Room. Lincoln stood with Robert at his side, the windows and ceiling—the entire White House—festooned with black crepe. Mary did not attend the funeral; she was in such a state of shock that she couldn't leave her room. But most of official Washington was present. There was Seward with a tremble in his face. There was Browning, Sumner, Trumbull, and dozens of others, including George McClellan, who was so moved by Lincoln's suffering that he later sent the President a tender note expressing his sorrow and thanking Lincoln for standing by him. When the services were over, pallbearers and a group of children—Willie's Sunday school class—carried the coffin out into a howling wind and placed Willie in a hearse. That morning a storm had lashed Washington, with gales so powerful that they blew off roofs, knocked down trees and chimneys, and ripped flags to shreds. In the carriage now, riding out to a Georgetown cemetery to see his son lowered into a grave, Lincoln scarcely noticed the whipping wind or the debris in the streets. For all people in this world are destined to die, even Willie. Yet he was such a gentle, intelligent boy, with a flair for poetry and a zest for life.

Hope and despondency are mixed together and follow one another like sun and storm. A smile and a tear, a song and a lament, wash after one another like waves. Yet for months after Willie's death there was no song and little laughter in the White House. As always, Lincoln found some escape in work, but his moments with Mary and Tad brought the pain back again. Tad, who until Willie's death had thought life a game, now broke into intermittent sobbing because Willie "would never speak to me any more." And Mary suffered a nervous breakdown and shut herself in her room for three eternal months. She tossed about in her bed in wild and inconsolable anguish, broke into fits of weeping, and cried out for Willie to come back to her. Lizzie Keckley recalled how tender Lincoln was to his distraught wife. But he worried about her, haunted by fears that she might lapse into insanity. One day he led her to a window and pointed to a distant building where mental patients were confined. "Try and control your grief," he said firmly, "or it will drive you mad, and we may have to send you there."

With Lizzie's help and Lincoln's firmness and care, Mary improved some. But the mention of Willie's name or the sight of a memento would send her into violent sobs. Unable to bear any memory, she gave away all his toys and anything else that might remind her of him. She never again went into the guest room where he'd died or into the Green Room where he'd lain in his coffin. She dropped all but the most imperative social functions and lived in virtual seclusion, trying desperately to hold on. Five months after his death she was still so shaken she could barely write her Springfield friends about "our crushing bereavement." And sometimes, in a chilling moment alone, she realized again that *"he is not with us"* and the anguish of the thought "often for days overcomes me."

Unable to endure on her own, she reached out to spiritualism to help her cope. She attended séances in and around Washington, feeling a rush of joy when some medium would pull back that "very slight veil" which "separates us from the loved and lost" and help Mary commune with Willie. As time went by, though, she found additional ways to ease her grief: she took to visiting military hospitals in Washington, distributing food and flowers among the wounded soldiers. Moreover, she developed a deep compassion, born of her own suffering and her friendship with Lizzie Keckley, for "all the oppressed colored people." She helped Lizzie care for "contraband" blacks who now filled up Washington—and even persuaded Lincoln to contribute two hundred dollars for them because "the cause of humanity requires it." Too, Mary became a "one woman employment bureau" for black people, doing what she could to find jobs for them.

Yet she remained profoundly unstable. Her mood swings, her headaches and explosions of temper, were worse now than ever. And she began seeing political conspiracies against her husband, especially on the part of that "hypocrite" Seward, that "dirty sneak" who'd tried and was still trying to take over Lincoln's job. Mary would put him in his place . . . in fact, she'd already told him off. Back in March of 1861, he'd all but insisted that he should hold the first Presidential reception, and Mary had told him that this is *our* function, Mr. Seward, that you are not President, my husband is! And she hated him all the more because he cheerfully ignored her hatred. Yet the other Cabinet members were just as evil, and in a carriage ride with Browning one day long after Willie's death, she remarked that "the Cabinet were all enemies of the President, working for themselves, and that they would have to be dismissed, and others called to his aid before he would succeed." Yet what upset her so was that her husband didn't seem to realize—didn't seem to understand—how they were all out to use or destroy him. He was so naïve, she thought. An innocent among thieves. She tried desperately to protect him. She talked about his enemies in her salon (her flatterers passing on her remarks), and sometimes she betrayed his confidences—but only because she wanted to help him. And she fretted about his safety, too, was terrified that something might happen to him. "Oh, God," she would say to Lizzie, what would she do then? She begged Lincoln to take guards along on his nocturnal walks to the War Department, begged him to be more careful, worried about him so, until it seemed to Lizzie that Mary "read impending danger in every rustling leaf, in every whisper of the wind."

And she worried about his own suffering over Willie. So that he too might find comfort, she persuaded him to attend an occasional séance with her, remarking that if she hadn't felt called to cheer her husband, "whose grief was as great as my own, I could never have smiled again." But Lincoln didn't believe in spiritualism and never took the séances seriously. No, he found his own ways to live with Willie's death. He treasured keepsakes and kept a framed picture of Illinois over his mantel, telling visitors that it was painted by "my boy, who died." And he became even closer to Taddie, took him everywhere—to neighborhood toy shops or to the front or on horseback rides. He rocked the boy at night and felt a twinge every time Taddie with his speech trouble called him "Papa day" for *Papa dear*. He left his office door open for the boy, and often Tad would fall asleep on the office sofa while Lincoln toiled at his desk; finally Lincoln would carry him off to bed and then return alone to his work. And he loved it when Tad would burst into his office, scramble into his lap and give him a hug, and then race off again.

Yet Willie's death left Lincoln with a permanent hurt inside. Sometimes he would dream that Willie was still alive, would see the boy playing in the leaves on the White House lawn and calling to him . . . only to wake up in the dark of his chamber and realize again that it was just a dream.

Ultimately it was Lincoln's own fatalism that eased his sorrow the most and helped him cope with his dreams and memories. After Willie's death, he talked more frequently about God than he had before—about how the

Almighty had taken Willie and how He controlled the fates and destinies of everyone. "I have all my life been a fatalist," he told a congressman one day, and said he agreed with Hamlet that "there is a divinity that shapes our ends." Yet his fatalism was more intense now and more deeply felt. As 1862 progressed, he saw himself increasingly as "an instrument of Providence," who'd been placed on this Earth, in the center of this war, for God's own designs.

McClellan woes again. In early March, over a month after Lincoln had approved the Urbana operation, McClellan still hadn't moved, contending among other things that his army was not yet battle-ready, that his lines of retreat were not yet secure. At the same time, he'd had more collisions with Wade and other members of the Joint Committee, who insisted that he fight the "traitor" army at Manassas. But McClellan steadfastly rejected a head-on attack. He had "reliable" information, he once told Wade, that more than 220,000 rebels were dug in there behind forts stronger than the Russians had at Sebastopol. And the Confederates had heavy guns, too, which could be seen by Union patrols. By now McClellan and Wade hated one another. The general was certain that Wade's committee was out to wreck his career simply because he was a Democrat (even though two Democrats served on the Joint Committee). And Wade and his colleagues thought McClellan was not only miserably incompetent but a traitor to boot. At one point, Wade and some of his distressed associates confronted Lincoln and demanded McClellan's dismissal.

"If I remove McClellan," Lincoln protested, "whom shall I put in command?"

"Well, anybody!" Wade said.

"Wade, *anybody* will do for you," Lincoln said, "but not for me. I must have *somebody.*" Then: "I must use the tool I have."

Still, Lincoln sympathized with the committeemen—they were right that "fighting, and *only* fighting," would win this war. But could he believe their imputations about McClellan's "disloyalty"? In this confusing conflict, it was sometimes difficult indeed to tell who your friends were. And there were persistent rumors that McClellan's Urbana plan was unmitigated treachery, that the general intended to sail away so that the rebels at Manassas could attack Washington without resistance. On March 8 Lincoln called McClellan to the White House for a confidential talk about all this. He confessed to the general that he still had terrible doubts about the Urbana operation. Then he brought up the charges that McClellan was a traitor. The general jumped to his feet and demanded that Lincoln retract that accusation. Much agitated himself, Lincoln explained that *he* wasn't calling McClellan a traitor, but was only repeating what others said. McClellan tearfully denied all such accusations and reaffirmed his loyalty to the Union. Finally Lincoln approved of Urbana a second time, but warned McClellan to leave Washington "entirely secure."

The next day, Lincoln received ominous news from Fort Monroe on the Virginia coast. A rebel ironclad, a hulking monster named the *Virginia,* had steamed out of Norfolk, descended on the Union fleet off Hampton Roads, and rammed and battered the wooden ships as though they were toys. Lincoln held an emergency Cabinet meeting, with some of the Secretaries beset with apocalyptic visions—of the rebel behemoth churning up the Potomac, obliterating McClellan's wooden troopships, shelling Washington and putting the government to flight, and then bombarding other Northern cities. Lincoln was anxious, too, going repeatedly to his office windows to see if the *Virginia* were approaching. But late that afternoon came a heartening dispatch: a little Union ironclad called the *Monitor,* with two guns mounted in a revolving turret, had dueled the *Virginia* to a draw that day. After the two vessels had bounced shells off one another for several hours, the *Virginia* had turned around and sailed back to Norfolk. The *Virginia* was neutralized; the blockade seemed safe for now. McClellan could proceed with his battle plans.

That evening came more startling intelligence. The rebel army had evacuated the Manassas area and fallen back to another defensive position south of the Rappahannock River, the better to defend Richmond in McClellan's impending campaign. At once McClellan threw his army into the abandoned enemy camps, only to find that many of those cannon that had intimidated the general were only imitation guns made of painted logs. What was more, observers could tell from camp remains that the rebel force was a lot smaller —it numbered about 36,000 in fact—than the massive hordes McClellan envisioned. Had the general attacked with his 120,000 troops, he might have won a decisive victory. As it was, McClellan was scandalized. Lincoln was disgusted, "almost in despair," and Congress was in a furor. "We shall be the scorn of the world," wailed Senator Fessenden of Maine. And Vice-President Hamlin, tall and grim-faced, begged Lincoln to remove McClellan from command.

On March 11 Lincoln consulted with the Cabinet about McClellan. Chase and Seward were bitter about his "imbecilic" behavior. Stanton fumed at the condition of the army, and he and Bates chorused that McClellan ' has no plans but is fumbling and plunging in confusion and darkness." Again Bates lectured Lincoln that he must "command the commanders."

Yes, Lincoln had to do something. McClellan was plainly "unaggressive" and the army a laughingstock. At last the President reached "a most painful" decision: he relieved McClellan as General in Chief, but left him in command of the Army of the Potomac as a favor, giving him "an opportunity to retrieve his errors." As Bates wanted, Lincoln would now function as Commander in Chief, watching all points and coordinating operations in East and West. He ordered all department commanders—including General Frémont, who now headed the new Mountain Department in western Virginia—to report directly to Stanton, Lincoln's right-hand man.

On March 13 Stanton brought in a report from McClellan. Though he had manfully accepted his demotion, the general was bitter nevertheless, angry at the politicians in Washington and embarrassed at how the rebels had fooled him with their "Quaker guns." Moreover, by falling back between Urbana and Richmond, they had ruined his plan of operations. But never mind. He had an alternative plan, perhaps better than Urbana, a magnificent campaign that promised to end the war in a master stroke. His army would sail down to Chesapeake Bay, land at the sandy peninsula between the York and James rivers, and then dash northwest into Richmond, thus outmaneuvering the Confederate force on the Rappahannock.

Lincoln was not impressed with the Peninsula operation. Yet he noted that all McClellan's corps commanders had endorsed it—so long as McClellan left forty thousand men to guard Washington. Maybe Stanton was right, maybe all the generals were afraid to fight the rebels in a stand-up battle but Lincoln still thought it unwise for a couple of civilians to overrule professional military men. So go ahead with the Peninsula plan, the President informed McClellan, as long as he secured Washington, meaning that he should leave forty thousand men there as his corps commanders advised. At all events, Lincoln said, attack the rebels "by some route."

On March 17 the Army of the Potomac began embarking at Alexandria; and a couple of weeks later McClellan himself sailed for the Virginia Peninsula. He was happy indeed to get away from Washington, away from "that sink of iniquity" and its "knaves" and "hypocrites," away from Lincoln and ex-friend Stanton, that "unmitigated scoundrel." After McClellan had gone, Lincoln confided in Browning that he'd studied his young general "and taken his measure as well as he could." McClellan was thoroughly loyal to the Union, Lincoln said, and a brilliant organizer who could make superb preparations for "a great conflict." But when the hour of battle approached, McClellan became "nervous and oppressed" and started piling up excuses for not fighting. Because he was that way, Lincoln had finally ordered him to move. And now at last he had.

Inevitably, controversy boiled up after the general. Stanton discovered that only nineteen thousand troops manned the forts and redoubts around Washington, rather than the forty thousand Lincoln expected and McClellan's own generals had called for. Stanton thought McClellan had arrogantly disobeyed the President's orders to safeguard Washington, and Lincoln believed so, too. Actually there were forty thousand men in various commands in northern Virginia, from the Manassas-Warranton area out to the Shenandoah Valley, and McClellan assumed that these were adequate protection. But Lincoln was extremely upset—he wanted forty thousand men around Washington itself, especially since a rebel force had now turned up in the valley. As a consequence, Lincoln detached McDowell's corps from McClellan's army and ordered it to defend the capital. Lincoln explained all this in dispatches to McClellan, pointing out that the general still had over 100,000 men and urging him to get on with his campaign "at once."

By the time McClellan learned about McDowell's corps, his army had landed in Virginia and the general was sitting on the ground near Yorktown, listening to the roar of his heavy guns as they fired off practice rounds. Of course he was enraged about McDowell. How in hell could he advance "at once" when those politicians in Washington took a corps of almost forty thousand men away from him? To Lincoln's dismay, McClellan refused to attack the thin rebel line in his front (for a time only eleven thousand Confederates stood between him and Richmond), but undertook a textbook siege of lightly defended Yorktown, thus giving the rebel army on the Rappahannock time enough to fall back and protect Richmond.

Why had Lincoln let McClellan go? Why had he allowed a general whose nerve he questioned take the Army of the Potomac far off on the other side of Richmond? Lincoln confessed to Sumner that McClellan "had gone to Yorktown very much against my judgment, but that I did not feel disposed to take the responsibility of overruling him." Sumner, for his part, was empurpled that Lincoln should violate his own instincts and leave an "utter incompetent" in command of the Union's finest army.

There was another reason why Sumner objected to McClellan. The two had clashed over whether the federal government could use the army to free the slaves in the rebel South. Echoing nearly all Northern Democrats, McClellan hotly opposed any such "revolutionary" move, inveighed against wartime emancipation and any other attempt to interfere with Southern property rights. By contrast, Sumner, Wade, Chandler, and many other Republicans considered the army a potent weapon for uprooting slavery. Should Lincoln issue an emancipation proclamation, the army would be obliged to smash the cornerstone of the South's planter-run social order and thus obliterate the mischievous planter class so that it could never foment another civil war. Moreover, as Sumner reminded everyone, military emancipation would also break the shackles of several million oppressed human beings.

And Lincoln's views that spring? Was he still just a month or so behind Sumner in the matter of wartime emancipation? To the senator's joy, Lincoln said "that he was now convinced that this was a great movement of God to end slavery and that the man would be a fool who should stand in the way."

Yet he still shied away from a Presidential decree, because he was still fretful about the border. And he shrank, too, from the derangement military emancipation would cause in the rebel South and the racial explosions it might ignite in the North as well. No, he still preferred gradual, compensated emancipation as an alternative policy, regarding this as the most bloodless way to expunge slavery and implement God's designs. So far his Delaware plan hadn't enjoyed any luck, since nobody in the Delaware legislature would sponsor it. But maybe he would get better results by starting on the national level and then working back to the states. If so, then emancipation would begin in the loyal border and then be extended into the rebel states as they were conquered.

In early March, in the midst of his troubles with McClellan, Lincoln sent a message to Congress which proposed a federal-state emancipation scheme, along the lines of his Delaware plan, whereby Congress would finance any gradual program the border states adopted. To ease white racial fears, he recommended colonization again. But Lincoln gave a warning. In his message last December, he'd said he would preserve the Union by "all indispensable means." He'd said, too, that he would not resort to any "radical" or "extreme" measures. Yet the pressures of this huge and awesome conflict were changing him and everybody else caught up in its flames. Should the rebellion grind on, he asserted, then other means "such as may seem indispensable, or may obviously promise great efficiency towards ending the struggle, must and will come."

Lincoln's message created a sensation—for it contained the first emancipation proposal a President had ever submitted to Congress in the history of the Republic—and most Republicans gave him enthusiastic support. Sumner, who'd criticized the message in manuscript, praised Lincoln's federal-state plan as an excellent step in the right direction. And Greeley's *New York Tribune* hailed the program as "the day-star of a new dawn," a hint from the President "THAT THE UNION BE PURGED OF SLAVERY." Among Republicans, one of the few sour notes came from Henry J. Raymond of the *New York Times,* who grumbled that Lincoln's plan would cost too much. On the contrary, Lincoln wrote him, the money required to fight the war for eighty-seven days would buy all the slaves in the entire border—in Delaware, Maryland, Kentucky, and Missouri—and would shorten the conflict by a lot more than any eighty-seven days. Raymond was so impressed that he now extolled Lincoln's program "as a masterpiece of practical wisdom and sound policy."

So he had powerful Republican voices behind him. Now to enlist the support of border congressmen, whose cooperation would be essential in prodding their states to act. On March 10 Lincoln met with the congressmen in the White House and beseeched them to endorse his plan, maintaining that it would shorten the war more swiftly than the greatest military victory. At the same time, he made no attempt to hide his strong antislavery convictions. He thought slavery was wrong and "ought never to have existed" and he would continue to think so. Nevertheless, he still respected the property rights of loyal Southerners and assured the border-state congressmen that he entertained no other antislavery designs for now. Yet, as Nicolay observed, they were full "of doubt, of qualified protest, and of apprehensive inquiry." When Congress subsequently approved Lincoln's plan, offering financial assistance to any state that enacted a gradual emancipation program, the border-state representatives voted against the resolution.

So they did not heed his warning, did not understand that his plan was designed to forestall a more revolutionary move against slavery, the root and branch of the rebellion. How could the border men be blind to the signs? Could they not perceive that the war was carrying them all into uncharted seas, where no one knew what measures would have to be adopted in order to save the Union? If in normal times the Constitution prohibited the President from touching slavery, these were hardly normal times. This was a national holocaust. And if his gradual plan should fail for lack of border-state support, God alone knew what Lincoln might have to do to preserve the government. As he phrased it later, one generally tried to save both life and limb. "Yet often a limb must be amputated to save a life," but a life must never be given to save a limb. What he wanted now was to amputate the limb as painlessly as possible.

Well, Congress at least had endorsed his gradual program, so perhaps there was still hope for that. Meanwhile, as the weeks passed, Lincoln and the Republican Congress collaborated in other ways to batter at the outer defenses of slavery in civil-war America. Congress sent him a bill which forbade Union officers to return fugitive slaves to the Confederacy. Lincoln signed it. The Republican Congress sent him a bill which abolished slavery in Washington, D.C., compensated the owners, and set funds aside for the voluntary colonization of freed blacks in Haiti and Liberia. Although Democrats cried that the bill was an entering "wedge of an abolition programme" and that "niggers" would now crowd white ladies out of Congressional galleries, the President signed the bill into law. Congress also sent him a bill which outlawed slavery in all federal territories, and he signed it, too, thus reversing the Dred Scott decision and implementing the Republican goal of slave containment. In addition, Lincoln joined with Congress in recognizing the black republics of Haiti and Liberia, a move that would facilitate colonization efforts in their lands. In time Negro diplomats would be strolling about in a slaveless Washington—something scarcely dreamed of when the war began.

Yet Lincoln continued to oppose military emancipation in the rebel states. And that is why he was stunned when he read in the newspapers what General David Hunter had done in his coastal department, on Union-held islands off South Carolina. On May 9 Hunter had issued a military order reminiscent of Frémont's in Missouri. Because slavery was inconsistent with a free country, Hunter declared, the slaves inside his lines were "free forever."

When Lincoln found out about the order, he revoked it at once, with Stanton supporting the President emphatically. "No commanding general shall do such a thing, upon *my* responsibility, without consulting me," Lincoln said. Though he regarded Hunter as an old friend and shared his wish "that all men everywhere, could be free," the President alone would decide when military emancipation was necessary to save the country.

The abolitionists, of course, cried out in protest. And so did Sumner, Wade, and Trumbull. Sumner in fact was mortified. Was Lincoln now going backward? Was he returning to his save-slavery policy of 1861? Despairing that Lincoln was not a month behind him after all, Sumner renewed his pressure on Lincoln to issue an emancipation proclamation and eradicate slavery in the rebel South as well. What was more, Trumbull, Sumner, and their Congressional colleagues now sponsored a new confiscation bill which would liberate the slaves of all those who supported the rebellion. The bill seemed a clear warning to Lincoln: either *you* strike at slavery or Congress will.

Lincoln was upset about all the antislavery salvos he was getting. He admitted that he did not fully understand the magnitude of the civil war, but he was put here and he must do the best he could. "I can only go just as fast as I can see how to go," he said. But if Democrats and conservative Republicans applauded his handling of Hunter, the President made it clear where his sentiments lay. "I gave dissatisfaction, if not offence, to many whose support this country can not afford to lose. And this," he said, "is not the end of it."

O n the military maps in his office, Lincoln tacked blue pins to mark Union gains that April. On the crucial Mississippi River, combined naval and army units had not only seized Island No. 10 in northwestern Tennessee, but had captured New Orleans in a blazing night attack that left the mouth of "the Father of Waters" in Union hands. And in southern Tennessee, at Shiloh Church near the Mississippi border, Grant had battled a rebel army for two bloody days before driving it from the field. Union casualties were staggering—over thirteen thousand killed, wounded, or missing. And rumors crawled through Washington that Grant had been drunk during the battle, prompting one critic to demand his dismissal. But Lincoln refused. "I can't spare this man," the President said. "He fights."

McClellan, meanwhile, was still stalled in front of Yorktown on the Virginia Peninsula. Instead of attacking the smaller rebel force there, the general threw up earthworks, dug trenches, graded military roads, installed mighty siege guns, and exhorted Lincoln to dispatch McDowell's corps without delay lest McClellan be overwhelmed by the rebels in his front. At last, on May 4, the Confederates pulled out of Yorktown and fell back toward Richmond; rebel forces defending the insurgent capital now numbered some seventy thousand men under the command of Joseph E. Johnston. McClellan was proud of his protracted siege—Yorktown had fallen virtually without casualties—but Lincoln was thoroughly disgusted. In his judgment, McClellan had wasted an entire month digging trenches around a city he could have taken with a single offensive punch.

On May 6 Lincoln sailed for the Peninsula with Chase and Stanton. The President wanted to get away from Washington for a few days and see for himself how the campaign was going. When he reached Fort Monroe, he found that the commander there—fusty old John Wool—hadn't even tried to seize Norfolk, which served as base for the *Virginia.* Damn! Lincoln threw his stovepipe hat on the floor. Was the army full of timid incompetents? All right, then, he would take command of Wool's troops and capture Norfolk himself. At Lincoln's orders, Union gunboats shelled rebel batteries protecting the city and Union soldiers crowded into transports for an amphibious assault. Lincoln even reconnoitered the Norfolk coast—he and Stanton in a tug and Chase in a revenue cutter, all looking for a place to land Wool's men. They went ashore and walked along the beach, with its ocean smells and lapping waves, until Chase located a perfect spot for a landing. At last Union troops swarmed ashore and drove against Norfolk, forcing the rebel garrison to blow up the *Virginia* and abandon the city. "So has ended a brilliant week's campaign of the President," Chase recorded in his diary, as Lincoln and his two Secretaries sailed back to Washington rather pleased with themselves.

By now McClellan had inched his way up the Peninsula and was nearing the gates of Richmond. As he moved, his telegrams to Washington became more urgent and apprehensive. He reported that he could only bring about eighty thousand men into battle, whereas the enemy had "perhaps double" that number. It was mandatory, therefore, that McDowell's corps and all other disposable troops in the Washington area move down the Potomac and reinforce McClellan at once.

Now that McClellan was actually on the march, Lincoln was sympathetic with him. So the President and Stanton contrived an ingenious plan to help the general: McDowell's 38,000 troops, then at Fredericksburg, would advance on Richmond by land, cooperating with McClellan as a separate force. By moving between Richmond and Washington, McDowell would be in a position to defend the latter place in case of a rebel attack from the Shenandoah Valley. But McClellan objected to the plan. Did the President not understand his telegrams? McDowell's corps should sail down the Potomac and join McClellan's army as a subordinate command; it was the only way. A separate force would only lead to more confusion, more chaos. But Lincoln thought his idea the more feasible one. With McDowell operating on one side of Richmond and McClellan on the other, they would pin the rebel capital between them while leaving Washington secure.

On May 24 and 25 came shattering news from the Shenandoah Valley, where rebel columns under Stonewall Jackson were on a rampage. Jackson had struck part of Frémont's command, routed a second Union force under Nathaniel Banks, and was now driving toward the Potomac in what seemed an all-out push against Washington. With Stanton backing him up, Lincoln ordered McDowell to stay in the Washington area and wired McClellan that the rebels had evidently mounted "a general and concerted" movement against the capital, forcing Lincoln to withhold McDowell again. "I think the time is near," the President added, "when you must either attack Richmond or give up the job and come to the defence of Washington." McClellan was incensed. The Confederates were massed in front of *him,* he said, not in front of Washington. How could he fight a battle when a "terribly scared" President kept promising him help and then withholding it?

Now word came that Jackson was retreating back down the valley—which meant that Lincoln and Stanton had been wrong, that Jackson's had been a diversionary movement all along, a feint to draw McClellan away from Richmond. Well, Lincoln said, he was going to spoil the rebel plan and bag Jackson before he got out of the valley. In the War Department, the President ordered Banks, Frémont, and McDowell to march at once and cut Jackson off before he got away. *Hurry,* Lincoln telegraphed his generals, *move.* "I see you are at Moorefield," he fumed at Frémont. "You were expressly ordered to march on Harrisonburg. What does this mean?" He drove and maneuvered his commanders like reluctant marionettes, warning them over and over, "Do not let the enemy escape you." But Jackson did escape, smashing Frémont and pummeling McDowell's advance units as he did so. Jackson accomplished all this with only sixteen thousand men.

Lincoln was crushed that Jackson had escaped and utterly disillusioned with his generals, all of whom seemed incompetent. Still, the fiasco in the Shenandoah was not entirely their fault. At the outset, their forces were so badly out of position that their chances of snaring Jackson were slender at best. Also, he was falling back toward his base of supplies while they marched away from theirs, which gave him a distinct advantage. And so the failures continued to mount. Now McDowell's corps was too depleted to help McClellan. And McClellan, edging to within five miles of Richmond and the very precipice of battle, was crying desperately for reinforcements and accusing the government of "irreparable fault" if he didn't get them. At the same time, Frémont was wailing for men, too, as was just about every other general from Washington to Saint Louis. On all fronts the story was the same: Union commanders telegraphed that they faced an overwhelming foe and must have reinforcements. Yet Lincoln had none to give.

Portentous news from Richmond: on May 31 and June 1 the rebel army fell on McClellan in the battle of Seven Pines, but McClellan repulsed the attack. Wild with excitement, McClellan wired Washington that he'd just fought "a desperate battle" against "greatly superior numbers." "Our loss is heavy, but that of the enemy must be enormous." Joe Johnston had been wounded, McClellan learned, and Robert E. Lee was now in command of the Confederate Army, which suited McClellan fine because he considered Lee a "weak" general, apt to be "timid and irresolute in action."

Lincoln expected McClellan to counterattack. But McClellan didn't budge. In truth, the battle of Seven Pines unnerved him. He couldn't bear the sight of all his dead and wounded men. This was not the way to fight a war. In his mind, war was a game in which you defeated your opponent by brilliant maneuvers with minimal loss of life. McClellan loved his soldiers, and the feeling was mutual. They called him "Little Mac." They looked up to him as to no other general in the army. How could he sacrifice their lives by hurling them insanely against a superior foe? So, no, he did not counterattack. Once again he dug in and called for reinforcements. Once again he upbraided the administration for not supporting "this Army." When Lincoln and Stanton sent him one of McDowell's divisions, McClellan found other reasons for delay. Continuous rains had lashed the marshy plains east of Richmond. McClellan reported that his artillery and wagon trains were bogged down in muddy roads, his army immobilized. Before he could move against Richmond, the general must build footbridges, must corduroy the roads. . . .

In Washington, Lincoln threw the dispatches aside. The rebels attacked in bad weather, Lincoln complained. Why couldn't McClellan? The general seemed to think that Heaven sent rain only on the just. Then on June 25 came an even more alarming letter from the front. McClellan declared that the rebel army now had 200,000 men (it actually numbered about 85,000) and was preparing to attack him. In righteous indignation, the general bemoaned his "great inferiority in numbers," chastised the government for scorning his pleas for help, and announced that he would die with his troops. And if the rebels did annihilate his "splendid Army," the responsibility must "rest where it belongs."

Lincoln had just about had enough of this. Your complaints "pain me very much," he informed McClellan. "I give you all I can." Anyway, Lincoln feared that McClellan's outburst was just another excuse for not advancing on Richmond. He really should never have let the general go down to the Peninsula. McClellan should've launched his big battle at Manassas, should've struck the rebel army while it was there. Now the enemy was entrenched in front of Richmond with a stronger force, McClellan was belligerently inert, Union commands in Virginia badly spread out, the chances of a victory increasingly dim.

More bad news from the Shenandoah Valley, where the rebels were building up again. To avoid another disaster there, Lincoln now did something he should have done a long time ago: he collected the disparate forces of McDowell, Frémont, and Banks into a single army—the Army of Virginia —and appointed General John Pope as its commander. A conceited, pugnacious Illinoisian, Pope had served with Halleck out in the West, where he'd masterminded the capture of Island No. 10. He was a big man, with burning eyes, a bewhiskered chin, and an abrasive personality. In Washington, he dazzled Wade and Chandler with his tough talk, asserting that the Union must wage an aggressive war, must attack and demolish the rebels without any more delays. If Pope were in charge of the Potomac Army, Wade and Chandler burbled, he would overpower Richmond at once and drive clear to the Gulf of Mexico.

On June 26, the same day Pope assumed command of the Army of Virginia, a fierce battle raged in front of Richmond. Lincoln hurried over to the War Department, but Stanton had only a vague idea what was happening. As far as Lincoln could make out, Lee had mounted an all-out offensive, and he and McClellan were fighting it out along Chickahominy Creek. At last came a report from McClellan. The general was ecstatic. He'd repulsed the rebel attack and won a complete victory against tremendous odds. A complete victory! Now Lincoln was ecstatic. If that were true, then Lee's army must be destroyed and McClellan battering at the very gates of Richmond. What an incredible change of luck.

Subsequent dispatches indicated that McClellan had crossed south of the Chickahominy and was on the move. There were no detailed reports yet, but Washington was "wild with rumors and suspense," Hay noted, as crowds gathered at Willard's Hotel to read the latest bulletins. Lincoln camped out in Stanton's office, lying on a sofa while General Pope implored him not to let McClellan retreat.

As it turned out, though, McClellan already had retreated. Convinced that his intelligence service was right, that Lee had assailed him with 180,000 troops, McClellan forgot about capturing Richmond and elected to change his base of supplies from the York River north of the rebel capital to the James River south of the city. His only consideration now was to get his army to a safe defensive position on the James, secure naval support, and call for reinforcements. As he struck out for the James, Lee pounded him with one attack after another, but McClellan skillfully parried every blow. For seven desperate days the two armies rolled southward, punching away at one another as they moved. But Lee could not win a decisive battle and McClellan would not counterattack, obsessed as he was with executing a strictly defensive operation. As he fell back McClellan telegraphed Washington: "I have seen too many dead and wounded comrades to feel otherwise than that the Government has not sustained this Army. If you do not do so now the game is lost."

Disregard his insubordinate blasts. Lincoln and Stanton were doing everything they could to reinforce the general without stripping Washington of its defenses. They ordered up units from the North Carolina coast and pressed Halleck to send what men he could spare (he had none to spare). Finally Lincoln had no choice but to call on the governors for more volunteers, knowing full well what a storm of criticism that would cause. Already the press was fulminating about McClellan's failure to capture Richmond—and now the general was marching away from the city instead of toward it. In an open letter to Seward, which the Secretary of State took to a governors' conference in New York, Lincoln explained that Union armies in all theaters suffered from chronic manpower shortages. If he sent McClellan reinforcements from the Washington area, the rebels would drive on the capital. If he brought in a force from the West, the rebels would recapture Tennessee, Kentucky, and Missouri. What Lincoln wanted was to hold Union positions in the West, defend Washington, and raise an additional 100,000 volunteers for McClellan. Then the general would take the offensive. Then he would seize Richmond and end the war. "I expect," Lincoln said, "to maintain this contest until successful, or till I die, or am conquered, or my term expires, or Congress or the country forsakes me."

Thanks to Seward's pleas and politicking, eighteen Union governors urged the President to call up all the men he needed. Consequently, on July 1 he summoned 300,000 more volunteers from the states and then wired McClellan to save his army at all costs, even if he had to retreat back to Fort Monroe. "We still have strength enough in the country," Lincoln declared, "and will bring it out."

When McClellan telegraphed back that he needed 50,000 men at once, Lincoln lost his patience altogether. We have no men to send you now, the President said. Apart from your own army, we don't have 75,000 men in the entire Eastern theater. The idea that we can send you 50,000 men immediately is "simply absurd." Whatever you do, save the army. I'll send you troops as soon as they come in from the states.

The next Lincoln heard from him, McClellan had reached Harrison's Landing on the James and was safely entrenched there. In his reports to Washington, he bragged about his brilliant retreat, stressing that changing bases across a fighting front was one of the most difficult maneuvers in warfare. Said the general: "It will be acknowledged by all competent judges that the movement just completed by this army is unparalleled in the annals of war." McClellan had not only saved the Army of the Potomac, but had preserved his guns, trains, "and, above all, our honor." The troops were now in review, bands were playing, "salutes being fired, and all things looking bright." Now if Lincoln would send him 100,000 reinforcements, he could go on the offensive in six weeks, capture Richmond, and win the war.

One hundred thousand reinforcements! Did McClellan not understand what Lincoln had been telling him? He offered the general "a thousand thanks" for saving the army, but there was no way on earth Lincoln could raise enough volunteers for McClellan to mount an offensive in six weeks.

Lincoln decided it was time to have a talk with McClellan. His army seemed extremely vulnerable down on the James. And that wasn't the only problem. The public outcry over McClellan's Peninsula campaign was shrill and menacing: he'd lost more than 23,000 killed, wounded, and missing and had won no victories, none at all. Republicans across the land were clamoring for McClellan's head, insisting that his failure in Virginia would cost the party the upcoming by-elections. Moreover, there was grave danger of European intervention this summer. If the Union did not reverse its military fortunes in the East—the showcase for European eyes—then Great Britain might well conclude that the Confederacy was invincible and go ahead and recognize her as an independent nation.

On July 7 Lincoln left Washington on the steamer *Ariel* and reached Harrison's Landing late the following afternoon. He had an earnest conversation with McClellan, "stout, short, and stiffly erect," and asked him whether he ought not to pull out of Virginia. The general eyed him suspiciously, certain that Lincoln was up to some "paltry trick," and replied that he wanted to remain right here on the James. In a conference on the *Ariel,* Lincoln asked McClellan's five corps commanders the same question, and three of them also wanted to stay. Lincoln noted that the army seemed in surprisingly good condition, so perhaps McClellan and the other three generals were right.

Before Lincoln left for Washington, McClellan handed him a letter that was supposed to discuss the military situation. In reality, though, the general said nothing about the Army of the Potomac or future military operations. Instead he strayed off into politics—which wasn't unusual for generals in those days—and offered Lincoln some sober advice on how to save "our poor country." The President must wage a strictly conservative war, one aimed not at the Southern people and their institutions, but at the rebel government and the rebel army. Above all, Lincoln must not emancipate the slaves, for such a "radical" departure would "rapidly disintegrate our present armies." Furthermore, the President must appoint another General in Chief—a hint that McClellan was sick of civilians like Lincoln and Stanton trying to direct professional soldiers.

Lincoln read the letter carefully, thanked McClellan, and sailed away without a word as to what he would do. The fact was that he'd already decided to restore the office of General in Chief. But McClellan would not get the job—Lincoln knew better than to travel that road again. No, he would appoint Henry Halleck to the post, because he liked what Halleck had done as overall commander in the West. In addition, Lincoln had just about reached a decision about slavery and the war, but it wasn't one young McClellan was likely to approve.

Throughout the Peninsula campaign, Lincoln had brooded constantly over slavery, with pressures to strike at the institution "thrust at him from various quarters." As always, abolitionists and liberal Republicans pressed him relentlessly about the absurdity of fighting a war without uprooting the one institution that had caused it. In Congress, a furious debate raged over the second confiscation bill and its controversial slave liberation provision. But with a colonization clause also included, the bill enjoyed support from a majority of Republicans on Capitol Hill and seemed sure to pass . . . something Sumner wouldn't let Lincoln forget in begging him to issue an emancipation proclamation. Also, with the threat of European intervention again in the wind, Sumner rehearsed his old arguments that freeing the slaves would check Britain's drift toward Confederate recognition—and that whatever powerful Britain did in the matter, the other nations were certain to follow.

On July 4, three days before Lincoln left for Harrison's Landing, Sumner called again and urged "the reconsecration of the day by a decree of emancipation." "You need more men," Sumner argued, "not only at the North, but at the South, in the rear of the Rebels: you need the slaves." Lincoln pretended to disagree with him, contending that a Presidential proclamation was "too big a lick," contending that it would cause half the officers to throw down their arms and three more states to rebel. But Lincoln's objections were largely rhetorical, for he was now coming around to Sumner's views. As the war dragged into its second year with no end in sight, some drastic action against slavery seemed imperative if the Union was ever going to win. Still, Lincoln wanted to give his gradual, compensated plan one more try. Once again he would appeal to the border-state congressmen, try and make them understand that his plan was the only alternative to federally enforced emancipation.

On July 12, two days after Lincoln returned from Harrison's Landing, Congress passed the second confiscation bill and would shortly send it to the President for his signature. On that same day, Lincoln assembled border-state representatives in the White House and read them a solemn statement. You must forget the *"punctillio,* and maxims" of a quieter time, Lincoln said, and look "to the unprecedentedly stern facts of our case." You say you want the Union restored with slavery preserved. That is now impossible. The war has doomed slavery—you cannot be blind to that fact. If you reject my plan, the war will kill off slavery in your states "by mere friction and abrasion," and you won't get a dollar for your loss. Don't you see that my plan is the best for you? I'm not asking you to free your slaves at once, but to make "a *decision* at once to emancipate *gradually.*" As patriots, I pray you, consider this proposition and commend it to your states and people. If you would perpetuate popular government, you must not fail to do this. "Our common country is in great peril, demanding the loftiest views, and boldest action to bring it speedy relief." If it is relieved, its form of government is saved for the world, its cherished history vindicated, and its happy future assured and "rendered inconceivably grand." "To you, more than to any others," the President told them, "the privilege is given, to assure that happiness, and swell that grandeur, and to link your own names therewith forever."

But most of the border men turned him down. They thought his plan would cost too much, would only fan the flames of rebellion, would sow dangerous discontent in their own states. Their intransigence was a sober lesson to Lincoln. It was proof indeed that slave owners—even loyal slave owners like them—were too tied up in the system ever to free their own Negroes and voluntarily transform their way of life. Too, the border men felt it unjust that they should emancipate their slaves while rebel traitors were spared that sacrifice, and Lincoln supposed they were right. In any case, it was now blazingly clear that emancipation could never begin as a voluntary program on the state level. If abolition was to come, it must commence in the rebel South and then be extended into the loyal border later on. Which meant that the President must eradicate slavery himself. Yes, Lincoln could no longer avoid the responsibility. The time had come for him to lay a "strong hand on the colored element." "Turn which way he would," Lincoln said, "this disturbing element which caused the war rose against him," compelling him to throw off his own maxims of a quieter era and strike at the very foundation of the rebel South.

The next day, in a carriage ride with Seward and Welles, Lincoln brought up the slavery problem and remarked that if Southerners persisted in their rebellion, it would be "a necessity and a duty on our part to liberate their slaves." The Secretaries turned and looked at him; they had never heard him talk like this before. Lincoln went on in an urgent voice. He was convinced that the war could no longer be won through forbearance toward the rebels. The time had come to take a bold new path and hurl Union armies at "the heart of the rebellion," using them to destroy the institution which caused and now sustained the insurrection. Southerners could not throw off the Constitution and at the same time invoke it to protect slavery. They had started the war and must now face its consequences.

He'd given this a great deal of grave and painful thought, he said, and he'd about concluded that Presidential emancipation was "the last and only alternative," that it was "a military necessity, absolutely essential to the preservation of the Union." Since the slaves were a tremendous source of strength to the rebellion, Lincoln must invite them to desert and "come to us and uniting with us they must be made free from rebel authority and rebel masters." His interview with the border men yesterday, he said, "had forced him slowly but he believed correctly to this conclusion." So far Seward and Welles were the only people he'd confided in about this. He asked them to think about his arguments and give him their opinions in a few days.

The next day, July 14, Orville Browning brought Lincoln a copy of the second confiscation bill and beseeched him to veto it. As the bill had moved through Congress, Browning and a few other conservative Republicans had broken from the party and fought with Democrats against full-scale confiscation of slaves and other property in the rebel South. Browning warned Lincoln that the bill violated the laws of "civilized warfare" and was unconstitutional. Moreover, it "would produce dangerous & fatal dissatisfaction in our army." Your administration has now reached a crossroads, Browning told Lincoln. Either you must control "the abolitionists and radicals" in the party or they will control you.

Lincoln revealed nothing about his own emancipation plans, merely promised to look the bill over and saw Browning to the door. When he turned to it later, Lincoln found much in the bill that he liked. Going beyond the first confiscation act, the measure stated that if the rebellion did not cease in sixty days, the executive branch was to confiscate the property of all people who supported, aided, or participated in the rebellion. Federal courts would determine which Southerners were guilty. Those convicted (among other penalties) would forfeit their estates and their slaves to the federal government, and their slaves would be set free. Section nine liberated other categories of slaves without court action: slaves of rebels who escaped to Union lines, who were captured by federal forces or were abandoned by their owners, "shall be deemed captives of war, and shall be forever free." On the other hand, the bill exempted loyal Unionists in the rebel South, allowing them to retain their slaves and other property. Another section empowered Lincoln to enlist Negroes in the military. And still another, aimed at easing Northern racial fears and keeping Republican unity, appropriated $500,000 to colonize blacks outside the country.

Lincoln agreed entirely with the spirit of the bill, insisting that "the traitor against the general government" should forfeit his slaves and other property as just punishment for rebellion. But he thought the wording about the slaves unfortunately vague, and assumed that the bill did not eradicate the institution of slavery within a state, something Congress had no authority to do. He assumed that the bill first transferred the ownership of captured and confiscated slaves to the national government and then Congress freed them, which he believed was legal. The vagueness in the bill, he pointed out, could easily be corrected by more precise wording. But in any case he raised no "substantial objection" to the sections on slave liberation. He did, however, disagree with other portions on technical grounds and indicated that he would veto the bill as a consequence. To avoid that, Congressional Republicans attached an explanatory resolution removing most of Lincoln's complaints. Satisfied, Lincoln signed the bill into law and commanded the army to start enforcing it after the sixty-day waiting period. To Browning's despair, Lincoln had sided with Sumner, Trumbull, Chandler, Wade, and other Republicans who wanted to wage a "hard" war against the rebel South. Perhaps this was a "radical" step, but Lincoln thought a majority of Republicans now favored it.

On July 21 Lincoln informed his Cabinet that he was ready himself "to take some definitive steps in respect to military action and slavery." In truth, he intended to go beyond the second confiscation act—intended to handle emancipation himself, abolish rebel slavery in a sweeping executive stroke, and avoid tangled case-by-case litigation over slavery in the courts. The next day he read the Cabinet a draft of a preliminary emancipation proclamation; in it, he vowed to enforce the second confiscation act (which he appended to his edict) and struck at the very heart of the rebellion as he said he would. Come January 1, 1863, in his capacity as Commander in Chief of the army and navy, the President would liberate *all* slaves in the rebel states—those of secessionists and loyalists alike. He would thus make it a Union objective to annihilate slavery as a state institution in the Confederate South. He believed that only the President with his war powers could do this, and he deemed it "a fit and necessary military measure" to ensure the salvation of the Union.

Only Welles and Seward had known about Lincoln's intentions beforehand; the other Secretaries were astonished. "The measure goes beyond anything I have recommended," Stanton scribbled on a piece of notepaper. Nevertheless, he urged Lincoln to issue the proclamation at once, and Bates concurred. Blair, however, warned that the country was not ready for an emancipation edict and that it would ruin the Republican party in the fall by-elections. And Chase, the Cabinet liberal who chummed around with liberal Republicans on Capitol Hill? Chase was surprised and apprehensive. Don't misunderstand him—he was all for arming the Negroes and enlisting them in the army (a move Stanton also favored). But Chase was afraid that a proclamation of this sort would cause profound financial instability and thereby damage the government. In his opinion, emancipation could best be accomplished by having the various theater commanders quietly liberate the slaves in local areas.

Lincoln was startled at Chase's timidity. And Seward? What had he decided? Seward wore a grave expression. He agreed with Chase and Stanton about the need to use Negro soldiers. But he was vehemently opposed to an emancipation proclamation at this time. Why? Because the Union had won no clear military victories, particularly in the showcase Eastern theater. As a consequence, Europe would view an emancipation proclamation as "our last shriek on the retreat," as a wild and reckless attempt to compensate for Union military ineptitude by provoking a slave insurrection. If Lincoln must give an emancipation order, Seward said, he must wait until the Union won a military victory.

Seward's argument impressed Lincoln. He thought it over, weighed Chase's and Blair's objections, and conceded that the proclamation would have to be postponed. But evidently he remained undecided, because at Seward's suggestion, Weed dropped by the White House that night and added his opposition to emancipation at this time. At that, Lincoln reluctantly gave in and filed his proclamation away in his desk. Yes, Seward was right —the proclamation would have to wait until the Union won a decisive victory, preferably here in the East. But how long would that take? How long before McClellan attacked and whipped Lee? To think that the proclamation must depend on George McClellan was a depressing irony indeed. Yet Lincoln guessed he had no choice in the matter. Above all, he didn't want emancipation misconstrued as an insidious attempt to cover up Union military blunders.

Also, Chase and Blair were undoubtedly right about the adverse effect Presidential emancipation would have in the North. Lincoln himself had said the same thing often enough in the past. Was it not a fact that tensions over the Negro were on the rise, that anti-Negro riots had flared in several Northern cities that summer? Something had to be done to ease racial tensions and prepare the way for emancipation here at home.

Sumner again. He'd learned that Lincoln had drawn up an emancipation proclamation, only to let Seward talk him into postponing it. The senator faced a tough fight for re-election in Massachusetts, where his political foes were combining to oust him from office, arguing that the "abolitionist" senator differed from the Lincoln administration when it came to slavery and accomplished nothing worthwhile for the state. Anxious to prove them wrong, Sumner pleaded with Lincoln to put forth his proclamation at once, "without any reference to our military condition." But Lincoln refused. We must wait, the President said, *"time* is essential." Whereupon Sumner stalked out of the White House in utter dismay over Lincoln's slowness, his "immense *vis inertiae,*" his extraordinary inexperience in public affairs.

Lincoln had a little more political skill than Sumner gave him credit for. As the summer wore on, the President took steps to calm white racial anxieties and prepare the Northern people for the proclamation he planned to drop on them. In August he made a great fuss about colonization—a ritual he went through every time he contemplated some new antislavery move. He approved a project, to be financed by the federal government, to resettle black people in the Chiriqui coal region of Central America. Then on August 14, with temperatures soaring over one hundred degrees in Washington, Lincoln met with several local Negro leaders and read them an address about colonization that would go out to the press.

"You and we are different races," Lincoln declared. "We have between us a broader difference than exists between almost any other two races. Whether it is right or wrong I need not discuss, but this physical difference is a great disadvantage to us both." He conceded that American blacks were suffering "the greatest wrong inflicted on any people." Yet it was impossible for them to be free in this country. Wherever they went here, whites would oppress them, refuse them equal rights, want them to go. It was a fact of life they all had to face. And yet whites were suffering too. "See our present condition—the country engaged in war!—our white men cutting one another's throats, none knowing how far it will extend; and then consider what we know to be the truth. But for your race among us there could not be war." He repeated that. "Without the institution of Slavery and the colored race as a basis, the war could not have an existence."

So, he told the Negro leaders, black people must be colonized outside the country. It was the only solution. You intelligent men must lead the way, must help recruit enough Negro volunteers to resettle in Liberia or in some place closer to the Union. He realized, of course, that there were free Negroes who did not want to leave, who contended that blacks were better off here than in some foreign land. He didn't want to be unkind, but that was "an extremely selfish view of the case."

The Negroes wanted to talk this over and report back later. "Take your full time," Lincoln said, "no hurry at all."

Two days later the Negro delegation wrote Lincoln that they "heartily" endorsed his position. But other Washington blacks denounced him furiously and accused the leaders of selling out their own people. And when Lincoln's address appeared in the newspapers, Frederick Douglass blasted the President for "his inconsistencies, his pride of race and blood, his contempt for negroes and his canting hypocrisy." In heated meetings across the North, Negroes rejected colonization—this was their country too!—and petitioned

Lincoln to deport slaveholders instead.

But never mind Negro protest. Lincoln went ahead with the Chiriqui project, because he was convinced that white Northerners would not accept emancipation unless a colonization scheme were under way. Meanwhile, in careful, legalistic, patriotic terms, he hinted to various whites that something was in the wind.

To Union men in occupied New Orleans: "I shall do nothing in malice. What I deal with is too vast for malicious dealing." "Still I must save this government if possible. What I *cannot* do, of course, I *will* not do; but it may as well be understood, once for all, that I shall not surrender this game leaving any available card unplayed."

To a New York financier: "This government cannot much longer play a game in which it stakes all, and its enemies stake nothing. Those enemies must understand that they cannot experiment for ten years trying to destroy the government, and if they fail still come back into the Union unhurt."

In an open letter to Horace Greeley, who in the *Tribune* had reproached Lincoln for not freeing the slaves: "My paramount object in this struggle *is* to save the Union, and is *not* either to save or to destroy slavery. If I could save the Union without freeing *any* slave I would do it, and if I could save it by freeing *all* the slaves I would do it; and if I could save it by freeing some and leaving others alone I would also do that. What I do about slavery, and the colored race, I do because I believe it helps to save the Union." He said in closing, "I have here stated my purpose according to my view of *official* duty; and I intend no modification of my oft-expressed *personal* wish that all men every where could be free."

Lincoln had good reason to speak of slavery strictly in terms of preserving the Union: they were the only terms the white public was likely to accept. And so his letter to Greeley was a calculated statement, part of his efforts to prepare Northern whites for the proclamation filed in his desk. You see, Lincoln was suggesting, I am keeping my personal hatred of slavery out of this. If I free some or all the slaves (and for the first time in public he claimed the authority to free them), I do so only to save our Union, our cause, our cherished experiment in popular government.

Meanwhile Lincoln was doing all he could to bring about a military victory. In late July he installed Henry Halleck as General in Chief, thus relieving Lincoln of that harrowing responsibility and establishing what he hoped would be a more efficient command system. The failures in the Valley and the Peninsula had convinced him that he needed a professional soldier in Washington who would not only direct the armies, but counsel Lincoln and translate civilian policy into language the military could understand. And Halleck seemed an excellent man for the job. Nicknamed "Old Brains," he was a Phi Beta Kappa from Union College and a West Point graduate who spoke French, wrote military manuals, possessed a mass of technical military information, and viewed war itself as a game of finesse and maneuver. No dashing knight like McClellan, he was a stout officer with an ample paunch, a double chin, flabby cheeks, and bulging, watery eyes. To some he seemed shifty and insincere, sitting in interviews with his head cocked sideways and looking "fishily" at people out of one eye. Predictably, dyspeptic Welles took an instant dislike to him, said he was a stupid man who sat, smoked, and stared, scratching and rubbing his elbow "as if that was the seat of thought."

Still, Lincoln had confidence in "Old Brains" and told him so—told him he was in charge of the army now and must make the generals fight and win. "We must hurt this enemy," Lincoln would say, "must whip these people now." With Stanton in the War Department, they surveyed the military situation on all fronts. In the West, Buell was moving with glacial slowness toward Chattanooga in south-central Tennessee. What a laggard Buell was. Lincoln wanted to fire him, but Halleck persuaded the President to give him another chance—and then wrote Buell that he'd better mount an attack or Lincoln and Stanton would sack him. In the East, meanwhile, Pope's Army of Virginia was massed on the Rappahannock south of Manassas Junction; and McClellan was still on the James River southeast of Richmond, still crying for reinforcements before he could move against the enemy capital. What else? Lincoln was certain that if he sent McClellan 100,000 men today, the general would telegraph tomorrow and assert that the rebels were about to annihilate him with 400,000 men, and demand still more reinforcements. With so much riding on military events in the East, Lincoln refused to tolerate McClellan any longer. He wasn't going to fight where he was, so let them fetch his army home, combine it with Pope's, and make a show in the Manassas area as Lincoln had wanted all along. Accordingly, on August 3, Halleck ordered McClellan to pull back to Washington without delay. And a bitter and resentful McClellan, boiling at how Washington had sold him out, slowly and reluctantly started his army home.

Ominous news from Pope. The general had pushed south of the Rappahannock, but on August 9 the rebels routed his advance units at Cedar Mountain and Pope retreated back across the Rappahannock, followed by Virginia Negroes fleeing from slavery. Subsequent dispatches reported rebel troop movements in the Rappahannock area—the beginning of another large battle? An attempt to crush Pope before McClellan arrived? In the War Department's telegraph office, Lincoln tried to sort out the reports while Halleck admonished McClellan to move, hurry, get his army back here and help Pope. On August 24 McClellan finally reached Aquia Creek on the Potomac and forwarded a few outfits on to Pope. But McClellan refused to send the rest of his army until Halleck clarified the command situation. Was McClellan in charge of Pope's army or Pope in charge of McClellan's? In all the confusion, Halleck went to pieces. What if Pope were already encircled and being cut to shreds? What should Halleck do? McClellan, now at Alexandria, advised that they should either break through to Pope or defend Washington and "leave Pope to get out of his scrape."

Lincoln was shocked at the latter crack. Was McClellan crazy? Did he want Pope to get whipped? And what was wrong with Halleck? Why didn't he combine both armies and *order* the rest of McClellan's troops to the front? The truth was that Halleck was immobilized with anxieties and unable to lead. Yet Lincoln was reluctant to step in and take over again. So with the Union command system again in chaos, Lincoln clung to the telegraph office, entreating Pope's headquarters to write him the news. At last, on August 30, Pope reported that he'd fought a desperate battle at Bull Run and had driven Lee's army from the field. Lincoln's spirits soared. That night he waited for telegrams of a complete Union victory. If so, then he could issue his emancipation proclamation and strike at the cornerstone of the Confederacy. Yes, and Europe would stay out of the war. Yes, and maybe he could see an end to this.

But the next day the terrible truth came about second Bull Run. Lee's army had turned and smashed Pope, sending him in headlong flight toward Washington. Back at the White House, a dejected Lincoln made his way into Hay' room. "Well, John, we are whipped again." Later he rode out to the Soldier Home so depressed he thought "we may as well stop fighting."

But the next morning he rode back to Washington determined to m? some changes. He knew they would be unpopular, but something had tc done about this military mess. Since Pope and McClellan could not coo ate, one of them had to go and Lincoln decided it would have to be P Lincoln talked this over with Halleck, who agreed with the President then they conferred with McClellan, ordering him to take charge o shington's defenses and merge Pope's battered troops with the Army Potomac. Was McClellan to have sole command? Yes, Lincoln McClellan was to have sole command. Because who else did Lincol Who else was better at regrouping dispirited men after a humiliating Who else was a better organizer? Yet he scolded McClellan for hi?

behavior during second Bull Run—said it seemed that he wanted Pope to fail. And while McClellan emphatically denied that, he was exultant that Lincoln had turned to him in this hour of peril. The general thought he'd been exonerated.

The Cabinet, though, was in turmoil. In a fiery meeting in the White House, Stanton, Chase, and most of the other Secretaries inveighed against Lincoln for keeping McClellan in command. They blamed the general for the disaster at Bull Run and wanted him punished and removed. Lincoln "seemed wrung by the bitterest anguish," Bates recorded, "said he felt almost ready to hang himself." Yet he stubbornly defended his decision about McClellan, and grudgingly, unhappily, the Cabinet yielded to his judgment, though Chase predicted that "it would prove a national calamity."

Alone in his White House office, Lincoln mulled over his ocean of troubles, mulled over the vast uncertainties of this war, and confessed that events had spun out of his control or that of any other man. He felt like someone who'd broken loose from his moorings and was hurtling forth on a ship he couldn't steer . . . as in a mysterious dream he'd had before Sumter and again before Bull Run. He was on a phantom vessel moving swiftly toward a distant shore. But what awaited him there? What did God have in store for him and his shattered country? "The will of God prevails," Lincoln wrote at his desk. In this great war, both parties claimed to act "in accordance with the will of God." Yet maybe the Almighty's purpose was different from that of either side. "I am almost ready to say this is probably true—that God wills this contest, and wills that it shall not end yet. By his mere quiet power, on the minds of the now contestants, He could have either *saved* or *destroyed* the Union without a human contest. Yet the contest began. And having begun He could give the final victory to either side any day. Yet the contest proceeds."

In the next few days it seemed racing toward some terrible climax. On September 5 Lee invaded Maryland, unleashing a fall offensive intended in part for British eyes. If the rebels could win a decisive victory on Union soil, maybe Great Britain would demand an armistice that would end the war and guarantee Confederate independence. With a rebel army loose in Maryland, Lincoln needed fighting generals more than ever. Yet all he had were McClellan and Halleck . . . Halleck who now had lost his nerve, who blamed himself for the defeat at second Bull Run and could no longer make decisions or give orders, for fear that he would blunder and fail. Lincoln observed that Halleck had lost his pluck and settled into an "habitual attitude of demur." Alas, Lincoln had to lead again, with Halleck functioning as a "first-rate clerk" who would advise and write out orders, but would not command.

As for McClellan, he would have to take the field and repel the invasion. Lincoln had no other choice, for he was not about to turn the Army of the

Potomac over to a new and inexperienced field commander, not against the likes of Lee and Jackson. So with Lincoln's solemn encouragement, McClellan led his army away from Washington—a proud young general, called on again to save his country, promising to give Bobbie Lee the drubbing of his life. But how Republicans wailed. Had Lincoln lost his mind? How dare the President entrust the fate of the Union to a coward and a Democrat. It was a desperate gamble, Welles said, for if McClellan failed to defeat Lee, "the wrath and indignation against him and the Administration will be great and unrestrained."

After McClellan had gone, Lincoln made a vow, a covenant. If the general won a victory, Lincoln would consider it "an indication of Divine Will" that God had "decided the question in favor of the slaves" and that it was Lincoln's duty to "move forward in the cause of emancipation."

But when a delegation of Chicago ministers called on him to free the slaves in the name of God, handing him a memorial from Chicago Christians of all denominations, Lincoln was skeptical and argumentative. You want me to liberate the slaves, he said, and others want me to leave them alone, and all claim to speak for God. Well, Lincoln wished to know God's will, too, but these were not days of miracles, so he guessed he wouldn't receive a direct revelation. No, he had to consider the "plain physical facts" of the matter and let them dictate what was "wise and right." Then he engaged in his old habit of quarreling with the very thing he hoped to do. Look, he asked the ministers, what good would it do to issue an emancipation proclamation? Wouldn't it have about as much authority as the Pope's bull against the comet? How could Lincoln enforce emancipation in states where he had no power? True, a proclamation might please some in the North, but not so many as the ministers thought. And what about the effect on the loyal border? And the crucial principle of constitutional government, which was also at stake? Yet, he said, "do not misunderstand me, because I have mentioned these objections." They merely reflected some of the difficulties involved. "I have not decided against a proclamation of liberty to the slaves, but hold the matter under advisement. And I can assure you that the subject is on my mind, by day and night, more than any other."

Dispatches were coming in from McClellan now. On September 14 he struck Lee's army at South Mountain and drove the rebels back. The general was almost hysterical. "The Enemy in perfect panic," he telegraphed Washington. "Gen Lee last night stated publicly that he must admit they had been shockingly whipped." In the telegraph office with Halleck and Stanton, Lincoln was enormously relieved. "General McClellan has gained a great victory," he wired his old friend Jesse Dubois. "He is now pursuing a flying foe." And then he telegraphed McClellan: "God bless you, and all with you. Destroy the rebel army, if possible."

Then Lincoln waited for McClellan to deliver a knockout blow and vanquish Lee's army. What news now? he telegraphed the front. Heavy fighting on the sixteenth near Antietam Creek. And the seventeenth? What goes? McClellan reported that he was fighting the most awesome battle of the war, maybe the most awesome battle of all history, and that he was up against great odds but that he would win. As the battle raged on at Antietam, Lincoln, Halleck, and Stanton hovered over the telegraph, snatching dispatches as they clicked in. At last McClellan sent a grand announcement that he'd conquered Lee and that his victory was "complete." Lincoln and Halleck, of course, interpreted this to mean that McClellan had overwhelmed Lee's army and was now eliminating it as a fighting force.

But subsequent communiqués demolished that hope. As it developed, complete victory for McClellan meant not the pursuit and destruction of Lee's army, but simply halting his invasion. On September 19, with McClellan doing little to impede him, Lee crossed safely back into Virginia. Yet McClellan was swept away with joy. Yes, he had saved the country a second time. He was truly the man of the hour. Because he'd "defeated Lee so utterly and saved the North so completely," McClellan would now insist that the government accord him the respect he was due. He would insist that Lincoln get rid of Stanton and Halleck and reinstate McClellan as General in Chief.

Lincoln was thoroughly disgusted that McClellan had let Lee's army escape, to nurse its wounds and fight again. But however indecisive and incomplete Antietam was, Lincoln still regarded it as a victory—and an unmistakable sign from God concerning emancipation. On September 22, therefore, Lincoln called a Cabinet meeting to discuss the final draft of his preliminary proclamation. His Secretaries were excited and tense, for this was an historic moment, but Lincoln opened the session with a story, reading them a chapter from Artemus Ward's "A High Handed Outrage at Utica." When Artemus showed up in the New York town with a cage of "wax figures of the Lord's Last Supper," the tale went, a burly fellow pulled Judas out and exclaimed, "What did you bring this pussylanermous cuss here fur?" "You egrejus ass," Artemus said, "that air's a wax figger—a representation of the false 'Postle." The fellow bellowed that Judas couldn't show himself in "Utiky with impunerty" and caved the figure's head in. Outraged, Artemus "sood" the man, and the jury returned a verdict of "Arson in the 3d degree."

Most of the Secretaries were not amused, Chase wearing a sick smile and Stanton frowning sourly. The President put the book aside and got down to business. He told the Cabinet about his vow to himself—and, he said hesitatingly, "to my Maker"—and then read his proclamation. When Congress convened in December, he would push again for gradual, compensated emancipation in the loyal states and would continue his efforts at voluntary colonization. Then he turned to the Confederate South. If the rebels did not stop fighting and return to the Union by January 1, 1863, the President would free "thenceforward and forever" all slaves in the rebel states; and the executive branch of the government, including the army and navy, would recognize and maintain the freedom of black people there. On the same day he would announce which states or parts of states would be affected. He also quoted the slave liberation sections of the second confiscation act and again enjoined the military to enforce it.

Lincoln would entertain criticism, he informed his Secretaries, but he would not rescind or postpone his proclamation. Bates, who'd endorsed the document back in July, now had reservations: he thought forced colonization ought to accompany emancipation and dreaded any step that might lead to Negro equality. Blair favored emancipation "from principle," but feared it would have dangerous repercussions in the loyal border. Lincoln replied that he was aware of the border problem, but that right now "the difficulty was as great not to act as to act." Given the violent and terrible dimensions the war had taken, the border could no longer dictate his policy.

Seward, for his part, privately questioned the wisdom of the proclamation, fretted that it might incite a slave rebellion behind Confederate lines and cause deep divisions in the North, argued that the war would kill off slavery without official government acts. But in Cabinet this day he supported the document regardless of his misgivings. Stanton was still all for it. And Chase? The Secretary of the Treasury had been talking over emancipation with Sumner and his Congressional colleagues, and Chase now approved the proclamation emphatically. In fact, he even wanted to enfranchise liberated blacks so they could protect themselves. Welles, who'd been silent about the matter in July, now voiced his support of the proclamation as well. "An extreme exercise of war powers," Welles admitted, but "under the circumstances and in view of the condition of the country and the magnitude of the contest," he too considered it a military necessity.

Upheld by most of the Cabinet, Lincoln published his decree the next day, proclaiming to people everywhere that the institution of slavery was to perish in the rebel South, and that human freedom "thenceforward and forever" was to become a Union war aim and rallying cry.

In the North the reaction was almost instantaneous. William Lloyd Garrison pronounced the edict "an important step in the right direction" and hoped that Lincoln would stick to his word next January 1. Other abolitionists cheered the proclamation enthusiastically —the President had succumbed to their moral crusade to make this a war for slave liberation after all. "We shout with joy," Frederick Douglass exclaimed, speaking for most Northern blacks, "that we live to record this righteous decree." Ralph Waldo Emerson asserted that it was something "most of us dared not hope to see, an event worth the dreadful war," and he ranked the proclamation alongside the Declaration of Independence and the English reform bill.

A majority of Republicans, of course, welcomed the proclamation as a bold and necessary war measure. And most liberals were as happy as the abolitionists—finally Lincoln had seen the light and hurled a thunderbolt at the cause of the rebellion as they had pressed him to do. It was indeed a "mighty act" and a glorious day. Charles Sumner, who'd prodded Lincoln harder than anyone, was especially elated. By now Sumner's home-state foes had launched a conservative "people's movement" to unseat the senator, on the grounds that he was at odds with Lincoln's policy. The proclamation scotched the movement and all but ensured Sumner's re-election in Republican Massachusetts. Though some Republicans took issue with the proclamation on technical matters, only a few conservatives like Browning really objected to military emancipation itself. Browning, up for re-election in Illinois, thought Lincoln's decree "unfortunate."

If Browning was apprehensive, most Democrats were enraged. As if the Republican-sponsored confiscation acts hadn't been "radical" enough, now the President was resorting to an unconstitutional fiat. In the beginning, Northern Democrats had been willing to side with Lincoln to maintain the old federal Union with slavery guaranteed. But they were not going to support him in a revolutionary struggle for "nigger" freedom. As a consequence, Democratic spokesmen summoned all conservatives to unite, throw "the shrieking and howling abolitionist faction" out of office, and return the war to its original purpose—"to save the Union as it was." One of the most incensed Democrats of all was George McClellan, whose victory at Antietam had made the proclamation possible. The general declared that he wouldn't fight to free the slaves. He would resign first. He would denounce the proclamation in public. But dissuaded by his advisers from doing anything rash, McClellan settled for an army order which called attention to the decree and reminded his troops that political errors could be corrected at the polls.

In the border slave states, the response to Lincoln's "mad act" was predictable. In the Confederacy, papers and politicians alike flailed the proclamation as a villainous attempt to stir up a slave revolt. Southerners did not misunderstand Lincoln's motives, asserted the Confederate press, for his proclamation was exactly what they had expected of him all along. "Fiendish" and "heinous" man, he dreamed of another Nat Turner uprising behind Confederate lines, of slaves butchering Southern women and children while Southern men were at the fronts. But he succeeded only in convincing Confederates to fight all the harder for their independence and their homes.

In England reactions were mixed. Union friends like the *London Star* extolled the proclamation as a humanitarian act. But the British pro-Southern press, led by the *London Times,* damned it as "the wretched makeshift of a pettifogging lawyer" who called to his aid "the execrable expedient of a servile insurrection." And *Punch* published a cartoon showing the President tossing in his last card, a black ace, in a game of rouge et noir. While some politicians thought the proclamation absurd, since it freed slaves where Lincoln had no authority, the British Cabinet gave the document some attention. That fall, the Cabinet was seriously considering Confederate recognition, but the news of the Union victory at Antietam and Lincoln's ensuing proclamation, combined with Prime Minister Palmerston's own nagging doubts about Southern invincibility, convinced the Cabinet to postpone recognition for now.

Back in Washington, meanwhile, Hay brought Lincoln a sampling of editorial opinion from the leading papers, but Lincoln wasn't too interested in their comments. He'd studied the emancipation problem so long, he said, that he knew more about it than the papers did. On September 24 Hay interrupted the President in his office and said a large crowd had come to serenade him on the White House lawn. Did he want to say anything? "No," Lincoln said flatly. But all the cheering and the band music finally brought him to the window for a few remarks. He supposed the serenade was because of the proclamation (cries of "Good" and applause from below). "What I did," Lincoln said, "I did after very full deliberation, and under a very heavy and solemn sense of responsibility." He added, "I can only trust in God I have made no mistake."

Later the serenaders gathered at Chase's home for speeches and refreshments. Everyone was in "a glorious humor," said Hay, who'd come along from the White House. After most of the crowd had gone, Chase and a few Republican "old fogies" sipped wine together and talked about the proclamation and the "insanity" of rebel slave owners. Had they remained in the Union, Chase said, they might have kept slavery alive for decades. But by fomenting civil war they had done what no public feeling and no political party in the North could ever have done—they had placed slavery on the "path of destruction." It was a delicious irony. As the evening passed and the wine had its desired effect and the full meaning of the proclamation swept over the group, "they all seemed to feel a sort of new and exhilarated life," Hay observed; "they breathed freer; the President's Proclamation had freed them as well as the slaves. They gleefully and merrily called each other and themselves abolitionists, and seemed to enjoy the novel sensation of appropriating that horrible name."

Their euphoria was short-lived. For the preliminary emancipation proclamation set off a powder box of racial discontent in much of the lower North, especially the Midwest, and portended a Republican disaster in the fall elections. In the Midwestern states, bipartisan support for Lincoln's war policy had all but disintegrated, as Democrats there counted off Lincoln's crimes—the military arrests, the catastrophes on the battlefield, and now this "monstrous" and "criminal" proclamation, which converted this into nothing but "a nigger war." What an "abolitionist" dictator Lincoln had become. From Illinois to New York, Democratic orators warned white audiences what they could expect under continued Republican rule: the destruction of their civil rights, endless casualties in a war for slave emancipation, Southern Negroes swarming into Northern communities to fornicate with white women and take away white jobs.

Republican candidates, on the other hand, defended Lincoln's war measures as indispensable for the salvation of the Union. And they tried hard to dispel Northern racial fears, contending that slaves would never leave the South because the climate there was more suitable to them, contending that after the war was won and slavery was eradicated in Dixie, Northern Negroes would also move there and leave the North an unsullied white man's country. And anyway, Republicans ultimately hoped to colonize all blacks abroad.

But white voters were sullen and war-weary, and in the fall elections they dealt the Republicans a smashing blow. The North's five most populous states—New York, Pennsylvania, Ohio, Indiana, and Illinois, all of which had gone for Lincoln in 1860—now returned Democratic majorities to Congress. The Democrats also carried New Jersey and made impressive gains in Michigan and Wisconsin. While Lincoln's party, in coalition with border Unionists, retained control of Congress by a narrow margin, the future looked bleak indeed for 1864. On the state level, New York and New Jersey chose Democratic governors, and the Midwest voted heavily for Democratic legislatures. In Illinois Congressional races, moreover, Lincoln's friend Leonard Swett, campaigning in favor of administration policies, went down to bitter defeat in the President's home district. And a Democrat would replace Browning in the Senate, since antiwar Democrats had won firm control of the Illinois legislature.

Lincoln had never felt more rejected and more alone. How to account for so appalling a defeat? He complained that Republican papers which had vilified his administration had given the Democrats a weapon to club him with. And "certainly the ill-success of the War had much to do with it." He admitted, too, that his proclamation had also played a large part. Yet he told a delegation of Kentuckians that he would rather die than retract a single word in it.

On the military fronts it was the same old story. In mid October, almost a month after Antietam, McClellan was still encamped on the battlefield, offering one excuse after another as to why he hadn't moved after Lee. Well, Lincoln must not tolerate McClellan's procrastination. Lincoln must trust his own judgment, must not defer any longer to timid generals who shrank in horror at their own shadows. He *must* command them. He *must* make them fight.

He ordered McClellan to move. Look at your maps, the President wrote him. Lee is in the Shenandoah with an insecure base. You are closer to Richmond than he is. You must cross the Potomac and get in between him and the rebel capital; if he starts for that place, you can beat him there "by the inside track." If he stands where he is, then hit him and destroy his army. If we can't defeat him where he is now, we'll never do it if he gets to the trenches around Richmond. Surely our men can march and fight as well as the enemy. "It is unmanly to say they can not do it."

He told Buell the same thing—Buell who after ten months of campaigning in Kentucky and western Tennessee was no closer to east Tennessee than when he began. Through Halleck Lincoln sent the general an ultimatum: You must enter east Tennessee before the autumn rains make roads impassable. Now *move*. But Buell did not move. He argued instead, insisting that it was best to operate against Chattanooga in south-central Tennessee. Yet he did not move against Chattanooga either. His supply problems seemed insurmountable. Why, Lincoln moaned, can't we live as the enemy lives, march and fight as he does? But Buell was not impressed with Lincoln's irritations. There were more delays. Delays in the East and the West and the people voting Republicans out of office in state after state. . . . Lincoln couldn't take it any longer. On October 23 he fired Buell, something he knew he should have done a long time ago, and put William S. Rosecrans in command of the Army of the Ohio. Another West Pointer, with blond hair and a ruddy complexion, "Old Rosey" was so obsessed with religion he could keep his aides up until four in the morning discoursing on theological questions. At the same time, he was a profane individual who preferred blonds for staff officers because in his opinion "sandy fellows" were smarter than brunettes. He had an excitable streak and sometimes lost his head in a crisis. But maybe he would move. Yes, maybe he would fight.

Lincoln was about to lower the ax on McClellan, too. It was now October 25 and McClellan still hadn't gone after Lee as the President ordered. Why? Because his horses had sore backs and were "broken down from fatigue," McClellan reported. Lincoln wired him back: "Will you pardon me for asking what the horses of your army have done since the battle of Antietam that fatigue anything?" McClellan hotly defended himself and his cavalry. But at last he moved. Slowly he put his army across the Potomac—complaining afterward about his supplies, Lincoln said—and then edged southward. Lincoln stormed about McClellan's "slows" and said he plainly "did not want to hurt the enemy." On November 5 Lincoln relieved the general of his command, because the President was tired of sticking "sharp sticks under McClellan's ribs," tired of dealing with a man who fought like "a stationary engine."

On the same day Lincoln put Ambrose E. Burnside in charge of the Army of the Potomac, because he had nobody else to choose. And so McClellan was gone, never to have another command, certain that he'd "tried to do what was right," bitter toward Lincoln and his cronies for the way they had treated him. And "Burn" was the commander now, "Burn" who stood six feet and sported prodigious sideburns, who looked virile and tough riding at the front of his troops with a pistol swinging at his side. Yet in reality he was a nervous insomniac and deeply insecure about himself. But I'm not fit to command an entire *army,* he sputtered when Lincoln's order arrived. Maybe so, but Lincoln thought anybody was better than McClellan. And anyway Burnside had stayed out of army quarrels, which Lincoln liked.

November passed with more delays. Rosecrans went to Nashville and squatted there for the entire month—building up supplies, he explained to Washington. Lincoln groaned that all this *"impedimenta"* was going to ruin the army, and he ordered Halleck to name another commander, but Halleck persuaded him to give Old Rosey a chance. In the East, meanwhile, Burnside had advanced south toward Fredericksburg on the Rappahannock, in hopes of driving from there on Richmond. But he neglected to take pontoon bridges along and his army couldn't cross the river without them. While he waited for the bridges, Lee hurried down to Fredericksburg to block Burnside's advance.

A delegation of women visited the White House and asked Lincoln for a word of encouragement. "I have no word of encouragement to give," he replied. Our people and our generals have not yet made up their minds that we are involved in an awful war. Our officers seem to think the war can be won by plans and strategy. That is not true. Only hard and tough fighting will win.

One woman noted that Lincoln's face was ghostly pale—as though he hadn't slept for weeks—and that he was stooped like an old man.

In December, in the midst of rising public indignation against him, Lincoln struck at slavery again. In his annual message to Congress, he recommended several Constitutional amendments to facilitate emancipation and racial adjustment. One amendment would offer federal funds to all states which abolished slavery any time between now and 1900. Another would guarantee the freedom of all black people liberated in the war, and would also compensate loyal Unionists who'd lost their slaves in the conflict. A third amendment would authorize additional federal appropriations to colonize blacks abroad. With an eye on white Negrophobia and the recent elections, Lincoln denied that emancipation would send "free colored people" flocking into the North. Before the war, blacks had come North to escape slavery. But if slavery were removed, they would have nothing to run from; they would remain in the South working for wages until the government could resettle them "in congenial climes."

Did his proposed amendments replace his emancipation proclamation? No, they were corollary moves, designed to push emancipation in the loyal border and to offer the rebels themselves a choice. Should they stop fighting and return to the Union before his proclamation took effect, they could avoid the "sudden derangement" of federally enforced abolition and enjoy the benefits of the President's gradual and compensated program. But if the rebellion persisted beyond next January 1, as Lincoln figured it would, then his proposed amendments would guarantee the permanency of his proclamation as well as the slave liberation sections of the confiscation acts. For Lincoln understood only too well that laws and decrees could be rescinded by a subsequent Congress or a later administration (and given the mood of the country as reflected in the recent elections, that was a distinct possibility). Now that he'd committed himself to emancipation, Lincoln was determined to accomplish it by any and all means. And so he asked Congress—and the Northern people beyond—to give him their support. "The dogmas of the quiet past," he reminded them, "are inadequate to the stormy present. The occasion is piled high with difficulty, and we must rise with the occasion. As our case is new, so we must think anew, and act anew. We must disenthrall our selves, and then we shall save our country.

"Fellow-citizens, *we* cannot escape history. We of this Congress and this administration, will be remembered in spite of ourselves. No personal significance, or insignificance, can spare one or another of us. The fiery trial through which we pass, will light us down, in honor or dishonor, to the latest generation. We *say* we are for the Union. The world knows we do know how to save it. We—even *we here*—hold the power, and bear the responsibility. In *giving* freedom to the *slave,* we *assure* freedom to the *free*—honorable alike in what we give, and what we preserve. We shall nobly save, or meanly lose, the last best, hope of earth."

Those words came from a visionary and deeply troubled man, waiting in his White House office while Congress heard his message through. Did he really expect the amendments to be adopted? Really expect Northerners to lay aside the habits and prejudices of a lifetime and join him in thinking anew? Orville Browning, exasperated by Lincoln's abolitionist posturings, thought the President was living in an "hallucination." In the White House later, Browning begged Lincoln to abandon his "reckless" emancipation schemes lest they shatter their already demoralized party and destroy what remained of their country.

I t was a month of unmitigated gloom. In mid December, Burnside launched a massive attack against Fredericksburg, only to suffer one of the worst defeats of the war. From reports and war correspondents, Lincoln put together what happened on that cold and ghastly battlefield. Burnside sent his giant army across the Rappahannock and threw one division after another against impregnable rebel positions behind a fortified stone wall, with cannon and muskets flaming up and down the line. Burnside sustained more than twelve thousand casualties at Fredericksburg. Afterward he broke down in tears—"Oh those men! Those men over there!" —and talked hysterically about leading his old corps on a suicidal charge. At last he pulled himself together and called his broken army back across the river, weeping as he did so.

For days after the battle, Lincoln struggled with a severe depression. If there was a man out of perdition who suffered more than he, Lincoln pitied him. How gladly would he change places with the soldier who'd died out there with the army. And the country! What a fusillade of criticism he was getting now, as both friends and foes blamed Fredericksburg on him and his "weak and timorous" Cabinet. We have "a cowardly imbecile at the head of the Government," a Cincinnati editor wrote Sumner. "Old Abe will do nothing decent till driven to it by a force which would save all the devils in hell." "I am heartsick," cried Fessenden of Maine, "when I think of the mismanagement of our army. . . . The simple truth is, there never was such a shambling, half and half set of incapables collected in one government before or since the world began." "Disgust with our present government is certainly universal," observed another man. "Even Lincoln himself has gone down at last. Nobody believes in him any more."

As the debacles mounted up, there were deepening schisms in Lincoln's official family, too. For their part, Chase and Welles became even more suspicious of Seward, even more convinced that he was an evil schemer who wooed Lincoln away from the other Secretaries and led the President into devastating misjudgments. Chase in particular felt snubbed and resentful, since Lincoln excluded him from military decisions these days and consulted almost exclusively with Stanton and Halleck. Too, Chase abhorred the President's lack of administrative system, his refusal to confer with the Cabinet regularly, his indecisiveness. But what bothered Chase the most was the

intimacy between Lincoln and Seward—their confidential chats, their carriage rides, their outrageous jokes and laughter. How Chase detested that
unctuous, conniving New Yorker. Yes, Welles was right: Seward was a
sycophant, a self-promoting politician who'd surrendered to border-state and
Negrophobic counsels and had dragged a naïve Lincoln along with him. In
talks with his liberal Congressional friends, Chase intimated that Seward was
a malignant influence on the President, that it was the Secretary of State who
was responsible for administrative bunglings.

So it was that Seward became a scapegoat for Republican discontent.
Inflamed by Fredericksburg, aroused by Chase's grumblings and Sumner's
own accusations that Seward was incompetent in foreign affairs and morally
obtuse, Congressional Republicans singled out the Secretary of State as a
symbol of executive impotence—as the biggest failure of a failed administration—and demanded that Seward be dismissed. On December 16 and 17 all
Republican senators but two caucused about Seward and chose a special
delegation to express their grievances to Lincoln. A Vermont conservative
led the delegation, which included Sumner, Wade, Trumbull, Fessenden,
and four others.

Lincoln agreed to see them on Thursday night, December 18, but he was
most upset. Because of the things Congress said about Seward, the Secretary
of State had offered his resignation—but of course Lincoln didn't want him
to go. Anyway, the senators really wished "to get rid of me," Lincoln told
Browning, "and I am sometimes half disposed to gratify them." "We are now
on the brink of destruction," Lincoln moaned. "It appears to me the Almighty is against us, and I can hardly see a ray of hope."

For three hours that Thursday night, Lincoln listened quietly as the senators aired their complaints. There was a widespread belief, they contended,
that Lincoln had failed to consult the Cabinet as a body and had made
important decisions without Cabinet approval. Then they opened up on
Seward, accusing him of indifference to Republican war aims, of causing
friction in the Cabinet and bringing on doom and drift throughout the
administration. When they finished, Lincoln said he would consider their
objections and respond the following night.

Alone, he decided that the senators were not out to destroy him after all.
No, they seemed "earnest and sad—not malicious nor passionate," and
seemed convinced of Lincoln's own honesty. Still, he didn't understand how
they could believe those "absurd lies" about Seward. The more Lincoln
thought about this, the more he believed that Chase "was at the bottom of
all this mischief." But what to do? Lincoln wanted to keep both Secretaries
—for Seward "comforts him," as Welles noted, and "Chase he deems a
necessity." At the same time, Lincoln had to stay on working terms with

Sumner and the other senators, whose support he would need in the trials ahead. He thought deeply about the problem, and by the next morning, December 19, he'd decided to lay a trap. He would "force certain men to say to the Senators *here* what they would not say elsewhere."

First he called a special Cabinet meeting—without Seward—and told his Secretaries about last night's grilling. He said he could not afford to lose any of his department heads, said he "did not see how he could get along with any new cabinet, made up of new materials." Maybe he'd called fewer meetings than some would have liked (an eye on Chase now), but that only demonstrated how he'd gained confidence in himself and needed less advice from people. Nevertheless, he'd consulted with them all on questions of fundamental policy and there had been no lack of unity on the measures they had adopted. And they all conceded that this was so—Chase included.

When the senators called that night, Lincoln had all the Cabinet with him except Seward. Right off he gave a speech about the harmony of his Cabinet, informing the delegation that there had been "no want of unity or of sufficient consultation" and summoning his Secretaries to back him up as they had in the Cabinet meeting. Chase was trapped. He couldn't change his position now, lest Lincoln and the other Secretaries think him the schemer, him the liar. Yes, he confessed, the Cabinet was generally in agreement on important matters. He even admitted that Seward was earnest in his conduct of the war, that in Cabinet he'd favored the emancipation proclamation and hadn't "improperly interfered" in anything. But Chase was offended. "He wouldn't have come," he said, "if he had expected to be arraigned." The senators, of course, were astonished. This was not what Chase had told them! In the course of the discussion, Fessenden remarked that they wanted only "to offer friendly advice." They didn't mean to dictate to Lincoln or infringe on his prerogatives. As they left, Lincoln asked whether they still thought Seward should be kicked out of office. Half of them now shook their heads.

The next day Chase tendered his resignation—and was mortified at Lincoln's reaction. "This cuts the Gordian knot," he exclaimed triumphantly. "I can dispose of this subject now. I see my way clear." Now that he had the resignations of both Cabinet rivals, he refused to accept either of them. The following day, December 21, he asked them both to stay on the job. Seward deeply appreciated what Lincoln had done for him and withdrew his resignation at once. And of course if Seward was going to remain, Chase must do so as well. But the entire episode left Chase bitter and alienated.

Nevertheless, Lincoln was proud of himself. At last he thought he'd done something right—had solved a dilemma "to entire satisfaction." He'd not only retained the services of two able men, but maintained a balance of conservative and liberal influences in his official family. Moreover, he'd refused to give in to Congressional pressure and yet had kept the lines of communication open with powerful Republican senators on Capitol Hill. And he'd exonerated Seward, too, and would tolerate no further attacks against him. In fact, when a couple of men did assail him later, Lincoln "rather gave my temper the rein" and "talked pretty damned plainly."

In those last days of December, Lincoln was absorbed with the details of his final proclamation and the fate of black people in America. As he pondered the Negro question, Browning and Sumner came one after another, Browning to protest against emancipation, Sumner to keep the President from wavering in his duty. Lincoln assured a worried Sumner that he would "stand firm." "There is no hope," Browning grieved. "The proclamation will come. God grant that it may not be productive of the mischief I fear."

What troubled Lincoln now was what to do with the liberated slaves. By now the Chiriqui colonization project had fizzled out, mainly because several Latin American countries objected to a U.S. colony in their midst and threatened to keep it out by force. So Lincoln crossed off Chiriqui, with growing doubts as to whether colonization was a workable solution to the race question anyway. If by some miracle the Union did win this war and four million Southern Negroes were ultimately freed, the cost of transporting all black people out of America would be astronomical, not to mention expenses for housing, shelter, and food to get them started in some new land. And in any case most Northern blacks seemed intractably opposed to colonization, which meant that it couldn't be brought off voluntarily as Lincoln insisted it must. Still, he wanted to leave all possibilities open—and above all to pacify Northern whites. So late in December he signed a contract with white promoters to resettle five thousand Negro volunteers on Haiti's Isle of Vache. Thus when Lincoln's proclamation took effect, he could show Northern whites that a colonization project was still on the drawing boards.

Nevertheless, Lincoln had just about decided that the fate of former slaves would have to be worked out in the South itself, that the white and black races in America would have to learn how to live with one another. He had a general idea that Southern "colored people" could be utilized as a free work force and be hired out to their former masters at decent wages, thus preventing the blacks from coming North as white people feared. He'd made an oblique reference to this in his recent message to Congress and would do so again in his final proclamation. But how to implement his idea? How to merge whites and free blacks without causing racial friction in South and North alike? He had no answer to that.

But he had reached a decision about able-bodied black men—something Sumner, Stanton, and others had urged on him for some time now. He would enlist blacks in the army—Southern slaves and Northern free Negroes—and would say so in his final proclamation. If he'd opposed this earlier in the rebellion, mainly out of concern for the border states, the terrible exigencies of the war had compelled him to change his mind. The fact was that the Union desperately needed black manpower, what with recruiting falling off sharply this winter and Union armies suffering troop shortages on all fronts. "The colored population," Lincoln said, "is the great *available* and yet *unavailed* of, force for restoring the Union," and he meant to avail himself of that force and to use the blacks as *soldiers,* not as menials. And though he was not quite ready to throw them into the front lines as combat troops, he would employ them extensively as garrison soldiers, thus liberating thousands of whites from garrison duty so that they could fight.

On New Year's Eve, with tension rising in Washington, Lincoln met with his Cabinet and discussed the wording of the final proclamation. Later he called at the War Department to check on military developments. Nothing much in the East, where Burnside's army was encamped north of the Rappahannock. In the West, Grant's Army of the Tennessee had moved out of Memphis and driven down the Mississippi. His target was Confederate Vicksburg, a strategic river fortress which disputed Union passage down the Mississippi. But Grant had run into difficulties. First rebel cavalry had raided his supply lines, burned his base, and forced him back to Memphis to reorganize. Then one of Grant's corps, under the command of red-haired William Tecumseh Sherman, had struck at Vicksburg from the northeast, in tangled bluff and bayou country, only to be thrown back. At year's end, Grant was stalled north of Vicksburg, and the prospects for a successful campaign seemed pretty dismal.

The news from Rosecrans was even worse. A few days ago the general had finally led his army out of Nashville to dislodge the rebels from middle Tennessee. But today's reports were disturbing. From all appearances, a Confederate army had fallen on Rosecrans at Murfreesboro and rolled back the Union right. Another catastrophe in the making? Halleck and Stanton didn't know, but they promised to keep Lincoln informed of the latest reports.

That night Lincoln tossed in fitful sleep, dreaming of corpses on a distant battlefield in Tennessee, of guns flashing in the night, of silent troops lying exhausted in the rain, of crowds reading casualty returns at Willard's Hotel. He woke in the darkness and lay there until the first gray of morning spread through his chamber. Tired and trembling from the night, he made his way down to the shop and lit the fireplace and gas lamps. So it was the dawn of a new year and the rebellion continued with no foreseeable end. What had begun as a ninety-day skirmish to restore the old Union had now swelled into a cataclysmic upheaval, forcing Lincoln to hurl an edict like a lance at the heart of the rebel South. As Lincoln told an Indiana senator, the war was the supreme irony of his life: that he who sickened at the sight of blood, who abhorred stridency and physical violence, should be cast in the middle of a great civil war, a tornado of blood and wreckage with consequences beyond prediction for those swept up in its winds.

At his desk, he put the final touches on his proclamation and then read it to make sure the words were right.

As of this day, the document said, all slaves in the rebellious states were "forever free." "For the present," the following areas were exempted from emancipation: those Louisiana parishes behind Union lines, certain occupied places in Virginia, the counties of West Virginia (which had recently separated from the rebels and would soon be admitted into the Union as a new state), the state of Tennessee, now under the military governorship of Andrew Johnson, and the entire loyal border. In issuing this decree, the President admonished the slaves to refrain from "unnecessary violence" and remain in the South working for "reasonable wages." Those slaves and free Negroes who so desired might now enlist in the Union army, "to garrison forts, positions, stations, and other places," and might sail on Union warships. "And upon this act," the President concluded, "sincerely believed to be an act of justice, warranted by the Constitution, upon military necessity, I invoke the considerate judgment of mankind, and the gracious favor of Almighty God."

He would sign an official copy of the proclamation later in the day. Meanwhile he went to fetch Mary for the usual New Year's reception, certain to be an ordeal. It was Mary's first reception since Willie had died, and she was not sure she was up to it. She looked lovely, though, bejeweled with diamonds and clad in a silver silk dress, with garlands in her hair and a black shawl about her head. Mary was extremely pleased about her husband's proclamation, for she favored emancipation as an act of humanity for "colored people."

At eleven the Lincolns went down to the Blue Room to greet a long procession of people—the diplomatic corps, army and naval officers in the order of their rank, politicians, and the public. Lincoln was remote, his mind on Murfreesboro, as he shook hands in the crowded room. But he recognized

Noah Brooks as he passed through the line. A native of Maine now working as Washington correspondent for a California newspaper, young Brooks was a frequent caller at the White House; in fact, he would soon be as close to the Lincoln family as Ellsworth had been. By Lincoln's side, Mary was most unsettled. "Oh, Mr. French," she whispered to a White House official, "how much we have passed through since last we stood here." The strain proved too much for Mary, and she left before the reception was over.

In the afternoon, Lincoln returned to his office for the signing of the emancipation proclamation. It was a casual affair, as Seward and several other Cabinet members and public officials wandered in. At the table, Lincoln dipped a gold pen in ink, but his hand trembled badly and he put the pen down. He assured everyone that he was never more certain of doing right. "If my name ever goes into history," he said, "it will be for this act." But he'd been shaking hands for hours and his right arm was "almost paralyzed." He worried that a nervous signature would invite his critics to say, "he hesitated." "But anyway it is going to be done." Then he took the pen and slowly and carefully wrote out his full name. There, he said, "that will do."

Out the proclamation went to an anxious nation. The reactions would come sweeping back soon enough, and Lincoln braced himself for the worst. Beyond the White House, a black preacher named Henry M. Turner ran down Pennsylvania Avenue with a copy of the proclamation and tried to read it to an assembly of Negroes. Out of breath, he gave the proclamation to a Mr. Hinton, who read it "with great force and earnestness," and the blacks broke into uninhibited demonstrations, shouted, clapped, and sang of jubilee while dogs barked at their sides. Presently an interracial crowd gathered in front of the White House and called for the President. When he appeared at the window and bowed to them, the blacks cheered Lincoln, cried out in ecstasy, and said if he would "come out of that palace, they would hug him to death." Preacher Turner exclaimed that "it was indeed a time of times," that "nothing like it will ever be seen again in this life."

"The occasion is piled high with difficulty," Lincoln told Congress in 1862, "and we must rise with the occasion. As our case is new, so we must think anew, and act anew. We must disenthrall ourselves, and then we shall save our country."

"If God now wills the removal of a great wrong, and wills also that we of the North as well as you of the South, shall pay fairly for our complicity in that wrong, impartial history will find therein new cause to attest and revere the justice and goodness of God."

"If the people over the river had behaved themselves, I could not have done what I have."

Mary Lincoln about a year after Willie's death. "I have sometimes feared," she wrote a friend, "that the *deep waters,* through which we have passed would overwhelm me."

Lincoln and Tad in 1864. After Willie's death, Lincoln became even closer to Tad, took him on journeys to the front, read him stories, and rocked the boy at night.

The strain of War—April, 1865. "We must both be more cheerful in the future," Lincoln said to Mary; "between the war and the loss of our darling Willie, we have both been very miserable."

WHITE HOUSE AIDES

John G. ("Nico") Nicolay (Lincoln's personal secretary).

John Hay (Lincoln's personal secretary).

Elizabeth Keckley, a former slave who became Mary Lincoln's seamstress and confidante.

William H. Seward, Secretary of State, with his daughter Fanny. A celebrated raconteur, Seward loved to pun and banter with the President, often laughing so hard at his own jokes that it left him hoarse.

Salmon P. Chase, Secretary of the Treasury, a solemn, three-time widower who spoke with a slight lisp and who hoped to replace Lincoln as President in 1864.

Simon Cameron, first Secretary of War. Before removing him, Lincoln complained that Cameron was not only ignorant and incompetent, but "openly discourteous" to the President himself.

Edwin M. Stanton, second Secretary of War, a talented administrator who became one of Lincoln's closest war-time friends.

Gideon Welles, Secretary of the Navy, a misanthropic individual who had a pathological mistrust of his colleagues, especially Seward.

Montgomery Blair, Postmaster General. Member of the powerful Blair clan and very loyal to Lincoln, Blair was a controversial, hot-tempered secretary who in 1864 mounted an all-out attack against Republican liberals.

REPUBLICAN COLLEAGUES IN THE SENATE

Senator Charles Sumner in 1864. More than anyone else, he pressured Lincoln to emancipate the slaves.

Benjamin F. Wade

Zachariah Chandler

THE GENERALS

John Charles Frémont, Republican Presidential candidate in 1856 and commander of the troubled Western Department in 1861.

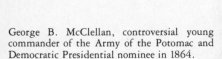

George B. McClellan, controversial young commander of the Army of the Potomac and Democratic Presidential nominee in 1864.

Don Carlos Buell, commander of the
Department of the Ohio, 1861-1862.

Henry ("Old Brains") Halleck,
Lincoln's General in Chief, 1862-1864.

William S. ("Old Rosey") Rosecrans, who replaced Buell in the West. Lincoln repeatedly admonished him to fight a decisive battle against the rebels in Tennessee, finally removing him for losing his nerve and acting as confused as "a duck hit on the head."

John Pope, ill-starred commander of the Army of Virginia who lost the second battle of Bull Run.

Ambrose E. Burnside, known for his prodigious sideburns and virile looks, who
led the Army of the Potomac to a disastrous defeat at Fredericksburg.

Joseph Hooker, commander of the Army
of the Potomac in the Chancellorsville
campaign, who wanted desperately to
impress Lincoln and talked obsessively
about what he would do once he captured
Richmond.

George Gordon Meade. Known among his men as a "damned goggle-eyed old snapping turtle," Meade led the Army of the Potomac from Gettysburg until the end of the War.

William Tecumseh Sherman, apostle of total war. "We are not only fighting hostile armies," he declared, "but a hostile people, and must make old and young, rich and poor, feel the hard hand of war."

Ulysses S. Grant, the hero of Vicksburg and Lincoln's favorite commander, who became General in Chief in 1864. He was "the joy of Lincoln's heart" because he fought without demanding reinforcements. Unlike some of Lincoln's other generals Grant did not believe "I have a right to question any policy of the government"; his job was to win the war.

Ford's Theater, Washington, D.C.

THE NEW RECKONING

Later that New Year's Day, Lincoln tried to relax in the telegraph office, resting his feet on a table and talking with Halleck and Assistant Navy Secretary Gustavus Fox. At that moment a battle was taking place at faraway Murfreesboro, and the outcome was much in doubt. Through that day and the next the telegraph chattered with reports from Rosecrans's headquarters. On January 2 the rebels attacked again, but this time the Union lines held. Yes, Old Rosey's army refused to crack and the rebels, exhausted and dispirited, withdrew from the field and retreated back to Chattanooga. At once Rosecrans pronounced Murfreesboro a Union victory—and it was of a sort, though hardly a crucial one since the general did not advance and the enemy retained control of south-central and eastern Tennessee. Nevertheless, Lincoln was massively relieved that another disaster hadn't resulted. But he was wrong if he thought the narrative had changed: soon Rosecrans was railing at Washington for not supporting him in his campaigns.

There was trouble with Burnside, too. Burnside, with his insomnia and self-doubts, had stayed up night after night trying to produce a plan that would please Lincoln and compensate for the horror of Fredericksburg. And he had one finally, yes, a winter offensive that would end in glory if Lincoln approved: "Burn" would cross his army north or south of Fredericksburg, outmaneuver Lee, and smash his way south to Richmond. Yet Lincoln had deep reservations. Most of Burnside's subordinate commanders, particularly "Fighting Joe" Hooker, condemned the plan and complained directly to Lincoln that they had no confidence whatever in Burnside. And Burnside, wounded to the core, told Lincoln that maybe he ought to quit the army and return to private life.

These family feuds were getting the best of Lincoln. Of course he didn't want Burnside to resign. And yet he worried about this plan of his —given the furies emancipation was bound to unleash, Lincoln could ill afford another military reversal in the East. Then let Halleck decide. Lincoln had brought him to Washington to solve precisely this kind of dilemma. Therefore he ordered Halleck to investigate Burnside's plan of operations and either approve or disapprove. "Your military skill is useless to me, if you will not do this," the President informed him in a note. Now it was Halleck's turn to be offended. He was willing to advise but not order Burnside what to do. Miffed at Lincoln's tone, Halleck asked to be relieved of command. In despair, Lincoln withdrew his note and ultimately left the matter of a winter offensive up to Burnside, with Halleck reminding the general that the destruction of Lee's army and not Richmond was his goal.

On January 20 Burnside led his army up the Rappahannock—130,000 men with artillery and wagon trains—and attempted to cross the river in a cold winter rain. It was calamitous. The entire army—troops, wagons, cannon, horses, all—bogged down in oozing roads and driving rains. At last the army extricated itself and returned to winter quarters. They called it the infamous "Mud March."

The army rocked with recriminations. Hooker, who'd already denounced Burnside as "incompetent," called Lincoln "a played out imbecile," and advocated a dictatorship, ridiculed his superior officer and informed reporters of what he would do as army commander. When Burnside tried to fire Hooker and other malcontents, Lincoln stepped in and relieved Burnside. It was either that or risk a total breakdown of army command. Then without consulting Halleck, Stanton, or anybody else, Lincoln put Hooker in charge of the Potomac Army. But in an interview at the White House, Lincoln handed the general a letter: "I believe you to be a brave and a skilful soldier, which, of course, I like. I also believe you do not mix politics with your profession, in which you are right. You have confidence in yourself, which is a valuable, if not an indispensable quality. You are ambitious, which, within reasonable bounds, does good rather than harm. But I think that during Gen. Burnside's command of the Army, you have taken counsel of your ambition, and thwarted him as much as you could, in which you did a great wrong to the country." Lincoln had also heard of Hooker's recent statement that both the government and the army needed a dictator. "Of course it was not *for* this, but in spite of it, that I have given you the command," the President wrote. "What I now ask of you is military success, and I will risk the dictatorship. . . .

"And now, beware of rashness. Beware of rashness, but with energy, and sleepless vigilance, go forward, and give us victories."

So Joseph Hooker, Massachusetts-born, hotheaded and hard-drinking, became the fourth commander of the Army of the Potomac. Yet he could not go forth, as winter storms dumped snow on Washington and much of northern Virginia, thus immobilizing the army. As Nicolay recorded, "The Army of the Potomac is for the present stuck in the mud, as it has been during nearly the whole of its existence."

W̲ith snow falling beyond his office windows, Lincoln studied the public reaction to his final proclamation, going over press clippings, letters, and memorials which his personal secretaries brought him in the mornings. The small abolitionist press generally forgave him for his exemptions, noting that they seemed temporary; and abolitionist circles took a great deal of credit for pushing and goading Lincoln at last to act. As he expected, most Republican journals gave the proclamation strong editorial support, assuring their readers once again that liberated slaves would not stampede into the North.

As Lincoln feared, the final proclamation shattered what remained of his broad bipartisan war coalition, with thousands of Democrats now in open opposition to his "revolutionary" war policy. From New York to Chicago, Democratic papers smeared the proclamation as "a wicked, atrocious, and revolting deed" which would unleash hordes of "negro barbarians." "It is impudent and insulting to God as to man," cried one Democrat, "for it declares those 'equal' whom God created unequal." With Democratic leaders up in arms, a storm of anti-Negro, anti-Lincoln protest rolled over the land, with rumbles of riot and disunion in parts of the Midwest. And there was trouble in the army as well. Correspondents who traveled with Union forces claimed that hardly one soldier in ten approved of emancipation; they reported that white soldiers cursed "niggers" with an "unreasoning hatred" and swore that they weren't going to fight to free any slaves. Some officers from the Midwest even resigned in protest.

Though they had expected a backlash, Republicans in Washington were alarmed all the same. "These are dark hours," Sumner said. "There are Senators full of despair—not I." Hale of Connecticut thought the party had made a terrible mistake and should have stayed clear of the Negro. Seward remarked that the proclamation was having a "pernicious influence," quite as he'd feared. And Browning and other conservatives caught a glimpse of the apocalypse. "We all agreed," Browning said, "that we were upon the brink of ruin, and could see no hope of an amendment in affairs unless the President would change his policy, and withdraw or greatly modify his proclamation." Even David Davis, whom Lincoln had recently called out of

Illinois and appointed to the U.S. Supreme Court, begged Lincoln to "alter" his emancipation policy and save the nation.

But Lincoln was as immovable as stone. "I am a slow walker," he said, "but I never walk back." He conceded that the final proclamation "has done about as much harm as good." And he confided in Sumner that he worried about "the fire in the rear" as much as the rebels at the front. Yet he'd made up his mind to shatter the cornerstone of the rebel South, to lay a strong hand on the blacks there, to eliminate the monstrous wrong of slavery, and no amount of public discontent was going to change his mind now. "To use a coarse, but an expressive figure," he wrote one unhappy Democrat, "broken eggs cannot be mended. I have issued the proclamation, and I cannot retract it."

Besides, if he could believe Charles Francis Adams, the proclamation was having a positive effect in England. Adams reported that "it has rallied the sympathies of the working classes" and produced mass meetings scarcely seen since Corn Law days. Lincoln, of course, had no way of knowing that Adams exaggerated British working-class sentiment for the proclamation, that in fact the weight of opinion in the cotton towns of Lancashire condemned the edict as a needless act of war. On the contrary, Lincoln thought Adams's observations were correct, especially when the President himself received a sympathetic memorial from a group of Manchester workers. Because Sumner insisted that public demonstrations would dissuade the British ruling class from helping the Confederacy, Lincoln did what he could to encourage them. He even drafted a resolution, to be adopted at English pro-Union meetings, that no "embryo state" based on slavery should ever be admitted into the family of civilized and Christian nations. Sumner then passed the resolution on to his English friends. And the British government? The proclamation may have aroused pro-Union sentiment in London and may have had some influence on British foreign policy—Great Britain did spurn Confederate recognition that winter and did remain neutral in the American war. Still, the major credit for that really belonged to Adams himself. Understanding that the British wanted above all to avoid another military entanglement in North America, he made it clear that if Great Britain recognized the South or gave her military assistance, it would mean war with his government.

As the winter passed, the War Department set about mobilizing black manpower and harvesting the fruits of Lincoln's proclamation. And Lincoln and Stanton worked out some sticky problems about military race relations. To ease white tensions, they agreed that black soldiers should receive less pay than whites and should serve in strictly segregated outfits under white officers. Even so, Lincoln believed that Negroes would make good Union troops. He even went so far as to assert that "the bare sight of fifty thousand armed and drilled black soldiers on the banks of the Mississippi, would end the rebellion at once."

And so in the occupied South—inside Union lines in Tennessee, Louisiana, Virginia, and South Carolina—Union commanders began enrolling blacks who'd come to them from the Confederacy, putting them in all-Negro units to serve at military garrisons and in some cases even to fight against their former masters. In the North, agents representing state and national governments rode through Negro communities mustering volunteers for Lincoln's armies. And those who enlisted did so to prove that blacks were not docile and moronic Sambos as most whites believed, but could fight as well as anybody, both to liberate their brothers and sisters in bondage and to save the Union. From Massachusetts came the Fifty-fourth "Colored" Infantry, under the command of a white colonel named Robert Gould Shaw; as the blacks marched through Boston to the tuck of drums, with guns bayoneted and banners high, Longfellow remarked that it was "an imposing sight, with something wild and strange about it, like a dream." Within eight months after Lincoln issued his proclamation, some fourteen Negro regiments were in the field from North and South alike and some twenty-four more were being organized.

As Union recruiters signed up black volunteers, Lincoln conferred with his advisers about how to control and utilize all the freed Negroes in the South. From the time he issued the proclamation, the slave grapevine across the Confederacy hummed with the news; and when Union lines drew near, more slaves than ever abandoned rebel farms and plantations and "demonstrated with their feet" their desire for freedom. By the spring of 1863, Grant and Rosecrans were complaining about the number of refugees who congested their lines and impaired their campaigns. Able-bodied black men, of course, could now be enlisted as soldiers. But what should be done with the rest? Obviously they couldn't be sent North, wrote a correspondent with Rosecrans's army, "in consequence of the heartless bigotry of the people of the free states." In fact, as Lincoln well knew, the fear of a Negro exodus into the North was one of the major reasons Northern whites objected to emancipation.

Before the emancipation proclamation, Lincoln had always promoted colonization as the most effective method of racial adjustment. But now he needed a more practical and expedient program, one he could put into operation without delay and hold up to the North as a successful device for race control. In March of 1863, after talking the matter over with Stanton and other advisers, Lincoln dispatched Adjutant General Lorenzo Thomas to the Mississippi Valley, to install a refugee program in occupied areas there. With indefatigable zeal, Thomas enrolled liberated blacks in the army, employed others as military laborers, and hired still others to work on farms and plantations for wages set by the government. Thomas's program was predicated on sound Republican dogma—it helped keep Southern Negroes out of the North, and it put the blacks to work, thus helping them to help themselves.

Lincoln was impressed with Thomas's program and embraced it as a more feasible solution to racial adjustment than colonization. In fact, by the spring of 1863 the Haitian colonization project had run into one problem after another—the promoters were either dishonest or inept and the black community generally hostile—so that only some 450 Negroes had volunteered to be transported, a far cry from the 5,000 people the operation originally called for. In May, 1863, the blacks moved down to Haiti, but the entire venture became a nightmare, as the colony suffered from corrupt white management, improper facilities, smallpox, starvation, mutiny, and hostility on the part of the Haitian government. In the end, Lincoln had to send a Union ship down to retrieve the survivors. After that Congress canceled all funds it had set aside for colonization purposes.

But even before the blacks left for Haiti, Lincoln had pretty much written colonization off as unworkable. Maybe in the abstract he still considered it the only way to avoid racial conflict in America. But after May of 1863 he never again championed colonization in public. Instead he let it be known that the administration now had a new policy—that of mobilizing blacks and using them in the South. As Lincoln's armies punched into rebel territory, liberating slaves as they moved, Thomas's refugee system would be employed to utilize the freedmen in military and civilian pursuits. The Lincoln administration hoped the program would not only assuage white racial anxieties, but prepare Southern blacks for life in a free society.

As the war approached its third year, Lincoln and his Republican colleagues conceded that another severe step was necessary to raise desperately needed manpower. While blacks were now joining up, Union forces on all fronts still suffered from shortages of white combat troops. So in March, 1863, the government produced yet another controversial measure—a Union draft. In theory, the conscript law was supposed to stimulate volunteering, though anybody who had the money could hire a substitute or buy a three-hundred-dollar exemption and still avoid the army. Nevertheless, Lincoln and Stanton saw to it that conscription was rigorously enforced, giving Stanton's ubiquitous provost marshals sweeping powers to enroll and conscript men in all corners of the Union.

There was a reason for this beyond getting the manpower. Both Lincoln and his Secretary of War hoped that enforcing the draft would win back the loyalty of soldiers already in the field—would demonstrate that the government intended to haul in reinforcements and stand behind its armies regardless of how unpopular the war became back home.

And back home, antiwar discontent was boiling. As though emancipation were not horrible enough, cried disaffected Northerners, white men would now be dragooned into fighting a war for slave liberation and dying in Lincoln's bungling armies. In the spring, dissident Democrats launched a broad "Peace Movement" to stop the war and bring the boys home. In all directions, Peace Democrats inveighed against Lincoln's tyrannies—the military arrests, the draft, and above all the emancipation proclamation—and exhorted the Northern people to end the butchery before Lincoln converted the country into a dictatorship. If the North made peace with the Confederacy, somehow the Union would be reunited with slavery preserved. Maybe Northerners and Southerners could even meet in a national convention and settle secession like reasonable men.

Antiwar protest flamed highest in the Midwest, where there was widespread war weariness and broad discontent with Lincoln's "half-witted" efforts to bring about "nigger equality" and "racial amalgamation" in the name of the Union. Here battalions of Peace Democrats stumped the countryside, staging boisterous antiwar rallies and ringing the bells of doom if the war and emancipation continued. Here the Peace Movement embraced Democrats of various kinds, from those who genuinely feared a Lincoln dictatorship to secret organizations like the Sons of Liberty, whose clandestine proceedings aroused the suspicions of Republicans everywhere.

In truth, Republicans tended to view all antiwar dissidents as disloyal, and they thundered back at Peace Democrats, calling them poisonous Copperheads one and all. In the states, various Republican agents, editors, politicians, and military commanders sent the administration embellished accounts of traitorous Copperhead operations. The reports charged that Copperheads obstructed the draft, discouraged enlistments in the army, raised guns and money for the Confederacy, and even engaged in night-riding terrorism, assassinating conscript officers and shooting it out with government agents who tracked them down. Unless the Copperheads were suppressed, the reports warned, the Union war effort would be dangerously undermined behind the lines.

Not ones to coddle treason, Lincoln and Stanton adopted a tough line against antiwar activities. With Lincoln's support, Stanton empowered provost marshals and army officers to jail anybody who interfered with the draft or otherwise helped the rebellion. And he got up dragnets in which state militia, home guards, police chiefs, and vigilantes all participated in rounding up alleged enemies of the Union. All told, Stanton's agents crowded more than thirteen thousand people—most of them antiwar Democrats—into Northern prisons.

But the outcry against arbitrary arrests became so strident that Lincoln and Stanton both tried to restrain excessive use of power whenever they could. Stanton speedily ordered the release of people unwarrantably jailed; and Lincoln himself spent a good part of workday afternoons writing out pardons for civilian prisoners. Also, when General Ambrose E. Burnside, now commander of the Department of the Ohio which encompassed the Midwest, suspended the *Chicago Times* for violent outbursts against the administration, Lincoln promptly revoked the order.

The most controversial political arrest was that of Clement L. Vallandigham, a tall, bewhiskered Ohio Democrat who led the Peace Movement in the Midwest. In May of 1863, Vallandigham crossed Ohio denouncing the draft and the despotism of Lincoln's government and advocating a negotiated peace with the Confederacy. During one of his orations, an officer in civilian dress, detailed from Burnside's headquarters, leaned against the platform taking notes. Three days later the army arrested Vallandigham, and a military commission sentenced him to imprisonment for the duration of the war. When Democrats decried all this, Lincoln snapped back, "Must I shoot a simple-minded soldier boy who deserts, while I must not touch a hair of a wiley agitator who induces him to desert?" Though Lincoln may actually have regretted the arrest, he refused to pardon Vallandigham, instead banishing him to the Confederacy.

Lincoln conceded that internal security in the midst of civil war was a complex problem and that errors and excesses had occurred. That was why he tempered military arrests with generous pardons, and why he refused to suppress popular assemblies and antiwar newspapers. But he would not rescind military law behind the lines. He explained that in this huge insurrection, when "combinations" in the North obstructed conscription and "assassinated" loyal officers, when the rebels themselves maintained a network of spies and informers there, only swift and efficient military law could preserve internal security. Without military law, Lincoln feared, the flames of rebellion would sweep into the North and consume the government from within. So, no, he would not close down the military courts or stop arresting those who interfered with the Union war machine. He would, he said, stand firm against "the fire in the rear."

Every time he had the opportunity, Lincoln studied the framed maps in his office and pondered the military predicament that spring. In western Tennessee, Rosecrans was inert and grumpy, still fuming about his lack of supplies. And Grant was still stalled in the tangled river country above Vicksburg. Since Sherman had failed to penetrate the bluffs and bayous northeast of Vicksburg, Grant had made one abortive attempt after another to get at the river garrison. He'd even tried to dig canals to facilitate troop movements, but the canals had also failed. Frankly Lincoln worried about Grant. Why didn't he forget about those foolish canals and simply march down the Louisiana side of the Mississippi and operate on Vicksburg from below? And there were other problems, too, like Grant's ill-considered effort to expel all Jewish peddlers from his lines, on grounds that they were rebel spies. Lincoln countermanded the order because it proscribed an entire religious class, some of whom were serving in Union ranks. In addition, reports came that Grant was drinking again. Chase passed on a letter from a Cincinnati editor who declared that Grant was "a poor drunk imbecile" who couldn't organize, control, or fight an army. "I have no personal feelings about it," the editor added, "but I know he is an ass."

Alarmed, Lincoln and Stanton sent out various people to look Grant over —among them Charles A. Dana, lately of the *New York Tribune.* Dana traveled with Grant's army and filed daily reports that were extremely complimentary about the general, praising him as industrious *and* sober. Dana's dispatches did much to restore Lincoln's confidence in the only army commander he had who could really fight.

In the East, Hooker seemed eager to fight Lee once the weather cleared. And "if the enemy does not run," Hooker snarled, "God help him." In early April Lincoln decided to pay Hooker a visit and look over the Army of the Potomac. The President assembled a small traveling party—Mary and Tad, Noah Brooks and Dr. Anson Henry, his old friend—and set out on a river steamer with a snowstorm blowing. At last they came to Aquia Creek, a village of crowded wharves and makeshift warehouses which served as Hooker's supply base. Snow still fell as they rode a rattling freight down to Falmouth, Virginia, where the army had spent the winter. Fredericksburg—and Lee's army—lay across the Rappahannock about a mile to the south. Lincoln went down to the picket lines along the river and peered at Fredericksburg through a field glass. He saw burned and bullet-torn homes and a church steeple that had been shredded by artillery fire. Smoke from the enemy camps rose from behind a ridge, and a rebel flag floated on the heights, just above the stone wall and the wide plain where thousands of Burnside's men had died. A blackened chimney stood alone near the riverbank, and a couple of rebel pickets huddled around a fire on the hearth, warming themselves in the cold.

Lincoln stayed five days at Falmouth: he lived in a large tent with his family, spoke with the wounded in nearby hospital tents, saw the ruins of a burned-out mansion "on our side of the river." As always, he attended gala troop reviews, riding past saluting regiments to the music of fifes and drums. The most impressive troops were the cavalry, now organized as a separate corps. Good, Lincoln said, now the cavalry could fight as the enemy fought, could sever communications and destroy supply bases as the rebel cavalry had done.

Between reviews, Lincoln climbed into a mule-drawn ambulance and bounced around the camp grounds visiting with the men. He was amused at his driver, who cursed everything with a flourish—the mules, the raw weather, the mud, the cavalry that sloshed by. Lincoln tapped him on the shoulder. "Excuse me, my friend, are you an Episcopalian?" No, the driver replied, he was a Methodist. "Well," Lincoln said, "I thought you must be an Episcopalian, because you swear just like Governor Seward, who is a churchwarden."

Lincoln was impressed with Hooker's army—over 130,000 drilled and spirited troops. Hooker called it "the finest Army on the Planet" as he showed the President around. The general was an imposing man, tall, with blue eyes and light brown hair—"by all odds," sighed Noah Brooks, "the handsomest soldier I ever laid eyes on." Hooker respected Lincoln as no other army commander had done and wanted desperately to please and impress him. He boasted constantly about what he would do after he captured Richmond, and Lincoln told Brooks that all the talk about the rebel capital "is the most depressing thing about Hooker." He seemed overconfident, with a cocky streak too reminiscent of McClellan to suit Lincoln. Besides, the general's objective was not Richmond, but Lee's army across the river, for Lincoln had come to understand that only the total annihilation of that army could win the war in the East.

When it was time to go, time to leave the Army of the Potomac and get back to the White House, Lincoln remarked that the trip had been a welcome relief, "but nothing touches the tired spot." By mid April he was back in Washington and submerged again in his multitude of cares: the Peace Movement, the draft, the blacks, the foreign situation. At least there was good news from Vicksburg, where Grant was on the move. Unable to take the city from the north, Grant threw military theories to the winds: he cut himself off from his Memphis base and led his force of twenty thousand down the Louisiana side of the river, heading southward through an area of humid bayous and mosquito-infested lakes. Meanwhile Union gunboats and transports ran past Vicksburg's batteries and steamed downriver. Good, Lincoln exclaimed, this was what he'd wanted Grant to do all along. Now if he would march on down the Mississippi and link up with Nathaniel Banks, who commanded the Department of the Gulf, they could mount a concerted operation against Vicksburg from the South. But Grant intended to campaign with the men he had. At the end of April, he used the transports to ferry his army back across the river, then drove inland to Port Gibson and prepared to strike out for Vicksburg. Lincoln feared it was a mistake not to join Banks, but he admired Grant's daring and refused to intervene.

Meanwhile Hooker, too, was on the move. Though hindered by continuous rains, he sent his cavalry on a raid around Lee's army, to cut his communications and menace Richmond. But Lincoln was confused about Hooker's exact plans. Where was the main army? Where would it attack Lee? On April 27 a staff officer rode pellmell up to the White House and handed Lincoln a letter from Hooker himself, a letter that described his objectives in full. The entire army was now on the march, Hooker said in the letter. He regretted withholding his final plan until now, but the countryside was full of spies and he feared a leak in security. With his cavalry rampaging behind Lee's lines, Hooker would divide his army, cross the Rappahannock, and pound Lee with combined frontal and flank assaults. Lincoln was pleased with the plan—it was excellent, the best he'd seen so far.

But his enthusiasm didn't last. Subsequent dispatches indicated that Hooker's advance had run into attacking rebels and that Hooker had fallen back to a village called Chancellorsville. Fallen back! Lincoln could not believe this. How could Hooker retreat when his was supposed to be the attacking force? May 3 brought even worse reports. Though outnumbered two to one, Lee had divided his army and struck Hooker with frontal and flank assaults, and a titanic battle roared around Chancellorsville. A cloudburst interrupted the telegraph for a time. But on May 6 came the awful news that Hooker had quit fighting and pulled back to Falmouth. Another defeat then? Yes, another defeat. Fighting Joe Hooker had let a smaller force whip the Potomac Army and had lost seventeen thousand casualties in the process. In the

White House, Lincoln paced back and forth with his hands behind his back. "My God!" he told Noah Brooks. "What will the country say! What will the country say!"

What was the condition of the army? Was the army all right? With Halleck, Lincoln set out for Falmouth to find out for himself. Hospital ships passed them going the other way, and rain fell from stormy skies. The army was still intact, but some of Hooker's subordinates complained bitterly about his inept generalship—what else?—and Lincoln returned to Washington in a black mood. By now the capital crawled with rumors that Stanton had resigned, that Halleck had been arrested, that McClellan had been restored to command. Newspapers screamed with the news about Chancellorsville, and the bar at Willard's was packed with men who drank, spat tobacco juice, and expostulated on what they would do if they commanded the army. Almost every night before he went to bed, Lincoln called at the telegraph office, expecting some word from Grant, from Rosecrans in Tennessee. He came and went alone, going out a side door of the War Department and crossing the brick walkway past the conservatory. A gas lamp flickered dimly in the dark, and shadows of trees lay like skeletons across the path. The White House loomed up ahead, blazing with lights.

Lincoln had said to Halleck and one of his corps commanders that he didn't blame anybody for Chancellorsville, but that its effect at home and abroad would be more injurious than any other defeat. What should the Army of the Potomac do now? At first he thought another offensive might restore morale all around, but soon changed his mind about that. No, he wrote Hooker, it was best for him to remain north of the Rappahannock, worry and fret the rebels with cavalry raids, and put his army into good condition again. "Still, if in your own clear judgment, you can renew the attack successfully, I do not mean to restrain you." But "I must tell you I have some painful intimations that some of your corps and Division Commanders are not giving you their entire confidence." Because of their complaints, Lincoln had doubts about Hooker, too, but remarked that he wouldn't throw a gun away because it hadn't fired the first time. He wanted to give Hooker another chance.

A cryptic telegram arrived from Grant. On May 11 he left Port Gibson—"you may not hear from me for several days," the dispatch read—and stormed northeast toward Vicksburg and

Jackson, subsisting entirely off hostile country. For two weeks nobody in Washington knew for sure what was happening. Then on May 25—a day to circle on the calendar—came a glorious report from Grant's chief of staff. The general had won five straight battles in Mississippi, captured the state capital of Jackson, split the rebel forces, and driven to the very trenches of Vicksburg itself.

Lincoln could not have been happier. In spite of his earlier reservations, he now pronounced Grant's campaign "one of the most brilliant in the world."

Yet Grant couldn't capture Vicksburg. When two assaults failed to break rebel lines, Grant settled in for a siege, with Union artillery shelling the city day and night. Though some contended that Vicksburg would fall in time, Lincoln was apprehensive. He'd been through too many campaigns, too many moments when Union armies seemed on the edge of victory, only to lose everything. He longed for a complete triumph at Vicksburg, so that he could tell a divided and dissident Union: here at last is a genuine conquest, the fall of a powerful rebel garrison and the elimination of its defenders as a fighting force. We *can* win. There *is* hope.

He turned to Rosecrans, still dallying in Tennessee. What was wrong with Rosecrans? Instead of cooperating with Grant and clearing the rebels out of middle Tennessee and Chattanooga, the general griped to reporters that somebody in Washington was out to get him (Stanton or Halleck, one or the other), and he bombarded Washington with accusations that the government was deliberately withholding supplies from him. Lincoln couldn't take his fussing any longer. He and Halleck telegraphed Rosecrans that he must fight the enemy in Tennessee, otherwise the rebels would transfer men from there to Vicksburg and try to crush Grant. Back came a letter from Rosecrans saying that it was dangerous to fight two large battles at the same time. Through Halleck Lincoln wired the general: Are you going to move immediately? Yes or no? Intimidated, Rosecrans finally advanced, but he wasn't happy about it.

Early in June, Lee's army was on the march, heading northwest along the Rappahannock with Hooker tracking him at a distance. In the War Department, Lincoln traced Lee's movements on a large military map and analyzed telegraph dispatches with Stanton and

Halleck. Beyond the War Office, crowds milled about Willard's Hotel, with "scary rumors abroad" that an invasion was imminent.

On June 14 Lee turned and drove into Maryland, sending shock waves across Pennsylvania and the rest of the North. The next day Lincoln ordered up 100,000 state militia to help repel the invasion, and he entreated Hooker to find a weakness in Lee's strung-out columns and break him. But Hooker, moving on a parallel line between Lee and Washington, did not strike.

Waiting anxiously for news, Lincoln, Stanton, and Welles remained in the War Department late into the night, with summer insects swarming about the lamps there. To ease the tension, Lincoln chatted about the American humorist Orpheus C. Kerr (his real name was Robert H. Newell). Kerr's outrageous satire gave Lincoln immense pleasure—except when Kerr aimed his pen at the President, which wasn't "too successful." But how he enjoyed Kerr's hits on Chase and Welles. And there was the wonderful poem about McClellan, Lincoln said, the one that described the general in mythological terms as the monkey who fought the serpent, with the serpent representing the rebellion. The monkey kept crying for "more tail! more tail!" which Jupiter kept giving him. The President shook with laughter.

In the next few days, telegraph reports indicated that Lee was now driving into Pennsylvania and that everything out there was pandemonium. Nobody could tell what Lee's objective was. Washington? Harrisburg? Philadelphia? Another dispatch said that the rebel cavalry was on a raid, racing somewhere south and east of Hooker's army . . . which meant that Lee had no cavalry for reconnaissance and could not know where Hooker was. If Hooker had generalship, Lincoln told Welles, he could attack and beat Lee now. "Hooker may commit the same fault as McClellan and lose his chance," the President said, "but it appears to me he can't help but win."

But Hooker still did not attack. Instead he pounded Washington with wild messages that the rebels outnumbered him and that he couldn't fight without large reinforcements. When he and Halleck clashed, Hooker "took umbrage" and offered his resignation. Lincoln accepted it. He did so because Hooker was obviously afraid to fight, because he showed the same fatal flaws as McClellan, and the President was not going to retain another McClellan in command this time. So out went Hooker—and in came George Gordon Meade to lead the Army of the Potomac in the middle of the invasion. Lincoln promoted Meade because he stayed out of army feuds and followed orders even if he disagreed with them. Another West Pointer, six feet tall and solemn, Meade rode with a slouchy hat pulled so low on his head, said an observer, that he resembled a helmeted knight. An irritable man with nerves taut as piano wire, he blew up so often that his troops called him "a damned goggle-eyed old snapping turtle."

June 30. The two armies—Lee with 70,000 effectives and Meade with 101,000—were moving on a collision course in southern Pennsylvania, and a battle seemed near. Lincoln skipped Cabinet meeting and hurried to the War Department, to wait it out as he'd done so often before. Through July 1 and 2 the telegraph chattered with continuous dispatches: the armies had come together at a town called Gettysburg and were grappling for positions on the ridges to the south. Later in the day of July 2, Lincoln returned to the White House and had reports brought to him there. Chandler called, and they spoke intently about the distant battle and about how the fate of the Union hung on the outcome. Chandler noted Lincoln's "restless solicitude" as he walked back and forth, "reading dispatches, soliloquizing, and often stopping to trace positions on a map."

Word from the Soldiers' Home—an evil omen? Mary had been hurt in a carriage accident and was lying in the home in shock, her head bruised and aching from the fall. Nobody knew for certain how it happened: evidently the seat came unbolted and threw the driver off, and the horses stampeded, throwing Mary to the ground. At first Lincoln thought her injury was slight and wired Robert so (later her head wound became infected and she lay in bed, sick and dazed, for three weeks). Lincoln saw to it that a nurse was present to care for her; he would like to stay at her bedside himself, as he always tried to do when she was ill, but he couldn't leave Washington in the midst of Gettysburg.

Early the next morning, July 3, he was back in the War Department, sweating in oppressive summer heat. The dispatches told of awesome fighting at Gettysburg yesterday with heavy losses on both sides. Today's reports were fragmentary, but apparently a showdown was at hand: early in the afternoon a furious artillery duel broke out, with masses of cannon thundering at one another from across the ridges. Late in the afternoon an officer in Baltimore telegraphed: "I learn that the suffering near the battlefield of Gettysburg and beyond is terrible, in the want of sufficient medical attendance." At last came a stream of bulletins and dispatches about what had happened that day. Lee had mounted a huge frontal assault, hurling 15,000 men at the center of the Union lines. But the Army of the Potomac had repulsed the charge and inflicted such crippling casualties that Lee's forces were incapable of another attack. On July 4 Lincoln released a press statement that the Army of the Potomac had fought "with the highest honor" and was on the verge of a mighty victory. The next day reports indicated that Lee was in full retreat back toward the Potomac, and the President was euphoric. Now Meade would "pitch into him" and obliterate his army.

But then Meade's congratulatory order hummed in over the wires, telling his men to "drive from our soil every vestige of the presence of the invader." Lincoln read this in anguish. "Drive the *invader* from our soil. My God! Is that all?" Meade was supposed to annihilate Lee's "traitor army," Lincoln cried, not drive him back to Virginia. This was McClellan all over again—Antietam all over again. And then the casualty returns started coming in—Union losses would amount to 23,000, the South's, 28,000—the worst of any battle so far. Lincoln grimaced at the losses, but insisted on facing the grim realities of war. Lee's smaller army was hurt badly, far worse than Meade's. Union soldiers "had expended all the skill, and toil, and blood" to plant an enormous crop, Lincoln said. They must harvest it. Meade must destroy Lee before he got away.

On July 7, with Lincoln fuming at Meade's slowness, Welles brought a report just in from Admiral David Porter, Union gunboat commander on the Mississippi. On July 4 Vicksburg had surrendered to Grant. Yes, the Union flag was flying over that rebel town at last. Lincoln shook Welles's hand and almost hugged him. "What can we do for the Secretary of the Navy for this glorious intelligence?" Lincoln beamed. "He is always giving us good news. I cannot tell you my joy." Bless Grant and bless his army, the President said, and he observed that the victory came on "the glorious old fourth." He grabbed his hat and headed for the telegraph office to wire Meade the news. Now if he finished his work and eliminated Lee's army, the rebellion, Lincoln said, would end.

But Meade still wasn't pursuing Lee, and Lincoln was beside himself. "The President is urgent and anxious that your army should move against him by forced marches," Halleck telegraphed Meade. But the general did not attack. By now Lee had reached the Potomac, but the waters were so swollen from heavy rains that he couldn't cross. What in hell was going on? Lee was just sitting out there and yet Meade did nothing to destroy him, nothing at all. Lincoln had "dreaded and expected this," knew something like this would happen. He said he ought to go up there and take command of the army himself. But it was too late. On July 14 came word that Lee had crossed safely into Virginia. "We had them within our grasp," Lincoln groaned. "We had only to stretch forth our hands & they were ours."

Noah Brooks called on him in the White House. Brooks had been at the front and had seen where the rebel army crossed the Potomac on pontoon bridges while Meade made little attempt to interfere. The President's "grief and anger," Brooks recorded, "were something sorrowful to behold."

"If I had gone up there," Lincoln told his son Robert, "I could have whipped them myself." He complained to Welles that "there is bad faith somewhere."

When Halleck informed Meade of Lincoln's "great dissatisfaction,"

Meade tendered his resignation. At that, Lincoln wrote him a blunt letter. "My dear general, I do not believe you appreciate the magnitude of the misfortune involved in Lee's escape. He was within your easy grasp, and to have closed upon him would, in connection with our other late successes, have ended the war. As it is, the war will be prolonged indefinitely. If you could not safely attack Lee last monday, how can you possibly do so South of the river . . . ? Your golden opportunity is gone, and I am distressed immeasurably because of it."

Once his temper subsided, Lincoln thought better of sending the letter, instead filing it away in his desk. He refused Meade's resignation and told Hay that while Meade had made a grave mistake, "I am very grateful . . . for the great service he did at Gettysburg."

So once again all was quiet on the Potomac. Yes, Welles quipped, Meade had dug in and was "watching the enemy as fast as he can."

Even so, Union military fortunes had never looked brighter. The simultaneous Union victories at Gettysburg and Vicksburg had not only given Northern morale a tremendous boost, but had delivered a stunning blow to the Confederacy both at home and abroad. After July, 1863, Confederate recognition was never again a serious threat, since the British had no reason to recognize a crippled belligerent and risk a military confrontation with the Union.

And there were other significant gains as well. After Lincoln made him advance, Rosecrans had skillfully driven the rebels out of middle Tennessee and was now—or so he claimed—preparing for a final push against Chattanooga near the Georgia border. What was more, Banks had captured Port Hudson, the last rebel stronghold on the Mississippi, so that the great river was now indisputably in Union hands.

In gratitude for Grant's own superb campaign, Lincoln sent him a personal note of thanks. "I do not remember that you and I ever met," Lincoln wrote, but he wanted to avow Grant's "almost inestimable service" to the country. Also, the President had a confession to make. When Grant got below Vicksburg, Lincoln had thought it a mistake for the general not to turn south and link up with Banks in Louisiana. "I now wish to make the personal acknowledgment that you were right, and I was wrong."

Lincoln could scarcely restrain his admiration for Grant. In truth, he was "the joy of Lincoln's heart," because the man fought without demanding reinforcements. Not once in the entire Vicksburg operation had Grant cried for help or balked at attacking the enemy. More significantly, he'd shown Lincoln and the rest of the army a new way of fighting, one that might well end the military deadlock on civil war battlefronts. Should other Union forces break from their supply bases and subsist off the country as Grant had done, they could maximize their freedom of movement and perhaps win a chain of Grant-style victories.

In spite of his conquests, though, Grant's critics kept circulating stories about his drunken sprees, and the tales have persisted through the years. Grant may have had his troubles with the bottle earlier in his life, but by the civil war he seems to have imbibed no more than most army officers in that hard-drinking time. He enjoyed his glass, and on rare occasion may have enjoyed one too many, but there is no documentary evidence that he was the chronic drunk his foes made him out to be. By now, Lincoln had little interest in the allegations about Grant's binges: the general conducted campaigns quite beyond the capacity of an inebriated lout, and Grant's victories and those alone were what concerned the President. As a consequence, an anecdote soon got around that a delegation of politicians marched into the White House and demanded that Lincoln fire Grant because he drank too much. But Lincoln—so the story went—replied that if they could find out Grant's brand of whiskey, he would "send every general in the field a barrel of it."

When asked if the story were true, Lincoln laughed. "That would have been very good if I had said it; but I reckon it was charged to me to give it currency."

Now that Union forces controlled the Mississippi, Lincoln was eager to extend Thomas's refugee program into newly captured territory in western Mississippi and Louisiana. In August he called on Grant to help Thomas enlist slave soldiers there, to send recruiters out among the farms and plantations along the Mississippi and muster all able-bodied Negro men into the Union army. "I believe it is a resource which, if vigorously applied now, will soon close the contest," Lincoln said. "It works doubly, weakening the enemy and strengthening us. We were not fully ripe for it until the river was opened. Now, I think at least a hundred thousand can, and ought to be rapidly organized along it's shores, relieving all the white troops to serve elsewhere." He added, "Mr. Dana understands you as believing that the emancipation proclamation has helped some in your military operations. I am very glad if this is so."

Grant's reply further endeared him to Lincoln. The general emphatically endorsed the use of black troops and contended that this, along with the emancipation proclamation, was "the heavyest blow yet given the Confederacy." Therefore Grant promised to give Thomas "all the aid in my power" in recruiting black soldiers in the Mississippi Valley. "I would do this whether arming the negro seemed to me a wise policy or not, because it is an order that I am bound to obey and I do not feel that in my position I have a right to question any policy of the government." God bless this man—he was truly Lincoln's kind of general. With Grant's support, Lincoln agreed with Stephen Hurlbut that "soon the banks of the Great River will bristle with the bayonets of colored Regiments taken from the former slaves of the soil."

Meanwhile there were ominous rumbles from the Confederacy about Lincoln's Negro soldiers—threats of reprisal and execution if blacks were caught in Union uniforms. Jefferson Davis himself announced that Negro prisoners would be turned over to state authorities—which meant certain death since it was a capital crime in Southern states for Negroes to carry arms. Also, the Confederate Congress decreed that white officers who led Negroes were guilty of inciting insurrection and should be "put to death or be otherwise punished."

Lincoln was determined to protect his men from rebel atrocities, and on July 30 he issued a retaliatory order: for every Union soldier killed in violation of the laws of war, Lincoln would have a rebel prisoner executed; and for every Union soldier sold back into slavery, he would have a rebel prisoner "placed at hard labor on the public works." Perhaps the order had some effect, since the Confederate government never really enforced its severe decrees.

By August, 1863, Lincoln felt thoroughly vindicated in his decision to employ Negro troops in the South. Black outfits were manning garrisons in various theaters of the war, and some were even serving as front-line combat troops—with Lincoln's blessing now and his praise. In a recent Union expedition against Charleston harbor, for example, the Fifty-fourth Massachusetts actually led an assault against Fort Wagner there. Colonel Shaw fell in that attack, dying in the wind-blown sand with many of his men. But the regiment fought on without him, the black soldiers charging through blasts of musketry and driving over the parapets of the fort to engage rebel whites hand to hand. But because supporting white regiments did not come up in time, the Fifty-fourth finally had to retreat. And so in a narrow technical sense the battle of Fort Wagner was a failure. But in a larger sense it was a tremendous victory for black Americans—it was their Bunker Hill, the *New York Tribune* declared—because it shattered the myth of Negro passivity and demonstrated to the army that blacks could be capable and courageous fighting men. After that, Negro units were to serve on the front lines in every theater from Texas to the Virginia and Carolina coasts. In all, some 180,000 blacks—a majority of them emancipated slaves—were to fight in Union military forces, adding enormous strength to Lincoln's war machine.

On August 10 Frederick Douglass came to the White House to talk about the Union's black soldiers. Lincoln was in his office, sitting in a low chair surrounded by books and papers, when Douglass came in and introduced himself. Lincoln rose—and kept rising, Douglass said, until the President stared down at him. "You need not tell me who you are, Douglass, I know who you are." Sit down, sit down. A handsome, bewhiskered man with full hair and stern, frowning eyes, Douglass was the most eminent black leader of his generation. In all honesty, he had several minds about Lincoln. He thought Lincoln "preeminently the white man's President, entirely devoted to the welfare of white men." For the first year and a half of the war, he "was ready and willing" to sacrifice blacks for the benefit of white people, and Douglass had been outspoken in his opposition to Lincoln's administration. But since the preliminary emancipation proclamation, Douglass said, American blacks had taken Lincoln's measure and had come to admire and even love this enigmatic man. Though Lincoln had taxed blacks to the limit, they had decided, in the roll and tumble of events, that "the how and the man of our redemption had somehow met in the person of Abraham Lincoln."

But that was not what Douglass had come to discuss. He was busily recruiting black soldiers himself—was on his way down to help Lorenzo Thomas on the Mississippi—and wanted to protest Lincoln's discriminatory policies against blacks in the army. Douglass argued that they deserved the same pay as whites and should get promoted for meritorious service just as white troops were. He also insisted that if the Confederacy executed Negro prisoners of war, Lincoln must kill rebel captives in retaliation.

Lincoln replied quietly, measuring his words. As for discrimination in the army, he pointed out that among whites there was great opposition to enlisting blacks at all. Many white men had even threatened to throw down their arms rather than fight beside them. Still others argued that Negroes were inferior and ought to receive less pay than whites. So "we had to make some concession to prejudice," the President explained. But—and he obviously had the Fifty-fourth Massachusetts in mind—"I assure you, Douglass, that in the end they shall have the same pay as white soldiers."

But Lincoln hedged about the treatment of black prisoners of war in the

South. For some reason he didn't mention his retaliatory order of July 30, with its warning that he would execute rebel prisoners if the Confederacy murdered captured blacks. According to Douglass, Lincoln now shunned "eye-for-an-eye" reprisals and said, "I can't take men out and kill them in cold blood for what was done by others."

That aside, Lincoln deeply appreciated Douglass's work in recruiting Negro soldiers and heartily approved of his plans to help Thomas on the Mississippi. After their interview was over, Douglass left the White House with a growing respect for Lincoln. He was "the first great man that I talked with in the United States freely," Douglass said later, "who in no single instance reminded me of the difference between himself and myself, of the difference of color."

Lincoln shuddered at the reports about civil disturbances that came to him that summer: white mobs in the Midwest killing conscript officers and ranting against Lincoln's war for Negro freedom . . . draft riots boiling up in Boston, Troy, Newark, and New York City, as whites who couldn't afford the three-hundred-dollar exemptions swarmed into the streets in savage protests against Lincoln's government. The New York draft riot was a macabre episode, a three-day orgy of violence which sickened Lincoln to read about. It began when a mob of white workers, most of them Irish, burned down the draft office and went on a rampage, breaking into saloons, looting jewelry stores, ganging up on policemen, and even attacking the mayor's house. Then in an explosion of racial fury, rioting whites stormed into New York's Negro area, where they set an orphanage afire, hung Negroes from lampposts and incinerated them, clubbed and whipped others to death, and assassinated policemen and white citizens who interfered. At last federal troops helped restore order in New York, but not before five hundred or more people had died. "Great God!" cried the *Christian Recorder.* "What is this nation coming to?"

It was heading for more outbursts of violence, complained New York Governor Horatio Seymour, a Democrat and an impassioned critic of Lincoln's administration, if the President did not suspend the draft and abandon emancipation and martial law. What did Lincoln want, Seymour raged, New York City ablaze with riots? The city cut off from the outside world and "given over to a howling mob?"

Of course Lincoln didn't want any more mob outbreaks—it was terrible, he said, for working people to maul and murder other working people as they had in New York City. But he told Seymour he would not suspend the draft, not when the enemy was forcing all available men into his ranks, "very much as a butcher drives bullocks into a slaughter pen," in hopes of attacking again and destroying all the Union had gained at Gettysburg. The Union needed men to preserve its military successes and see this war through to victory; therefore the draft must continue in New York and everywhere else. And if he had to, Lincoln would enforce it with federal troops and state militia.

In mid September Lincoln prepared a two-fisted defense of the draft, arguing that it was not only Constitutional, but based on sound historical precedent as well. Did not the Founding Fathers resort to conscription in the Revolution and the War of 1812? Are we not now to use what our own Fathers employed? "Are we degenerate? Has the manhood of our race run out?" He was resolutely determined, he informed the Cabinet, to stand behind the draft—and to deal with officials who obstructed it as he'd dealt with Vallandigham: he would banish them all to the Confederacy.

Hay was amazed at how tough Lincoln was becoming. "The Tycoon is in fine whack," Hay said of the President. "He is managing this war, the draft, foreign relations, and planning a reconstruction of the Union, all at once. I never knew with what tyrannous authority he rules the Cabinet, till now. The most important things he decides & there is no cavil." "He will not be bullied —even by his friends."

As the fall state elections drew near, Lincoln became increasingly uneasy, for he regarded them as a test of his administration—proof of whether the voters had come around to emancipation, the draft, Negro troops, and martial law as necessary war measures. Already Peace Democrats were out on the stump, haranguing white rallies with Negrophobic, anti-Lincoln, antiwar diatribes. Clement Vallandigham, now in Canada, was running for the governorship of Ohio *in absentia* and showering the state with letters and written speeches against Lincoln's war. To offset Democratic propaganda, Republican campaigners not only played up Gettysburg and Vicksburg for all they were worth, but trumpeted Thomas's program in the Mississippi Valley, declaring it a superlative device for keeping Southern "coloreds" out of the North. In an attempt to attract bipartisan support, Republicans included prowar Democrats in their ranks and labeled theirs the Union party, plainly implying that the regular Democrats were against the Union, were indeed treasonable Copperheads.

In his concern about the elections, Lincoln composed a public letter in which he lashed out at Democrats "who are dissatisfied with me." You say you are for the Union, Lincoln wrote, addressing himself rhetorically to the Democrats. Yet you oppose Lincoln because he uses too much force to prevent dissolution. Then you must favor "some unimaginable *compromise.*" Rebel leaders have shown no willingness to stop fighting and compromise their independence, so we have no choice but to fight on.

"But, to be plain, you are dissatisfied with me about the negro. Quite likely there is a difference of opinion between you and myself upon that subject. I certainly wish that all men could be free, while I suppose you do not." So you resist emancipation. "You say you will not fight to free negroes." Well, Lincoln said, "some of them seem willing to fight for you; but no matter. Fight you, then, exclusively to save the Union. I issued the proclamation on purpose to aid you in saving the Union. . . .

"I thought that in your struggle for the Union, to whatever extent the negroes should cease helping the enemy, to that extent it weakened the enemy in his resistance to you. Do you think differently? I thought that whatever negroes can be got to do as soldiers, leaves just so much less for white soldiers to do, in saving the Union. Does it appear otherwise to you? But negroes, like other people, act upon motives. Why should they do any thing for us, if we will do nothing for them? If they stake their lives for us, they must be prompted by the strongest motive—even the promise of freedom. And the promise being made, must be kept."

And anyway, he said, the signs are looking better. The Mississippi River now "goes unvexed to the sea." We have won brave victories at Gettysburg and Murfreesboro. Our job "is a great national one," and we are all fighting together—the great Northwest, New England, the Empire and Keystone states, Jersey, and the border South are all lending a hand. And so are black men—so are they fighting with us for the salvation of the Union. And if the Union can be preserved, "there will be some black men who can remember that, with silent tongue, and clenched teeth, and steady eye, and well-poised bayonet, they have helped mankind on to this great consummation; while, I fear, there will be some white ones, unable to forget that, with malignant heart, and deceitful speech, they have strove to hinder it."

Out the letter went, to be widely published in the press and read aloud to a cheering Union rally in Springfield, Illinois. At the same time, Lincoln consulted with Stanton about the soldier vote, and Stanton argued persuasively that most soldiers detested antiwar "Copperheads" and would vote down the line for Union candidates. With Lincoln's approval, Stanton furloughed soldiers home to vote in Ohio, New York, Delaware, and other states with crucial elections coming up.

Though he did all he could for Union candidates, Lincoln remained apprehensive. Welles said he worried more about the state elections in 1863 than he did about the Presidential contest in 1860. In the White House, Lincoln spoke with "a good deal of emotion" about the likes of Vallandigham—and about the grim possibility that the voters might deal Republicans another political defeat.

But this time the electorate turned against the Democrats. In Ohio, Union candidate John Brough trounced Vallandigham by more than 100,000 ballots, with some 40,000 soldiers voting for Brough as well. "Glory to God in the highest," Lincoln exclaimed, "Ohio has saved the Nation." When all the fall returns were in, state Union candidates had carried the rest of the Midwest, Pennsylvania, and every other Northern state except New Jersey —proof indeed that Republican propaganda had been persuasive, easing racial tensions and discrediting antiwar Democrats, and that Lincoln's own Springfield letter had had a powerful impact. Charles Sumner, who loved Lincoln's "true and noble" letter, thought it had proved unanswerable. And Lincoln himself was elated. Yes, he viewed the elections as a victory for the Republicans' tough war policy, and a sign that the country was changing so far as emancipation was concerned. He told Chandler how glad he was "that I have not, by native depravity, or under evil influences, done anything bad enough to prevent the good result." And Chandler for once spoke well of the President. While Lincoln moved more slowly than some people liked, "when he puts his foot down, he is there."

As Lincoln scanned the whole military picture that summer and autumn, he tried to prod his armies into the kind of concerted operations he'd desired since 1861. On his orders, Union forces finally mounted something of a joint offensive out in Tennessee, with Rosecrans moving south through the mountains and driving the rebels out of Chattanooga. At the same time, Burnside took a small army into east Tennessee and captured most of the Union-sympathizing mountain region. Lincoln, who'd wanted to liberate east Tennessee as though "my own home, and family were in Knoxville," sent Burnside "a thousand thanks" for his successful campaign.

Meanwhile Lincoln kept urging Meade to strike at Lee before bad weather set in. In September the two armies faced one another in northern Virginia, about midway between Washington and Richmond, each army defending its capital and menacing the other's. Meade reported that he could probably make Lee retreat, but doubted that he could pursue the rebel army clear to Richmond and siege the enemy capital.

Siege the enemy capital. That, Lincoln complained, was the main trouble with Meade—he couldn't comprehend that Lee's army was his target, not Richmond. Somehow Meade must be made to understand that if the Potomac Army couldn't defeat Lee where he was, it could never whip him in the trenches around Richmond. When Meade telegraphed that he wasn't sure where Lee was, Halleck suggested that he attack and find out. But Meade wouldn't budge until he discovered Lee's exact position.

"It is the same old story of this Army of the Potomac," Lincoln told Welles. "Imbecility, inefficiency—don't want to *do.*" "It is terrible, terrible, this weakness, this indifference of our Potomac generals, with such armies of good and brave men." "What can I do with such generals as we have? Who among them is any better than Meade?" The whole thing "oppresses me."

On September 19 came alarming news from Rosecrans. The rebels had attacked him along Chickamauga Creek south of Chattanooga, and the fighting was "terrific." At once Lincoln ordered Burnside to march to Chickamauga and reinforce Rosecrans. Then Lincoln took up his usual battle station in the War Department, where for two days he was "awake and watchful." He felt something bad was happening out there, could "feel trouble in the air." As he feared, subsequent dispatches from Rosecrans confessed that Chickamauga was a Union rout, that the rebels, reinforced by men from Lee's army, had crashed through Rosecrans's lines and driven him back to Chattanooga. From there Rosecrans telegraphed an apocalyptic message: the enemy was now threatening his entire front—Burnside would arrive too late to help. We are "brave and determined," Rosecrans said, "but our fate is in the hands of God."

Lincoln was shocked at the news from Chattanooga. Where was Burnside now? Could he get to Rosecrans in time? As it happened, Burnside believed that the rebels had retreated at Chickamauga and that Rosecrans no longer needed him. Therefore he had turned around and marched away from Chattanooga, heading off to Jonesboro. "Damn Jonesboro!" Lincoln cried. This report was "incomprehensible." It made him "doubt whether I am awake or dreaming." He wrote Burnside a harsh rebuke, only to change his mind about sending it. Upbraiding him wouldn't accomplish anything—the damage had already been done. Burnside's army was too far away to reach Rosecrans in time.

On September 23, though, Rosecrans wired that he'd secured Chattanooga and believed he could hold it unless attacked by a superior rebel force. Now Lincoln didn't know what to think. Was the danger really past —or was Rosecrans just too confused to tell? That night, Lincoln rode out to the Soldiers' Home to try and get some rest. But no sooner had he climbed into bed than a rider came galloping into the grounds. It was John Hay. Stanton had called a special war council; Lincoln had better come at once. Now he was really distressed—Stanton had never summoned him from his sleep, not even during the worst days of 1862.

Lincoln and Hay rode back to Washington in brilliant moonlight. At the War Office, Stanton, Halleck, Seward, Chase, and several other officials greeted Lincoln as he walked in. Yes, Stanton? The Secretary was grave and urgent. The reports are that Rosecrans ran away from the battlefield at Chickamauga—that he left his men and raced with two other commanders back to Chattanooga. Then he ordered General George Thomas, whose troops had held the Union left during the battle, to fall back as well.

Lincoln conferred with his advisers until dawn. Above all, Chattanooga must be held. It was the key to middle Tennessee and an indispensable base for future Union operations against the Deep South. Accordingly they decided to detach twenty thousand men from the Army of the Potomac and rush them to Chattanooga by railroad. At the same time, Grant and Burnside would send all the help they could spare. Lincoln would decide later what to do about Rosecrans.

After the conference, Hay showed Lincoln a dispatch from the rebel newspapers. Among the Confederates killed at Chickamauga was Ben Hardin Helm, Mary's brother-in-law, husband of sister Emilie. Mary was away in New York, so Lincoln sent her a telegram from the War Department that Helm was dead. Poor Emilie. Well, Lincoln would talk with Mary about it when she returned.

More reports about Rosecrans—one from Dana claimed that he was losing control. Lincoln grumbled that since Chickamauga Rosecrans had acted "confused and stunned like a duck hit on the head." In mid October the President gave Grant command of all armies in the West and ordered him down to Chattanooga with the authority to fire Rosecrans if he wanted to. Grant, then at Louisville, relieved the general by telegram and hurried south to Chattanooga, riding horseback over gullied mountain roads. Lincoln was relieved indeed when Grant took charge of the city, secured his supply lines, and reorganized his defenses. In Grant's hands, Chattanooga was sure to be safe. In fact, the way he fought, it wouldn't be long before he counterattacked.

Early in November, Lincoln received an invitation to speak at the commemoration of a new National Soldiers' Cemetery on the Gettysburg battlefield. Attorney David Wills of Gettysburg had conceived the idea for the cemetery, and the enterprise now enjoyed the support of the Pennsylvania governor and a spate of other public officials. The commemoration, to take place on November 19, was expected to attract thousands of people, Wills said, with Edward Everett of Boston as orator of the day and many other luminaries in attendance.

Though he'd turned down other requests to speak (he couldn't spare the time), Lincoln accepted the Gettysburg invitation, because he thought it an appropriate setting to say something significant about the meaning of the war, to explain how Union armies were fighting not just to subdue a rebellion, but to save democracy and America's liberal institutions. Yes, something like that.

In rare moments alone, in his office or on a crisp ride in the country, Lincoln reflected on his speech and arranged phrases in his mind. He managed to write down several lines on executive stationery, but was too busy to finish the speech before it was time to go. He confided in Noah Brooks that it was going to be "short, short, short."

Mary was home now and anxious about his leaving her. She'd taken Helm's death pretty hard and worried about sister Emilie. Lincoln was troubled about Helm, too, well remembering how he'd offered the man a coveted post in the Union army back in early 1861 and urged him not to join the Confederacy. Lincoln had even argued that Mary needed Emilie here in Washington. But Helm was inflexible in his Southern loyalties and went away to enlist in the rebel army. And he wasn't the only person in Mary's family who'd sided with the rebels. So had her mother, her full brother George, all her half-sisters, and all her half-brothers except one. By now three of her half-brothers had died in battle, and each death had deeply affected Mary, until she said she'd drawn a shell around her heart and hoped all her rebel brothers would be killed or captured. But why? a friend protested. How can you say that? "They would kill my husband if they could," Mary replied, "and destroy our Government—the dearest of all things to us."

But Ben and Emilie were exceptions. Mary loved her sister and adored attractive and charming Ben. And now he was dead and God knew what had become of Emilie. Then, as if that were not enough, Tad came down with a serious illness which upset Mary even more. What if Taddie died? What if he died like Willie? Lincoln didn't want to leave her—he didn't want to leave Tad either—but he'd given his word to the people in Gettysburg. He had to go.

He left on November 18, but the train was too crowded and noisy for him to work on his speech. As the coach droned through the Maryland and Pennsylvania countryside, Lincoln chatted with Seward, Blair, Hay, Nicolay, and John Usher, who by now had replaced Caleb Smith as Secretary of the Interior. Their conversation turned to Missouri, where Unionists were divided into squabbling "radical" and "conservative" factions. Inevitably Lincoln became embroiled in their pestilent bickering, as both sides prevailed on him for support. The so-called radicals wanted in particular to extend the emancipation proclamation to Missouri, thus freeing slaves there at once. The "radicals" not only endorsed all of Lincoln's war measures, but requested that he suppress the "Copperhead" conservatives and put only true loyalists like themselves in power. The conservatives, by contrast, favored gradual emancipation in Missouri and a more lenient civilian and military administration there. For his part, Lincoln went for gradual emancipation in loyal Missouri because he thought it best for both races. And he refused to suppress conservative Unionists, reminding the "radicals" that regardless of factional differences, their rivals were loyalists, too. Nevertheless, Lincoln's sympathies lay with the "radicals." As he told Hay later, "I know these Radical men have in them the stuff which must save the State and on which we must mainly rely." "If one side *must* be crushed out & the other cherished there could be no doubt which side we would choose as fuller of hope for the future. We would have to side with the Radicals."

The train pulled into Gettysburg just after dusk. Already thousands of people were in town, most of them relatives of slain soldiers "who had come from distant parts to look at and weep over the remains of their fallen kindred," a reporter said. The hotels were brimming with "notables and nondescripts," and the streets crowded with soldiers, serenaders, businessmen, governors, and mayors. Lincoln and his Cabinet Secretaries made their way to Wills's house on the square, where they and other guests—including Edward Everett—sat down to an elegant dinner. This was the first time Lincoln had met Everett, a prim, white-haired Bostonian and one of the foremost orators in America. He'd had a distinguished and polygonal career, first as a Boston minister, then as a professor of Greek, president of Harvard, governor of Massachusetts, U.S. senator, minister to England, and Secretary of State. He'd just toured the battlefield and spoke sadly of the shell-torn trees, the shallow graves with their wooden markers, the Southern corpses covered with rocks at Devil's Den. Outside, a military band struck up a serenade for the President and called for a speech. Lincoln went out to greet them, but declined to say anything formal. "In my position," he said, searching for his words, "it is somewhat important that I should not say any foolish things."

The band music and serenades went on into the night. Those who'd lost kinfolk at Gettysburg shared stories in their rooms; and Nicolay and Hay, after an oyster dinner at the college, gathered with an acquaintance to sip whiskey and sing "John Brown's Body." In the Wills home late that night, Lincoln sat in a guest bedroom toiling on his speech. Before he went to bed, a telegram arrived from Stanton: All is quiet on the battlefronts, and "Mrs. Lincoln informs me that your son is better this evening."

The next morning Lincoln finished his speech, took a wagon ride with Seward, and then joined a slow, chaotic procession out to the battlefield cemetery south of town. Flags flew at half-mast that day, minute guns spoke, and soldiers stood in salute as the President passed. At last the procession reached the top of Cemetery Hill and Lincoln took his place on a wooden platform there, with some fifteen or twenty thousand people milling about and band music wavering in the wind.

A half hour passed before Everett arrived; he'd been wandering over the battlefield all morning to get inspired. The orator of the day spoke for two hours, with people drifting in and out of the audience. It had rained in the early morning, but the sky was clear now and the sun shone brilliantly on the rolling countryside, a wall of blue mountains rising in the distance. Cemetery Hill afforded a sweeping view of the battlefield and peach orchards and wheat fields south of Gettysburg. Off to the west lay Seminary Ridge, where half of Lee's army had drawn its lines. Cemetery Ridge, where the Army of the Potomac had entrenched, ran south from Cemetery Hill for a mile or more to Little and Big Round Tops. Here elements of the two armies had clashed in desperate and bloody combat for control of the Union left flank. On the third day of battle, across the open meadow there and up the slope, George Pickett had led fifteen thousand of Lee's finest in a charge of awful grandeur, a frontal assault against the center of the Union line at a place called the Angle, a charge conducted in perfect battle formation with flags and banners flying high. But Union forces met the attack with murderous artillery and musket fire, blowing enormous holes in Pickett's lines and hurling his shredded regiments back down the slope until all the flags were gone. The next night, in a torrential rainstorm, Lee started his battered army south, and for days after the wind cried with bugles, and with dirges for the dead.

In his oration, Everett described the three-day battle with cadenced gestures and drew lessons from European history. From where he sat, Lincoln could see coffins lying in silent formations in trees and open areas, could see skeletons of army horses in the distance, trees decapitated by artillery fire, and souvenir hunters prowling over the fields. He shifted in his chair, took his manuscript from his coat pocket, put on his steel-rimmed glasses, and looked over his speech. When Everett sat down, the President unwound himself, stepped to the front of the platform, and recited in a high and singing voice his hymn to the Union dead.

"Four score and seven years ago our fathers brought forth on this continent, a new nation, conceived in Liberty, and dedicated to the proposition that all men are created equal." He waited for the applause to die away, then went on: "Now we are engaged in a great civil war, testing whether that nation, or any nation so conceived and so dedicated, can long endure. We are met on a great battle-field of that war. We have come to dedicate a portion of that field, as a final resting place for those who here gave their lives that that nation might live. It is altogether fitting and proper that we should do this.

"But, in a larger sense, we can not dedicate—we can not consecrate—we can not hallow—this ground. The brave men, living and dead, who struggled here, have consecrated it, far above our poor power to add or detract. The world will little note, nor long remember what we say here, but it can never forget what they did here. It is for us the living, rather, to be dedicated here to the unfinished work which they who fought here have thus far so nobly advanced. It is rather for us to be here dedicated to the great task remaining before us—that from these honored dead we take increased devotion—that we here highly resolve that these dead shall not have died in vain—that this nation, under God, shall have a new birth of freedom—and that government of the people, by the people, for the people, shall not perish from the earth."

The audience gave him a sustained ovation. Then "the music wailed," Hay recorded, "and we went home through crowded and cheering streets." That evening Lincoln returned to Washington by train; he was worn out, said little, lay in a side seat in the drawing room with a wet towel across his eyes. In his Springfield letter in the recent elections, he'd defended his proclamation of freedom for black people and urged dissident whites to rejoin him in fighting to save their Union. At Gettysburg this day, he'd called for a national rededication to the proposition that all men were created equal, a new resolve to fight for that proposition and salvage America's experiment in democracy for all mankind. Let Union people of all colors and conditions come together in a new commitment to freedom and a new national crusade. Let them cease their petty quarrels, put aside their differences, and vow together that those who died in battle, those who lay in coffins at Gettysburg and a thousand other cemeteries across the land, had given their lives for a true and noble ideal—for the liberation of the human spirit in a government by and for all the people.

Back in Washington, Lincoln came down with what doctors diagnosed as "varioloid," a mild form of smallpox. So now he and Tad both were sick—though Tad at least was improving. Where were the office seekers? Lincoln quipped. Now he had something he could give everybody.

As he lay in his chamber, Stanton brought him a procession of electrifying dispatches from Tennessee. In late November, Grant attacked the rebels south of Chattanooga and defeated them in the battles of Lookout Mountain and Missionary Ridge. A report from Dana moved Lincoln deeply, for it told how eighteen thousand Union men charged without orders from Grant or Thomas, how they crawled, pulled, and hacked their way to the summit of Missionary Ridge and put the rebels to flight.

With the Confederates driven back into Georgia, Grant sent Sherman marching for Knoxville, where Burnside was fighting off an attack by James Longstreet. But by the time Sherman reached Burnside, Longstreet was already in full retreat back toward Virginia. "This is one of the most important gains of the war," Lincoln told Nicolay. "It secures us East Tennessee." "Now if this Army of the Potomac was good for anything—if the officers had anything in them—if the Army had any legs, they could move thirty thousand men down to Lynchburg and catch Longstreet. Can anybody doubt, if Grant were here in command that he could catch him?" But of course Meade wasn't going to do anything. It was disgraceful "how hard we have tried to get this Army to move towards the enemy and we cannot succeed."

In between military communiqués, Lincoln propped himself up on pillows and worked on his message to Congress, scheduled to convene on December 8. Though his principal concern now was reconstruction, he had some things to say about immigration and Indians, too. He wanted the government to encourage immigration, because the country had a shortage of laborers in all fields of industry, particularly in mining and agriculture. As for the Indians, Lincoln had carried on the policy of "concentration" begun in 1851, another step in the long process of oppression and removal which had characterized federal Indian policy since Jefferson's Presidency. While involved in the smoke and steel of battle, Lincoln noted, his administration had negotiated sundry treaties for "extinguishing the possessory rights of the Indians to large and valuable tracts of land" and for concentrating the tribes in specified areas on the plains and deserts west of the Mississippi. Lincoln fervently hoped that these treaties would result in cordial relations with the Indians, especially those who'd suffered bloody collisions with whites out in the Southwest and up in Minnesota. Though Lincoln himself had intervened in the Minnesota Indian war of 1862 and had prevented vengeful whites from executing a number of innocent Sioux, he had a paternalistic attitude toward the Indians, as did most enlightened white leaders of his generation (unenlightened whites simply wanted to exterminate them). "Sound policy and our imperative duty to these wards of the government," Lincoln wrote in his bed, "demand our anxious and constant attention to their material well-being, to their progress in the arts of civilization, and, above all, to that moral training which, under the blessing of Divine Providence, will confer upon them the elevated and sanctifying influences, the hopes and consolation of the Christian faith."

Then Lincoln turned to the liberation of black people in the South. He gave a careful and reasoned defense of his proclamation and his decision to employ Negro troops, who'd proved "as good soldiers as any." Though his policy on Negroes had been greatly vilified in those "dark and doubtful days" that followed his proclamation, Lincoln declared the recent state elections "highly encouraging" for those who must guide the country "through this great trial."

"Thus," he wrote, "we have the new reckoning. The crisis which threatened to divide the friends of the Union is past."

And now for reconstructing conquered Southern states, a complex and potentially inflammable problem. Lincoln firmly believed that the President and not Congress should restore loyal civilian rule in the rebel South, that it was an executive responsibility to be carried out swiftly and efficiently by the War Department and the army. Now that the Union had liberated Tennessee and occupied sections of Arkansas, Louisiana, Texas, Florida, and Virginia, he was determined to reconstruct these areas along lines he thought were correct, because he didn't know whether he would be re-elected or even renominated in 1864. Sure, his party had done well in the state elections, but Lincoln himself remained an enormously unpopular President, stridently criticized from all directions. He had no idea what would happen next year, who his replacement might be or what he might do. Therefore time was crucial if he was to get a sound and systematic reconstruction program adopted.

Already he'd made piecemeal attempts to restore occupied Louisiana, Arkansas, and Tennessee. He'd established military regimes there whose task it was to rally the Southern Unionist minority—whites who'd opposed secession and war, many of them former Whigs—and help them set up state governments loyal to Washington and committed to the liberation of the slaves. He'd installed cranky Andrew Johnson as military governor of Tennessee and exhorted him to prevent the state from falling back into the hands of "our enemies." Johnson must use whatever power was necessary to organize a Unionist government and draft a new state constitution that ratified the emancipation proclamation.

Lincoln had admonished General Banks to do the same in Louisiana, but reconstruction there had run into a sea of difficulties. From the start, the Unionist minority had split into warring factions, with "conservatives" struggling to retain the old 1852 constitution which guaranteed slavery, while a "radical" faction demanded a new constitution which would eradicate the institution altogether. Although Lincoln agonized over such destructive quarreling, he nevertheless sided with the "radicals." Through Banks, he warned Louisiana Unionists that they must not only accept the emancipation proclamation, but must expunge slavery in the parishes he himself had exempted. In short, black people must be freed everywhere in the state. Still, he would be flexible about racial adjustment. He would not object if Louisiana Unionists, working in cooperation with his administration, devised "a reasonable temporary arrangement in relation to the landless and homeless freed people," so that the two races "could gradually live themselves out of their old relation to each other, and both come out better prepared for the new." Now get on with reconstruction, Lincoln wrote Banks and the Louisiana Unionists, and cooperate with one another in a common and hazardous task: the creation of a free state government in an occupied area filled with hostile Confederate sympathizers. But to Lincoln's bitter disappointment, Louisiana reconstruction bogged down in continued feuds and delays.

As if squabbling Southern Unionists were not enough to contend with, trouble was brewing with Sumner and other Congressional liberals, who argued that Congress and not the President had jurisdiction over reconstruction. If reconstruction were left to the executive branch, Sumner feared that Seward would scheme and connive until he controlled reconstruction policy —God forbid!—whereupon he might try to negotiate a peace that would restore the South with slavery still alive. Also, Sumner distrusted Southern white Unionists and frowned on any attempt to build loyal governments around them. The only loyal people in the South, the senator said, were the slaves. Accordingly, he wanted Congress to assume control of reconstruction and guarantee the freedom of Southern blacks by granting them citizenship and the right to vote.

Lincoln was well aware of Sumner's arguments in favor of Congressional reconstruction. But he chose to ignore them. "I can do nothing with Mr. Sumner in these matters," Lincoln remarked. "While Mr. Sumner is very cordial with me, he is making his history in an issue with me on this very point. . . . I think I understand Mr. Sumner; and I think he would be all the more resolute in his persistence . . . if he supposed I were at all watching his course."

Maybe Lincoln ignored Sumner, but conservative Republicans did not. They pointed out that if former slaves could vote as citizens, they could also run for office—which meant that Southern "niggers" would soon be sitting here in Congress and influencing the destiny of the United States! Montgomery Blair, who loathed Negroes, went even further than Sumner's conservative critics. In a hot and bigoted speech in Maryland, Blair accused the senator and his colleagues of trying to achieve *"amalgamation, equality* and *fraternity"* with the Negro. They "would make the manumission of the slaves the means of infusing their blood into our whole system." As Postmaster General, Blair condemned Sumner's Congressional approach as flagrantly unconstitutional and announced that the President alone had jurisdiction over reconstruction in the South.

These family conflicts. While Lincoln regretted Blair's speech, thought he was using the executive branch to make war on Republican liberals, deplored his Negrophobia, Lincoln convinced himself that the controversy between Blair and Sumner was "one of mere form and little else." Lincoln didn't think Blair wished to restore the Southern states as they used to be, so that whites there could revive their old tricks and "bedevil us" as they did before the war. As for Sumner, the President didn't believe he really wanted to prevent loyal Southern whites from directing their own affairs. True, Sumner desired to take away the power Lincoln now exercised over insurrectionary districts. But about the vital question—the right and privilege of Southern people to govern themselves—"I apprehend there will be little difference among loyal men," Lincoln said.

No, the paramount problem was not Sumner or even political infighting among Southern Unionists. The major problem was how to prevent the rebel majority in a given state from outvoting and overwhelming the loyalist minority and thus restoring the old Southern ruling class to power. Should that happen, the war would truly have been fought in vain.

As he lay in bed that December, ruminating over reconstruction and all its potential woes, Lincoln drafted a plan of restoration, appended to his Congressional message as a special proclamation, which virtually outlawed the old Southern rulers, so that loyal Unionists—the former Whigs and antisecessionists—could become the new leaders in Dixie. Specifically, he refused to pardon the following classes, thus disqualifying them from voting or holding political office in the occupied South: all men who'd held Confederate civilian and diplomatic posts, all who'd served as rebel officers above the rank of colonel in the army and lieutenant colonel in the navy, all who'd resigned from the U.S. armed forces or left the Congress or judicial positions to assist the rebellion, and all who'd treated Union soldiers other than as prisoners of war. Apart from these, he granted full pardons to all other Southerners who'd engaged in rebellion so long as they took an oath of allegiance to the United States. This, Lincoln contended, was a fair and liberal test "which accepts as sound whoever will make a sworn recantation of his former unsoundness." Once a number of people equal to ten percent of those who'd voted in 1860 had taken the oath, these people could establish a loyal civilian government and elect U.S. representatives, and their states would be restored to the Union with full federal protection.

All reconstructed governments, however, must accept and obey the emancipation proclamation and all Congressional laws bearing on slavery. "To now abandon them would be not only to relinquish a lever of power," Lincoln said, "but would also be a cruel and an astounding breach of faith." But he repeated what he'd told Louisiana Unionists. He would accept temporary state control of Negroes "as a laboring, landless, and homeless class," so long as the state recognized their permanent freedom and provided "for their education."

Lincoln indicated that this was only one plan for reconstructing the rebel South, and while it was the best he could think of for now, he would gladly consider others and possibly adopt them. He might even modify his own classes of pardons, if that seemed warrantable. His whole approach to reconstruction was empirical; what plan he used depended on the circumstances and exigencies of each place and moment. Furthermore, he solicited Congress's approval and cooperation (which was why he appended his ten percent plan to his annual message). For he understood that either house could refuse to seat the representatives from the states he reconstructed, thus preventing his work from being fully consummated.

On December 8 Lincoln sent his message and appended proclamation up to Capitol Hill, and the next day the clerk read them to a joint session of Congress. Afterward almost everybody but die-hard Democrats seemed happy with the ten percent plan. Hay and Brooks circulated through the

Capitol and reported that "men acted as if the millennium had come." "We only wanted a leader to speak the bold word," one fellow burbled. "It is done and all can follow." A Michigan congressman praised Lincoln as a great leader now, "the great man of the century." And Owen Lovejoy, whom Lincoln considered "the best friend I had in Congress," declared that he would actually live to witness the end of slavery in this country. In the Senate, both liberal and conservative Republicans pronounced Lincoln's plan "highly satisfactory." Chandler "was delighted," and Sumner himself was "not displeased," conceding that in the absence of a Congressional program for the present, Lincoln's approach seemed proper enough. Later Sumner came to the White House and personally congratulated Lincoln. So did Wade, Lovejoy, and many others.

Lincoln basked in all the praise. He almost felt serene that day. Never, it seemed, had Congressional Republicans approved of his work with such unanimity. Yes, on the subject of reconstruction, Lincoln saw little essential difference among good Union men.

No, he was wrong. Chase objected to his proposals, and Lincoln thought he was about the only Republican who did. Increasingly sullen and remote from the President, Chase seemed to object more and more these days. A sign of coming trouble? A move on his part to challenge Lincoln for the nomination next year? There were rumors that he was scheming in the shadows and making "capital" out of military troubles like those with Rosecrans. In front of others Lincoln laughed at the rumors. But in private he decided that his Treasury Secretary would bear some watching this winter.

PART·TEN

MIGHTY SCOURGE
OF WAR

In December, Lincoln received a wire from Fort Monroe that Emilie Helm was there with her daughter. The story goes that she refused to take the oath of allegiance, so Lincoln telegraphed back, "Send her to me." When she reached the White House, tired, anguished, and pregnant with her third child, both Lincolns embraced her with "the warmest affection," insisting that she stay and let them comfort her. She'd been through a nightmarish ordeal—had seen her husband buried in Atlanta and then made her way to federal lines in Virginia, in hopes that "Brother Lincoln" might help her get home to Kentucky. She was proud and sadly beautiful—and still very loyal to the Confederacy, so much so that she wouldn't sign an oath of allegiance Lincoln prepared for her. Though Emilie's rebel sympathies hurt him, Lincoln wrote out a pass for her anyway, certain that her code of honor would prevent her from embarrassing him.

Lincoln knew that Emilie's presence in the Executive Mansion would be troublesome. When Browning called, the President confessed that his sister-in-law was here, "but he did not wish it known." The news leaked out anyway, spawning ugly rumors around Washington about the Confederate wife in the White House. The "little rebel" was Mrs. Lincoln's half-sister too —proof indeed of the First Lady's traitorous sympathies, the gossips said. For was it not so? That her entire family in Kentucky was infested with rebels? Lincoln, for his part, tried to ignore the slander—though some of it was getting into the newspapers now—and to make Emilie feel as welcome as he could. Initially he avoided talk about Helm's death, but finally he said, "You know, Little Sister, I tried to have Ben come with me. I hope you do not feel any bitterness or that I am any way to blame for all this sorrow."

No, Emilie didn't blame him. It was "the fortunes of war," she said, and while Ben had been very grateful for Brother Lincoln's offer, he had "to follow his conscience." Lincoln put his arms around her and both of them cried.

As the December days passed, Lincoln left the two sisters to dine by themselves in the evenings, to reminisce together about Lexington and

Springfield, families and friends. Emilie regretted, though, that "this frightful war comes between us like a barrier of granite," so we cannot "open our hearts to each other as freely as we would like." Still, the sisters wept together about their dead. And Emilie found an endearing tenderness in the White House, where both Lincolns "pet me as if I were a child." She noticed, too, how concerned Mary was about Lincoln's health. "Emilie, what do you think of Mr. Lincoln, do you think he is well?" He was run-down from his recent bout with varioloid, and Emilie replied, "He seems thinner than I ever saw him." "Oh, Emilie," Mary said, "will we ever awake from this hideous nightmare?"

It became apparent to Emilie that something was wrong with Mary: she seemed tense, extremely unsettled by the war and Willie's death. One night in Emilie's room, in fact, Mary started saying strange and disturbing things about Willie. "He lives, Emilie." Yes, "he comes to me every night, and stands at the foot of my bed with the same sweet, adorable smile he always had; he does not always come alone; little Eddie is sometimes with him. . . . You cannot dream of the comfort this gives me." A chill passed through Emilie. "It *is* unnatural and abnormal," she wrote in her diary; "it frightens me. It does not seem like sister Mary to be so nervous and wrought up. She is on a terrible strain and her smiles seem forced."

A day or so later Lincoln took Emilie aside. "Little Sister, I hope you can come up and spend the summer with us at the Soldiers Home. You and Mary love each other—it is good for her to have you with her. I feel worried about Mary, her nerves have gone to pieces." She couldn't hide it from him, Lincoln said. He could tell that the strain she was under had been too much for her mental and physical health.

Yes, Emilie said, "she seems very nervous and excitable and once or twice when I have come into the room suddenly the frightened look in her eyes has appalled me. She seems to fear that other sorrows may be added to those we already have to bear. I believe if anything should happen to you or Robert or Tad it would kill her."

"Stay with her as long as you can," Lincoln said.

Later that same day, Lincoln and Mary had a disagreement about Robert, who wanted to leave Harvard and enlist in the army. Lincoln was willing to let him go, for he found it embarrassing that other young men Robert's age were fighting in the army while the President's son remained in college.

"Of course, Mr. Lincoln, I know that Robert's plea to go into the Army is manly and noble and I want him to go," Mary said. But "I am so frightened he may never come back to us."

"Many a poor mother, Mary, has had to make this sacrifice and has given up every son she had—and lost them all."

There was a tremble in Mary's voice now. "Before this war is ended I may be like that poor mother, like my poor mother in Kentucky, with not a prop left in her old age."

Finally Lincoln let it go. There was no use in tormenting her about Robert and aggravating her nerves. Still, it troubled him that he exempted his own son while drafting thousands of other sons into the army. But what could be done.

Soon after Christmas Emilie came to Lincoln and said she had to go. She missed her mother, missed her home and her family in Kentucky. She hoped Lincoln understood—and he did, of course. Later he and Mary saw her off with embraces and repeated goodbyes.

The New Year brought only more woe for the Lincolns. By now the rumors about Mary's Confederate relatives had swelled into grotesque accusations that the First Lady was a rebel spy who passed on state secrets to the enemy. In fact, she received so many abusive letters that she asked one of the White House secretaries to screen her public mail before bringing it to her. Mary could not understand how people could think her disloyal to her country or in any way friendly to the Confederates. "Why should I sympathize with the rebels," she once told Lizzie Keckley, "are they not against me? They would hang my husband to-morrow if it was in their power, and perhaps gibbet me with him. How then can I sympathize with a people at war with me and mine?"

Over the winter, Lincoln worked hard at getting his ten percent plan implemented in occupied Louisiana, which he hoped to make a model reconstructed state. To avoid conflict and confusion there, he designated General Banks as the sole "master" of reconstruction, instructing him to preserve whatever labor had already gone into restoring civilian rule and get on with registering pardoned voters, calling elections, and finding work for the freedmen. As he did so, Banks was to cooperate with loyal civilians and entertain their views and recommendations. If necessary, he could even modify Lincoln's reconstruction plan, so long as the basic principles were not violated. Above all, Banks must stop the pernicious feuding that had impeded reconstruction thus far and rally all friends of the Union—conservative, moderate, and radical—behind a loyal Louisiana regime. And the Southern leaders Lincoln himself had disqualified? In a supplementary proclamation in March, 1864, the President announced that such people could apply directly to him for clemency.

Though Louisiana was still rife with political dissension, Banks set about restoring the state as Lincoln directed. For blacks who could not find their own employment, Banks administered a Negro free-labor system similar to what Thomas had installed in the Mississippi Valley: under the system, a federal agent hired out freedmen to white planters or government "lessees" —those who managed abandoned rebel estates—at government-regulated conditions and wages. Lincoln approved of the system, as long as "the colored people" were paid and treated fairly. Under Banks's military supervision, Louisiana loyalists elected "moderate" Michael Hahn as governor and chose delegates to a constitutional convention; the delegates met in April and drafted a new state constitution that outlawed slavery and established a segregated public school system for black and white children. As Lincoln noted, the delegates chose not to establish a Negro apprentice system and thus gave Louisiana blacks economic independence. Later Louisiana voters ratified the constitution by a margin of more than four to one and elected state and national representatives. Lincoln accepted reconstruction in Louisiana as the best that could be done and took similar steps in Arkansas, Tennessee, and Virginia.

Meanwhile his program had run into stiff opposition from Sumner and some of his liberal colleagues on Capitol Hill. Following Lincoln's work in Dixie with a critical eye, they had grave doubts that the President was really changing anything there. They were especially unhappy with the administration's treatment of the former slaves, claiming that they were not adequately cared for or protected. There were reports that the administration's Negro labor system was unfair to the blacks, that the rate of pay was much too low— lower in fact than what free Negroes used to earn in antebellum days—and that some white employers were guilty of abusing and exploiting their Negro workers. How, then, had conditions materially improved for Southern black people? Were they not still oppressed? Were they not still dependent on their former masters? Moreover, by Lincoln's reconstruction program, would not the planters still control Southern state politics, able to retain their political clout merely by repenting for their sins and obtaining pardons? Should Lincoln restore the entire South along the lines he applied in Louisiana, what would the war have accomplished beyond the killing of several hundred thousand men? Sumner in particular was incensed, because he thought Lincoln was betraying "the only Unionists of the South"—the former slaves themselves. Time and again Sumner demanded that Congress enfranchise Southern blacks so they could protect themselves and secure their freedom.

Negro suffrage? This was an extremely unpopular measure in 1864, and Lincoln knew it was political dynamite. Since blacks couldn't vote in most of the North, nearly all Republicans shied away from the issue of Negro suffrage like the plague itself. Even Wade and many other liberals, mindful of their white constituents, thought Negro enfranchisement impractical at this time.

Yet Lincoln was sympathetic to the Negroes. Even before the constitutional convention met in Louisiana, he studied petitions from New Orleans black people, who belonged to a relatively well-educated and outspoken black community and who beseeched him to grant them the right to vote. In March, when Louisiana Unionists were considering the matter of voter qualifications, Lincoln wrote Governor Hahn: "I barely suggest for your private consideration, whether some of the colored people may not be let in —as, for instance, the very intelligent, and especially those who have fought gallantly in our ranks. They would probably help, in some trying time to come, to keep the jewel of liberty within the family of freedom." Yet Lincoln wouldn't force Negro suffrage on Louisiana, wouldn't make it a condition of reconstruction for that or any other Southern state at this time. Because he was afraid of alienating white Unionists there, afraid of losing the very people he hoped would rule the new South, afraid of destroying in one stroke all he'd tilled and cultivated in occupied Dixie.

As it turned out, the Louisiana constitutional convention rejected Lincoln's suggestion and categorically refused to give Negroes the suffrage. But evidently Lincoln and Banks persuaded the lawmakers to reconsider their decision and strike a compromise in the constitution they drew up: while the document did not enfranchise Louisiana blacks, it did empower the white legislature to do so. Still, this wasn't likely to happen any time soon, since most white Unionists in the state were hostile to black political rights.

Lincoln defended the white loyalists in Louisiana and stood by the all-white government they established there. Maybe it was imperfect, but he thought it a lot better than no civilian government at all. And in time, when prejudice and passion had subsided, its more glaring defects could be repaired. It was, in short, a foundation to build on for the future—for blacks as well as whites. But the black people of New Orleans did not agree with him. Bitterly they announced that if Negroes wanted justice, they would have to find another President.

Yes, Sumner was angry with him again, fussing that the President didn't care enough about black people. Through the spring of 1864, the Massachusetts senator campaigned tirelessly for black civil rights, eloquent in his appeal for homestead legislation and voting rights to protect Southern Negroes. Yet Lincoln "does not know how to help or is not moved to help," Sumner complained. True, Lincoln did not push for black political rights at this time. But from his view there were doubtlessly political reasons for his caution in this volatile area. After all, his emancipation proclamation—the most revolutionary measure ever to come from an American President thus far—had already caused tremendous turmoil in the North. Should he now champion Negro suffrage, he would almost surely enrage the country's capricious white constituency and split his own party, and this was a Presidential election year.

So, no, Lincoln did not summon Congress to enact a body of equal rights legislation for black people. But in subtle, unpublicized ways he did offer Negroes a Presidential hand: he helped Sumner guide through Congress a bill which gave widows and orphans of slave soldiers the same benefits as whites. And he readily signed bills into law that opened federal courts to black witnesses, granted equal pay to black soldiers, and outlawed racial discrimination in Washington's streetcar system.

When pressed about a second term back in 1863, Lincoln was usually noncommittal, for no President since Jackson had served more than four years, and a single term had become something of a tradition. "I wish they would stop thrusting that subject of the Presidency into my face," Lincoln snapped on one occasion. "I don't want to hear anything about it." But by early 1864 it was no secret that he wanted a second term badly. For re-election would demonstrate to his critics that the American people approved of his war policies, approved of his decision to fight the rebels with everything the Union could muster, approved of emancipation, Negro troops, and Presidential reconstruction. A change of administrations, he confided in Noah Brooks, would be "virtually voting him a failure."

Still, he had mixed feelings about remaining in office, as though he were doomed no matter what he did. "This war is eating my life out," he told Owen Lovejoy. "I have a strong impression that I shall not live to see the end." Yet he must stay in office, to vindicate himself and his work. And Lovejoy, who now lay dying in bed, sent out word that he favored Lincoln for a second term. "If he is not the best conceivable President," Lovejoy informed Garrison, "he is the best possible."

Not all Republicans agreed. In fact, there was formidable opposition to a second Lincoln administration throughout Republican ranks, as conservatives thought him a radical and liberals thought him a failure. In Congress, liberal Republicans derided Lincoln as a "simple Susan" who'd dragged his feet on every significant war measure. "This vacillation and indecisiveness of the President," asserted a Republican-sponsored pamphlet, "has been the real cause why our well-appointed armies have not succeeded in the destruction of the rebellion. . . . The cant about 'Honest Old Abe' was at first amusing, it then became ridiculous, but now it is absolutely criminal." A correspondent on Capitol Hill reported that not a single senator favored Lincoln's re-election, not a single one.

Who to replace him with? Conservative James Gordon Bennett promoted Grant for the Presidency, but Grant axed the move, contending that the country needed Lincoln and that his defeat would be "a great national calamity." On the other hand, Horace Greeley and many other liberals favored Chase for the Republican nomination. And in truth, Chase was ready and eager to run if the party wanted him. In January he declared himself "available" and subtly pushed his candidacy around Washington, regaling his liberal colleagues with accounts of Lincoln's blundering, his ignorance about fiscal matters, his catering to Seward. Chase candidly and honestly thought himself a better man than Lincoln. If elected, Chase promised to run an efficient administration, make no deals with anybody, and preserve America's free institutions. By February the Chase boom was gathering momentum, with Senator Samuel Pomeroy of Kansas heading a Chase central committee and Chase clubs springing up around the country. Perhaps his most ardent supporter, though, was his daughter Kate Chase Sprague, a striking woman with bewitching charm. Since her father was a widower, Kate would reign as First Lady in a Chase White House. And she was quite certain she would be more successful than Mary Lincoln.

Mary, of course, could not abide Kate Sprague and her ingrate of a father. But at first Lincoln made light of Chase's Presidential ambitions, chuckling that Chase was like "a horsefly on the neck of a plowhorse." Yes, his campaign was "a devilish good joke," wasn't it? Another chapter in that "voracious desire for office . . . from which I am not free myself." But as Chase picked up strong liberal support, Lincoln watched him closely, trying to separate fact from rumor about his activities. For Lincoln's friends brought him ugly reports about Chase's intrigues—reports that he was using his own department to promote himself, filling up Treasury posts with pro-Chase men who electioneered in his behalf. How could the President retain a scoundrel like him in the Cabinet? Lincoln's friends asked. But Lincoln said he flat didn't want to hear about these things. Sure, Chase was "a little insane" about the Presidency and it strained their relationship, but Lincoln had no intention of removing him so long as he ran his department with competence. And Lincoln thought he was doing that.

On the surface Lincoln ignored the Chase boom, but in private he admitted that he was concerned. He consulted with Henry Raymond of the *New York Times*, now Lincoln's "Lieutenant General in politics," and mapped out the President's own preconvention strategy. Evidently he wooed Charles Sumner away from the drop-Lincoln movement, until Sumner declared himself neutral about Lincoln, neither for nor against his renomination. "My relations with the President are of constant intimacy," Sumner said, "and I have reason to believe that he appreciates my reserve." Lincoln did indeed. But the opposition of other Republican liberals hurt him deeply.

In late February came a bombshell. For several days Pomeroy had been privately circulating an anti-Lincoln letter among prominent Republicans. On February 22 the letter appeared in the newspapers, announcing that Lincoln's election was "practically impossible," that he was given to "compromise and temporary expediency of policy," that no President since Jackson had gone on to a second term, that another Lincoln administration would damage "the cause of human liberty and the dignity and honor of the nation," and that the best man for the White House was Salmon Chase.

The Pomeroy Circular created a sensation, with Lincoln's friends and many of his foes denouncing it as a rabid and scandalous assault gotten up by Chase himself. And though Pomeroy argued that Chase knew all about the letter, Chase seemed mortified. He assured Lincoln that he had "no knowledge" of Pomeroy's letter and offered to resign.

Lincoln refused to let him go. He would judge Chase on his public service, would keep or fire him on that issue and no other. Lincoln told Chase that he'd heard about the letter before it appeared in print, but that he hadn't read it and didn't think he would. He believed Chase's disavowal and said neither of them should be held accountable for what their friends did without their consent. With that, Lincoln dropped the Pomeroy business and refused to talk any more about it.

As it developed, the Pomeroy Circular galvanized Lincoln's political backers into action. Out in Illinois, hard-drinking Governor Richard Yates defended Lincoln's re-election, and so did Republican papers there. In Washington, Frank Blair, Jr., opened up on Chase, proclaiming on the floor of Congress "that a more profligate administration of the Treasury Department never existed, that the whole Mississippi Valley is rank and fetid with the frauds and corruptions of its agents." Lincoln disclaimed any responsibility for what Blair said, but made no attempt to shut him up either. Meanwhile, in state after state, Lincoln's managers persuaded Union leagues and local conventions to announce for the President, until the Chase boom simply collapsed. The worst blow came when a caucus of Ohio Republicans, indignant over the Pomeroy Circular, rejected their favorite son and threw their support to Lincoln. On March 5 a humiliated Chase withdrew his candidacy. And when the Republican National Committee subsequently lined up with Lincoln, his renomination seemed all but guaranteed. A lot of Republicans, however, remained bitterly unhappy with him and grumbled on about his crass ineptitude.

In late March a delegation of Kentuckians met with Lincoln about border-state troubles, reiterating all the old objections to his slave policy, particularly his use of Negro troops. Lincoln gave an impassioned defense of his Presidency, declaring that a year and a half of trial had more than vindicated his decision to free the slaves and enroll them in the army. Was the Union not winning abroad, at home, and on the battlefronts, thanks in no small part to the thousands of black soldiers now fighting in Union ranks? Still, it was God and not Lincoln who had brought about the revolutionary changes of this great contest, especially so far as slavery was concerned. "I claim not to have controlled events," Lincoln said, "but confess plainly that events have controlled me. Now, at the end of three years struggle the nation's condition is not what either party, or any man devised, or expected. God alone can claim it. Whither it is tending seems plain. If God now wills the removal of a great wrong, and wills also that we of the North as well as you of the South, shall pay fairly for our complicity in that wrong, impartial history will find therein new cause to attest and revere the justice and goodness of God."

I t was time for a major shake-up in command, time to get somebody in charge of Union forces in the East who could demolish Lee and show Republican dissidents that Lincoln was out for total victory, that he wasn't slow or timid or half-witted as they charged. In March he did something he'd been contemplating since early in the year: he called Grant out of the West and appointed him General in Chief of all Union armies. In addition, Lincoln promoted him to lieutenant general, a rank no commander had held since Winfield Scott.

Grant arrived in Washington on March 8. There were the usual receptions, ceremonies, interviews. Then he and Lincoln met together in private. Going on forty-two now, Grant was a shy, slight man who stood about five feet eight and weighed around 135 pounds. He had stooped shoulders and mild blue eyes, walked with a lurch, chewed cigars, and surrendered words with much reluctance. Ohio-born, he'd graduated twenty-first in a class of thirty-nine at West Point, fought in the Mexican War, and married a trim and vivacious Missouri girl named Julia Dent. Later he went off to outpost duty on the West Coast, but loneliness and whiskey got the best of him. In 1854 he quit the army, rejoined his family in Missouri, and tilled a small dirt farm he called "Hardscrabble." Eventually he moved into Saint Louis, but failed there as a real estate salesman. Suffering from ague and rheumatism and feeling useless, Grant moved his family to Galena, Illinois, where he clerked in his father's leather goods store. When the war broke out, he led an Illinois outfit to Missouri and in early 1862 won national fame when he stormed Fort Henry and Fort Donelson in Tennessee. And now two years later he was commander of some 860,000 total troops in all Union armies, about 500,000 of them effective fighting men ready to campaign against the Confederacy. At first Grant thought he would remain in the West, but Lincoln quickly disabused him of that idea. Sherman could command his old army out there. Lincoln needed Grant here in the East.

With Grant as General in Chief, Halleck now functioned officially as chief of staff, coordinating information and giving out advice as he'd always done. Since Grant wanted to stay out of the cauldron of Washington politics, he elected to set up headquarters in the field and travel with Meade and the Potomac Army, coordinating its movements with those of armies in other theaters. In reality, Grant would now act as the strategic commander of the Potomac Army and Meade as its tactical commander, fighting as Grant ordered him to fight—a role that suited Meade best.

In late March, Grant established his headquarters at Culpeper Court House down in Virginia. But once a week he rode back to Washington and conferred with Lincoln in the White House, with snow and rain falling intermittently outside Lincoln's office windows. Together they produced a grand strategy that called for simultaneous offensive movements on all battlefronts. In the East, Grant and Meade would worry Lee without letup, trying to draw him out into a showdown battle and liquidate his army. At the same time, the Army of the James, under ubiquitous Benjamin Butler, would advance up from Fort Monroe and cut off the railroads south of Richmond which supplied Lee's army. If Lee fell back to Richmond, he would have to battle both Union forces without supply lines and would eventually be starved out.

In the West, meanwhile, Sherman's Army of the Cumberland, now 112,-000 strong, would slash into Georgia, capture Atlanta and its crucial railroad nexus, and burn up the economic resources in the Atlanta area. As Sherman advanced, Banks's Louisiana forces would drive on Mobile, Alabama, and eventually link up with Sherman in the Deep South. In sum, the Union war machine would now utilize its vastly superior manpower and smash the Confederacy with concerted blows on all battlefronts. Grant observed (as Lincoln had noted earlier) that Union armies so far had never pulled together in the East and West, thus allowing the rebels to shift their troops across interior lines from one theater to another. But now that would be impossible, as Union forces would pound at the Confederacy from all directions and thrust toward "a common center."

Lincoln loved the grand plan. It entailed exactly the kind of concerted action he'd advocated since 1861. And though the plan basically was Grant's, Lincoln helped him work it out in their weekly meetings in Lincoln's White House office. When Grant promised to utilize the entire federal line, Lincoln remarked happily, "Those not skinning can hold a leg." Grant was so enamored with Lincoln's figure of speech that he repeated it in a letter to Sherman.

Now that strategic plans had been perfected, Lincoln gave Grant a fairly free hand in executing them. Though Lincoln would make final decisions on civilian and military policy and would give Grant advice, he would not interfere in Grant's battles as he'd had to do with previous Eastern commanders. "Grant is the first general I have had," Lincoln said. "You know how it has been with all the rest. They wanted me to be the general. I am glad to find a man who can go ahead without me."

And so on all fronts Union armies prepared for the mightiest offensive of the war, with Sherman poised in Tennessee like a coiled spring, ready to strike at Georgia; and Butler mobilizing his army at Fort Monroe; and Grant assembling an awesome force of 120,000 north of the Rapidan River in Virginia. In Washington, people knew something big was in the wind, as artillery and infantry regiments passed through for the front. Here came Burnside's new army corps, an amalgam of white and black regiments, marching up Pennsylvania Avenue bound for Grant's army. Lincoln stood on the balcony of Willard's Hotel that rain-swept day, watching Burnside's men as they passed in review—white veterans of Fredericksburg and Chancellorsville who snapped along in booted cadences and black recruits who cheered and waved their hats when they saw him there. The rain fell harder now, but Lincoln refused to go inside. "If they can stand it, I guess I can," he said, as the burned and bullet-torn flags whipped by in the street below.

By April 30 Grant was ready to attack, and the White House was in a state of high tension. Lincoln wired Grant a word of encouragement, then steeled himself for the coming campaign. With the election approaching and Republican ranks split and wavering, Lincoln could not survive another Peninsula fiasco or another Chancellorsville; no, another disaster would be the end for him. But he must have faith in Grant; if the hero of Vicksburg and Missionary Ridge couldn't whip Lee, nobody could. Night came, but Lincoln couldn't sleep. He opened a volume of Thomas Hood and surrendered himself to Hood's delicious caricatures, especially "an Unfortunate Bee-ing." Presently Lincoln went out and down the White House corridor in flopping slippers, going down to read "an Unfortunate Bee-ing" to Nicolay and Hay. When Lincoln wandered into their room, the secretaries found him "infinitely funnier" than anything in Hood's book. He stood there oblivious to his loose slippers and his short nightshirt, which hung above his legs and set out behind him "like the tails of an enormous ostrich." What an eccentric, extraordinary man he was, Hay thought. Here he was absorbed all day with cataclysmic problems, "deeply anxious" about the fate of Grant's army, his own fate "hanging on the events of the passing hour." And yet he had enough bonhomie "that he gets out of bed & perambulates the house in his shirt" to share with us "one of poor Hood's queer little conceits."

On May 3 Grant flung his huge army across the Rapidan and hit Lee's force of 65,000 almost at once, and a confused and terrible battle raged in the dense pine woods and tangled thickets south of the river, an area known as the Wilderness. In the White House, Lincoln spotted Grant's movements on his Virginia map and waited tensely for news. But few reports came from Grant's headquarters, so that Lincoln had no idea what was happening down there. "Grant has gone into the Wilderness, crawled in, drawn up the ladder, and pulled in the hole after him," the President said, and nobody would know who'd won until he came out. Still, the silence was nerve-racking. For four days Lincoln scarcely slept. He paced in his office, paced in the corridors, paced in the War Department. "I *must* have some relief from this terrible anxiety," he said, "or it will kill me." He went off to the opera wrapped in a shawl, black rings like bruises under his eyes, and sat in his box "either enjoying the music," observed a newsman, "or communing with himself."

At last, on May 8, a telegram came from Grant. He was pounding Lee again and again, his casualties were heavy, he was inching forward. But still no showdown battle yet—no final, apocalyptic engagement that would end with Lee's annihilation. But at least the Army of the Potomac was attacking now instead of waiting to be attacked. You know, Lincoln told Hay, that any other general but Grant would have fallen back across the Rapidan by now. "It is the dogged pertinacity of Grant that wins," Lincoln said.

More dispatches now: Grant was moving southeast through the Wilderness, hitting Lee's army with one attack after another, but Lee parried his assaults from behind earthworks and refused to come out for that showdown fight Grant hoped for. Still, he would not give up. "I propose to fight it out on this line if it takes all summer," Grant wired Washington. Lincoln read those ringing words to a Washington crowd, and they produced a sensation across the land.

Yet Grant's casualties were staggering. Every day now whistling steamers crowded the Sixth Street wharves in Washington and unloaded their cargo of dead and wounded. Ambulances and hearses clattered endlessly through the city, and wounded men marched forlornly to the depot bound for home. Lincoln saw them in the twilight as he rode out to the Soldiers' Home, muttering about "those poor fellows," "this suffering," "this loss of life." And the news grew ominous as the days passed, until it seemed that the whole grand plan might go awry. In Louisiana, Banks had mounted an expedition up the Red River, which ended in failure. And Sherman had punched across into Georgia, only to encounter stiff resistance from the rebels so that his progress was distressingly slow. In the East, Butler's army had advanced up the James River, then stumbled back into a trap between the James and Appomattox rivers, where the rebels sealed him like "a bottle tightly corked." And then came the worse report of all: on June 1, in a final attempt to destroy Lee, Grant launched a massive attack at Cold Harbor, but failed to break through the rebels' earthworks and lost ten thousand men in the effort. Because Lee would not come out from behind his defenses and risk his army in open battle, Grant questioned whether his current plan of operations would work.

And so after a month of fighting Grant was stalled nine miles northeast of Richmond with Lee's army entrenched in his front, and total casualties for the Wilderness campaign stood at 54,000 killed and wounded. "The immense slaughter of our brave men," Welles recorded, "chills and sickens us all." From all corners of the Union came waves of indignation against Lincoln, that he could sanction such senseless carnage, that he could put a butcher like Grant in command. And outcries against him broke out anew among Republicans, who held the administration responsible for Grant's stunning losses.

With passions running high against Lincoln, a group of liberal Republicans —men like Governor John Andrew of Massachusetts and William Cullen Bryant of New York—started a new movement to dump Lincoln and replace him with liberal stalwart John Charles Frémont. On May 31 the Frémont men assembled in Cleveland and formally nominated him for the Presidency, with Frémont railing at Lincoln's bankrupt and slaughterous policies and warning party regulars that Lincoln's re-election would "be fatal to the country."

Since the Republican convention was scheduled for Baltimore on June 7, Lincoln's advisers fretted that the Frémont movement would split the party. Though Lincoln too was concerned, he noted that Sumner and most other party bigwigs had stayed away from Cleveland. As for Frémont himself, Lincoln agreed with Hay that the Pathfinder "would be dangerous if he had more ability & energy." Frémont "is like Jim Jett's brother," Lincoln said. "Jim used to say that his brother was the damndest scoundrel that ever lived, but in the infinite mercy of Providence he was also the damndest fool."

Anyway, Lincoln had the Baltimore convention well in hand; he and his managers had so skillfully controlled Republican state delegations that even mutinous liberals resigned themselves to a Lincoln victory. As the convention drew near, a veritable parade of delegates, party leaders, and assorted "wire-pullers" called on Lincoln to divine his wishes. He didn't care whom they chose as Vice-President, but he insisted on a plank in the platform calling for a Constitutional amendment that would outlaw slavery everywhere in the land, thus safeguarding his proclamation so that no future President could overthrow it. Recently Congressional liberals had introduced such an emancipation amendment and guided it through the Senate, only to see it fail in the House. Lincoln was eager now to put the amendment before the electorate with his own unequivocal support. Yes, the amendment would eradicate slavery in the loyal border as well, and Lincoln was in favor of that now, for fear that gradualists and immediatists in the border states would fuss and fight so much that emancipation would be lost there altogether. In addition, his support of the amendment might rally liberal Republicans back to his side.

When the convention opened in Baltimore on June 7, Brooks and Nicolay were on hand as Lincoln's observers. Nicolay reported back that the delegates indulged in "little drinking—little quarreling" and exhibited "an earnest intention to simply register the expressed will of the people and go home. They were intolerant of speeches—remorselessly coughed down the crack orators of the party." As everybody expected, the delegates nominated Lincoln unanimously. But they dropped Hamlin in favor of Andrew Johnson as Lincoln's running mate, contending that Johnson's credentials as a border man and a War Democrat would strengthen the "Union" ticket.

So it was official now: Lincoln and Johnson, two native-born Southerners, would run on a Union ticket whose platform called for an unconditional surrender of Southern rebels and a Constitutional amendment that would permanently extirpate slavery in the United States. Lincoln said he was glad to have "Andy" Johnson on board and proclaimed the emancipation amendment as "a fitting, and necessary conclusion to the final success of the Union cause." Then to all Republicans: "In the joint names of Liberty and Union, let us labor to give it legal form, and practical effect."

Garrison came to the White House to praise the projected amendment and offer Lincoln his firm support. So this balding and bespectacled fellow was the celebrated Boston abolitionist. Why, he looked more like a scholar or a prim New England deacon than the outspoken crusader who'd enraged a generation of Southern slaveholders and Northern Negrophobes. Lincoln thanked Garrison for his support and pointed out that the emancipation plank in the Republican platform had been put there at Lincoln's request.

Garrison was "much pleased" with Lincoln's spirit, certain that he was doing all he could to uproot slavery "and give fair-play to the emancipated." And in the *Liberator* Garrison slashed away at the Frémont movement—which abolitionists like Douglass and Wendell Phillips had endorsed—and warned that unless abolitionists and Republicans united behind Lincoln, the hated Copperheads would win the government.

Harriet Beecher Stowe defended Lincoln, too, for she thought him a noble and effective President with a singular talent for literary expression; there were passages in his state papers, she declared, that ought "to be inscribed in letters of gold." She had met Lincoln in the White House, and he'd supposedly jested in reference to *Uncle Tom's Cabin:* "So you're the little woman who wrote the book that made this great war." They had a warm and candid conversation, and Lincoln remarked that whichever way the war ended, he didn't think he would last long after it was over. Stowe went away impressed with his suffering, though she noted that his was "a dry, weary, patient pain, that seemed to some like insensibility." In 1864, in a biographical assessment published in two Boston journals, she extolled Lincoln as a man of peculiar strengths, not a strong, aggressive individual so much as a passive one with the durability of an iron cable, swaying back and forth in the tempest of politics, yet tenacious in carrying his "great end." "Surrounded by all sorts of conflicting claims, by traitors, by half-hearted, timid men, by Border States men, and Free States men, by radical Abolitionists and Conservatives, he has listened to all, weighed the words of all, waited, observed, yielded now here and now there, but in the main kept one inflexible, honest purpose, and drawn the national ship through." Because he was cautious, because he stayed within Constitutional limitations and time-honored landmarks, Stowe considered him the safest leader the country could have in war. "A reckless, bold, theorizing, dashing man of genius," she said, "might have wrecked our Constitution and ended us in a splendid military despotism."

For Lincoln it was a time of unrelenting discouragement. In northern Georgia, Sherman had to fight his way foot by bloody foot against Joseph Johnston's stubborn rebel army, and Lincoln had no idea when or whether Sherman would ever reach Atlanta. In Virginia, Grant despaired of ever whipping Lee north of Richmond, so with Lincoln's approval he hurried the Potomac Army east of the capital and raced south for Petersburg, hub of all railroads connecting Richmond to the rest of the South. If Grant could sever the roads, Lee would have to fight for his communications, and Grant would crush him.

As with everything else, it didn't work. Grant attacked Petersburg, but couldn't take the city before Lee arrived there and dug in behind a labyrinth of trenches and redoubts. On June 18 Grant settled in for a protracted siege, and Lincoln was deeply disappointed. He didn't need another siege. He needed a tremendous victory at Petersburg that would revive his sagging fortunes.

But Grant assured Lincoln that he would get Lee in time. The general would bombard Petersburg with heavy artillery, engage Lee all across his line, and drive around and destroy the railroads that supplied his army, thus flushing him out for a showdown. Lee's army was so crippled now, so decimated by the "bloody and terrible" battles of the Wilderness, that Grant would demolish him in open battle.

Grant's determination made Lincoln feel a little better. He wired the general to do his best at Petersburg, to "hold on with bull-dog grip, and chew & choke, as much as possible."

On the political front, meanwhile, Lincoln suffered one crisis after another in his own Republican house. In late June came an open rupture with Chase that was bound to rile Republican liberals even more. By now there was such antagonism between Lincoln and his Secretary that they communicated only through chilly notes. When Lincoln interfered with one of Chase's Treasury appointments in New York, indicating that it would offend New York Senator Edwin Morgan, Chase accused the President of infringing on the integrity and independence of the Treasury Department. Once again Chase offered his resignation, certain that Lincoln would decline it as he had in the past. But this time Lincoln accepted Chase's offer. You are a faithful and able man, Lincoln wrote him, "yet you and I have reached a point of mutual embarrassment in our official relation which it seems can not be overcome, or longer sustained." Chase, of course, was shocked and deeply hurt, moaning that Lincoln seemed to sympathize "more with those who assailed and disparaged than with those who asserted and maintained the views" of his administration.

As Lincoln feared, the Senate shook with recriminations against him. Chase's liberal friends there, led by Fessenden and his Finance Committee, hurried to the White House and demanded to know why Chase had been relieved. Lincoln told them why, citing Chase's obstinacy about Treasury appointments and explaining that the strain between them had become so intolerable that Lincoln had no choice but to let him go. If the Senate didn't like what he'd done, Lincoln declared, then it could have *his* resignation and see Hannibal Hamlin as President.

Lincoln knew he was treading on a political minefield. That night he lay on a library sofa, his hands on his chest, trying to figure out a way to mollify the liberals and prevent any further defections to Frémont. He was well aware that a reconstruction bill was about to pass in Congress, one he planned to veto, so he had to do something now to soften that blow. It occurred to him that he ought to appoint Fessenden himself as Secretary of the Treasury. Of course! As chairman of the Senate Finance Committee, he "knows the ropes thoroughly." Moreover, he had a national reputation and was a popular liberal, though without the petulance and fretfulness of many of his associates. Lincoln thought the idea of Fessenden in the Treasury was proof that "the Lord hadn't forsaken me yet." With another liberal in his official family, maybe he could forestall any further rifts with that wing of the party.

The next day, July 1, Lincoln startled everyone on Capitol Hill by naming Fessenden to head the Treasury Department. Liberal congressmen were pleased indeed about the appointment—if about little else in Lincoln's "fumbling" administration. With Fessenden in his Cabinet, Lincoln now prepared himself for a fiery confrontation over the Congressional reconstruction bill. It was the brainchild of Maryland Congressman Henry Winter Davis, a pale, mustached, contentious man and an inveterate rival of the Blairs for control of Maryland politics. Last December, Davis had tangled with Lincoln himself over Maryland's patronage and had stormed out of the White House vowing to have no further "political intercourse" with the President. After that, Davis lashed out at Lincoln's ten percent plan and produced a tougher program of his own, incorporating it into a bill which he guided through the House and Wade pushed through the Senate. Since it enlarged Congress's role in reconstruction, Sumner endorsed the measure, too, though he failed to get Negro suffrage included because Republicans of all persuasions still opposed so radical a move. Like Lincoln's reconstructed governments, the

Wade-Davis bill restricted the Southern vote to whites only.

As finally passed, the Wade-Davis bill clarified and stiffened the reconstruction policy Lincoln had begun, and nearly all Republicans in both houses
gave the measure their support. Among other things, it prohibited slavery
in all reconstructed states and made slave owning a federal crime punishable
by fines and imprisonment. Moreover, the bill threw out Lincoln's ten percent test and decreed that a majority of voters in a conquered rebel state must
take the oath of allegiance before they could establish a new government. As
outlined in the bill, the restoration process would now work like this for
every rebel state: the President would appoint and the Senate confirm a
provisional governor whose job was to administer the oath and call a constitutional convention charged with creating a republican form of government.
So far as the convention was concerned, the bill required that an "iron-clad"
oath be taken in order to exclude all ex-Confederates from participation:
before Southerners could vote for delegates, attend the convention, or ratify
its results, they must swear that they had never aided or fought for the
Confederacy. Once a state had produced an acceptable constitution, it could
hold regular elections for state and national offices. But like Lincoln's proclamation of last December, the Wade-Davis bill disqualified all Southerners
who'd held important military and civil posts in the Confederacy, thus denying them the right to vote or hold political office.

On July 4, as Lincoln sat in a room in the Capitol signing other documents,
Sumner and Chandler brought the Wade-Davis bill for the President's signature. Since Congress was scheduled to adjourn in less than an hour, Sumner
was "in a state of intense anxiety," Hay noted, for the word was out that
Lincoln would use a pocket veto. "Aren't you going to sign it?" Chandler
asked when Lincoln set the bill aside. "No," the President replied. "If it is
vetoed," Chandler warned, "it will damage us fearfully in the Northwest. It
may not in Illinois; it will in Michigan and Ohio. The important point is that
one prohibiting slavery in the reconstructed states."

"That," Lincoln told him firmly, "is the point on which I doubt the
authority of Congress to act."

But, Chandler protested, it was no more than Lincoln had done! True,
Lincoln said, but in a national emergency he could do things on grounds of
military necessity which Congress could not do.

When the senators stalked angrily off, Lincoln turned to Hay. "I do not
see how any of us can deny and contradict all we have always said, that
Congress has no constitutional power over slavery in the states." Fessenden,
who'd just come into the room, said he agreed entirely with Lincoln. Maybe
Congress could provide for the confiscation of slave property in wartime,
but Lincoln had always contended that Congress had no authority to abolish
slavery as a state institution. No, only the President with his war powers—

or a Constitutional amendment—could do that.

So he pocket-vetoed the Wade-Davis bill. And Congressional Republicans were extremely embittered. "I am inconsolable," Sumner cried. And Winter Davis, pale with wrath, predicted a civil war in the North unless Republicans revolted against Lincoln's nomination. In a desperate effort to prevent more schisms in his troubled party, Lincoln defended his veto in a special proclamation. He rejected the Wade-Davis bill, he explained, because he wasn't willing to commit himself to any single plan of reconstruction, or to set aside the loyal government he'd established in Louisiana by his ten percent plan —and a similar regime now installed in Arkansas. Nor would he concede, by signing the bill, that Congress had the power to extinguish slavery in the reconstructed states. Nevertheless, he entirely approved of the stringent program contained in the bill. And he gladly offered executive and military assistance to any loyal Southerners who desired to restore their states according to the Wade-Davis plan.

His proclamation only infuriated liberal Republicans all the more. What audacity the President has, roared Thaddeus Stevens. He vetoes the bill and then issues a proclamation about how far he will conform to it! And Sumner was simply livid. What this country needs, the senator hissed, is "a President with brains, one who can make a plan and carry it out." Wade and Davis, for their part, were so furious that they penned a searing manifesto which appeared in newspapers around the Union. They blasted Lincoln's proclamation as the most "studied outrage" ever visited on Congress and accused him of exercising "dictatorial usurpation" over reconstruction. If he wanted their support, Wade and Davis fulminated, he must content himself with suppressing the rebellion and leave reconstruction to Congress, whose authority was "paramount."

It was a remarkable salvo against a President by members of his own party. The *New York Times* thought it "the most effective Copperhead campaign document" the *Times* had seen so far. And in truth the manifesto was so severe that even some liberals refused to endorse it. And Lincoln? "To be wounded in the house of one's friends," he told Noah Brooks, "is perhaps the most grievous affliction that can befall a man."

In the middle of the Wade-Davis battle, a rebel force under Jubal Early materialized in the Shenandoah, seized Harpers Ferry, and drove across the Potomac to menace Washington and

Baltimore. Lincoln was incredulous that a large rebel force could be rampaging on Union soil. What would the country say about his leadership now? On July 10, with Early driving on the capital, Lincoln wired Grant and advised him to bring part of his army to Washington and destroy Early's force so it could never invade again. Grant telegraphed back that he wouldn't come personally, but he would send a crack corps at once.

By the time Grant's troops arrived, Early had cut Washington's telegraph lines, burned Monty Blair's home at Silver Spring, pushed to within two miles of the Soldiers' Home, and sent out skirmishers to probe Washington's defenses. On July 12 Lincoln stood on the parapet at Fort Stevens, watching as Union and rebel forces fired away at one another. He saw a body of Union soldiers move across shimmering summer fields and drive the rebels from a house and an orchard with muskets smoking. By now rebel sharpshooters were sniping at Fort Stevens, but Lincoln stood there oblivious to the bullets whizzing around him, and a soldier fell at his side. Finally an officer made the President get down before he was killed.

The next day Early was gone, falling back somewhere toward the Shenandoah. Lincoln was enraged. How could Early have just marched away? Why hadn't Union forces given chase and obliterated him? Now Early would only come back again, threaten Washington again. But the blame for Early's escape lay not so much with timid generalship as with Washington's chaotic command system: no single general was in charge of the various forces in the area, no one directing and coordinating the action against Early. In all the confusion, nobody was made explicitly responsible for chasing Early down.

On July 31 Lincoln met with Grant down at Fort Monroe to talk about Early and Washington—and confer, too, about the delicate political situation and the role of the army in it. Grant understood that Lincoln's political fortunes were linked closely to military developments, and the General in Chief was anxious to produce victories, protect Washington, eliminate Early, and help Lincoln in his bid for re-election. After their conference, Grant dispatched Philip Sheridan, cavalry leader of the Potomac Army, to take field command of all troops in the Washington area, march into the Shenandoah, and follow Early "to the death." Thirty-three now, Sheridan was a bandy-legged little Irishman, a West Pointer, and a hard-fighting cavalryman. But Early was a tough and elusive foe, so that all Sheridan accomplished that summer was inconsequential maneuvering.

In the meantime, Lincoln had called up 500,000 more men to fill depleted Union armies, realizing that this would only fuel the inferno of criticism against him. As the summer wore on, the military situation became even more discouraging, as Grant repeatedly failed to dislodge Lee from Petersburg and the siege dragged on. Out in Georgia, Sherman had finally driven to Atlanta, but failed to carry the city by assault. In fact, one-legged John B. Hood of Texas, the new rebel commander, struck Sherman with fiery counterattacks. Unable to knock out Hood's army, Sherman began a long and dispiriting siege of Atlanta.

With Union forces bogged down in both major theaters and demoralization spreading across the North, Republican regulars were in despair. Greeley begged Lincoln to negotiate for peace, arguing that the people wanted the war to end and that "our bleeding, bankrupt, almost dying" country couldn't stand new conscriptions and "new rivers of human blood." "I find everywhere a conviction that we need a change," said another Republican, "that the war languishes under Mr. Lincoln and that he *cannot* and *will* not give us peace." "There are no Lincoln men," wrote a New York Republican. "We know not which way to turn." Defeatism settled across the party that August, as Lincoln's friends and foes alike were certain he could not be re-elected. Weed bluntly told him so; and Swett, Raymond, and Browning all agreed.

In the privacy of his office, Lincoln admitted that it was just about the end of the line for him. In a week or so, the Democrats would nominate their candidate, and everybody was sure it would be George McClellan. That McClellan would likely defeat him and become the next President was painfully ironic, but Lincoln resigned himself to that probability. He took out a sheet of stationery and dated it August 23, 1864. "This morning, as for some days past, it seems exceedingly probable that this Administration will not be re-elected. Then it will be my duty to so co-operate with the President elect, as to save the Union between the election and the inauguration; as he will have secured his election on such ground that he can not possibly save it afterwards."

He stared at his memorandum, reflecting on what he would tell McClellan when that depressing day arrived. "General," he would say, "the election has demonstrated that you are stronger, have more influence with the American people than I. Now let us together, you with your influence and I with all the executive power of the Government, try to save the country. You raise as many troops as you possibly can for this final trial, and I will devote all my energies to assisting and finishing the war."

At last he signed the memorandum, then folded and sealed it. Later he had all his cabinet members sign their names on the back of the document, without telling them what it said. After that he put it away in his desk.

Some conservative Union men argued that there was only one way Lincoln could save his Presidency. He must abandon emancipation and offer the rebels peace conditions that promised to leave slavery alone. But Lincoln answered with a resounding no. The Union needed its thousands of Southern black soldiers now more than ever. And anyway he would not—could not —throw them back to their former masters. He would "be damned in time & eternity" if he did.

But in truth he began to waver. With what seemed massive opposition to him and his war and racial policies, he confessed that maybe the nation would no longer sustain a war for slave liberation. Maybe he should give up pulling the country down a road it didn't want to travel. On August 24 he wrote a letter which appointed Henry Raymond as a special peace commissioner and instructed him to go down to Richmond and offer Davis the following terms: Lincoln would stop fighting if the rebels would return to the Union and recognize the national authority. Slavery and all other questions would be left for "adjustment by peaceful means." If Davis rejected these terms, Raymond was to find out what ones he might accept.

But Lincoln changed his mind and didn't send the letter. The next day he met with Stanton, Seward, and Fessenden—"the stronger half of the Cabinet," Nicolay said—and the President spoke with awakened resolution. They all decided against a peace overture to the rebels, for that would be worse than losing the election, they declared; "it would be ignominiously surrendering it in advance."

August 30 brought yet another blow to Lincoln's candidacy. A group of disgruntled party warriors called for another convention to gather in Cincinnati on September 28 and if necessary choose another nominee, one who had a chance of victory. Greeley, Winter Davis, Benjamin Butler, William Cullen Bryant, and several others were involved in the project, and Chase too gave it his guarded support. But Wade and Sumner held back from so drastic a move. Sumner thought Lincoln should resign voluntarily instead of being forced out. With Republicans in open revolt against their own candidate, the Democratic *New York World* editorialized that it was hard to tell what hurt Lincoln the most—the frontal blows from his "manly opponents" or the stabs in the back from his friends.

Out in Chicago, meanwhile, the Democrats gathered in a new wigwam and nominated McClellan on a platform that advocated peace "at the earliest possible moment." Welles thought "there is fatuity in nominating a general and a warrior in time of war on a peace platform." But McClellan made it clear that he wouldn't stop the war at any price, as some Democrats insisted. If he did that, the general said, he couldn't look his former soldiers in the eye. No, he would restore the Union—and slavery with it.

So as August closed, the campaign of 1864 got under way in the very midst of rebellion and civil war. As always, the rival propaganda mills labored at white heat, with Republicans portraying McClellan as a "cowardly traitor" and insisting that he would end the war on any terms and sell the Union out.

At the same time, Democrats reiterated all their old accusations against Dictator Lincoln, smeared him again as the bastard son of one "Inlow," and distributed posters and pamphlets about Republican racial policies. One broadside showed Republican men and black women dancing and hugging at "the Miscegenation Ball." A scurrilous pamphlet warned readers what could be expected under a second Lincoln regime—a cover sketch had a thick-lipped Negro man kissing a young white girl. Said a Democratic paper in Wisconsin: "If he is elected to misgovern for another four years, we trust some bold hand will pierce his heart with dagger point for the public good."

The campaign had barely begun when electrifying news flashed in from Georgia. On September 2 Atlanta surrendered to Sherman as Hood's army left the city and fell back to the south. Atlanta, strategic railroad center, symbol of Deep Southern resistance, queen city of the cotton states, was now in Union hands. After the summer of gloom that had just passed, Lincoln played up Atlanta for everything it was worth. He pronounced it a gift of God, designated Sunday as a national day of thanksgiving, and ordered one-hundred-gun salutes to sound in Washington, in New York, Boston, Philadelphia, Pittsburgh, Baltimore, Saint Louis, and New Orleans. Let news of Atlanta sweep the Union from ocean to ocean. Let it sing in the newspapers, sing in a thousand telegraphs in every city in the nation, that this administration was not whipped yet, that Lincoln's armies were winning again.

September brought good news from the Shenandoah, too. After weeks of indecisive skirmishing, Sheridan caught Jubal Early and struck him at Berryville, defeated him at Winchester, and pounded him again at Fisher's Hill. Lincoln read the dispatches with rising hopes, for the victories of Sherman and Sheridan—along with a recent Union naval triumph in Mobile Bay—seemed to kick the bottom out of the Democratic platform, as Republican orators claimed.

Meanwhile Lincoln was hard at work in his own behalf. Though neither he nor McClellan gave formal campaign speeches, the President wrote letters, spoke unofficially to troops who passed through Washington, and dropped political remarks to commissions, delegations, and serenaders. There was no doubt in his voice now, no wavering in his resolve to stand his ground. If the Union signed an armistice at this time, he declared, it would be "the end of the struggle" and the loss of all that Union men had fought and died for. If his policy on slaves and Negroes were abandoned, it would mean the loss of tens of thousands of black soldiers—and that would be catastrophic to the Union cause. Keep his policy, he said, "and you can save the Union. Throw it away, and the Union goes with it." So long as he was in office, he would not violate his promise of freedom to black people. He would accept no measure that re-enslaved them. "It can *not* be."

There was another reason to fight on to victory, and that was to preserve something that lay at the heart of the American promise, something Lincoln had cherished and defended almost all his political life. As he explained it to an Ohio regiment: "I happen temporarily to occupy this big White House. I am a living witness that any one of your children may look to come here as my father's child has. It is in order that each of you may have through this free government which we have enjoyed, an open field and a fair chance for your industry, enterprise and intelligence; that you may all have equal privileges in the race of life, with all its desirable human aspirations. It is for this the struggle should be maintained, that we may not lose our birthright."

As he vowed to see the war through, Lincoln made adroit moves to rally his party behind him. He manipulated the patronage with consummate skill, ordering government employees to support the Union ticket or lose their jobs. And he tried his best to woo back mutinous liberals. He even made a deal with Chandler, who offered to get Frémont to quit if Lincoln would remove Monty Blair from the Cabinet. Lincoln agreed that Blair would go; the President could no longer keep so controversial and distempered a man in his official family anyway. For a long time now Blair had made war on Republican liberals from Lincoln's Cabinet. He'd castigated Chase, stormed at Stanton, raged on street corners about how "this man is a liar, that man is a thief." Blair had been loyal to the President, but he could no longer keep a man of such indiscretion and personal antagonisms.

As it turned out, Frémont withdrew unconditionally from the Presidential race, not because he bore Lincoln any love, but because McClellan would save slavery. A couple of days later, on September 23, Lincoln replaced Blair with former governor William Dennison of Ohio. Blair, of course, growled that he'd been "decapitated" to appease Frémont partisans. But his dismissal had its desired effect, as dissident liberals rejoiced over Lincoln's decision to get rid of Blair. Henry Winter Davis now gave the President his support. And Wade also came around, informing Chandler that he would do all for Lincoln that he would do for a better man.

With Frémont out and the military situation looking brighter, Republicans one after another closed ranks behind Lincoln for President. Chase came to the White House on a peace mission and went out to stump for his former boss. Greeley promised Nicolay that he would "fight like a savage" for

Lincoln because "I hate McClellan." At Union rallies, men who'd favored the Cincinnati call now appeared on the platform in defense of Lincoln. And many black and white abolitionists climbed on Lincoln's train as well. As Douglass said, if he had to chose between McClellan and Lincoln, he wouldn't hesitate to support the President.

In October, early state elections were scheduled in Indiana, Ohio, and Pennsylvania. Both sides regarded the elections as wind signals for the Presidential race and poured in crack speakers and rival soldier clubs—the McClellan Legion and Lincoln's Veteran Union Club—to win these critical states. Lincoln, for his part, was anxious to secure the soldier vote, hoping that Republican campaigners were right—that most troops favored unconditional surrender and would stand with the President, their Commander in Chief. Unlike Ohio and Pennsylvania, Indiana did not allow her soldiers to vote in the field. So at Lincoln's request, Sherman granted wholesale furloughs to the twenty-nine Indiana regiments serving in his army. But Lincoln advised Sherman that the Indianans need not remain for the Presidential election. They could return to duty at once.

On election day, October 11, Union candidates narrowly carried Pennsylvania, but scored impressive wins in Ohio and Indiana, with the soldier vote helping to swell Republican majorities. Predictably, Republicans pronounced this a repudiation of McClellan and predicted a mighty Republican sweep in November. But Lincoln was not so optimistic as they. In his office he tallied up how he thought the Presidential contest would go. The Copperheads would carry New York, Pennsylvania, New Jersey, Delaware, Maryland, Missouri, Kentucky, and his home state of Illinois. His party would win the New England states, Michigan, Wisconsin, Minnesota, Oregon, California, Kansas, Indiana, Ohio, West Virginia, and the new state of Nevada (to be officially admitted on October 31). That gave Lincoln 120 electoral votes, McClellan 114. But Lincoln feared some of his states could still go the other way.

More superlative dispatches came in from the Shenandoah. On October 19 Sheridan crushed Early at Cedar Creek, with Sheridan personally rallying his men and leading them to a dramatic and decisive victory that made him a national war hero. After that Sheridan's army visited total war on the fertile Shenandoah, burning fields and driving off livestock in order to help starve the Confederacy into submission. On the campaign trails, Republican orators sang on about the successes of Sheridan and Sherman and flailed away at McClellan's "peace-at-any-price" candidacy. In the White House, Lincoln joked that it was a good thing Sheridan was a little man: there was no telling what he would do to the rebels if he were bigger. But at bottom Lincoln was still apprehensive about the election.

Election day, November 8, came in with dark and rainy skies. As usual Lincoln was up at daybreak and later breakfasted with his family. Mary seemed more nervous about the election than he was. He didn't know the full reason why—didn't know that her New York shopping forays, so therapeutic for her headaches and yet so costly, had run her personal debts up to $27,000. She'd confessed her debts to Lizzie Keckley—and her fears that Lincoln might find them out, which was why she worried so about the election. If he lost, all her bills would be sent in "and he would know all." In desperation she'd demanded that Republican politicians help pay her debts. "Hundreds of them are getting immensely rich off of the patronage of my husband," she reasoned, "and it is but fair that they should help me out of my embarrassment." During the campaign, she approached Cabinet members and executive office workers and wooed, cajoled, and wept, begging them to give her money for her debts. She tried desperately to hide all this from Lincoln, complaining to one of her male confidants how "Poor Mr. L" was "almost a monomaniac on the subject of honesty."

So she was exceedingly nervous on election day—and guarded in what she said to her husband. After she returned to her chamber, Lincoln went to his shop and chatted with Hay. The White House seemed deserted, with Stanton sick in bed and most of the other Cabinet members gone home to vote. Lincoln told Hay that it was curious about the political contests he'd been in. While he wasn't a vindictive man, almost all his elections were marked with "great rancor" and bitterness. Around noon Noah Brooks joined Lincoln in the office, and they discussed the election today. "I'm just enough of a politician to know that there was not much doubt about the result of the Baltimore convention," Lincoln said; "but about this thing I am very far from being certain. I wish I were certain."

At seven that evening, Lincoln went with Hay and Brooks to the War Department to catch the returns. "The night was rainy, steamy, and dark," Hay said, as they splashed across the grounds to a side door of the War Department, where a "soaked and smoking sentinel" stood huddled in a rubber cloak. In the telegraph office, Stanton's clerks handed Lincoln a sheaf of returns: Philadelphia reported a 10,000 Union majority. The Union ticket was also running ahead in Baltimore. "That is superb," Lincoln said, and he sent word about the early returns back to Mary. "She is more anxious than I," he said.

Gustavus Fox came in with word that Winter Davis had gone down to defeat in Maryland. "It served him right," Fox declared. Lincoln conceded that Davis had been "very malicious against me," but he bore Davis no personal ill will. In truth, Lincoln tried not to carry personal resentments. "A man has not time to spend half his life in quarrels," he said.

The telegraph was ticking again. The signs looked very good now—Lincoln was making steady gains in Pennsylvania and running well ahead in Indiana and most other crucial states. Around midnight Major Eckert of the telegraph office brought supper into the room, and Lincoln awkwardly dished out fried oysters to everyone. Though weary to his bones, he seemed mellow tonight, almost happy.

Back at the White House, he informed Mary that she could relax now—his victory seemed assured. Then he retired to his own chamber and tried to sleep. Unknown to him, Ward Hill Lamon came to Hay's room armed with pistols and bowie knives, borrowed some blankets, and lay down in the hall outside Lincoln's door. Lamon was afraid that somebody might try to assassinate the President now that he was re-elected. The next morning, he left before Lincoln awoke, leaving the blankets at Hay's door.

When the final election returns came in, Lincoln had defeated McClellan by almost half a million popular votes out of some four million cast, and had carried every Union state except Delaware, New Jersey, and Kentucky, giving him a thumping victory in the electoral college of 212 to 21. Lincoln had won the soldier vote overwhelmingly, with 116,887 casting their ballots for him and only 33,748 for McClellan. Lincoln, for one, interpreted the election as a popular mandate for him and his emancipation policy. But in reality the election provided no clear referendum on slavery, since Republicans played emancipation down and concentrated on the "treasonable" peace plank in the Democratic platform, which doubtlessly hurt McClellan now that Union armies were winning again. Sumner was probably right when he asserted that the election was a "vote *against* McClellan rather than *for* Lincoln."

In the week following the election, Lincoln looked old beyond his fifty-five years, for the strain of 1864 had taken its toll on him. Though Mary was most happy about the election triumph, she was concerned about "Mr. Lincoln," for he seemed "so broken hearted, so completely worn out." In the mornings he reviewed court-martial cases—one of his most pressing and painful chores these days. As always, he pardoned convicted deserters on any reasonable pretext. "There are already too many weeping widows in the United States," he said. "For God's sake do not ask me to add to the number."

One late night, serenaders crowded around the White House with lanterns and banners, and Lincoln stood at a window reading them a statement about what the election meant to him. It meant above all that America's free institutions could work, that the country could carry out a free canvass in the middle of civil war. Had he abandoned or postponed the election, the rebellion could have well claimed a victory. But by holding the contest, Northerners had saved their free government and proven how strong they still were. And he was glad, he said, that most people had voted for him, the candidate most dedicated to the Union and opposed to treason.

He turned away from the window and muttered to Hay, "Not very graceful, but I am growing old enough not to care much for the manner of doing things."

Tired though he was, Lincoln pushed himself through a maelstrom of work that fall and winter, his days consumed with pardons and promotions, with public levees, interviews, meetings, and serenades. And now that he was re-elected, the office seekers were back again. "It seems that the bare thought of going through again what I did the first year here, would *crush* me," he told a New Hampshire Senator. There was a Cabinet change, too, as Bates retired as Attorney General and Lincoln chose James Speed, Joshua's brother, to take Bates's place. Also, the President appointed Salmon Chase as Chief Justice of the Supreme Court, to replace Taney, who'd died in October. Though still concerned about Chase's political ambitions, Lincoln selected him because he was a prominent jurist and a dedicated Republican who could be counted on to sustain Republican war measures.

On the battlefronts, Grant was still sieging Petersburg, but out in Georgia Sherman was preparing to burn his way across the state, destroying fields and wrecking railroads. "If the people raise a howl against barbarity and cruelty," Sherman wired Halleck, "I will answer that war is war, and not popularity seeking." "We are not only fighting hostile armies, but a hostile people, and must make old and young, rich and poor, feel the hard hand of war." Lincoln had misgivings about Sherman's plan of operation, not because the President opposed total war, but because he thought Hood's army ought to be eliminated first. When Grant raised the same objection, Sherman reported that he couldn't conquer the rebels from Atlanta, but if he drove south across Georgia, Hood would have to pursue and Sherman would turn and smash him. Moreover, Sherman planned to borrow a chapter from Grant's Vicksburg campaign and subsist entirely off hostile country, thus facilitating his speed and freedom of movement. Grant finally endorsed Sherman's plan; and Lincoln chose not to interfere, though he still worried about the operation—worried that the rebels might trap Sherman somewhere in the interior of Georgia.

On November 16 Sherman marched out of Atlanta to slash his way to the sea. As he moved, Hood struck out for Tennessee in a desperate attempt to draw Sherman out of Georgia and defeat him in the mountains. But Sherman let him go. George ("Old Pap") Thomas had an army in Tennessee and could take care of Hood. Lincoln, though, remained uneasy about all this— and his worries increased when Sherman cut off his communications and headed deep into Georgia. For several weeks Lincoln hurried to the War Department whenever he could, to see if there was news from Sherman, if anybody knew where he was now. "I know what hole he went in at," Lincoln said, "but I can't tell what hole he will come out of."

There was plenty of news from Tennessee, however, and it was all bad. Hood's tired and ragged army made it into the hills south of Nashville, where it seemed an easy target for annihilation. Yet Thomas, who held Nashville with a powerful force of seventy thousand men, refused to attack, instead sitting on his haunches inside the city—and doing what? Making preparations, the general explained, getting every detail right before he moved out against Hood. His procrastination upset Lincoln—it was all too reminiscent of Buell and Rosecrans, of all the endless delays and excuses for not fighting that had impeded Union operations in the first three years of the war. Lincoln was beginning to think that Thomas too was afraid of battle.

Excellent news from Georgia. On December 10, after cutting a swath of destruction through the heart of the state, Sherman reached the sea and started shelling Savannah. Now if only Thomas would attack Hood in Tennessee . . . but Thomas was still in Nashville making his preparations. By December 15, Grant was so exasperated that he vowed to go out there and relieve Thomas personally. But Lincoln balked at such a drastic act. After all, Thomas was on the ground in Nashville; perhaps he knew the situation better than Grant and Lincoln did here in the East.

Feeling sick and melancholy, Lincoln went home to bed. But some time after eleven that night, his secretaries woke him and said that Stanton and Major Eckert were downstairs with telegrams from Tennessee. Lincoln lit a candle and plodded out to the second-story landing in his nightshirt. Yes, Stanton? What has happened? Greatly excited, Stanton exclaimed that Thomas had attacked this morning, shattered Hood's line, and would surely win a decisive victory tomorrow. Lincoln said he was delighted at the news. And Major Eckert, looking up at him, would never forget the sight of that tall and ghostly man, standing at the top of the stairs and holding a candle overhead.

The next day Thomas crushed Hood with massive attacks and sent him reeling back to Mississippi with Thomas's cavalry chasing him. It was a devastating pursuit, as the cavalry pounded and slashed at the retreating rebel army until only remnants were left. What Lincoln had desired all along had finally been accomplished—Hood's army had been wiped out as a fighting force. And Old Pap Thomas had done it, too. If he was a stickler for detailed preparations, once he got going he fought with bulldog fury.

On Christmas Day, Lincoln received a telegram from Sherman: "I beg to present you as a Christmas gift the city of Savannah." On December 21 Savannah had capitulated, and Sherman now had a base on the sea. Lincoln wired him "many, many thanks" and confessed that at first he'd been fearful about Sherman's campaign. How glad he was now that he hadn't intervened. This and Thomas's victory in Tennessee brought "those who sat in darkness, to see a great light. But what next?" "The Great Next," as Lincoln found out, called for Sherman to march through the Carolinas, smashing up the railroads that supplied Lee's army in Petersburg. Lincoln thought it a masterful plan: with Lee's supply lines destroyed, he would have to come out and fight. Yes, and then Grant could destroy him, too, and this struggle would end.

As his armies defeated the rebels in Georgia and Tennessee, Lincoln sought to consummate a cherished political goal—the adoption of the Thirteenth Amendment. He deeply feared that Congress, the courts, or a later administration might overturn the emancipation proclamation as an illegal use of power. Also, there was the argument that the proclamation affected only those slaves in the rebel South who reached Union lines and that "it did not meet the evil." But a Constitutional amendment, as he phrased it, would be "a King's cure for all the evils." Accordingly, when Congress assembled in early December, Lincoln reminded the House of Representatives that the Senate had already approved the amendment, and he enjoined the House to do so as well. True, this was the same House that had rejected the amendment last May, before the Republican convention. But since then a national election had taken place which Lincoln insisted was a mandate for permanent emancipation. If the present House failed to pass the amendment, the next one "almost surely would." So "at all events, may we not agree that the sooner the better?"

Whereupon he set about using his powers of persuasion and patronage to get the amendment through. As December passed, he invited congressmen to the White House and plotted ways to pressure conservative Republicans and recalcitrant Democrats who opposed the amendment. On January 6 a heated debate began over the measure, with its chief sponsor quoting Lincoln himself that *"if slavery is not wrong, nothing is wrong."* When it seemed that the amendment would fail to get the necessary votes, Lincoln buttonholed former Whigs and exhorted them in the name of Henry Clay to support the amendment, "my chief hope to bring the war to a speedy close." And he singled out "sinners" among the Democrats who were "on praying ground," letting them know they had a better chance for the federal jobs they desired if they voted for the amendment. Presently two Democrats swung over in favor of it. With the outcome still "very doubtful," Lincoln was involved in "certain negotiations" about federal patronage—negotiations that were never made public—to bring wavering Republicans and opposition Democrats into line.

The vote came on January 31, 1865, with spectators filling the corridors and galleries of the Capitol and Lincoln awaiting the outcome in the White House. As the roll call proceeded, every Republican member of the House voted yes, until at last the chamber hushed and the clerk read out the results: 119 for the amendment and 58 against—just three votes more than the required two-thirds majority. At once "a storm of cheers" broke out among the Republicans, who jumped around, embraced one another, and waved their hats and canes overhead. There were shouts and applause in the galleries, too, where women's handkerchiefs were fluttering in the air. In all the rejoicing, Representative George Julian of Indiana said he felt as though he were in a new country. As the telegraphs carried the news over the Union, Negroes gathered in mass meetings and clapped and sang: "Sound the loud timbrel o'er Egypt's dark sea, Jehovah has triumphed, His people are free." Soon a popular cartoon circulated in the newspapers, showing a laughing black man who said, "Now I's nobody's nigger but my own." Up in Boston, William Lloyd Garrison, who'd struggled most of his adult life to free the slaves, gave Lincoln the credit for getting the amendment passed.

In the White House, Lincoln pronounced the amendment "a great moral victory." He happily noted that Maryland and Missouri had already abolished slavery by state action—and so had his reconstructed governments in Louisiana and Arkansas. Now once the amendment was ratified, slavery would be permanently eradicated everywhere in America. He pointed across the Potomac. "If the people over the river had behaved themselves, I could not have done what I have."

Lincoln and Congressional Republicans may have collaborated on the Thirteenth Amendment, but they clashed over Lincoln's reconstruction regimes in Louisiana and Arkansas, a collision that had been shaping up since the Wade-Davis troubles last summer. Through December and January, Lincoln defended his work in Louisiana and Arkansas and fought doggedly to get their representatives seated in Congress. He argued that Louisiana's free-state constitution ratified under his ten percent plan was "better for the poor black man than we have in Illinois." Thanks to the military protection Lincoln had provided, Louisiana now had a nucleus to build around in order to get her back in the proper relation to the federal government. He pointed out that Southern secessionists would love to see his Louisiana government fail—would love to see Congress reject it and refuse to admit its representatives. But it must not fail. It must be coddled by Lincoln and protected by the army until it grew into a truly free and democratic state. "We shall sooner have the fowl by hatching the egg than by smashing it," Lincoln believed.

He called Sumner to the White House and tried to make him see the light. But the senator continued to impugn Lincoln's entire approach to reconstruction. Not only did he contend that Congress had sole responsibility over restoring the South; Sumner also rejected Lincoln's argument that the military could establish civilian regimes there. "The eggs of crocodiles can produce only crocodiles," he retorted, "and it is not easy to see how eggs laid by military power can be hatched into an American state." As always, Sumner also objected to Lincoln's reconstructed governments because they did not enfranchise blacks, "whose votes are as necessary as their muskets."

Still, Lincoln and Sumner remained warm personal friends and wanted to avoid an outright political split. They were eager, too, to maintain harmony between their branches of the government. And so with the support of several other Congressional Republicans they forged a compromise: Sumner and his colleagues would accept Lincoln's present Louisiana regime and Lincoln, in return, would approve for all other rebel states a projected bill enfranchising all citizens "without distinction of color." Though Lincoln didn't like to impose the black vote on white loyalists, he didn't think the bill laid down any ironclad reconstruction policy or wrenched control of restoration away from the President, so he agreed to endorse it.

But the compromise was doomed because most Congressional Republicans still opposed Negro suffrage and would not sanction any such measure as Sumner and his friends had in mind. Moreover, news of the compromise with Lincoln leaked out and brought obstreperous denunciations from liberals and conservatives alike, with the latter shrieking that Lincoln had surrendered himself to "radical" influences and the "nigger" vote. At that the compromise fell apart. When it did, Sumner and his cohorts moved to kill off Lincoln's governments in Louisiana and Arkansas lest they become models for reconstructing all other rebel states. In the debates that February, Sumner, Wade, and other liberals castigated Lincoln's "seven months abortion" in Louisiana and fought it by parliamentary maneuver and a threatened filibuster. And they won in the end, as Congress declined to seat the representatives from either Louisiana or Arkansas, thus leaving their status—and the entire question of restoring the Confederate South—an unresolved mess. At once rumors flew about that the personal intimacy between Lincoln and Sumner had ended, since the senator, as the *New York Herald* phrased it, had "kicked the pet scheme of the President down the marble steps of the Senate chamber."

Lincoln was distressed about Louisiana, extremely so, but he didn't break off his friendship with Sumner because of it. Though the President abhorred family quarrels, he realized that differences were bound to exist among principled men. So however much it pained Lincoln that Louisiana hadn't been readmitted, he saw no reason to terminate a relationship because he and Sumner were at odds over his government there. Too, he and Sumner agreed on other crucial issues—both were out to abolish slavery totally and to muzzle the South's rebellious white majority—and they needed one another. So in spite of their disagreements about reconstruction, they maintained close political ties and went right on seeing one another. In fact, only a few days after Sumner had won his fight against Louisiana, Lincoln invited the senator to accompany him and Mary to the inaugural ball.

For Lincoln there was another reason to keep Sumner coming to the White House. He was Mary's friend, too, and Lincoln understood how much the senator meant to her. In truth, Sumner was Mary's favorite caller; she delighted in him, found his brilliance and fastidious ways irresistible. And he was always elegantly dressed, his favorite attire consisting of a brown tailored coat, maroon vest, lavender or checked trousers, and a gold cane. Mary was flattered that this distinguished bachelor should care for her, that he could visit her in her drawing room and discuss diplomacy and read her letters from powerful Englishmen, that he could escort her regularly to the opera or the theater when Lincoln was too busy to go. Sumner made her feel like a cultivated woman, and she adored him and sent flowers for his table. When he threw off his aloof and haughty manner, Mary said, he could "make himself very *very* agreeable" and they could have "such delightful conversations." By now she subscribed to his views on black people—she admired him so—and urged Lincoln himself "to be an *extreme* Republican." And Sumner enjoyed her, too, for she was an adoring disciple, gossipy, spirited, and marvelously indiscreet.

Gossips, of course, attributed dishonorable motives to Sumner's White House visits. Some whispered that Sumner used Mary to control Lincoln, others that Lincoln used Mary to control Sumner. There was not a scintilla of truth to either speculation. Sumner made Mary happy, and Lincoln respected their relationship. Moreover, Lincoln welcomed Sumner's company, too, even if the senator did grumble about the "Lincolnisms" that studded his speech. Mary recalled how they could talk and laugh together "like *two* schoolboys."

By contrast, there was little genuine communication between Lincoln and Mary these days. They had drifted steadily apart during the last couple of years, so that by January of 1865 they pretty much went their separate ways. As Mary wrote an old friend, she felt lucky if her "tired and weary" husband dropped by at 11 P.M. to chat with her at all. And even when they were together they shut one another off. She lived in terror that *"forbidden subjects, might be introduced"*—like her $27,000 in debts or her begging for money in the recent campaign, which she took pains to conceal from him. On the other hand, Lincoln no longer confided in her either, because she couldn't be trusted to keep critical matters to herself. She hungered to discuss his affairs—wanted to share in them—but he gave evasive answers. And hurts and resentments grew between them. Mary, who in the past had never publicly criticized Lincoln, had never spoken disrespectfully about him around other people, disparaged one of his 1864 addresses as "the worst speech I ever listened to in my life." Moreover, when Mary mysteriously dismissed old Edward McManus, long-time White House doorkeeper, she and Lincoln had an ugly scene. In her misery, Mary told one of her male friends about the quarrel, but later wrote the man to keep it a "sacredly guarded" secret that she'd talked with him about her marital troubles. Because she knew how Lincoln grieved "over any coolness of mine," she went to him, and they made up and had "quite a little laugh together." But no matter how much they still loved and cared for one another, there was considerable tension between them now, a guarded and painful reticence that hurt them both. As a consequence, Mary turned to Sumner and other men for the companionship she desperately needed. Lincoln, for his part, found what comfort he could with Seward, or Stanton, or his personal secretaries, or Tad.

In January the problem of Robert and the army surfaced again. Now in Harvard Law School, Robert wanted to enlist more than ever, but Mary still said no, still feared he would die and add to "the *deep waters*" that threatened her. One day, though, a New York senator asked her bluntly: "Why isn't Robert in the Army? He is old enough to serve his country. He should have gone to the front some time ago." When Mary made excuses, the senator pointed out that his only son was in the army fighting for his country. Finally, with Lincoln and Robert both pressing her, Mary relented and said he could go. Lincoln asked Grant whether he might find a place for Robert with some nominal rank, whereupon Grant appointed young Lincoln a captain and assistant adjutant general of volunteers, and in February Robert left for Petersburg.

As the inauguration approached, Lincoln's health was failing. Browning, who saw him in late February, recorded that Lincoln "looked badly and felt badly—apparently more depressed than I have seen him since he became President." Joshua Speed came by for a visit and found Lincoln contending with the usual crowds. "I am very unwell," Lincoln admitted to his old friend. "My feet and hands always cold—I suppose I ought to be in bed." While Speed was there, two women entered the office and pleaded with the President to release their kin, imprisoned for resisting the draft. Lincoln supposed those fellows had suffered enough now and signed an order for their release. One of the women broke into tears and tried to kneel. "Get up," Lincoln told her firmly. "Don't kneel to me— Thank God and go."

"Lincoln," Speed said, "with my knowledge of your nervous sensibility it is a wonder that such scenes as this don't kill you."

Lincoln said it was the only thing today that had given him pleasure. He wanted Speed to say of him that Lincoln had good judgment, that he always tried to plant a flower when he thought one would grow.

March came with a flurry of rumors that Lincoln would be abducted or assassinated on Inauguration Day. Alarmed, Stanton took every precaution to secure the capital, to make certain there would be "no disturbances, no fires, no raids or robberies." He assigned extra guards to protect Lincoln, ignoring his irritated protests. And he deployed detectives around the capital, to keep an eye on all the Northern riffraff and rebel deserters who roamed Washington these days.

On the morning of March 4, heavy clouds moved over Washington, raking the capital with tails of rain. At noon Lincoln made the traditional carriage ride up Pennsylvania Avenue, teeming with bands and spectators and clattering cavalry. Ahead the new iron dome of the Capitol, crowned with a statue of liberty, moved against a somber sky. The inaugural platform stretched out from the east front of the Capitol; here Lincoln took his place, still sick and dispirited, as clouds scudded across Washington's horizon, rolling and turning in the wind. Below him, he noticed Frederick Douglass in the crowd. Behind him, on the platform and along the front of the Capitol, were various officials and spectators. Up behind the railing of the right buttress, looking down at the President, was the actor John Wilkes Booth, a dashing man with raven hair and a black mustache, wearing a fashionable stovepipe hat. Lincoln had seen Booth perform in *The Marble Heart* at Ford's Theater; it was on the night of November 9, 1863, just over a week before Lincoln spoke at Gettysburg.

Presently Lincoln rose and stepped to the speaker's stand; as he did so the sun broke through the clouds, flooding the entire gathering with brilliant light. Then the clouds closed in again. Lincoln began reading in "ringing and somewhat shrill tones," said Noah Brooks, which "sounded over the vast concourse." It was a terse speech, succinct and lyrical, like his address at Gettysburg. Yet his words today blazed with religious eloquence as he reiterated what he'd told a delegation of Kentuckians in the darkening spring of 1864: in his search for the meaning of this vast struggle, he'd come to view it finally as a divine punishment for the sin of slavery, as a terrible retribution visited by God on a guilty people, in North as well as South.

When the war began four years ago, he sang out, neither side expected it to last as long or grow to such magnitude as it had. Neither side anticipated that slavery might perish in the flames, "that the *cause* of the conflict might cease with, or even before, the conflict itself should cease." No, Lincoln said, "each looked for an easier triumph, and a result less fundamental and astounding. Both read the same Bible, and pray to the same God; and each invokes His aid against the other. It may seem strange that any men should dare to ask God's assistance in wringing their bread from the sweat of other men's faces; but let us judge not that we be not judged. The prayers of both could not be answered; that of neither has been answered fully. The Almighty has His own purposes." And God's purpose perhaps was to will "this terrible war" on both North and South to remove "the offence" of slavery.

"Fondly do we hope—fervently do we pray—that this mighty scourge of war may speedily pass away. Yet, if God wills that it continue, until all the wealth piled by the bond-man's two hundred and fifty years of unrequited toil shall be sunk, and until every drop of blood drawn with the lash, shall be paid by another drawn with the sword, as was said three thousand years ago, so still it must be said 'the judgments of the Lord, are true and righteous altogether.' "

And if it were God's will, the country must see this grim purgation through to its Providential conclusion, when at last both sides might be cleansed and regenerated: "With malice toward none; with charity for all; with firmness in the right, as God gives us to see the right, let us strive on to finish the work we are in; to bind up the nation's wounds; to care for him who shall have borne the battle, and for his widow, and his orphan—to do all which may achieve and cherish a just, and a lasting peace, among ourselves, and with all nations."

Lincoln turned now, took the oath from Chief Justice Salmon Chase, and kissed an open Bible. Chase noted the spot—Isaiah 5:27 and 28. "None shall be weary nor stumble among them; none shall slumber nor sleep . . . Whose arrows are sharp, and all their bows bent, their horses' hoofs shall be counted like flint, and their wheels like a whirlwind."

The President bowed to the cheering assembly, and an artillery salvo exploded on the wind. Then a huge procession escorted his carriage back to the White House, bleak against the clouds. Later Lincoln encountered Noah Brooks. "Did you notice that sunburst?" the President asked. "It made my heart jump."

In a reception that evening, Lincoln stood in line in the White House, shaking hands with hundreds of well-wishers. In a moment somebody informed him that Frederick Douglass was at the front door, but that the police wouldn't let him in because he was a Negro. Douglass had come to congratulate Lincoln, contending that since black men were dying in battle for the Union, they had a right to come and shake the President's hand like any other citizen: "if the colored man would have his rights," Douglass said, "he must take them."

Lincoln had Douglass shown in at once. "Here comes my friend Douglass," the President announced when Douglass entered the room. "I am glad to see you," Lincoln told him. "I saw you in the crowd today, listening to my address." He added, "There is no man in the country whose opinion I value more than yours. I want to know what you think of it." Douglass said he was impressed: he thought it "a sacred effort." "I am glad you liked it!" Lincoln said, and he watched as Douglass passed down the line. It was the first inaugural reception in the history of the Republic in which an American President had greeted a free black man and solicited his opinion.

PART · ELEVEN

"MOODY, TEARFUL NIGHT"

The signs seemed auspicious now, the war seemed rushing toward its destined end. In February, Sherman's army stormed into South Carolina like a horde of avenging angels—South Carolina with its militant Southernism, where secession and the rebellion had begun. Civilians fled their homes and evacuated their cities before Sherman's advancing columns. Columbia, the state capital, went up in smoke, the city set afire either by rebels or by Sherman's men. In the Shenandoah, meanwhile, Sheridan burned a broad path of devastation clear to the Rapidan River. With rebel cities and fields in flames, Lincoln met with three Confederate emissaries on a steamer near Hampton Roads, Virginia, and gave them his terms for peace: the rebels must recognize the national supremacy and accept the complete and permanent eradication of slavery before hostilities would cease. He did, however, favor compensation for rebel slaveholders, if that was any consolation. The emissaries hedged and vacillated, and the conference ended without result. Afterward the Cabinet unanimously rejected Lincoln's compensation proposal, arguing that Congress in its present mood would never approve such a measure, and Lincoln abandoned the plan. Later the President learned that Lee had contacted Grant and suggested that the two generals settle the war. Through Stanton Lincoln sent Grant a clear, uncompromising order: *I* will deal with political questions and negotiate for peace. Your job is to fight.

As spring approached, with Union armies driving into the heart of the Confederacy, some of Lincoln's friends were concerned about his safety, fearful that rebel sympathizers might try to harm him—or that the rebels themselves might try to kill or kidnap the President in a desperate attempt to save the Confederacy. But Lincoln had heard this before—in 1864 rumors had circulated in the press of a rebel plot to assassinate him—and he had persistently dismissed all such rumors. "Even if true," he said, "I do not see what the Rebels would gain. If they kill me, the next man in line will be just as bad for them."

And the hate mail came in with depressing regularity, but Lincoln said it didn't bother him much any more. "Soon after I was nominated in Chicago," he told a friend, "I began to receive letters threatening my life. The first ones made me feel a little uncomfortable; but I came at length to look for a regular installment of this kind of correspondence in every mail." Nevertheless, he scrupulously filed the threat letters in an envelope marked "Assassination."

"I know I'm in danger," Lincoln confessed to Seward, "but I am not going to worry about it." In point of fact, somebody might already have tried to kill him. Lincoln told Lamon about the incident, Lamon said, and it made his skin crawl. According to Lamon, it happened on an August night in 1862: Lincoln was riding out to the Soldiers' Home on a horse called Old Abe. Suddenly there was a gunshot; the horse stampeded and raced pell-mell all the way to the Soldiers' Home. Lincoln recounted the event in "a spirit of levity," Lamon recalled, and then the President broke into a laugh about the "ludicrous" and "farcical" scene of the two Old Abes—him and the horse —breaking speed records back to the Soldiers' Home. The gunshot? Lincoln admitted that the bullet whistled close by, but denied that anybody had really tried to shoot him. He thought it was probably an accident—somebody out hunting most likely.

Lamon never forgot that story. He was certain that Lincoln's life was in danger, certain that his enemies had tried to murder him that night and would try again. Ever since Lincoln's re-election, Lamon had fretted about his safety, haunted by fears that desperate and hateful men were out to assassinate him. He fussed at Lincoln for walking alone at night, scolded him for going off to the theater without guards, pleaded with him to be careful. You must understand, Mr. President, that *"You are in danger."* "You know, or ought to know, that your life is sought after, and will be taken unless you and your friends are cautious; for you have many enemies within our lines."

"Lamon," Lincoln said to a Cabinet member, "is a monomaniac on the subject of my safety. . . . What does any one want to assassinate me for?" Besides, Lincoln agreed with Seward. "Assassination," Seward declared, "is not an American practice or habit, and one so vicious and desperate cannot be engrafted into our political system." "As to crazy folks," Lincoln told Hay with a shrug, "I must take my chances." And anyway what could he really do to save himself? As he said to Noah Brooks, "I long ago made up my mind that if anybody wants to kill me, he will do it. If I wore a shirt of mail, and kept myself surrounded by a body-guard, it would be all the same. There are a thousand ways of getting at a man if it is desired that he should be killed."

That was what he kept telling Stanton, too, but Stanton was just as bad as Lamon, constantly lecturing Lincoln about protecting himself. After all, hate for the President had boiled up all over the Union for the past four years; such a violent atmosphere could produce an assassin at any time—a face in the crowd, a gun, a knife. In his worries about Lincoln's safety, Stanton had done everything he could think of to guard the President: he'd ordered a company of Pennsylvania troops to encamp on the White House lawn. During the past winter, he'd worked with the Washington police chief to provide Lincoln with constant protection, and the chief had detailed four plain clothesmen—among them Thomas Pendel and William Crook, both excellent lawmen—to serve as Lincoln's personal bodyguards. Moreover, Stanton had assigned a military escort for Lincoln's nocturnal walks to and from the War Department, in addition to surrounding him with detectives on special occasions like the inauguration.

As always, Lincoln griped about all this protection. It is important, he argued, "that the people know I come among them without fear." And sometimes he would elude Stanton's soldier escort in his nightly walks—or just ask the men to stay behind. Which infuriated Stanton, who wanted to know why in hell the President was so stubborn about a guard? Did he not care about his own life? The Secretary begged Lincoln to look out for himself, and he ordered the soldiers never under any circumstances to let the President go into the night alone. Was that clear? Finally Lincoln consented to the escort, mostly out of concern for the men. "If Stanton should learn that you had let me return alone," he told them once, "he would have you court-martialed and shot."

By mid March, his heavy work schedule and sleepless nights had almost completely worn him down, and Mary and his friends worried about his failing health. On March 14 he felt so ill that he had to conduct a Cabinet meeting in his bedroom. Still, he was encouraged about the war, as reports from Grant indicated that the end indeed was near. On an invitation from Grant himself, Lincoln decided to visit the general's headquarters at City Point, at the confluence of the Appomattox and James rivers some twenty miles south of Richmond. As Mary said, the rest and fresh air would be good for him. More than that, though, he hoped to be there when the final victory came.

On March 23 he sailed for Virginia on the *River Queen,* accompanied by Mary, Tad, detective Crook and Captain Charles B. Penrose, detailed by Stanton to help guard the President. The following night they anchored at City Point, with the lights of Grant's headquarters flickering on a nearby bluff and shadows of warehouses looming along the shoreline. Grant and his staff greeted them at the wharf, and everybody went up to Grant's headquarters huts for a social hour, with Julia Grant as hostess and several officers and their wives in attendance. General Meade was there, and so was General Edward O. C. Ord, who'd replaced Butler as commander of the Army of the James.

The next morning Lincoln followed a scampering Tad back up to the bluff to see the sights. There was an exhilarating breeze, and the view was spectacular: across the James and off to the north was Bermuda Hundred, its tents and warehouses visible in the haze; near there Butler had been bottled up until Grant liberated him on his sweep southward to Petersburg. Beyond Bermuda Hundred were the timbered slopes of Malvern Hill where McClellan had repulsed Lee in the Seven Days fighting. And below that lay Harrison's Landing on the James, where Lincoln had met with McClellan at the end of the unhappy Peninsula campaign. And there was the river itself, teeming with gunboats and troop transports, their whistles sounding in the distance. A medley of jumbled noises rose from the wharves of City Point, where laborers unloaded supplies for the Potomac Army. And back around to the southwest lay Petersburg, with its tangle of trenches and redoubts carved in the earth. Across the eastern front of Petersburg were the Union siege lines, stretching for miles toward the south.

Captain Robert Lincoln, a member of Grant's staff, joined his father for breakfast and reported "a little rumpus up the line this morning." In the chill dawn, Lee had attempted to break Union lines at Fort Stedman, just northeast of Petersburg, but Union troops had counterattacked. Eager to see the battle site, Lincoln left City Point and set out for the front on a slow, jolting military train, staring out the window at the scenes of war passing by. There was Fort Stedman now, where Confederate and Union troops had fought that morning in surging hand-to-hand combat before the latter repulsed the rebel attack. Lincoln noticed a line of rebel prisoners captured in the fighting and commented on their "sad condition." He saw dead Union men and dead Confederates, saw the wounded of both armies lying bandaged, bloodied, and forlorn, looking his way as the train ground by. He said "he had seen enough of the horrors of war, that he hoped this was the beginning of the end." But mostly he "was quiet and observant," said a man who rode with him, as other battlefields moved by in the windows of the coach.

Back at City Point, Lincoln joined Grant and his staff around a crackling fire. Lincoln brushed smoke away from his face, an officer noted, and "spoke of the appalling difficulties encountered by the administration, the losses in the field, the perplexing financial problems." Later, looking "worn and haggard" from his journey to the front, he retired to the *River Queen* for the night. The next day, he and Mary sailed up to Aikin's Landing for a grand review of Ord's Army of the James. On the way Lincoln waved at some soldiers watering their horses at the riverbank—they were part of Sheridan's army, come down from the Shenandoah to join Grant in the final campaign. At Aikin's Landing, Sheridan himself came on board to pay his respects. "General Sheridan," Lincoln said with a grin, "when this peculiar war began I thought a cavalryman should be six feet four high, but"—and he gazed down now—"I have changed my mind—five feet four will do in a pinch."

From the landing, Lincoln rode out to the field on horseback, while Mary and Julia Grant followed in an ambulance, bouncing over a rough corduroyed road until Mary's head was aching. At the review site, the Army of the James was drawn up in parade formation, line after line of regiments with banners and flags rippling in the breeze. Lincoln moved down the lines with attractive Mrs. Ord riding at his side. When Mary arrived, she was shocked to find her husband reviewing the troops with another woman. Lincoln and Mrs. Ord trotted over to greet her, but Mary was in a jealous rage and gave Mrs. Ord such a tongue-lashing that the general's wife burst into tears. As the officers turned away in acute embarrassment, Lincoln tried to calm Mary, apologized and called her "Mother." But she turned on him now, attacking him with tears in her eyes, crying that all the soldiers would think that Ord woman was his wife instead of her. Lincoln turned and walked off.

For several days afterward, Mary remained "indisposed" aboard the *River Queen.* Then she left Tad with his father and returned to Washington. Later, though, she wired Lincoln that she missed him "very very much" and that she might return to City Point with a little party of her own. Was she sorry, then? Remorseful? Perhaps so, but the incident over Mrs. Ord left a painful wound between Lincoln and Mary.

As March closed, Sherman left his army in North Carolina and came up to confer with Grant, and Lincoln participated in their deliberations as much as possible. He'd met red-haired "Bill" Sherman a couple of times in the first year of the war and admired his fierce, go-ahead spirit. An unkempt man, tall, lean, and loose-jointed, Sherman chain-smoked cigars, spoke in picturesque phrases, mocked military orthodoxy, and defended the kind of total war he'd waged in the Deep South. As Lincoln listened in, Grant seemed to think one more large battle would be necessary, but the President was largely in the dark about his plans. Staff officers noticed that Lincoln was sad and nervous, asking repeatedly when Grant's army would move, when the final engagement would come.

On March 29 Grant set out for Petersburg, and Lincoln paced on the *River Queen* waiting for news. That night, as he stood on the deck, an artillery barrage opened up near Petersburg, followed by the rattle of heavy musketry. Grant starting his attack now? It was a dark and rainy night, and

Lincoln could see the flashes of the guns against the clouds. Later he wired Stanton that he really ought to get back to Washington, but was reluctant to leave with the end so close. Stanton telegraphed back that all was well in Washington and that Lincoln should stay where he was. "A pause by the army now would do harm," Stanton said; "if you are on the ground there will be no pause."

April 1. Heavy fighting all day around Petersburg. That evening a war correspondent came aboard the *River Queen* with some rebel battle flags captured that day. Grant sent them back to Lincoln with his compliments. "Here is something material," Lincoln rejoiced, "something I can see, feel, and understand. This means victory. This *is* victory." Inside the cabin, Lincoln asked the correspondent to relate everything he'd seen at the front. The man observed that every table was covered with huge military maps dotted with red and black pins.

Before dawn the next morning, hundreds of guns opened up all across Grant's front, with concussions that could be felt back at City Point. For a time, Lincoln stood out in the trenches and watched the battle, then he manned the City Point headquarters where he relayed messages on to Washington as rapidly as they came in from Grant. The next morning, April 3, came the news Lincoln had been waiting for. Lee had evacuated Petersburg, and Meade and Sheridan were racing now to cut off his retreat. At that, Lincoln rode into smoldering Petersburg and pumped Grant's hand for this glorious victory. At first, Lincoln said, he'd thought Grant might wait for Sherman to come up from North Carolina and then launch his attack. No, Grant said, he'd decided that the Potomac Army—Lee's old foe—ought to destroy the Army of Northern Virginia without help from Western troops. Lincoln remarked that his anxieties were so great he didn't care where help came from, "so that the work was perfectly done."

Back at City Point, there were more portentous dispatches. Richmond too had fallen, with Davis and his government fleeing into the Virginia interior. Unable to stay here, Lincoln wired Stanton that he was going up to have a look at Richmond. "I will take care of myself," he added.

With Tad, Captain Penrose, and detective Crook, Lincoln headed up the James on a gunboat, passing dead horses, broken ordnance, and half-sunken boats in the currents. At last Richmond came into view, with columns of smoke billowing up against the sky. When Lincoln stepped onto the docks, followed by a dozen sailors armed with navy carbines, black workers recognized the tall, gaunt man with the stovepipe hat. "Glory!" cried a black woman. "Glory glory!" Accompanied by Crook, Penrose, and the naval guard, Lincoln took Tad by the hand and set out for downtown Richmond a couple of miles away. By now black people were flocking around him, yelling his name and reaching for his hand. "God bless you," said one woman, and he smiled at her and nodded at the others. The city was under martial law and patrolled by a number of black Union troops, so the white inhabitants stayed in their homes, watching from their windows as Lincoln and a column of cheering Negroes passed in the dusty streets. "Every window was crowded with heads," said Crook. "But it was a silent crowd. There was something oppressive in those thousands of watchers without a sound, either of welcome or hatred." Lincoln scanned the windows, as though expecting a jeer or some sort of defiance. But nothing. Only a white woman shaking her head in disgust.

He stepped along through the business section now, surrounded by burned-out buildings. Before evacuating the city, the rebels had set the bridges and warehouses afire, but a high wind whipped the flames into downtown Richmond, and some structures were still blazing. A cavalry escort came and took him to the rebel executive mansion, now Union military headquarters. Inside, Lincoln looked around the deserted rooms. He seemed "pale and utterly worn out," said one witness, as he asked for a glass of water. In the executive office, Lincoln sat down in Jefferson Davis's chair, and the Union troops broke into cheers.

He took a whirlwind carriage ride through the city, passing Libby Prison and dozens of fire-ravaged buildings. He came to the capitol where the rebel Congress had met and roamed about in the wreckage inside, with papers and broken furniture scattered everywhere. Before he left Richmond, he met with the rebel Assistant Secretary of War and dictated terms of complete and unconditional surrender. He did not promise amnesty to rebels, but did say he had the power to pardon them and "would save any repenting sinner from hanging." Later he tentatively authorized the rebel state legislature to convene and help disband Virginia troops and send them home.

Back at City Point, he pored through dispatches for word of the destruction or surrender of Lee's army. Sheridan had seized the Danville Railroad, thus cutting off any chance for Lee to escape into North Carolina; and on April 6 Union cavalry and infantry columns smashed half of his army south of the Appomattox. Sheridan wired Grant: "If the thing is pressed I think that Lee will surrender." Lincoln read that and telegraphed Grant: "Let the *thing* be pressed."

Word came from Washington that Seward had been severely injured in a carriage accident there. Alas, Lincoln had to return now. But before he left, he toured Union hospital camps and shook hands with some seven thousand wounded soldiers. He said he would probably never see them again and he wanted them to know how much he appreciated what they'd done for the Union.

Mary rejoined Lincoln at City Point with a "choice little party" that included Sumner and Lizzie Keckley. They'd come down a few days ago and toured Richmond themselves; and the sight of the rebel capital had transformed Sumner "into a lad of sixteen." On the journey back to Washington, they had a long discussion about Shakespeare, and Lincoln entertained the group by reading the scene in Macbeth where Duncan is assassinated.

The *River Queen* reached Washington early on the evening of April 9, and Stanton greeted Lincoln with a momentous telegram from Grant: "General Lee surrendered the Army of Northern Virginia this morning," at a place called Appomattox Courthouse. Lincoln and Stanton threw their arms around one another, and Stanton, "his iron mask torn off, was trotting about in exhilarated joy," said an onlooker. Lincoln made his way through the torch-lit streets, already thronging with people, and called at Seward's home. He blanched when he saw his Secretary of State: he was suffering from facial lacerations, a fractured jaw, and a broken arm, lying there covered with bandages, his neck in a steel brace. "You are back from Richmond," Seward whispered. Lincoln nodded and told him about the news of General Lee. Seward managed a smile, but reminded Lincoln that a force under Joseph Johnston was still at bay in North Carolina, so that technically the war wasn't over yet. Lincoln expected word at any moment that Johnston too had given up. He stayed an hour with Seward, describing his Richmond visit and giving his Secretary words of sympathy and cheer.

The next day, April 10, official news of Lee's surrender threw Washington into bedlam. By Stanton's orders, five hundred guns rocked the city, breaking out windows on Lafayette Square, and newspapers and official proclamations blazed with headlines about Appomattox. "The nation seems delirious with joy," Welles wrote in his diary. "Guns are firing, bells ringing, flags flying, men laughing, children cheering, all, all are jubilant."

In the White House, Lincoln breakfasted with Noah Brooks, while "hurrahing legions" marched about outside. Later in the day a brass band led some three thousand people up to the White House, where they clamored for a speech from the President. Tad appeared at the window and waved a captured rebel flag, and the crowd roared with delight. Lincoln promised a formal speech tomorrow night—and then summoned the band to strike up

"Dixie." It was always one of his favorite tunes, he said. The rebels took it as theirs, but now the Union had recaptured it—there were happy shouts from below—and so "it is our lawful prize." At that the band played "Dixie," with its cornet trills, and then launched into "Yankee Doodle" as an encore.

While Washington celebrated that day, Lincoln worked on reconstruction. He told his Cabinet he was glad and relieved that the war was over, but he had little time to rejoice, for his postwar burdens were among the most difficult he'd ever faced. He had to rebuild and restore the conquered South, maintain the loyalty of white Unionists there, protect Negro freedom, and contend with an increasingly hostile Congress. So the country wanted a speech, did it? He would give one, all right—but it wasn't going to be a patriotic hymn. No, he wanted to talk about the problem of reconstruction that faced them now, and how he thought it should be solved.

Stanton came in later, looking exhausted and utterly worn down from illness. He was here to offer his resignation. He'd promised himself he would leave his post when Richmond fell, and now Lee had capitulated, and he felt he wasn't needed any longer. Lincoln put his hands on Stanton's shoulders. "Stanton, you have been a good friend and a faithful public officer and it is not for you to say when you will be no longer needed here." Reluctantly Stanton agreed to remain. In the crucial matter of reconstruction, Stanton decided, the President required a tough and experienced adviser at his side. So Stanton would stay with Lincoln a little longer.

On the night of April 11, hundreds of people assembled on the White House lawn, as Lincoln prepared to address them from an upstairs window. It was misty out, but even so one could see the new illuminated dome of the Capitol. Off in the distance, across the Potomac, Lee's Arlington plantation was aflame with colored candles and exploding rockets, as hundreds of ex-slaves sang "The Year of Jubilee." With Brooks holding a candle from behind a curtain, Lincoln stepped to the window and unrolled his speech. Stretched out below, Brooks noted, "was a vast sea of faces, illuminated by the lights that burned in the festal array of the White House."

"We meet this evening, not in sorrow, but in gladness of heart," the President began. The fall of Richmond and the surrender of Lee's army gave hope of a speedy peace "whose joyous expression can not be restrained." But now the nation faced the trial of reconstructing the conquered South. "It is fraught with great difficulty," Lincoln said. "Unlike the case of a war between independent nations, there is no authorized organ for us to treat with. No one man has authority to give up the rebellion for any other man. We simply must begin with, and mould from, disorganized and discordant elements. Nor is it a small additional embarrassment that we, the loyal people, differ among ourselves as to the mode, manner, and means of reconstruction."

With Sumner and other Republican critics in mind, Lincoln observed that he was much censored for the state government he'd created in Louisiana. A few Republicans were especially displeased because black men there could not vote (in fact, that very day Salmon Chase had complained to Lincoln about this, insisting that it was criminal to deny Southern Negroes the suffrage and thus leave them in the political control of their former masters). Well, Lincoln too was unhappy that blacks couldn't vote in Louisiana. He himself preferred that "very intelligent blacks" and those who'd served in Union forces should have the suffrage. But he wasn't going to throw out Louisiana's government because it failed to enfranchise Negroes. No, it was wiser to accept Louisiana with all its current imperfections and "help to improve it." After all, he said, Louisiana already had a fine constitution which outlawed slavery all over the state, granted black people economic independence, provided public school benefits equally for both races, and even empowered the legislature to enfranchise Negroes if it wanted to. Wasn't it better to work from this nucleus than to dismantle the state government and start all over again? How would the latter course more effectively restore Louisiana to the Union? Help "the colored man" there? "Grant that he deserves the elective franchise," Lincoln said, "will he not attain it sooner by saving the already advanced steps toward it, than by running backward over them?" Moreover, if Republicans rejected Louisiana, they rejected one more vote for the Thirteenth Amendment.

Once again he assured his critics that he was flexible about reconstruction. His ten percent plan by which Louisiana had been restored was one mode —but not the only mode—for reconstructing the other rebel states. He was willing to bend, modify, and adjust according to the peculiar problems and conditions of each locality. An inflexible plan, he contended, would just become a new entanglement.

Still, the enormous problems of Southern reconstruction, of how to control the rebellious white majority and protect Southern loyalists whether black or white, were far from being solved. And so he closed with a warning: "In the present *'situation'* as the phrase goes, it may be my duty to make some new announcement to the people of the South. I am considering, and shall not fail to act, when satisfied that action will be proper."

There was a patter of polite applause, but it was obvious that most of the audience was extremely disappointed in the speech. They weren't interested in the technical problems of Louisiana and reconstruction, not in the glow of military victory. They wanted a rousing oration about the triumph of Union arms and the glory of the cause. As Lincoln read his address, a lot of people became bored and wandered off. Others glowered when Lincoln endorsed limited Negro suffrage, fuming that he was as rabid and radical as the Democrats had always charged and that the country owed all its troubles to him.

Lincoln was "very anxious" about the cool reaction of the crowd. And Sumner and Chase—what did they think? Sumner remained categorically opposed to Louisiana, so Lincoln could count on continued salvos from Capitol Hill when Congress convened again. Chase, for his part, was glad that Lincoln "openly avows" the vote for intelligent Negroes and black soldiers, but was "sorry that he is not yet ready for universal or at least equal suffrage." Nevertheless, Lincoln's position that April put him in advance of most Republicans, who still shrank from Negro suffrage lest their own white constituents turn them out of office.

Meanwhile there was static in the Cabinet. When it got out that Lincoln had authorized the Virginia legislature to convene, Stanton and Speed vigorously protested, arguing that this was tantamount to recognizing the rebel legislature as a legal body. And Welles, too, was apprehensive. At first Lincoln was a little defensive about his action. But finally he admitted that "he had perhaps made a mistake," and on April 12 he rescinded the order.

In his exhausted condition, Lincoln still wasn't sleeping well, troubled lately by strange and ghostly dreams. One night in the second week of April, with Mary, Lamon, and one or two others in the White House, Lincoln started talking about dreams, and Mary commented on how "dreadful solemn" he seemed. "I had one the other night which has haunted me ever since," Lincoln said. After he woke from the dream, he opened the Bible and everywhere he turned his eye fell on passages about dreams, visions, and supernatural visitations.

"You frighten me!" Mary exlaimed. "What is the matter?"

Maybe he'd "done wrong" in even mentioning the dream, Lincoln said, "but somehow the thing has got possession of me."

What had possession of him? Mary asked. What had he dreamed?

Lincoln hesitated, then began in a voice sad and serious: "About ten days ago I retired very late. I had been up waiting for important dispatches from the front. I could not have been long in bed when I fell into a slumber, for I was weary. I soon began to dream. There seemed to be a death-like stillness about me. Then I heard subdued sobs, as if a number of people were weeping. I thought I left my bed and wandered downstairs. There the silence was broken by the same pitiful sobbing, but the mourners were invisible. I went from room to room; no living person was in sight, but the same mournful sounds of distress met me as I passed along. It was light in all the rooms; every object was familiar to me; but where were all the people who were grieving as if their hearts would break? I was puzzled and alarmed. What could be the meaning of all this? Determined to find the cause of a state of things so mysterious and so shocking, I kept on until I arrived at the East Room, which I entered. There I met with a sickening surprise. Before me was a catafalque, on which rested a corpse wrapped in funeral vestments. Around it were stationed soldiers who were acting as guards; and there was a throng of people, some gazing mournfully upon the corpse, whose face was covered, others weeping pitifully. 'Who is dead in the White House?' I demanded of one of the soldiers. 'The President,' was his answer; 'he was killed by an assassin!' Then came a loud burst of grief from the crowd." As he recounted the dream, Lamon observed, Lincoln was "grave, gloomy, and at times visibly pale."

"That is horrid!" Mary said. "I wish you had not told it."

"Well," Lincoln said, "it is only a dream, Mary. Let us say no more about it, and try to forget it."

Lamon, of course, was just as frightened as Mary. But Lincoln sported with him as always. "For a long time you have been trying to keep somebody— the Lord knows who—from killing me," Lincoln said. "Don't you see how it will turn out? In this dream it was not me, but some other fellow, that was killed. It seems that this ghostly assassin tried his hand on someone else." In a moment Lincoln grew solemn again. He sighed, as though talking only to himself: "Well, let it go. I think the Lord in His own good time and way will work this out all right."

On Good Friday morning, April 14, he woke in a pleasant mood. For once he'd slept fairly well—no insomniac worries about reconstruction and no bad dreams. No, last night he'd had the other dream again, the one that had come to him several times on the eve of significant military events—especially Union victories like Antietam, Gettysburg, and Vicksburg. In the dream, he was on a phantom ship moving swiftly toward a dark and indefinite shore. It must portend favorable news today—no doubt that Johnston had surrendered in North

Carolina, thus ending organized rebel resistance.

In his office, Lincoln lit the fireplace to take off the morning chill. Outside, it was a lovely spring day—the Judas trees and dogwoods were in bloom, and there was a scent of flowers in the air. In a few days, guests in Washington's hotels would move their chairs out to the sidewalks, and young people would ride to the Great Falls for picnics.

At eight he enjoyed an intimate breakfast with Mary. Trying to put what happened at City Point behind them, they were affectionate with one another now. On Wednesday Lincoln had even sent her a little note about going on a carriage ride, "playfully & tenderly worded," Mary said, "notifying, the hour, of the day, he would drive with me." And today they would ride together again—"by ourselves," he said, just the two of them. This evening they were going out to Ford's Theater, where the celebrated English actress Laura Keene was starring in *Our American Cousin,* billed as an "eccentric" English comedy. The Lincolns had invited the Grants to go with them, and today's papers officially announced their plans.

Robert joined his parents for breakfast. He'd come back to Washington with General Grant, proud of his stint on Grant's staff, proud of the fact that he'd been at Appomattox Courthouse that famous day, standing out on the porch while Lee surrendered inside. This morning, he and Lincoln chatted about his future, Lincoln drawling that he wanted Robert to finish Harvard Law School and read law for three years. Then "I hope that we will be able to tell whether you will make a lawyer or not."

At 11 A.M. there was a meeting with Grant and the Cabinet. "I never saw Mr. Lincoln so cheerful and happy," remarked one Secretary, as the President asked Grant about Sherman and commented on his dream last night about the phantom ship and the good news it surely betokened. Then he turned to the conquered South and confessed that he'd "perhaps been too fast in his desires for early reconstruction." Lincoln had asked Stanton to draft a tentative plan for Cabinet consideration, and Stanton now read a proposal for restoring Virginia and North Carolina. He recommended that the two states be combined into a single military department, to be administered by the army and the War Department under military law. Lincoln advised the Secretaries to consider Stanton's plan carefully. "This is the great question before us," the President said in reference to reconstruction, "and we must soon begin to act." He added that he was glad Congress was not in session now to impede and impair their work. There was some disagreement about certain technical points in Stanton's report, but Lincoln and all the other Secretaries endorsed Stanton's military approach and agreed that an army of occupation might be necessary to safeguard the reconstruction process. This no doubt was what Lincoln had in mind at the close of his April 11 speech.

After all, he'd always believed the military indispensable in restoring civilian rule in the South, insisting that the Unionist minority there must have federal military protection in establishing loyal state regimes. Without the army, he feared that Southern rebels would overwhelm white loyalists and maybe even re-enslave the blacks, thereby obliterating all the gains of the war. This was the position he'd taken in Louisiana and Arkansas, and it was the position he adhered to now.

What Lincoln disliked about Stanton's plan was that it merged Virginia and North Carolina into a single department. When Welles argued that the two states should be dealt with separately, Lincoln entirely agreed and directed that Stanton redraft his plan, applying military reconstruction separately to the two states, and that the revised proposals be discussed at the next meeting.

On other reconstruction matters, they deferred the question of Negro suffrage, knowing that it would require extended debate. As for punishing the rebels, Lincoln made it clear that he wanted "no bloody work," no war trials, hangings and firing squads—not even for rebel leaders. But he would like to "frighten them out of the country," he said, "open the gates, let down the bars, scare them off." He waved his hands as though he were shooing chickens.

After the meeting, Grant regretted that he and Julia couldn't go to the theater tonight, mumbling that they planned to visit their sons in New Jersey and were catching an early train today. No doubt Lincoln understood. At City Point, Mary had not only screamed at Mrs. Ord, but had insulted Julia Grant as well, raging at her in the presence of officers: "I suppose you think you'll get to the White House yourself, don't you?" Since then, Julia had developed such an intense dislike for Mary that she couldn't bear an evening out with her.

So what couple to invite? As it turned out, the Stantons too were unable to go, mainly because Ellen Stanton and Mary didn't get along either. Also, Stanton worried about Lincoln's theatergoing, worried that some deranged rebel sympathizer might take a shot at him from the streets, and beseeched him to stay home at night. At lunch, Lincoln told Mary he had half a mind not to go, but she thought it important to stick with their plans, since they had been announced in the newspapers. Later in the day they finally selected a handsome young couple to accompany them—Major Henry R. Rathbone and Clara Harris, his fiancée, the daughter of Senator Ira Harris of New York.

Lincoln spent the afternoon in his shop, laboring over pardons and reprieves and contending with the usual callers. At last he dismissed them and went to fetch Mary for their ride. In the corridor, he encountered two women and stopped to chat. One of them was the wife of C. Dwight Hess, manager of Grover's Theater, where a dramatic new play called *Aladdin! or, The Wonderful Lamp* had just opened. Hess had invited the Lincolns to attend his theater tonight, and Lincoln told Mrs. Hess how sorry he was that he and Mrs. Lincoln couldn't accept, but they had decided on other plans. Lincoln preferred something farcical tonight anyway, something that would make him laugh.

Around five the Lincolns rode out the White House gates and headed for the Navy Yard. It was raw and gusty out now, and they sat close together as the carriage rolled through the streets. Lincoln was in good spirits. You seem "so gay," Mary said, "so cheerful." "And well might I be," he said. "I consider this day the war has come to a close." Then more tenderly now, recalling the troubles they'd had: "We must both be more cheerful in the future; between the war and the loss of our darling Willie, we have been very miserable."

He talked about the years ahead, about what they would do when his second term was over. They would travel to Europe with their sons and maybe even visit Jerusalem—he'd always wanted to see it. Then they would journey out West, out to California, where soldiers discharged from the army could find jobs in gold and silver mines. After that, he would return to his law practice, open an office in Springfield or Chicago, and make enough to provide for them. It must have been a comforting thought after the hell he'd gone through here: getting back to quiet, unhurried legal work, back to the Illinois Supreme Court where he loved to argue technical cases, back to the circuit and the discussions with friends around hotel fireplaces. . . .

Between six and seven they returned to the White House for dinner. Afterward Mary complained of a headache and may have changed her mind about attending the theater now. Lincoln was a little tired himself, but said he needed to "have a laugh over the country cousin" and didn't want to go to Ford's alone. Besides, he would get no rest if he remained in the White House—visitors would pester him all evening. Seeing how much he wanted to go out, Mary relented. And anyway she'd never "felt so unwilling to be away from him."

After dinner, Lincoln and detective Crook made a quick trip to the War Department, but still no news from North Carolina. On the way back, Crook "almost begged" Lincoln not to attend the theater tonight. But the President assured him that everything would be all right. Then couldn't Crook stay on duty and go along as an extra guard? "No," Lincoln said, "you've had a long, hard day's work, and must go home." They parted at the portico of the White House, Lincoln calling out goodbye as he started up the steps.

Because of last-minute visitors, Lincoln and Mary were late in getting away. At last, at eight-fifteen, they climbed into the Presidential carriage, Mary wearing a gray silk dress and a bonnet to match, Lincoln a black overcoat and white kid gloves. The carriage rolled out into the street and headed for Senator Harris's place. Only the coachman and Lincoln's personal attendant rode with him and Mary; the guard for the night, one John F. Parker, had gone on ahead to Ford's Theater. A patrolman on the Washington police force, Parker was a lazy oaf with an appalling record of drunkenness, insubordination, and inefficiency. Through bureaucratic confusion and negligence, he'd been assigned to the White House while Lincoln was at City Point.

It was a foggy night, so gloomy that the Lincolns could scarcely make out the buildings they passed. At street corners, gaslights glimmered eerily in the drifting mist. By the time they picked up Rathbone and Miss Harris, they were so late that the play had begun without them. At eight-thirty the carriage pulled up in front of Ford's, blazing with lights, and the two couples hurried inside—Mary on Lincoln's arm, pretty young Clara on Rathbone's. The major was an ebullient fellow with a walrus mustache and a sloping nose. The two couples made their way up the winding stairway and crossed the dress circle at the back of the first balcony. The theater was packed tonight with high army brass and assorted Washington socialites; when they spotted the President, the audience gave him a standing ovation and the orchestra promptly struck up "Hail to the Chief." The Presidential party swept around the back row of chairs now, passed through a door and down a short hallway to the "state box," which directly overlooked the stage. Lincoln sank into a rocking chair provided by the management, with Mary seated beside him and Rathbone and Miss Harris to their right. The front of the box was adorned with drapes and brilliant regimental and Union flags. On stage, Harry Hawk, the male lead, ad-libbed a line: "This reminds me of a story, as Mr. Lincoln would say." The audience roared and clapped, and Lincoln smiled, uttering something to Mary. Behind him, the box door was closed but not locked; in all the excitement of Lincoln's arrival, nobody noticed a small peephole dug out of the door. As the play progressed, guard John Parker left his post in the hallway leading to the state box, and either sat down out in the gallery to watch the play or went outside for a drink.

On stage, the players hammed it up in absurd and melodramatic scenes, and waves of laughter rolled over the audience. As the Lincolns picked up the story, Hawk was a homespun American backwoodsman named Asa Trenchard, and Laura Keene, a stunning woman with high cheekbones and braided auburn curls, was his young English cousin Florence Trenchard. A scheming English matron named Mrs. Mountchessington, convinced that Trenchard was a rich Yankee, was out to snare him as a husband for her daughter Augusta. In the state box, Lincoln tried to relax, tried to get his mind off reconstruction, tried to lose himself in the play's preposterous humor. Mary rested her hand on his knee, called his attention to situations on stage, and clapped happily at the funniest scenes. One of the actresses noticed that Lincoln never clapped, but that he did laugh "heartily" from time to time. At one point he felt a chill, as if a cold wind had blown over him, and he got up long enough to put on his overcoat.

During the third act, Mary slipped her hand into Lincoln's and nestled close to him. "What will Miss Harris think of my hanging on to you so?" she whispered. "She won't think anything about it," Lincoln replied.

On stage, Mrs. Mountchessington finally discovered the shocking truth about Trenchard: he was "church mouse poor" and no catch at all for her daughter. In stiff British rage, she sent Augusta to her room, reproached the American for his ill-mannered impertinence, and flounced haughtily into the wings, leaving Trenchard alone on stage. Behind Lincoln, the door opened and a figure, a man, stepped into the box and aimed a derringer at the back of Lincoln's head, not six inches away. Mary was still sitting close to Lincoln and Rathbone and Miss Harris were looking rapturously at Trenchard. "Don't know the manner of good society, eh?" Trenchard called after Mrs. Mountchessington. "Wal, I guess I know enough to turn you inside out, old gal—you sockdologizing old mantrap." A gunshot rang out in the state box, and Lincoln's arm jerked up convulsively. For a frozen instant nobody moved —Mary and Miss Harris sat rigid in their seats, and the man stood there enveloped in smoke. As Lincoln slumped forward, Mary reached out instinctively and struggled to keep him from falling. She screamed in deranged, incomprehensible terror. The man jumped out of the smoke brandishing a dagger, a wild-looking man in a black felt hat and high boots with spurs. He yelled something and stabbed at Rathbone, gashing his arm open to the bone. Then he leaped from the box, only to catch his spur in a regimental flag and crash to the stage, breaking his left shinbone in the fall.

The audience was stunned, incredulous. Why, it was the actor John Wilkes Booth. Was this part of the play? An improvised scene? Witnesses heard him shout something in defiance—either *"Sic semper tyrannus!"* (Thus be it ever to tyrants) or "The South shall be free!" Then he dragged himself out the back stage door, while Rathbone and Miss Harris both screamed, "Stop that man! Stop that man!" "Won't somebody stop that man?" Miss Harris pleaded. "The President is shot!"

By now the theater was pandemonium: screams, a medley of voices. "Is there a doctor in the house!" someone cried. People were shoving into the aisles and rushing for the exits, with Laura Keene yelling at them from the stage, "For God's sake, have presence of mind and keep your places, and all will be well." In all the commotion, a young army doctor named Charles A. Leale fought his way to the President's box, where a tearful Clara Harris tried to console Mary, who was holding Lincoln in the rocker and weeping hysterically. Leale lay the President on the floor and removed the blood clot from the wound to relieve the pressure on the brain. The bullet had struck him just behind the left ear, tunneled through his brain, and lodged behind his right eye. The President was almost dead, Leale thought. He was paralyzed, his breath shallow, his wrists without a pulse. Quickly Leale reached into his mouth, opened his throat, and applied artificial respiration in a desperate attempt to revive him. By now another doctor had reached the box, and Leale had him raise and lower the President's arms while Leale himself massaged the left breast with both hands. Then he placed his face against Lincoln's and gave him mouth-to-mouth resuscitation, drawing in his own breath and forcing it again and again into the President's lungs. At last he was breathing on his own, his heart beating with an irregular flutter. But he was still unconscious. Leale shook his head, unable to believe what was happening. "His wound is mortal," he said softly; "it is impossible for him to recover."

They carried him out into the mist, across the street to the Petersen House, and down a narrow hallway to the room of a War Department clerk, where they lay him across a four-poster bed. Miss Harris and Rathbone, his arm bleeding profusely, brought Mary over. When she saw Lincoln in the back room, Mary ran to him, fell sobbing to her knees, and called him intimate names and whispered endearments and begged him to speak to her. Taddie! She must fetch Taddie. He would speak to Taddie, he loved him so. The doctors led her to the front parlor, where she broke into convulsive weeping.

By now the word had flashed through Washington, and a procession of government officials came running in the night—Welles, Speed and Fessenden, Stanton, Dana, Halleck and Major Eckert, all crowding into the dim little room where Lincoln lay, a battery of doctors at his side. Robert arrived with John Hay, scarcely hearing the family physician who told them at the doorway that it was no use. When he saw his father lying diagonally across the bed, his brain destroyed, his eye swollen now and discolored, Robert broke down in despair and disbelief. Finally, in shock himself, he joined his mother in the front parlor and tried to comfort her, but Mary was so drugged from grief she hardly knew he was there.

As Robert left the clerk's room, Sumner came in, his face contorted in anguish; the senator took Lincoln's hand and spoke to him, but a doctor said "he can't hear you. He is dead." "No, he isn't dead," Sumner protested. "Look at his face; he is breathing." But all the doctors assured him that Lincoln would never regain consciousness. At that, Sumner clasped Lincoln's hand tightly, bowed his head close to the pillow, and sobbed.

With the President dying and the government at a standstill, Secretary of War Stanton took over. Close to breaking down himself, tears burning his eyes, Stanton set up headquarters in the back parlor, where a federal judge and two other men helped him take down testimony from witnesses, who identified John Wilkes Booth, a man of militant Confederate sympathies, as the President's assassin. At the same time, word came that another assassin had attempted but failed to murder Seward and that Washington was in a reign of terror. At once Stanton put the city under martial law and organized dragnets to bring in Booth and all other suspects. As Stanton came and went, barking out orders, calling for Andrew Johnson, mobilizing troops and the police, Mary lay on a sofa in the front parlor, alternately quiet and overcome with spells of frantic weeping. When she recalled Lincoln's dream of mournful voices and a dead body in the White House, Mary cried miserably, "His dream was prophetic," and she begged God to take her, too, let her go with her husband. Beyond the windows, a crowd gathered in the fog, keeping a constant vigil in front of the Petersen House, asking the leaders of government who came and went if there was any word or any hope.

And so it went through the night, as the hours ticked by and the doctors released half-hourly press bulletins which went out over the telegraphs. When the first gray of dawn spread through the Petersen House, a heavy rain beat against the window panes. Sumner was still holding Lincoln's hand when Mary came one last time to see him. She kissed his face and whispered to him, "Love, live but one moment to speak to me once—to speak to our children." Then she looked at him, looked at his bruised and swollen eye, and realized perhaps for the first time that he was beyond help. "Oh, my God," she wailed as they led her away, "and have I given my husband to die?"

With the end close now, Lincoln's friends and colleagues gathered at his bedside. Stanton and Robert came in from the front rooms; and Robert, giving way to uncontrollable anguish, put his head on Sumner's shoulder. Then many of the others also began to cry. At last, at 7:22 A.M., April 15, Lincoln died. The surgeon general pulled a sheet over his face, and the pastor of the New York Avenue Presbyterian Church uttered a prayer. Stanton, fighting back his tears, muttered that he now belonged to the ages.

Moments later Robert led his mother out into the rain and they took a carriage back to the White House, where a confused and frightened Taddie waited for them. And watching the carriage disappear in the muddy street, men and women, black people and white, stood in the rain and wept. Somewhere a church bell began to toll. And the news went out to a shocked and grieving nation that Abraham Lincoln, sixteenth President of the United States, had been shot and killed in Washington, one of the final casualties of a war that had broken his heart and had now claimed his life, gone to join the other Union dead he himself had so immortalized.

Never had the nation mourned so over a fallen leader. Not only Lincoln's friends, but his legion of critics— those who'd denounced him in life, castigated him as a dictator, ridiculed him as a baboon, damned him as stupid and incompetent—now lamented his death and grieved for their country. As if civil war had not been atonement enough, for the first time in the history of the Republic a President had been assassinated, and the haunting echo of Booth's derringer troubled Americans in all corners of the Union.

On Wednesday, April 19, Lincoln lay in the East Room of the White House, his coffin resting on a flower-covered catafalque—his "temple of death." His head lay on a white pillow, a faint smile frozen on his lips, his face pale and distorted in death. The room was hushed and dim, the adjoining rooms festooned in black crepe. Upstairs, Mary lay imprisoned in her room, almost deranged from grief and hysterical weeping, unable to attend the services below. Elizabeth Keckley and Tad both tried to soothe her. Though stricken himself, Tad would throw his arms around Mary's neck and plead with her: "Don't cry so, Momma! don't cry, or you will make me cry, too! You will break my heart."

Services began around eleven that morning, with some six hundred people crowded into the East Room. Robert, ashen and grave, stood at the foot of the coffin. General Grant, a black mourning crepe on one arm, sat alone at the other end, staring at a cross of lilies. He began to weep, unable to believe or to bear what had happened, contending that it was the saddest day of his life. By now nearly all official Washington was there—President Johnson and the Cabinet, Sumner and his Congressional colleagues, Chase, the diplomatic corps, numerous generals and naval officers, Lincoln's personal cavalry escort and bodyguards, and Nicolay, Hay, and Brooks.

Four ministers spoke and prayed for Lincoln. After that twelve veteran reserve corps sergeants carried his coffin out to the funeral car. Then with minute guns shattering the sunlit day, with bells tolling and bands playing dirges for the dead, the funeral procession started up Pennsylvania Avenue in slow, measured cadences, a detachment of black troops in the lead, then the funeral car followed by a riderless horse, then columns of mourners all moving to the steady muffled beat of drums. The lines swelled with wounded soldiers, who left their hospital beds and marched along now, some hobbling on crutches, torn and bandaged men who followed after their fallen chief. There was a procession of black citizens, walking in lines of forty from curb to curb, four thousand of them in high silk hats and white gloves, holding hands as they moved.

At last the procession reached the Capitol, and the sergeants bore Lincoln into the rotunda, where he lay in state on another catafalque. All the next day, thousands of people filed through the rotunda to pay Lincoln their last respects. From the stairs leading to the dome, Noah Brooks described the scene: "Directly beneath me lay the casket in which the dead President lay at full length, far, far below; and like black atoms moving over a sheet of gray paper, the slow-moving mourners, seen from a perpendicular above them, crept silently in two dark lines across the pavement of the rotunda, forming an ellipse around the coffin and joining as they advanced toward the eastern portal and disappeared."

On April 21 a nine-car funeral train, decorated with Union flags, waited at the depot as a hearse carried Lincoln down from the Capitol. Earlier some men had removed Willie Lincoln's coffin from its Georgetown grave and put it in the funeral car. Now they placed Lincoln with Willie, and the train crawled forward with ringing bells, setting out on a 1,600-mile journey back to Illinois. All along the route, men and women gathered with their children and watched in silence as the train passed against the sky. In New York City, 85,000 people accompanied a funeral cortege through the streets, and a sobbing Walt Whitman would mourn Lincoln's death ("O moody, tearful night") with every returning spring, when lilacs blooming in the dooryard would remind Whitman "of him I love." He would never forget the coffin passing in the street, the pomp of the inlooped flags, the flaming torches and sea of faces, "the dim-lit churches and the shuddering organs," and the tolling, tolling bells. In the statehouse in Albany, people came all through the night to view the open coffin, and a state senator who walked by thought again what he'd decided the first time he met the President—he had the saddest face the senator had ever seen. Later, as the train puffed west for Ohio, newspapers carried reports that Booth had been cornered in Virginia and killed. In Cleveland, the coffin rested in a pagoda in Monument Park, where more than 150,000 pilgrims from northern Ohio, Michigan, and Pennsylvania paid Lincoln homage. On the night run to Indianapolis, bonfires lit up the route, and mute crowds stood in the rain as the train rolled by. In Chicago, thousands of Lincoln's fellow Illinoisians—columns of officials, ordinary citizens, Ellsworth Zouaves, immigrants, carpenters, Chicago actors, Jews, Negroes, and ten thousand schoolchildren wearing black arm-bands—all marched with the coffin in final tribute. From Chicago the train ran south across the prairies, taking Lincoln and Willie home now, home at last to Springfield.

REFERENCE NOTES

Because of the organization adopted in this biography, numbered citations would have been unmanageably and repellingly large. Therefore I abandoned them in favor of a collective reference, identified by page numbers, to each section in the titled parts of the biography. So far as possible, sources are listed in the order I used them in preparing the text of each section.

My references, of course, do not contain a definitive indexing of the vast literature on Lincoln. Nor do they mention all the scores of manuscript collections, published documents, books, and journal articles I consulted but did not specifically use in constructing this volume. My citations list only those materials from which I extracted quotations and factual matter or derived my interpretations. Those concerned with Lincoln historiography should turn to D. E. Fehrenbacher, *The Changing Image of Lincoln in American Historiography* (Oxford, 1968), David M. Potter, *The South and the Sectional Conflict* (Baton Rouge, 1968), 151–176, and Benjamin P. Thomas, *Portrait for Posterity: Lincoln and His Biographers* (New Brunswick, N.J., 1947).

The basic source for any biography of Lincoln is *The Collected Works of Abraham Lincoln* (9 vols., New Brunswick, N.J., 1953–1955), edited by Roy P. Basler, Marion Dolores Pratt, and Lloyd A. Dunlap. Subsequently, Basler edited and published *The Collected Works of Abraham Lincoln—Supplement, 1832–1865* (Westport, Conn., 1974). In my references, I've abbreviated the titles to Lincoln, *CW* and *CWS*.

Among the many manuscript collections germane to the Lincoln story, the most valuable are the Robert Todd Lincoln Collection of the papers of Abraham Lincoln and the Herndon-Weik Collection, both located in the Library of Congress, Washington, D.C., and both containing thousands of manuscript items about Lincoln's life and career. Throughout my notes, I've referred to these sources as the RTL Coll. and the H-W Coll. All other manuscript and published sources are cited without abbreviation.

PART ONE: RIVERS OF TIME
Pages 3–5

Scripps quotations ("chief difficulty I had" and "seemed painfully impressed") in Scripps to Herndon, June 24, 1865, H-W Coll., and in Ward Hill Lamon, *The Life of Abraham Lincoln* (Boston, 1872), 18, and William H. Herndon, *Herndon's Life of Lincoln* (paperback ed., edited by Paul M. Angle, N.Y., 1961), 45–48; Lincoln, *CW,* IV, 60–67, 68. See also Grace L. S. Dyche, "John Locke Scripps, Lincoln's Campaign Biographer," *Journal of the Illinois State Historical Society* (October, 1924), 333–351.

Pages 5–7

Lincoln's recollections and remarks about his family and environment in Lincoln, *CW,* I, 456, III, 511, and IV, 60–62; quotation ("leader of grocery-store dialogue") in Dennis Hanks to Herndon, June 13, 1865, and A. Grigsby to Herndon, September 12, 1865, H-W Coll. For a discussion of the lives and ancestries of Nancy Hanks and Thomas Lincoln, see Betty L. Atkinson, "Some Thoughts on Nancy Hanks Lincoln," *Lincoln Herald* (Fall, 1971), 127–131; Richard N. Current, *The Lincoln Nobody Knows* (N.Y., 1958), 22–31; and Louis A. Warren, *Lincoln's Parentage and Childhood* (N.Y., 1922). Quotations ("tall spider of a boy" and "piece of human flotsam") in Albert J. Beveridge, *Abraham Lincoln, 1809–1858* (2 vols., Boston and N.Y., 1928), I, 30, 37; Louis A. Warren, *Lincoln's Youth: Indiana Years* (Indianapolis, Ind., 1959), 10–14.

Pages 7–9

Lincoln, *CW,* III, 511, IV, 62; quotation ("Sing Much") in Hanks to Herndon, June 13 and December 24, 1865, H-W Coll.; Lincoln quotation ("withered features") in Lincoln, *CW,* I, 117; Warren, *Lincoln's Youth,* 60–70; Current, *Lincoln Nobody Knows,* 24–25, 31; and Beveridge, *Lincoln,* I, 70.

Pages 9–12

Warren, *Lincoln's Youth,* 24–30, 81ff; John Romine to Herndon, September 8, 1865, and Sarah Lincoln to Herndon, September 8, 1865, H-W Coll.; Francis F. Browne, *Every-Day Life of Lincoln* (N.Y. and St. Louis, 1886), 53; Beveridge, *Lincoln,* I, 50–55, 69, 83–84; John Locke Scripps, *Life of Lincoln* (ed. by Roy P. Basler and Lloyd A. Dunlap, Bloomington, Ind., 1961); Lincoln's boyhood poetry, *CW,* I, 1–2, comments on education and Indiana, *ibid.,* III, 511, IV, 62, 235–236, and poem about Matthew Gentry, *ibid.,* I, 384–386.

Pages 12–13

Lincoln's emotional profile: Joseph C. Richardson to Herndon, September 14, 1865, and David Turnham and John Hanks to Herndon [n.d.], H-W Coll.; Warren, *Lincoln's Youth,* 154–155, 173–175; and Beveridge, *Lincoln,* I, 80. Lincoln and his father: Lincoln, *CW,* IV, 61; Dennis Hanks to Herndon, June 13, 1865, and Sarah Lincoln to Herndon, September 8, 1865, H-W Coll.; quotation ("Old Tom couldn't read") in Beveridge, *Lincoln,* I, 78. See also Current, *Lincoln Nobody Knows,* 23, 25, 29–31, and Warren, *Lincoln's Youth,* 84–85, 128.

Pages 14–17

Lincoln, *CW,* IV, 62–64; Warren, *Lincoln's Youth,* 176ff; John J. Duff, *A. Lincoln, Prairie Lawyer* (N.Y., 1960), 5–7; Beveridge, *Lincoln,* I, 77–109; statements of William Butler and William G. Green in David C. Mearns (ed.), *The Lincoln Papers* (2 vols, in 1, reprint ed., N.Y., 1961), 151, 152; Herndon, *Herndon's Lincoln,* 98, 100–101. See also Charles H. Coleman, *Lincoln and Coles County Illinois* (New Brunswick, N.J., 1955); Ida Tarbell, *In the Footsteps of the Lincolns* (N.Y., 1924), 162.

Pages 17–19

Benjamin P. Thomas, *Lincoln's New Salem* (revised ed., Chicago, 1966), 5–67, 121–126; quotation ("tyrant of spirits") in Lincoln, *CW,* I, 272–279; Green's statement in Mearns, *Lincoln Papers,* 80; Beveridge, *Lincoln,* I, 82–83, 534. For Lincoln and Ann Rutledge, consult J. G. Randall, *Lincoln the President* (paperback ed., N.Y., 1945), II, 321–342; David Donald, *Lincoln's Herndon* (N.Y., 1948), 218–241; and editors' note on Rutledge in Lincoln, *CW,* IV, 104.

Pages 20–22

Lincoln, *CW,* I, 5–9, IV, 62; quotation ("something knotty") in Green's statement in Mearns, *Lincoln Papers,* 153; Thomas, *Lincoln's New Salem,* 68–71; Paul Simon, *Lincoln's Preparation for Greatness* (Norman, Okla., 1965), 4–6; Reinhard H. Luthin, *The Real Abraham Lincoln* (Englewood Cliffs, N.J., 1960), 22–24.

Pages 22–23

Lincoln, *CW,* I, 509–510, IV, 64; quotation ("The red light of the morning sun") in Beveridge, *Lincoln,* I, 123.

Pages 24–27

Lincoln quotations ("deeper and deeper in debt" and "politics an objection") in Lincoln, *CW,* IV, 64–65, also I, 16–20n; Thomas, *Lincoln's New Salem,* 88–118; Zarel Cratic Spears and Robert S. Barton, *Berry and Lincoln, Frontier Merchants* (N.Y., 1947); William E. Baringer (ed.), *Lincoln Day by Day* (editor in chief Earl Schenck Miers, Washington, 1960), I, 30; Green's statement in Mearns, *Lincoln Papers,* 153; Beveridge, *Lincoln,* I, 127–132; Duff, *Prairie Lawyer,* 23–27; Herndon, *Herndon's Lincoln,* 122–124.

Pages 27–29

William E. Baringer, *Lincoln's Vandalia* (New Brunswick, N.J., 1949), 3–62; Simon, *Preparation for Greatness,* 5, 22–31; Beveridge, *Lincoln,* I, 133–140, 163–169; Duff, *Prairie Lawyer,* 30–32; Herndon, *Herndon's Lincoln,* 122–124; William J. Wolf, *The Almost Chosen People: A Study of the Religion of Abraham Lincoln* (Garden City, N.Y.,

1959), 40ff; Elton Trueblood, *Abraham Lincoln, Theologian of American Anguish* (N.Y., 1973), 14–16; Lincoln quotations ("fall of fury," "doctrine of necessity," and "intensity of death") in Lincoln, *CW*, I, 265, 279, 382; Simon, *Preparation for Greatness*, 225–247; Thomas, *Lincoln's New Salem*, 53, 69–70, 121–126, and Thomas, *Abraham Lincoln* (N.Y., 1952), 42–43.

Pages 30–33

Lincoln quotations ("legitimate object of government" and "sheet anchor") in Lincoln, *CW*, II, 220–221, 266; Lincoln's 1836 platform and speeches, *ibid.*, I, 48, 61–69, 108–115, 159–179, 196; Lincoln's retrospective comments on political beliefs, *ibid.*, IV, 168–169, 235–236, 240; quotation ("liar and scoundrel"), *ibid.*, VIII, 429; Lincoln on Mary Owens, *ibid.*, I, 117; Mary Owens Vineyard to Herndon, July 22, 1866, H-W Coll.; Beveridge, *Lincoln*, I, 154–155; Thomas, *Lincoln's New Salem*, 126–127. For Lincoln's economic thought, consult G. S. Borit, "Lincoln and the Economics of the American Dream: the Whig Years, 1832–1854" (Ph.D. dissertation, Boston University, 1968).

Pages 33–35

Lincoln, *CW*, I, 54–55, 65–66. Simon, *Preparation for Greatness*, 76–105, drawing on abundant documentary evidence, dispels as myth the popular belief that the Long Nine voted as a unit in 1836–1837 and logrolled extensively to get Springfield chosen as capital.

Pages 35–40

Lincoln's "Protest" and quotations ("persuasion" and "gallon of gall") in Lincoln, *CW*, I, 74–75, 272–273, and ("briefly defined his position"), *ibid.*, IV, 65; *Illinois House Journal, 1836–1837*, 134, 241–244, 309–311, 824; Simon, *Preparation for Greatness*, 122–134; Beveridge, *Lincoln*, I, 188–195. In his addresses in the mid and late 1850s, Lincoln sometimes spoke of views he'd long held regarding slavery; see, for example, Lincoln, *CW*, II, 230–283, 492. For racial tension and descrimination in the North, consult Leon F. Litwack, *North of Slavery: the Negro in the Free States, 1790–1860* (Chicago and London, 1961); Eugene H. Berwanger, *The Frontier Against Slavery* (Urbana, Ill., 1967), 7–59; and Leonard L. Richards, *"Gentlemen of Property and Standing": Anti-Abolition Mobs in Jacksonian America* (N.Y., 1970).

PART TWO: WHY SHOULD THE SPIRIT OF MAN BE PROUD?
Pages 43–45

Lincoln's remark about Springfield and letters to Mary Owens and Mrs. Browning in Lincoln, *CW*, I, 78–79, 94–95, 117–119; Mary Owens Vineyard to Herndon, May 22 and June 22, 1866, H-W Coll.; Beveridge, *Lincoln*, I, 154–158; R. Gerald McMurtry, *Lincoln's Other Mary* (Chicago, 1946); Paul M. Angle, *"Here I Have Lived": A History of Lincoln's Springfield* (new ed., Chicago and Lincoln's New Salem, 1971), 21, 83–108.

Pages 45–47

Duff, *Prairie Lawyer,* 41–46, 51–61; quotation ("never saw so gloomy a face") in Herndon, *Herndon's Lincoln,* 170–171, and Speed to Herndon, December 6, 1866, H-W Coll.; quotation ("by nature a literary artist") in Beveridge, *Lincoln,* I, 302, also 209, 298–301; Richards, *Gentlemen of Property and Standing,* 16–17; Lincoln, *CW,* I, 108–115.

Pages 48–49

Beveridge, *Lincoln,* I, 212–218; Angle, *Here I Have Lived,* 64–71; Simon, *Preparation for Greatness,* 74, 107–108, 171; Robert S. Harper, *Lincoln and the Press,* 2ff; *Sangamo Journal,* June 24, July 8, 1837; Lincoln's letters and documents in the Adams controversy, *CW,* I, 19–93, 95–106; Lincoln on Douglas, *ibid.,* 107; quotation ("commencement") in Duff, *Prairie Lawyer,* 68. My sketch of Douglas is based on Robert W. Johannsen, *Stephen A. Douglas* (N.Y., 1973), chaps. I–IV and *passim.*

Pages 49–52

Lincoln's speeches, remarks, and resolutions, 1838 and 1839, *CW,* I, 122ff; quotation ("not worth a damn"), *ibid.,* 143; Lincoln's campaign documents of 1840, *ibid.,* 180–181, 201–205; *Sangamo Journal,* December 20, 1839, January 3, April 10 and 14, May 8, July 17, August 2, September 18, 1840; *Illinois State Register,* November 23, 1839; Simon, *Preparation for Greatness,* 135–139, 211–217; Baringer, *Lincoln's Vandalia,* 111ff; Beveridge, *Lincoln,* I, 232–274; Emanuel Hertz, *The Hidden Lincoln: From the Letters and Papers of William H. Herndon* (N.Y., 1938), 435–436; Willard L. King, *Lincoln's Manager: David Davis* (Cambridge, Mass., 1960), 38; Johannsen, *Douglas,* 75–82; quotation ("He stumped . . . the state") in Luthin, *Lincoln,* 55.

Pages 52–56

Beveridge, *Lincoln,* I, 304–310, and *Lincoln Lore,* November 30, 1931. Mary's life and background and courtship with Lincoln: Justin G. Turner and Linda Levitt Turner, *Mary Todd Lincoln: Her Life and Letters* (N.Y., 1972), 3–28; Ruth Painter Randall, *Mary Lincoln, Biography of a Marriage* (Boston, 1953), chaps. I–IV; Mary's comment on Stephen A. Douglas in *ibid.,* 10; quotation ("very creature of excitement") in Conkling to Levering, September 21, 1840, Illinois State Historical Library; quotations ("could not hold a conversation with a lady" and "glitter and pomp and power") in Elizabeth Edwards's first statement [n.d.] and second statement, September 27, 1887, H-W Coll.; Todd family opposition to Lincoln in Katherine Helm, *The True Story of Mary, Wife of Lincoln* (N.Y., 1928), 82, and Frank E. Stevens, *A Reporter's Lincoln* (St. Louis, 1916), 75–76; quotation ("crime of matrimony") in Turners, *Mary Lincoln,* 19–22; Lincoln's remark about Todd and God cited in Luthin, *Lincoln,* 90, also 85–89; Carl Sandburg and Paul M. Angle, *Mary Lincoln, Wife and Widow* (N.Y., 1932), 329–350.

Pages 56–58

Simon, *Preparation for Greatness,* 225–247; Lincoln, *CW,* I, 213–214, 215–216. The "fatal first" and Lincoln's depression: Lincoln, *ibid.,* 228, 229–230; Browning's remarks in John G. Nicolay and John Hay, *Abraham Lincoln, A History* (10 vols., N.Y., 1890), I, 187; Speed's remarks in Beveridge, *Lincoln,* I, 314–315, and Speed to Herndon, January 6, 1866, H-W Coll.; Randall's analysis in *Mary Lincoln,* 36–51; Conkling's comment on Lincoln's emaciated appearance in Angle, *Here I Have Lived,* 95–96; quotation ("two duck fits") in Martin McKee to John Hardin, January 22, 1841, Illinois State Historical Library; quotations ("crazy for a week" and "crazy as a loon") in Beveridge, *Lincoln,* I, 315, and Elizabeth Edwards's first statement [n.d.] in H-W Coll.; Turners, *Mary Lincoln,* 24–25; Lincoln to Stuart in Lincoln, *CWS,* 6. Lincoln signed a Whig protest against the judiciary bill and served on a Whig committee which drafted a circular denouncing it; see *CW,* I, 234–237, 244–249.

Pages 58–61

Quotation ("deems me unworthy of notice") in Turners, *Mary Lincoln,* 26–27; quotation ("Poor A") in Levering to Conkling, February 7, 1841, photostat in Illinois State Historical Library; Conkling's remarks about Lincoln in Angle, *Here I Have Lived,* 96, and in Sandburg and Angle, *Mary Lincoln,* 180–181; Duff, *Prairie Lawyer,* 76–95, 231–232; Donald, *Lincoln's Herndon,* 18; John P. Frank, *Lincoln as a Lawyer* (Urbana, Ill., 1961), 12–14; quotation ("a pretty good lawyer") in "Stephen T. Logan Talks About Lincoln," *Bulletin of the Abraham Lincoln Association* (June 1, 1928), 3, 5; Beveridge, *Lincoln,* I, 318–322; Woldman, *Lawyer Lincoln,* 37–38. Lincoln and Speed: Lincoln's visit with Speed in Lincoln, *CW,* I, 259–261; Lincoln's description of the slave coffle on the Ohio River in *ibid.,* 260, and II, 320; and Lincoln's remarkably revealing letters to Speed in *ibid.,* I, 265–266, 267–268, 269–270, 280–281, 282, 288–289. See also Robert Lee Kincaid, *Joshua Fry Speed* (Harrogate, Tenn., 1943).

Pages 61–63

The Shields affair: Lincoln, *CW,* I, 291–297, 299–303. Benjamin P. Thomas, *Abraham Lincoln,* 83, suggests that Lincoln couldn't have been serious about his dueling instructions; and Beveridge, *Lincoln,* I, 353, calls the Shields episode "the most lurid personal incident" in Lincoln's life. See also Harry E. Pratt, *Concerning Mr. Lincoln* (Springfield, Ill., 1944), 18. Lincoln's marriage: Lincoln to Speed, *CW,* I, 269, 303; Lincoln quotation ("a matter of profound wonder") in *ibid.,* 305; quotation ("To hell, I reckon") in Beveridge, *Lincoln,* I, 355; Randall, *Mary Lincoln,* 65–74; and Turners, *Mary Lincoln,* 29–31.

Pages 63–66

Lincoln, *CW,* I, 323–325, 328, 389–391. My profile of the Lincoln marriage derives from Turners, *Mary Lincoln,* 30–39, 68–69, 114, and *passim;* Harriet A. Chapman to Herndon, November 21, 1866, and James Gourley's statement [n.d.] in H-W Coll.; Beveridge, *Lincoln,* I, 502–508; Randall, *Mary Lincoln,* chaps. VII–XIII, XXX;

Mary's remarks about her "tall Kentuckian" in Helm, *Mary, Wife of Lincoln,* 140, and Turners, *Mary Lincoln,* 52, and how Lincoln was everything to her in Mary to Sally Orne, December 16, 1869, *ibid.,* 534. See also David Donald, *Lincoln Reconsidered* (paperback ed., N.Y., 1956), 49–56.

Pages 67–70

Lincoln quotations ("madness" and "house divided") in Lincoln, *CW,* I, 314, 315; quotations ("candidate of pride" and "belonged to no church") *ibid.,* 320; also *ibid.,* 322–323, 324–325; Donald W. Riddle, *Congressman Abraham Lincoln* (Urbana, Ill., 1947), 4–5; Beveridge, *Lincoln,* I, 357–368. Lincoln's Indiana visit: Lincoln, *CW,* I, 377–379, 384–386; and quotation ("If the fruit of electing Mr. Clay") *ibid.,* 347–348.

Pages 70–71

Lincoln's verse and correspondence about it in Lincoln, *CW,* I, 366–370, 378–379, 384–386, and II, 90. See also David J. Harkness and R. Gerald McMurtry, *Lincoln's Favorite Poets* (Knoxville, Tenn., 1959).

Pages 71–75

My account of the Lincoln-Herndon partnership draws from Donald, *Lincoln's Herndon,* 6–49 and *passim;* Duff, *Prairie Lawyer,* 94–117; Herndon, *Herndon's Lincoln,* 261–293; Lincoln, *CW,* VIII, 424; Herndon to "Friend Weik," October 21 and November 19, 1885, in Hertz, *Hidden Lincoln,* 95, 105; Randall, *Mary Lincoln,* 116–117; Turners, *Mary Lincoln,* 33–34.

Pages 75–77

Lincoln quotations ("fair play," "moccasan track," and "utter injustice" and "yet think better") in Lincoln, *CW,* I, 350, 352–353, 360–366, and Lincoln, *CWS,* 9; Riddle, *Congressman Lincoln,* 5–6; Beveridge, *Lincoln,* I, 372–383; Lincoln, *CW,* I, 382–384, 391.

Pages 77–78

Riddle, *Congressman Lincoln,* 8–12; William H. Townsend, *Lincoln and His Wife's Home Town* (Indianapolis, 1929), 140–155; Baringer, *Lincoln Day by Day,* I, 295.

Pages 78–84

Turners, *Mary Lincoln,* 35; quotation ("attending to business") in Lincoln, *CW,* I, 465; Beveridge, *Lincoln,* I, 404–430; Frank, *Lincoln as Lawyer,* 105–110; Riddle, *Congressman Lincoln,* 14–161; Lincoln on Alexander Stephens in Lincoln, *CW,* I, 448; Lincoln's "spot" resolutions and Mexican War speech, *ibid.,* 420–422, 431–442; also *ibid.,* IV, 66. Reactions to his Mexican War stand: *Illinois State Register,* February 25, March 10, June 13, 1848; Lincoln quotations ("tell the truth or tell a lie") in Lincoln, *CW,* I, 446–447, 451, 457–458; Lincoln to Mary, *ibid.,* 465–466. To retain party harmony and abide by the rotation principle, Lincoln had decided not to run for re-election before his Mexican War speech. So the adverse reaction to it in Illinois did not ruin his Congressional career, as some have contended (see Lincoln, *ibid.,* 430–431, IV, 66–67). Lincoln and the 1848 election: Lincoln quotations on Clay and the election in Lincoln, *ibid.,* I, 463, 468, 476–477, 478; quotation ("every one play the part he can play best") *ibid.,* 491; Lincoln's speech on Taylor, *ibid.,* 501–516; Lincoln's New England speeches, *ibid.,* II, 1ff; Lincoln quotation ("We have got to deal with this slavery question") in Frederick W. Seward, *William H. Seward* (3 vols., N.Y., 1891), II, 80; Lincoln on Logan's loss, *CW,* I, 518–519.

Pages 84–87

Riddle, *Congressman Lincoln,* 162–180; Johannsen, *Douglas,* 206–261; Lincoln quotation on Georgia pen in *CW,* II, 237–238, 253; Lincoln's projected District of Columbia emancipation bill, *ibid.,* II, 20–22, 22n; Calhoun's "Address" in Beveridge, *Lincoln,* I, 481–486; Allan Nevins, *Ordeal of the Union* (2 vols., N.Y., 1947), I, 225–227.

Pages 87–89

Lincoln, *CW,* II, 25, 31ff; Lincoln quotations on the Land Office fight in *ibid.,* 41, 46, 49, 58, 67, 79; also *ibid.,* 61, 65; Riddle, *Congressman Lincoln,* 181–235; Turners, *Mary Lincoln,* 39–40; Lincoln quotation ("my great devotion") in Lincoln, *CW,* II, 91–92.

PART THREE: ON THE PILGRIMAGE ROAD
Pages 93–95

Lincoln quotation ("Upon his return") in Lincoln, *CW,* IV, 67; Turners, *Mary Lincoln,* 40–41; Ruth Painter Randall, *Lincoln's Sons* (Boston, 1955), 28–32; Lincoln's eulogy on Taylor, *CW,* II, 90.

Pages 95–96

Randall, *Lincoln's Sons,* 8; Lincoln on his father in *CW,* II, 96–97, 111–113.

Pages 96–97

Turners, *Mary Lincoln,* 41–42; quotation ("her little sunshine") in Randall, *Lincoln's Sons,* 8, also 33, 41, 49–53, 68; statements of Gourley and Mrs. Wallace in Beveridge, *Lincoln,* I, 505; quotation ("My Father's life") in Mearns, *Lincoln Papers,* 94, also 5–7; quotation ("Bob is little, proud") in Hertz, *Hidden Lincoln,* 261.

Pages 97–105

Profile of Lincoln as a self-made man and lawyer: quotation ("Mr. Lincoln dressed") in Paul M. Angle (ed.), *Abraham Lincoln By Some Men Who Knew Him* (reprint ed., Freeport, N.Y., 1969), 50, also 23, 48, 109–110; Current, *Lincoln Nobody Knows,* 1–21; Pratt, *Concerning Mr. Lincoln,* 18–19; Harry E. Pratt, *The Personal Finances of Abraham Lincoln* (Springfield, Ill., 1943), 26, 122, 71–82, 123, 164, and *passim*

(Frank, *Lincoln as Lawyer,* 40 and 40n, contends that Lincoln's income from 1855 to 1860 may have been higher than most authorities have assumed); Duff, *Prairie Lawyer, passim.;* quotations ("tell us about the books" and "preferring the Germans") in Donald, *Lincoln's Herndon,* 124–125, 128–129, also 32–38. Lincoln's reticence and melancholy: Herndon's remarks in Hertz, *Hidden Lincoln,* 110–111, 121, 159, 204; Henry C. Whitney, *Life on the Circuit with Lincoln* (Boston, 1892), 139, 171, and *passim;* Lamon, *Life of Lincoln,* 466–504; Current, *Lincoln Nobody Knows,* 11–13; J. G. Holland, *Life of Abraham Lincoln* (Springfield, Mass., 1866), 241; Louis A. Warren, "Abraham Lincoln—A Melancholy Man," *Lincoln Lore,* September 24, 1934. Lincoln's humor: quotation ("to whistle down sadness") in David Davis's statement in Whitney, *Life on the Circuit,* 171; joke about the traveler in Carl Sandburg, *Abraham Lincoln: the War Years* (4 vols., N.Y., 1939), III, 312; "Bass-Ackwards" in Lincoln, *CW,* VIII, 420.

My account of the range and nature of Lincoln's legal work draws from Duff, *Prairie Lawyer,* 118ff; King, *Lincoln's Manager,* 71–95; Frank, *Lincoln as Lawyer,* 51ff; quotation ("sickly sentamentalism") *Illinois Citizen,* May 29, 1850; Usher F. Linder, *Reminiscences of the Early Bench and Bar of Illinois* (Chicago, 1879), 183; Whitney, *Life on the Circuit, passim;* quotation on Dubois in Lincoln, *CW,* VIII, 422; Lincoln's denial that he lobbied for the Illinois Central in *ibid.,* III, 244; Stanton's behavior in the McCormick Reaper case in Benjamin P. Thomas and Harold M. Hyman, *Stanton: the Life and Times of Lincoln's Secretary of War* (N.Y., 1962), 63–66; Herndon to Weik, January 6, 1887, H-W Coll.; Lincoln's notes for a law lecture in *CW,* II, 81–82.

Pages 105–108

Lincoln, *CW,* II, 121–132. His reactions to the Kansas-Nebraska Act in *ibid.,* II, 282, IV, 67.

PART FOUR: REVOLT AGAINST THE FATHERS
Pages 111–120

The Kansas-Nebraska Act and free-soil response: Johannsen, *Douglas,* 374–464; Stephen B. Oates, "The Little Giant Reconsidered," *Reviews in American History* (December, 1973), 534–540; Nevins, *Ordeal of the Union,* II, 122–159; Don E. Fehrenbacher, *Prelude to Greatness: Lincoln in the 1850s* (Stanford, Calif., 1962), 23–39; Eric Foner, *Free Soil, Free Labor, Free Men: The Ideology of the Republican Party Before the Civil War* (N.Y., 1970), 73–102; Hans L. Trefousse, *The Radical Republicans: Lincoln's Vanguard for Racial Justice* (N.Y., 1969), 66–95. Lincoln in the anti-Nebraska movement: Lincoln, *CW,* II, 226–239, IV, 67; Lincoln's "Peoria Speech" in *ibid.,* II, 247–283; Lincoln, *ibid.,* 288; Beveridge, *Lincoln,* II, 242–287; Mark M. Krug, *Lyman Trumbull, Conservative Radical* (N.Y., 1965), 87ff; Donald, *Lincoln's Herndon,* 76–77. Lincoln's bid for the Senate in 1854–1855: Lincoln, *CW,* II, 286ff, and Lincoln, *CWS,* 25–26; Elihu Washburne to Lincoln, November 14, 1854, and Lamon to Lincoln, November 21, 1854, RTL Coll.; King, *Lincoln's Manager,* 107; Lincoln quotation ("great consolation") in Lincoln, *CW*, II, 306; Turners, *Mary Lincoln*, 43–44, 274; Whitney's remarks about Lincoln in Beveridge, *Lincoln*, II, 287.

Pages 120–122

Lincoln quotation ("no objection to fuse with anybody") in Lincoln, *CW*, II, 316–317; quotation ("Can we, as a nation, continue . . . half slave") in *ibid.*, 318; Lincoln to Speed, *ibid.*, 320–323. See also Fehrenbacher, *Prelude to Greatness*, 40–43, and David M. Potter and Don E. Fehrenbacher, *The Impending Crisis, 1848–1861* (N.Y., 1976), 225–260.

Pages 122–123

Lincoln quotation ("physical rebellions & bloody resistances") in Donald, *Lincoln's Herndon*, 82; quotation ("resist the laws of Kansas") in Herndon, *Herndon's Lincoln*, 309; Lincoln, *CW*, IV, 131.

Pages 123–125

Fehrenbacher, *Prelude to Greatness*, 44–46, 86; Donald, *Lincoln's Herndon*, 83–88; Lincoln, *CW*, II, 273–274, 333; Beveridge, *Lincoln*, II, 362–365; Lincoln quotation ("man couldn't think, dream") in Lincoln, *CW*, II, 340–341. The Bloomington convention: Lincoln, *CW*, II, 341; King, *Lincoln's Manager*, 112; *Illinois State Journal*, June 3, 1856; Baringer, *Lincoln Day by Day*, II, 169–170. Specialists have discredited Whitney's version of Lincoln's "Lost Speech" in Whitney, *Life on the Circuit* (see Randall, *Lincoln the President*, I, 99).

Pages 125–127

Donald, *Lincoln's Herndon*, 93, 128–129; Herndon to Trumbull, February 15, 1856, Lyman Trumbull Papers, Library of Congress; Beveridge, *Lincoln*, II, 436–439. Lincoln's two fragments on slavery and fragment on sectionalism in *CW*, II, 222–223, 353. The editors of the *CW* hesitantly assigned July, 1854, as the date the fragments on slavery were written, conceding that they could easily belong to a later period. In my judgment, they fit the logic of events in 1856–1858, not those in 1854, and so I elected to use them in the later period.

Pages 127–130

Lincoln, *CW*, II, 342–343, 379; Nevins, *Ordeal of the Union*, II, 487–514; Avery Craven, *The Coming of the Civil War* (2nd ed., Chicago, 1957), 365–381; Herbert Aptheker, *American Negro Slave Revolts* (paperback ed., N.Y., 1969), 84–85, 111; Lincoln's speeches and correspondence in the 1856 campaign, *CW*, II, 347, 356, 358, 361–366, 368ff; racist reactions to Lincoln in *Illinois State Register*, August 19 and September 4, 1856, and Luthin, *Lincoln*, 186; quotation ("weak woman's heart") in Turners, *Mary Lincoln*, 44–46; Lincoln quotation ("we were constantly charged") in Lincoln, *CW*, II, 390–391; quotation ("all countries and colors") and Lincoln's fragment on Douglas, *ibid.*, 382–383.

Pages 130–135

Turners, *Mary Lincoln,* 48–49; Donald, *Lincoln's Herndon,* 100; Taney quotation in Potter and Fehrenbacher, *Impending Crisis,* 275; Johannsen, *Douglas,* 567–573; Lincoln's "Notes for a speech at Chicago" and his Springfield speech, *CW,* II, 390–391, 398–410; *Illinois State Register,* June 30, 1857.

Pages 135–137

Beveridge, *Lincoln,* II, 521–525; quotation ("next husband shall be rich") in Turners, *Mary Lincoln,* 50; Baringer, *Lincoln Day by Day,* II, 198–199; Pratt, *Lincoln's Finances,* 53. *Effie Afton* Case: Lincoln, *CW,* II, 415–422; Duff, *Prairie Lawyer,* 332–343; Frank, *Lincoln as Lawyer,* 85–87.

Pages 137–141

Douglas and Lecompton: Trefousse, *Radical Republicans,* 112–118; Johannsen, *Douglas,* 590–634, *passim;* Oates, "Little Giant Reconsidered," *Reviews in American History* (December, 1973), 535–540; Fehrenbacher, *Prelude to Greatness,* 53–61. Lincoln and Republican flirtation with Douglas: quotation ("What does the New York Tribune mean") in Lincoln, *CW,* II, 430, also 427; Trumbull to Lincoln, January 3, 1858, RTL Coll.; Krug, *Trumbull,* 137–138; Lincoln quotation ("sentiment in favor of white slavery") in Lincoln, *CW,* II, 341, also III, 53–54; Lincoln's fragment on "A House Divided," *ibid.,* II, 448–454; Beveridge, *Lincoln,* II, 494–495, 562–564; quotation ("unanimous for Lincoln") in Horace White, *Life of Lyman Trumbull* (Boston, 1913), 87; Herndon to Trumbull, April 12 and 24, 1858, Trumbull Papers, Library of Congress; *Chicago Tribune,* June 14, 1858; Fehrenbacher, *Prelude to Greatness,* 61–64, 90–93; Lincoln quotation ("if we do not win") in Lincoln, *CW,* II, 443.

Pages 141–142

My account of the Armstrong murder case is from Duff, *Prairie Lawyer,* 351–359.

Pages 142–145

Lamon, *Life of Lincoln,* 398; Baringer, *Lincoln Day by Day,* II, 218; quotation ("By —God, deliver it") in Donald, *Lincoln's Herndon,* 118; Lincoln's "House Divided" speech in *CW,* II, 461–469; quotation (" 'old gentleman' Greeley's notice") in Herndon to Trumbull, June 24, 1858, Trumbull Papers, Library of Congress; Beveridge, *Lincoln,* II, 585; quotation ("go ahead and fight") in Greeley to Medill, July 24, 1858, RTL Coll.; Douglas quotation on Lincoln in John W. Forney, *Anecdotes of Public Men* (2 vols., N.Y., 1877–1881), II, 179. See also Fehrenbacher, *Prelude to Greatness,* 72–94.

PART FIVE: YEARS OF METEORS
Pages 149–152

Lincoln quotations ("indifference to the honors" and "battle upon principle") in Lincoln, *CW*, II, 482, 506. Chicago: Paul M. Angle (ed.), *Created Equal? The Complete Lincoln-Douglas Debates of 1858* (Chicago and London, 1958), 12–25; Lincoln's Chicago speech, *CW*, II, 484–501; *New York Times*, July 12, 1858; Fehrenbacher, *Prelude to Greatness*, 112–114; Donald, *Lincoln's Herndon*, 123; quotation ("this thing is settled") in Beveridge, *Lincoln*, II, 590; quotations ("new combinations" and "never sell old friends") in Lincoln, *CWS*, 29–30; Lincoln, *CW*, II, 457, 472; Johannsen, *Douglas*, 641–644. The debates: *ibid.*, 662–665; Lincoln, *CW*, II, 522, 527n, 528–530, 531; Douglas to Lincoln, July 24, 1858, RTL Coll.; Fehrenbacher, *Prelude to Greatness*, 99–112.

Pages 152–160

Lincoln on Douglas's strategy in Lincoln, *CW*, II, 530; Mary's remark about Lincoln and the Presidency in Henry Villard, *Lincoln on the Eve of '61* (ed. by Harold G. and Oswald Garrison Villard, N.Y., 1941), 6. The Ottawa debate: Lincoln, *CW*, III, 1–37; Angle, *Created Equal*, 102–137; Johannsen, *Douglas*, 660; Carl Schurz, *Reminiscences* (3 vols., N.Y., 1907–1908), II, 89–95, with a description of Lincoln's voice and speaking manner. My discussion of the general nature of the campaign draws from William E. Baringer, *Lincoln's Rise to Power* (Boston, 1937), 23ff; *Chicago Times*, August 22 and September 2, 1858; Herbert Mitgang (ed.), *Abraham Lincoln, A Press Portrait* (new ed., Chicago, 1971), 92–125; and the sources cited above. See also Harry V. Jaffa, *Crisis of the House Divided* (Garden City, N.Y., 1959), and Saul Sigelschiffer, *The American Conscience: the Drama of the Lincoln-Douglas Debates* (N.Y., 1973).

Lincoln confers with Medill and others: Lincoln, *CWS*, 32–33; Whitney to Lincoln, August 26, 1858, and Medill to Lincoln, August 27, 1858, RTL Coll.; Fehrenbacher, *Prelude to Greatness*, 125–127. The Freeport debate: Lincoln, *CW*, III, 38–76; Angle, *Created Equal*, 138–176; *Illinois State Journal*, August 30, 1858; Fehrenbacher, *Prelude to Greatness*, 137, 142; Johannsen, *Douglas*, 670–671. Lincoln's complaint about Douglas in *CW*, III, 83; speeches of Lincoln and Douglas at Jonesboro, *ibid.*, 102–144; Baringer, *Lincoln's Rise to Power*, 28–29; speeches of Lincoln and Douglas at Charleston in Lincoln, *CW*, III, 145–201; Angle, *Created Equal*, 232–234; Baringer, *Lincoln's Rise to Power*, 30–31; Lincoln's fragments on Negroes and on the Southern conspiracy in Lincoln, *CW*, III, 204–205; speeches of Lincoln and Douglas at Galesburg, Quincy, and Alton, *ibid.*, 207–325; Edwin E. Sparks (ed.), *The Lincoln-Douglas Debates of 1858* (Springfield, Ill., 1908), 333–496; Angle, *Created Equal*, 285–402; Baringer, *Lincoln's Rise to Power*, 37ff. The election and the aftermath: Fehrenbacher, *Prelude to Greatness*, 115–119; David Davis to Lincoln, November 7, 1858, RTL Coll.; *Chicago Tribune*, November 10, 1858; Lincoln quotation ("glad I made the late race") in Lincoln, *CW*, III, 339; quotation ("too badly hurt to laugh") in Angle, *Lincoln*, 50.

Pages 161–167

Lincoln quotations ("on expenses," "fight must go on," "we shall beat them") in Lincoln, *CW*, III, 337, 339, 340, 341, 346, IV, 121. Presidential possibilities: Baringer, *Lincoln's Rise to Power*, 48–88; scene with Fell from Fell's statement in O. H. Oldroyd (ed.), *The Lincoln Memorial* (Springfield, Ill., 1890), 473–476; Paul M. Angle, *Lincoln, 1854–1861* (Springfield, 1933), 260–261; Lincoln quotation ("not fit to be President") in Lincoln, *CW*, III, 377; *ibid.*, 395, 505; quotation ("claiming no greater exemption") in *ibid.*, IV, 43. Lincoln's promotional campaign: *ibid.*, III, 368–369, 375–376, 380, 387–388, 390–391. Lincoln and Douglas: *ibid.*, 345, 394–395, 397–398; Douglas's *Harper's* article in Johannsen, *Douglas*, 670–671, 680–698, 706–712. Lincoln's Ohio tour: Turners, *Mary Lincoln*, 58–59; Lincoln's Columbus and Cincinnati speeches, *CW*, III, 400–425, 438–463; Baringer, *Lincoln's Rise to Power*, 96–108. Lincoln's defense of the free-labor system: Lincoln, *CWS*, 43–45, and *CW*, III, 462–463, 477–481. For an analysis of Republican ideology, see Foner, *Free Soil, Free Labor, Free Men*, especially 11–32, 301–317.

Pages 167–171

John Brown: *Chicago Press & Tribune*, October 19–21, December 3, 1859; Baringer, *Lincoln's Rise to Power*, 113–114; Stephen B. Oates, *To Purge This Land with Blood: A Biography of John Brown* (N.Y., 1970), 310–324, 353; Lincoln's remarks about Brown in Lincoln, *CW*, III, 496–497, 498–502, 503. Lincoln's preconvention moves: Fehrenbacher, *Prelude to Greatness*, 145, 194; Lincoln, *CW*, III, 505; Lincoln quotation ("there is not much of it") in *ibid.*, 511–512; Baringer, *Lincoln's Rise to Power*, 131–132, 141–148; Orville H. Browning, *Diary* (ed. by Theodore Calvin Pease and James G. Randall, 2 vols., Springfield, Ill., 1927–1933), I, 395, 407, 409; Donald, *Lincoln's Herndon*, 131–137; Lincoln quotation ("your end of the vineyard") in Lincoln, *CW*, III, 517; *Chicago Press & Tribune*, February 16 and 24, 1860; Judd to Lincoln, February 27, 1860, RTL Coll. For detailed accounts of the factional feuds among Illinois Republicans, see Fehrenbacher, *Prelude to Greatness*, 148–151, and King, *Lincoln's Manager*, 127–133.

Pages 171–176

Cooper Institute: Lincoln's speech, CW, III, 522–550, 522n; *Illinois State Journal*, March 3 and 8, 1860; Baringer, *Lincoln's Rise to Power*, 150–164. New England tour: Lincoln, *CW*, III, 550–55., IV, 2–30, and Lincoln to Mary, *CWS*, 49–50; Randall, *Lincoln's Sons*, 67–68; Baringer, *Lincoln Day by Day*, II, 274–275; John Niven, *Gideon Welles, Lincoln's Secretary of the Navy* (N.Y., 1973), 288–289; Fehrenbacher, *Prelude to Greatness*, 151–154. Presidential maneuvering and Decatur convention: Lincoln quotations ("taste in my mouth," "give no offense," and "write no letters") in Lincoln, *CW*, IV, 34, 36, 45–46, 47; Lamon, *Life of Lincoln*, 444–446; *Illinois State Journal*, May 11, 1860; Lincoln's remarks at Decatur, *CW*, IV, 48–49; Baringer, *Lincoln's Rise to Power*, 170–187.

Pages 176-179

My profile of Seward derives from Glyndon G. Van Deusen, *William Henry Seward* (N.Y., 1967), 162-219; David M. Potter, *Lincoln and His Party in the Secession Crisis* (New Haven and London, 1942), 23-24; Randall, *Lincoln the President*, I, 107-108, 146-147. Lincoln's convention strategy: King, *Lincoln's Manager*, 133-136; Baringer, *Lincoln's Rise to Power*, 193-219; Reinhard H. Luthin, *First Lincoln Campaign* (Cambridge, Mass., 1944), and Luthin, *Lincoln*, 206-208. Lincoln during the convention: Davis and Dubois to Lincoln, May 15, 1860, and Davis to Lincoln, May 17, 1860, RTL Coll.; Knapp to Lincoln as quoted in Baringer, *Lincoln's Rise to Power*, 231; Lincoln's instructions to Davis in Lincoln, *CW*, IV, 50; King, *Lincoln's Manager*, 138; Baringer, *Lincoln's Rise to Power*, 289-290, 305; Angle, *Here I Have Lived*, 236; telegrams to Lincoln in Mearns, *Lincoln Papers*, 237; *New York Tribune*, May 25, 1860; *Illinois State Journal*, May 19, 1860; Champaign *Central Illinois Gazette*, May 23, 1860; Lincoln's response to serenade, *CW*, IV, 50.

Most historians have argued that Lincoln's managers secured his nomination by making bargains with Republicans from other states, promising them Cabinet posts and other offices if they would swing their delegations to Lincoln. See Baringer, *Lincoln's Rise to Power*, 266ff, and Erwin Stanley Bradley, *Simon Cameron, Lincoln's Secretary of War* (Philadelphia, 1966), 148-150. In my judgment, King, *Lincoln's Manager*, 137ff, successfully refutes this interpretation. Analyzing the contemporary evidence, King demonstrates that while Lincoln's men may have made conditional offers and overtures (as any managers would do), they followed Lincoln's own instructions and did not bind him to any convention deals. See Lincoln's own letter to Davis in Lincoln, *CWS*, 54-55. Moreover, as Fehrenbacher has observed (*Prelude to Greatness*, 155-159), Seward's men had as many offices to promise as Lincoln's managers. What won Lincoln the nomination was not the peddling of spoils but a hard decision by the convention "that the leading candidate could not win and must give way to someone who could."

Pages 179-185

Henry Steele Commager (ed.), *Documents of American History* (8th ed., N.Y., 1968), I, 363-365; Fehrenbacher, *Prelude to Greatness*, 156-157; Lincoln on the tariff in Lincoln, *CW*, III, 487, IV, 49; Helen Nicolay, *Lincoln's Secretary: A Biography of John G. Nicolay* (N.Y., 1949), 3-38, 42; Randall, *Mary Lincoln*, 181-184; quotation ("slouchy, ungraceful") in Thomas Webster to John Sherman, November 15, 1860, John Sherman Papers, Library of Congress; quotation ("high-toned . . . gentleman") in Charles M. Segal (ed.), *Conversations with Lincoln* (N.Y., 1961), 33-34. On the campaign biographies: Lincoln, *CW*, IV, 60-67, 68; Scripps to Lincoln, June 18, July 17, 1860, Mearns, *Lincoln Papers*, 259, 260; Lincoln quotation ("I authorize nothing") in Lincoln, *CW*, IV, 79-80; Allan Nevins, *Emergence of Lincoln* (2 vols., N.Y., 1950), II, 273-278; Luthin, *Lincoln*, 233-235. My analysis of the liberal wing of the Republican party (the so-called Radicals or Jacobins) draws information from Trefousse, *Radical Republicans*, 4-136; Foner, *Free Soil, Free Labor, Free Men*, 103-148; and

Fawn Brodie, *Thaddeus Stevens, Scourge of the South* (N.Y., 1959), 63ff. Lincoln's visit with Weed: Lincoln, *CW,* IV, 71; Thurlow Weed, *Autobiography* (Boston, 1883), 602–603; Lincoln, *CWS,* 54–55; King, *Lincoln's Manager,* 144–147. Lincoln quotation ("most audacity") in Lincoln, *CW,* IV, 90; *ibid.,* 81–82; Segal, *Conversations with Lincoln,* 34; Lincoln quotation ("can not fail") in Lincoln, *CW,* IV, 87; quotation ("bear up under defeat") in Turners, *Mary Lincoln,* 63–64, 66.

Pages 185–189

The 1860 campaign in the North: Luthin, *First Lincoln Campaign,* passim; Nevins, *Emergence of Lincoln,* II, 261ff; Potter and Fehrenbacher, *Impending Crisis,* 434–441; Randall, *Lincoln the President,* I, 186–189; Angle, *Here I Have Lived,* 236–246; Lincoln's remarks at the Springfield rally, *CW,* IV, 91–92; James M. McPherson, *The Struggle for Equality: Abolitionists and the Negro in the Civil War and Reconstruction* (Princeton, N.J., 1964), 3–23, 272–273; Current, *Lincoln Nobody Knows,* 28. The 1860 campaign in the South: Michael Davis, *The Image of Lincoln in the South* (Knoxville, Tenn., 1971), 7–15; Nevins, *Emergence of Lincoln,* II, 287–298; Steven A. Channing, *Crisis of Fear: Secession in South Carolina* (N.Y., 1970), 299ff; Marshall J. Rachleff, "Racial Fear and Political Factionalism: A Study of the Secession Movement in Alabama" (Ph.D. dissertation, University of Massachusetts, Amherst, 1974), chap. VI; Donald E. Reynolds, *Editors Make War: Southern Newspapers in the Secession Crisis* (Nashville, 1970); Randall, *Lincoln the President,* I, 190–192; quotation ("social monstrosities") in *Southern Advocate,* December 12, 1860; quotation ("the South . . . would never submit") in Atlanta *Southern Confederacy* as quoted in the *New York Times,* August 7, 1860. Lincoln and the campaign: Lincoln, *CW,* IV, 95; quotation ("for the commercial gain") in Segal, *Conversations with Lincoln,* 35–37; quotation ("as if the Government is about to fall") in Lincoln, *CW,* IV, 126–127.

Pages 189–191

My account of Lincoln and Springfield on election day is based on reports in the *New York Tribune,* November 10 and 12, 1860; the *Illinois State Journal,* November 7, 1860; Angle, *Here I Have Lived,* 251–253; quotation ("perfectly wild") in Pratt, *Concerning Mr. Lincoln,* 27–29; Stevens, *A Reporter's Lincoln,* 48; William E. Baringer, *A House Dividing: Lincoln as President Elect* (Springfield, 1945), 5–7; Randall, *Mary Lincoln,* 186–190; Segal, *Conversations with Lincoln,* 37–38; telegrams to Lincoln in RTL Coll.; Lincoln's recollection of election night in Gideon Welles, *Diary* (ed. by John T. Morse, Jr., 3 vols., Boston, 1911), I, 82.

PART SIX: MY TROUBLES HAVE JUST BEGUN
Pages 195–198

Quotation ("Well, boys") in Segal, *Conversations with Lincoln,* 38; Lincoln's recollection of the day after his election in Welles, *Diary,* I, 82; Baringer, *House Dividing,* 7, 26–28, 76–79, 138; Van Deusen, *Seward,* 240; Villard, *Lincoln on the Eve,* 17; *New York Herald,* November 22, 1860. Threats and hate mail: *New York Tribune,*

November 20, 1860; *Illinois State Journal,* November 24, 1860; Randall, *Mary Lincoln,* 190; Mearns, *Lincoln Papers,* 296, 336, 402–412. Lincoln's looking-glass vision: Ward Hill Lamon, *Recollections of Abraham Lincoln* (ed. by Dorothy Lamon Teillard, Washington, D.C., 1911), 112–113; Noah Brooks, *Washington, D.C., in Lincoln's Time* (ed. by Herbert Mitgang, Chicago, 1971), 198–200. Lincoln's secession policy: Nicolay's notes in Nicolay, *Lincoln's Secretary,* 54, and in Segal, *Conversations with Lincoln,* 44–45; Webster to Sherman, November 15, 1860, Sherman Papers, Library of Congress; Lincoln, *CW,* IV, 130, 132–133, 138, 139–140; quotation ("self respect demands") in *New York Tribune,* November 20, 1860; *Illinois State Journal,* November 21, 1860; quotation ("just as I expected") in Lincoln, *CW,* IV, 145–146; quotation ("eyes but does not see") in Villard, *Lincoln on the Eve,* 34; Potter, *Lincoln and His Party,* 9–19, 139–142; Baringer, *House Dividing,* 13–55. Lincoln's Cabinet: Mearns, *Lincoln Papers,* 317; Lincoln quotation ("give up their offices") in Donn Piatt, *Memories of the Men Who Saved the Union* (N.Y., 1887), 33–34; quotation ("from all I can learn") in Segal, *Conversations with Lincoln,* 47; Baringer, *House Dividing,* 82–89, 181; Lincoln, *CW,* IV, 148–149 (see the analyses of Lincoln's Cabinet offer to Seward in Van Deusen, *Seward,* 240, and King, *Lincoln's Manager,* 166–167); Edward Bates, *Diary* (ed. by Howard K. Beale, Washington, D.C., 1933), xi–xiii, 128–129, 164–165, 166n; Marvin R. Cain, *Lincoln's Attorney General: Edward Bates of Missouri* (Columbia, Mo., 1965), 1–126.

Pages 198–201

Lincoln's letters to Trumbull, Kellogg, Washburne, John A. Gilmer, and Seward, *CW,* IV, 149–150, 151–154, 183; Henry Adams, *Letters* (ed. by Worthington Chauncey Ford, Boston and N.Y., 1930), 68–69; Potter, *Lincoln and His Party,* 112–133, 156–161; Current, *Lincoln Nobody Knows,* 80–82. Weed's visit: Weed, *Autobiography,* 605–614; King, *Lincoln's Manager,* 167–169; Baringer, *Lincoln Day by Day,* II, 302; Lincoln to Stephens, *CW,* IV, 160; Lincoln quotations ("they ought to hang him" and "good ground") in Nicolay memorandum, December 22, 1860, John G. Nicolay Papers, Library of Congress; Lincoln, *CW,* IV, 157–159.

Pages 201–203

On Cameron: Lincoln's memorandum and correspondence regarding Cameron, *CW,* IV, 165–167, 169–170, 174, 179–180; W. C. Bryant to Lincoln, January 4, 1861, Trumbull to Lincoln, December 30, 1860, and January 20, 1861, RTL Coll.; Herndon to Trumbull, January 27, 1861, Trumbull Papers, Library of Congress; White, *Trumbull,* 146; Welles as quoted in Randall, *Lincoln the President,* I, 264. King, *Lincoln's Manager,* 162–167, demolishes the popular belief that Cameron's appointment came about because of an alleged pledge at Chicago. Bradley, *Cameron,* 173–174, agrees that bargains didn't force Lincoln to appoint Cameron. Quotation ("more trouble than balance of Union") in Luthin, *Lincoln,* 248. On Chase: Lincoln, *CW,* IV, 168, 171; Salmon Chase, *Inside Lincoln's Cabinet: the Civil War Diaries of Salmon P. Chase* (ed. by David Donald, N.Y. and London, 1954), 1–8; Reinhard H. Luthin, "Salmon P. Chase's Political Career before the Civil War," *Mississippi Valley Historical*

Review (March, 1943), 517–540; Chase to Washburne, January 14, 1861, and Ray to Washburne, January 7, 1861, Elihu Washburne Papers, Library of Congress; Nicolay and Hay, *Lincoln,* III, 359–360; Trefousse, *Radical Republicans,* 140; J. W. Shafer to Washburne, January 29, 1861, Washburne Papers, Library of Congress; Lincoln quotation ("teased to insanity") in Lincoln, *CW,* IV, 173.

Pages 203–205

Montgomery Mail as quoted in the *Nashville Republican Banner,* November 11, 1860; Lincoln quotation ("whatever springs of necessity") in Lincoln, *CW,* IV, 183, also 172, 175–176; Potter, *Lincoln and His Party,* 170–187, 247–314; Van Deusen, *Seward,* 248–249; Herndon to Trumbull, January 27, 1861, Trumbull Papers, Library of Congress.

Pages 205–208

Mearns, *Lincoln Papers,* 346–347, 378–379, 431–432, 435; Pratt, *Concerning Mr. Lincoln,* 68, 69; Segal, *Conversations with Lincoln,* 91–92; quotation ("dingy . . . back room") in Herndon, *Herndon's Lincoln,* 376; Turners, *Mary Lincoln,* 69–72; Randall, *Mary Lincoln,* 192–193; quotation ("winter gayeties") in Villard, *Lincoln on the Eve,* 54–55. Lincoln's visit with Sarah Lincoln: Lamon, *Life of Lincoln,* 462–464; *Illinois State Journal,* February 2, 1861; Baringer, *House Dividing,* 252–254; J. G. Randall, *Mr. Lincoln* (ed. by Richard N. Current, N.Y., 1957), 130. Last days in Springfield: Angle, *Lincoln, 1854–1861,* 370; Browning, *Diary,* I, 453; Herndon, *Herndon's Lincoln,* 379–382; Segal, *Conversations with Lincoln,* 69–71; Nicolay, *Lincoln's Secretary,* 61; Randall, *Mary Lincoln,* 195, 199; quotation ("plucky wife") in Villard, *Lincoln on the Eve,* 53, 66–67; Lincoln's Springfield speech, *CW,* IV, 190.

Pages 208–213

Journey East: Randall, *Mary Lincoln,* 202–206; Nicolay, *Lincoln's Secretary,* 102–104; *New York Herald,* February 13 and 18, 1861; Villard, *Lincoln on the Eve,* 75–105; Lincoln quotations at Columbus, Pittsburgh, Cleveland, Westfield, and Albany, in Lincoln, *CW,* IV, 204, 208–212, 215–216, 219, 225; quotation ("long and short of it") in C. Percy Powell (ed.), *Lincoln Day by Day* (editor in chief Earl Schenck Miers, Washington, D.C., 1960), III, 20; Baringer, *House Dividing,* 276–281; Thomas, *Lincoln,* 241; Powell, *Lincoln Day by Day,* III, 19; William H. Russell, *My Diary North and South* (ed. by Fletcher Pratt, N.Y., 1954), 16. Baltimore plot and Lincoln at Philadelphia, Harrisburg, and Washington: Norma B. Cuthbert, *Lincoln and the Baltimore Plot* (San Marino, Calif., 1949), which contains Pinkerton's Record Book of 1861 and his, Judd's, and Lamon's accounts of the plot; Mearns, *Lincoln Papers,* 442–443; Randall, *Mary Lincoln,* 205–206; Baringer, *House Dividing,* 292–296; George S. Bryan, *The Great American Myth* (N.Y., 1940), 21–43; Lamon, *Life of Lincoln,* 520–525; Winfield Scott to Seward, February 21, 1861, RTL Coll.; Lincoln's speech at Philadelphia, *CW,* IV, 240. Lincoln in Washington: Russell, *My Diary,* 17–18, 26, 195; Brooks, *Washington, D.C.,* 13–20, 46; Bryan, *Great American Myth,* 3–12, 51–52; quotation ("if you don't resign") in A. G. Frick [?] to Lincoln, February

14, 1861, Chicago Historical Society (Randall, *Mary Lincoln,* 207); R. Gerald McMurtry, "Scotch Cap and Military Cloak a Fabrication," *Lincoln Lore,* October, 1956; Harper, *Lincoln and the Press,* 91; Turners, *Mary Lincoln,* 77–79.

Pages 213–217

Lincoln quotation ("bad enough in Springfield") in Henry Villard, *Memoirs* (2 vols., Boston, 1904), I, 156; Nicolay and Hay, *Lincoln,* III, 370; Powell, *Lincoln Day by Day,* III, 22–24; quotation ("in God's name") in Johannsen, *Douglas,* 841–842. Lincoln and Southern delegations: John Hay, *Lincoln and the Civil War in the Diaries and Letters of John Hay* (selected by Tyler Dennett, reprint of 1939 ed., N.Y., 1972), 30; Charles S. Morehead's eyewitness account in Segal, *Conversations with Lincoln,* 85–90; Potter, *Lincoln and His Party,* 353–354; Baringer, *House Dividing,* 315–318, 321, 329. The Cabinet: *ibid.,* 319–329; King, *Lincoln's Manager,* 178–179; Harry J. Carman and Reinhard H. Luthin, *Lincoln and the Patronage* (N.Y., 1943), 48–50; quotation ("Violence is not to be met"), Monty Blair to John A. Andrew, January 23, 1861, in Nevins, *Emergence of Lincoln,* II, 452; William E. Smith, *The Francis Preston Blair Family in Politics* (2 vols., N.Y., 1933), I, 508–516, and *passim.* Lincoln and Seward: Nicolay and Hay, *Lincoln,* III, 370–372; Lincoln to Seward, *CW,* IV, 273; Van Deusen, *Seward,* 253–254. Lincoln on secession: Lincoln, *CW,* IV, 154, 421–440; Nicolay and Hay, *Lincoln,* III, 248; Nicolay, *Lincoln's Secretary,* 55; Current, *Lincoln Nobody Knows,* 100–103.

Pages 217–219

Bryan, *Great American Myth,* 53–55; Powell, *Lincoln Day by Day,* III, 24–26; Randall, *Mary Lincoln,* 208–209; Nicolay, *Lincoln's Secretary,* 71–73; R. Gerald McMurtry, "The Inauguration of Abraham Lincoln," *Lincoln Lore,* January, 1952; Ben: Perley Poore, *Recollections of Sixty Years in the National Metropolis* (2 vols., Philadelphia, 1886), II, 69; *New York Herald,* March 5, 1861; Lincoln's inaugural address, *CW,* IV, 262–271; Potter, *Lincoln and His Party,* 319–337; quotation ("face of a galvanized corpse") in Sandburg, *Lincoln, War Years,* I, 122; quotation ("thank God") in Julia Taft Bayne, *Tad Lincoln's Father* (Boston, 1931), 17–20.

Pages 219–227

My account of Lincoln and Fort Sumter draws generally from Richard N. Current, *Lincoln and the First Shot* (N.Y., 1963), 43 and *passim;* Van Deusen, *Seward,* 276–285; Potter, *Lincoln and His Party,* 337–366; Potter and Fehrenbacher, *Impending Crisis,* 570–583; Randall, *Lincoln the President,* I, 320ff; Niven, *Welles,* 325–339; Kenneth M. Stampp, *And the War Came: the North and the Secession Crisis* (paperback ed., Chicago, 1964), 263–286. Specific documentation is as follows: Anderson's report and Joseph Holt and Winfield Scott to Lincoln, March 5, 1861, and Scott to Lincoln, March 11, 1861, RTL Coll.; Lincoln, *CW,* IV, 279, 284, 285n; replies of Cabinet members to Lincoln's interrogatory in RTL Coll.; quotation ("little Illinois lawyer") in Van Deusen, *Seward,* 336; Lincoln's memorandum on Sumter, *CW,* IV, 288–290;

Lincoln quotation ("all the troubles and anxieties") in Browning, *Diary,* I, 476. Quotation ("loaves and fishes") in Luthin, *Lincoln,* 288; quotation ("man letting lodgings") in Nicolay and Hay, *Lincoln,* IV, 69; Segal, *Conversations with Lincoln,* 43–61; *New York Times,* April 3, 1861. Hurlbut to Lincoln, March 27, 1861, RTL Coll.; Gideon Welles, *Lincoln and Seward* (N.Y., 1874), 64–65; Nicolay and Hay, *Lincoln,* III, 395, IV, 44; Powell, *Lincoln Day by Day,* III, 31; Seward's opinion of Lincoln in Van Deusen, *Seward,* 280–281; Seward's April 1 memorandum to Lincoln in RTL Coll.; Lincoln's reply, *CW,* IV, 316–317; quotation ("President is the best of us") in Thomas, *Lincoln,* 254; Lincoln's meeting with Baldwin in Segal, *Conversations with Lincoln,* 102–107; Nicolay and Hay, *Lincoln,* IV, 70–71; quotation ("immediate dissolution") in Lincoln, *CW,* IV, 426; Lincoln and Douglas in George Ashmun's account, *Washington Daily Morning Chronicle,* October 26, 1864, and Johannsen, *Douglas,* 859–860; Lincoln and Virginia secession in Lincoln, *CW,* IV, 427.

PART SEVEN: STORM CENTER
Pages 231–234

Lincoln and Lee: Nicolay and Hay, *Lincoln,* IV, 97–101; Douglas Southall Freeman, *Robert E. Lee, A Biography* (4 vols., N.Y., 1934–1935), I, 421, also 416–436; Margaret Sanborn, *Robert E. Lee, A Portrait* (2 vols., Philadelphia, 1966–1967), I, 157, 255, 302; quotation ("Magruder came to me in this room") in Nicolay's notes, Nicolay, *Lincoln's Secretary,* 95; Lincoln's reference to Lee, Magruder, and others as traitors in Lincoln, *CW,* VI, 265, also IV, 438. Baltimore riot: Nicolay and Hay, *Lincoln,* IV, 127; Nicolay's notes in Nicolay, *Lincoln's Secretary,* 93–94; Lincoln, *CW,* IV, 340–342. Washington in fear: Hay, *Diaries* (Dennett ed.), 2–6; Nicolay and Hay, *Lincoln,* IV, 148–153; Randall, *Lincoln the President,* I, 363; quotation ("midst of traitors") in Nicolay, *Lincoln's Secretary,* 98; Cabinet meeting of April 21, 1861: Lincoln, *CW,* V, 241–242; Nicolay, *Lincoln's Secretary,* 95–96; quotation ("only Northern realities") in Hay, *Diaries* (Dennett ed.), 11.

Pages 234–235

Hay, *Diaries* (Dennett ed.), 12, 17, 26; quotation ("everywhereness of uniforms"), Nicolay's letter in Nicolay, *Lincoln's Secretary,* 98.

Pages 235–237

Quotation ("great trouble") in Brooks, *Washington, D.C.,* 267; quotation ("corps of spies") in Lincoln, *CW,* VI, 263–264; *ibid.,* IV, 372, 428–432; Van Deusen, *Seward,* 289–290; quotation ("Are all the laws, but one") in Lincoln, *CW,* IV, 430–431. See also Harold M. Hyman, *A More Perfect Union: the Impact of the Civil War and Reconstruction on the Constitution* (N.Y., 1973), 81–170, and J. G. Randall, *Constitutional Problems Under Lincoln* (Urbana, Ill., 1951), 140–168. Lincoln and the border: quotation ("whole game") in Lincoln, *CW,* IV, 532; quotation ("tangible evidence") *ibid.,* 523, also 344, V, 24; Van Deusen, *Seward,* 290; Randall, *Lincoln the President,* I,

366–367, II, 3–11; Lincoln's interview with Garret Davis in Segal, *Conversations with Lincoln,* 115–116; Lincoln, *CW,* IV, 368–369, 497.

Pages 238–240

Mobilization woes: Lincoln, *CW,* IV, 369n, 370n, 370–376, 402ff; Nicolay, *Lincoln's Secretary,* 113; Hay, *Diaries* (Dennett ed.), 18–19; Randall, *Lincoln the President,* I, 361, 372, II, 86; Lincoln quotation ("plain matter of fact") in Lincoln, *CW,* V, 20. See also Fred A. Shannon, *The Organization and Administration of the Union Army* (2 vols., Cleveland, 1928), and A. Howard Meneely, *The War Department, 1861* (N.Y., 1928). Negroes and slave contraband: James M. McPherson, *The Negro's Civil War* (N.Y., 1965), 20–23; Trefousse, *Radical Republicans,* 205–206; David M. Nellis, "Between a Crown and a Gibbet: Benjamin F. Butler and the Early War Years" (M.A. thesis, University of Massachusetts, Amherst, 1973), 15–19; Bell I. Wiley, *Southern Negroes, 1861–1865* (Baton Rouge, 1938), 175–178; quotation ("I'se contraband)" in Sandburg, *Lincoln, the War Years,* I, 279.

Pages 240–241

Washington Evening Star, May 24, 1861; *New York Herald,* May 25 and 26, 1861; Nicolay and Hay, *Lincoln,* IV, 310–314; Turners, *Mary Lincoln,* 92; Powell, *Lincoln Day by Day,* III, 43, 46.

Pages 241–245

Seward's "manifesto": David Donald, *Charles Sumner and the Rights of Man* (N.Y., 1970), 18–25, 39, 41–42; Van Deusen, *Seward,* 295–299, 301, 371; Lincoln, *CW,* IV, 376–380. Portrait of Sumner from Brooks, *Washington, D.C.,* 32–33, and Edward Dicey, *Spectator of America* (ed. by Herbert Mitgang, Chicago, 1971), 101–102. Portrait of Seward from Van Deusen, *Seward,* 255–268, 274, 335, 339–341; Brooks, *Washington, D.C.,* 35–36; Russell, *My Diary,* 19–20. Cabinet conflict: Van Deusen, *Seward,* 275; Welles, *Diary,* I, 6–8, 135; Chase, *Inside Lincoln's Cabinet,* 11, 12, 16; Burton J. Hendrick, *Lincoln's War Cabinet* (reprint of 1946 ed., Gloucester, Mass., 1965), 154ff. Welles: Welles, *Diary,* I, *passim;* Niven, *Welles,* 318ff; Thomas and Hyman, *Stanton,* 150. Quotation ("pride and dignity") in Donald, *Sumner and the Rights of Man,* 23; quotation ("afterthought only") in Broughaum Villiers and W. H. Chesson, *Anglo-American Relations, 1861–1865* (London, 1919), 193.

Pages 245–250

Quotation ("capricious" sleep) in Nicolay, *Lincoln's Secretary,* 102; Lincoln's "shop" in F. B. Carpenter, *Six Months at the White House with Lincoln* (Century House edition, ed. by John Crosby Freeman, Watkins Glen, N.Y., 1961), 66; Lincoln's writing style described in Albert Chandler, "Lincoln and the Telegrapher" (ed. by E. B. Long), *American Heritage* (April, 1961), 32–34. Lincoln's secretaries: Nicolay, *Lincoln's Secretary,* 75–88, 122–123; quotation ("Hellcat") in Hay, *Diaries* (Dennett ed.), 40–41 and *passim.* Morning crowds: Lincoln's memoranda, April, 1861, and August 23, 1862, RTL Coll.; Nicolay, *Lincoln's Secretary,* 82–83, 117–118; Brooks, *Washington,*

D.C., 79; Agnes Macdonnel, "American Then and Now: Recollections of Lincoln," *The Contemporary Review* (May, 1917), 566–568. Lincoln's language in George Templeton Strong, *Diary, 1835–1875* (ed. by Allan Nevins and Milton H. Thomas, 4 vols., N.Y., 1952), III, 188, 204–205; quotation ("damned rascal") in Thomas, *Lincoln,* 459; quotation ("go away") in Nicolay, *Lincoln's Secretary,* 85; quotation ("flabbiness") in Allen Thorndike Rice (ed.), *Reminiscences of Abraham Lincoln by Distinguished Men of His Time* (N.Y., 1886), 428–429, and Noah Brooks, "Personal Recollections of Abraham Lincoln," *Harper's* (July, 1865), 226–227. Tools of war: Hay, *Diaries* (Dennett ed.), 22, 82; Lincoln, *CW,* V, 365, and *CWS,* 88, 92; Robert V. Bruce, *Lincoln and the Tools of War* (Indianapolis, 1956), 99ff. Cabinet meetings: Thomas and Hyman, *Stanton,* 255; Welles, *Diary,* I, 136–137, 194–202; Chase, *Inside Lincoln's Cabinet,* 16–17; Hendrick, *Lincoln's War Cabinet,* 227–228. Quotation ("wanting to work") in Lincoln, *CW,* IV, 556. Carriage rides and guards: Bryan, *Great American Myth,* 59, 60, 62–63; Brooks, *Washington, D.C.,* 266. Evening receptions: *ibid.,* 68–69; Melville to Elizabeth Melville, March 24, 1861, *Letters of Herman Melville* (ed. by Merrell R. Davis and William H. Gilman, New Haven, 1960), 209–210. Lincoln's humor: quotation ("cheeriest of talkers") in Thomas, *Lincoln,* 475; quotation ("wrinkled his nose") in Horace H. Furness to his wife, November 10, 1862, in *Letters of Horace H. Furness* (2 vols., Boston, 1922), I, 126; quotation ("You speak of Lincoln stories") in Brooks, *Washington, D.C.,* 255–256; quotation ("stories are not so nice") in John F. Farnsworth's testimony as quoted in Sandburg, *Lincoln, the War Years,* III, 305; Nicolay, *Lincoln's Secretary,* 100–102; Randall, *Mr. Lincoln,* 213–239. Opera and theater: Lincoln, *CW,* VI, 392; Brooks, *Washington, D.C.,* 72–73; Carpenter, *Six Months in the White House* (Century House ed.), 29; quotation ("grotesque or absurd") in Hugh McCulloch's recollections of Lincoln in the *New York Tribune,* June 14, 1885; quotation ("hefty flirtation") in John Hay, *Letters of John Hay and Extracts from Diary* (3 vols., Gordian Press ed., N.Y., 1969), I, 199. Quotation ("twenty thousand drowned cats") in *ibid.;* Nicolay, *Lincoln's Secretary,* 115. Soldiers' Home: Bryan, *Great American Myth,* 62; Turners, *Mary Lincoln,* 94; Brooks, *Washington, D.C.,* 78–79, 260–261; Lincoln, *CW,* VI, 392; David Homer Bates, *Lincoln in the Telegraph Office* (N.Y., 1907), 214–215, 223–224, 236; Lamon, *Recollections of Lincoln,* 110–122.

Pages 250–254

Picture Book War: Russell, *My Diary,* 188; Frank Moore (ed.), *The Rebellion Record: A Diary of American Events* (12 vols., N.Y., 1862–1871), II, 385; Robert Underwood Johnson and Clarence Clough Buel (eds.), *Battles and Leaders of the Civil War* (facsimile reprint of 1887–1888 ed., 4 vols., N.Y., 1956), I, 93–98; Margaret Leech, *Reveille in Washington, 1860–1865* (N.Y., 1941), 66–90; Henry Steele Commager (ed.), *The Blue and the Gray* (Indianapolis and New York, 1950), 66–76; Bruce Catton, *Mr. Lincoln's Army* (Garden City, N.Y., 1951), chap. I ("Picture Book War"). Bull Run advance: Nicolay and Hay, *Lincoln,* IV, 360; quotation ("you are green") in E. D. Townsend, *Anecdotes of the Civil War* (N.Y., 1884), 57, and T. Harry Williams, *Lincoln and His Generals* (N.Y., 1952), 21; Warren W. Hassler, Jr., *Commanders of the Army of the Potomac* (Baton Rouge, 1962), 3, 11. Lincoln, Congress, and Republican liberals: Lincoln's message, *CW*, IV, 426, 438–440; profiles of Wade and Chandler in Trefousse, *Radical Republicans*, 7–9, 30–32, 158–159, 172–173, 203–208; quotation ("bulldog obduracy") in Brooks, *Washington, D.C.*, 34; Segal, *Conversations with Lincoln*, 144, 153; quotation ("deeply convinced") in Donald, *Sumner and the Rights of Man*, 17; V. Jacque Voegeli, *Free but Not Equal: the Midwest and the Negro During the Civil War* (Chicago, 1967), 4–8, 14–15. There is an enduring historical myth that Lincoln didn't enforce the confiscation act. On the contrary, Lincoln not only ordered General Frémont, commander in Missouri, to abide by the act, but insisted in his Annual Message, December 3, 1861, *CW*, V, 49, that he'd faithfully enforced it.

Pages 254–259

Bull Run: Bates, *Lincoln in the Telegraph Office*, 88; Browning, *Diary*, I, 484–485; Nicolay's memorandum in Nicolay, *Lincoln's Secretary*, 109; Williams, *Lincoln and His Generals*, 22–23; quotation on clouds in Russell, *My Diary*, 231, also 232–233; Nicolay and Hay, *Lincoln*, IV, 355; Trefousse, *Radical Republicans*, 173–175; Moore, *Rebellion Record*, II, 284; Johnson and Buel, *Battles and Leaders*, I, 167–193; *New York Herald*, July 25 and 26, 1861. Aftermath: Lincoln, *CW*, IV, 457–458; Williams, *Lincoln and His Generals*, 23–24, 33–35. Lincoln and McClellan: *ibid.*, 24–31; Brooks, *Washington, D.C.*, 25; Hassler, *Commanders of the Army of the Potomac*, 26–31; George B. McClellan, *McClellan's Own Story* (N.Y., 1887), 82–83, 85, 172, 229; William Starr Myers, *A Study in Personality: General George Brinton McClellan* (N.Y., 1934), 212–214. Lincoln's depression: Greeley to Lincoln, July 29, 1861, RTL Coll.; Nicolay's comment in Nicolay, *Lincoln's Secretary*, 105; Browning to Lincoln, August 19, 1861, RTL Coll.; Browning, *Diary*, I, 488–489.

Pages 259–263

Frémont's proclamation: Carman and Luthin, *Lincoln and the Patronage*, 194–197; Frank Blair, Jr., to Montgomery Blair, September 1, 1861, RTL Coll.; Speed to Lincoln, September 1 and 3, 1861, *ibid.;* quotation ("over the mill dam") in Thomas, *Lincoln*, 275; Lincoln to Frémont, *CW*, IV, 506. Lincoln and Jesse Frémont: Frémont's letter of September 8, 1861, and Jesse Frémont to Lincoln, September 12, 1861, RTL Coll.; Jesse Frémont's account in Segal, *Conversations with Lincoln*, 131–134; Nicolay and Hay, *Lincoln*, IV, 415; Lincoln quotation ("taxed me so violently") in Hay, *Diaries* (Dennett ed.), 133; Lincoln, *CW*, IV, 517–518, 519. Reaction to Lincoln's order: Donald, *Sumner and the Rights of Man*, 26; Trefousse, *Radical Republicans*, 177; McPherson, *Struggle for Equality*, 72–81; Lincoln quotation ("range of military law") in Lincoln, *CW*, IV, 531–532; Nicolay's conversation with the President, October 2, 1861, in Nicolay, *Lincoln's Secretary*, 125–126. Corruption in Frémont's department: *House Report 2*, 37th Cong., 2nd Sess., xxxvi, xlii, lxi–lxiv, 37–99; Report of the Commission on War Claims, *House Executive Document 94*, 37th Cong., 2nd Sess., 11–18, 25–26, 34, and *passim*. Thomas to Cameron, October 21, 1861, in U.S. War Department, *The War of the Rebellion: A Compilation of the Official Records of the Union and Confederate Armies* (70 vols. in 128, Washington, D.C., 1880–1901), ser. I, vol. III, 540–549. This hereafter is cited as *OR*. Lincoln quotation ("cardinal mistake") in

Lincoln, *CW,* IV, 513; Trefousse, *Radical Republicans,* 177. For a defense of Frémont, see Allan Nevins, *The War for the Union* (4 vols., N.Y., 1959–1971), I, 307ff.

Pages 263–267

Ball's Bluff: quotation ("Don't let them hurry me") in Hay, *Diaries* (Dennett ed.), 27; John Hay, "Edward Baker," *Harper's* (December, 1861), 108; Randall, *Mary Lincoln,* 235; quotation ("smote") in Luthin, *Lincoln,* 303; Nicolay, *Lincoln's Secretary,* 101; Russell, *My Diary,* 257–258; Willie's poem in the *Washington National Republican,* November 4, 1861. Lincoln, liberal senators, and McClellan: Trefousse, *Radical Republicans,* 179–180; Williams, *Lincoln and His Generals,* 29–32; Hassler, *Commanders of the Army of the Potomac,* 31–33; Lincoln quotation ("have a good notion") in Hay, *Diaries* (Dennett ed.), 31–32. McClellan as General in Chief: Randall, *Lincoln the President,* I, 393–395; quotation ("greatest geese") in McClellan, *McClellan's Own Story,* 86, 167, 169–170, 175–176; Bates, *Diary,* 196; quotation ("a vast labor") in Hay, *Diaries* (Dennett ed.), 33; McClellan's snub in *ibid.,* 34–35, and Nicolay, *Lincoln's Secretary,* 141–142; Williams, *Lincoln and His Generals,* 32, 42–50; Kenneth P. Williams, *Lincoln Finds a General* (5 vols., N.Y., 1950–1959), I, 103–121.

Pages 267–270

Lincoln, *CW,* V, 29–31; Browning, *Diary,* I, 512; Lincoln's message, *CW,* V, 48–49. Cameron's report: Nicolay and Hay, *Lincoln,* V, 125; Randall, *Lincoln the President* (paperback ed., N.Y., 1945), II, 56–58; quotations ("greatly aggrieved" and "severe strain") in Alexander K. McClure, *Lincoln and Men of War-Times* (Philadelphia, 1892), 162–163; Nevin, *Welles,* 393–396; Thomas and Hyman, *Stanton,* 133–135; Bradley, *Cameron,* 201–205. Sumner and Lincoln on emancipation: Trefousse, *Radical Republicans,* 210; Donald, *Sumner and the Rights of Man,* 27, 46, 48; Sumner's speech on Baker in T. Harry Williams, *Lincoln and the Radicals* (Madison, Wis., 1941), 61; Charles Sumner, *Works* (15 vols., Boston, 1870–1883), VI, 152; Edward Everett Hale, *Memories of a Hundred Years* (2 vols., N.Y., 1902), II, 191.

Pages 270–272

My account of the *Trent* affair derives from many sources, among them: Bruce Catton, *Terrible Swift Sword* (Garden City, N.Y., 1963), 108–116; quotation ("white elephants") in Benson John Lossing, *Pictorial History of the Civil War* (3 vols., Philadelphia, 1866–1868), II, 156–157; Donald, *Sumner and the Rights of Man,* 35–44; Randall, *Lincoln the President,* II, 37–49; Jay Monaghan, *Diplomat in Carpet Slippers: Abraham Lincoln Deals with Foreign Affairs* (Indianapolis, 1945), 173–193; E. D. Adams, *Great Britain and the American Civil War* (2 vols., N.Y., 1925), I, 231–234; Nevins, *War for the Union,* I, 384ff; Browning, *Diary,* I, 516–519; Bates, *Diary,* 213–214; Chase, *Inside Lincoln's Cabinet,* 53, 55; quotation ("bitter pill") in Horace Porter, *Campaigning with Grant* (N.Y., 1897), 406–409.

Pages 272–276

Turners, *Mary Lincoln,* 84–89, 93, 96–100, 110–113, 120, 137; Randall, *Mary Lincoln,* 244–268; quotation ("bliss") in Randall, *Lincoln's Sons,* 126–127; quotation ("gossip of courts and cabinets") in Forney, *Anecdotes of Public Men,* I, 367; Elizabeth Keckley, *Behind the Scenes: Thirty Years a Slave and Four Years in the White House* (reprint of 1868 ed., N.Y., 1968), 17–95; quotations ("common to overrun appropriations" and "it can never have my approval") in Benjamin French to Pamela French, December 24, 1861, Benjamin French Papers, Library of Congress; Angle, *Lincoln by Some Men Who Knew Him,* 110–111; quotation ("stand around") in Keckley, *Behind the Scenes,* 124–126.

Pages 276–280

War Department abuses: *House Report 2,* 37th Cong., 2nd Sess., iv–xi, 34–54, and *passim;* Meneely, *War Department,* 252–269; Shannon, *Union Army,* I, 61–62; Lincoln, *CW,* V, 96–97; quotation ("Send word to the Csar") in Albert Gallatin Riddle, *Recollections of War Times* (N.Y., 1895), 179–180; Lincoln's defense of Cameron, *CW,* V, 240–243. For charges of corruption against the Navy and Treasury Departments, see Stephen B. Oates, "Abraham Lincoln," in C. Vann Woodward (ed.), *Responses of the Presidents to Charges of Misconduct* (N.Y., 1974), 116–117. Stanton: Thomas and Hyman, *Stanton,* 4–118, 124, 141–168, 170, 230, 375–378, 390–393; quotation ("a perpetually irritated look") in Bryan, *Great American Myth,* 129–130; Strong, *Diary,* IV, 266; Brooks, *Washington, D.C.,* 36–37; Nevins, *War for the Union,* II, 458–459; quotation ("champagne and oysters") in Thomas and Hyman, *Stanton,* 170.

PART EIGHT: THIS FIERY TRIAL
Pages 283–287

Trefousse, *Radical Republicans,* 184–185; Lincoln, *CW,* V, 88; quotation ("nation requires it") in Bates, *Diary,* 218–220; Nicolay and Hay, *Lincoln,* V, 155; Lincoln to Halleck and Buell, *CW,* V, 87, 90, 91; Williams, *Lincoln and His Generals,* 54–57, 62–66; Chase, *Inside Lincoln's Cabinet,* 57–58; January 6 meeting with the Committee in Williams, *Lincoln and the Radicals,* 83–84, and Thomas, *Lincoln,* 291; Lincoln to Buell and McClellan, *CW,* V, 91–92, 94; Lincoln quotation ("exceedingly discouraging") in *ibid.,* 95; quotation ("what shall I do") in Montgomery Meigs, "General Montgomery Meigs on the Conduct of the War," *American Historical Review* (January, 1921), 292–293, 302; William Swinton, *Campaigns of the Army of the Potomac* (N.Y., 1866), 80–82; Browning, *Diary,* I, 523. January 13: Lincoln, *CW,* V, 98–99; conference with McClellan in Randall, *Lincoln the President,* II, 76; accounts of Meigs and McClellan in Segal, *Conversations with Lincoln,* 153–156. Lincoln's General and Special War Orders, *CW,* V, 111–112, 115; Hay, *Diaries* (Dennett ed.), 36; Nicolay and Hay, *Lincoln,* V, 159; Lincoln, *CW,* V, 120n–124n; Lincoln defends McClellan in Keckley, *Behind the Scenes,* 132 (Lincoln said much the same to Browning, *Diary,* I, 525–526). Hassler, *Commanders of the Army of the Potomac,* 34–37, defends McClellan and paints an unflattering portrait of Stanton and Lincoln.

Pages 287–288

Randall, *Lincoln's Sons,* 93–127; Brooks, *Washington, D.C.,* 249–250; Turners, *Mary Lincoln,* 120.

Pages 289–290

Nicolay, *Lincoln's Secretary*, 132; Bates, *Diary*, 233, 235, 236; quotation ("My poor boy") in Keckley, *Behind the Scenes*, 100–105; quotation ("Well, Nicolay") in Nicolay's memorandum, February 18 and 20, 1862, Nicolay Papers, Library of Congress; Browning, *Diary*, I, 530–531; Randall, *Lincoln's Sons*, 130–132.

Pages 290–293

My account of Mary's suffering draws from Turners, *Mary Lincoln*, 121ff; quotations ("Try and control" and "impending danger") in Keckley, *Behind the Scenes*, 104–105, 120–121; Brooks, *Washington, D.C.*, 66–68; quotation ("hypocrite") in Van Deusen, *Seward*, 336–337; Randall, *Mary Lincoln*, 239, 285–302; Browning, *Diary*, I, 608–609. Lincoln's suffering: Nicolay, *Lincoln's Secretary*, 133; Randall, *Lincoln's Sons*, 102ff; Wolf, *Almost Chosen People*, 122ff; Trueblood, *Lincoln*, 29ff; Current, *Lincoln Nobody Knows*, 62–65, 71–74.

Pages 293–296

McClellan's remark about Sebastopol in *American Historical Review* (April, 1918), 551–552; Lincoln quotation ("Wade, anybody") in Nicolay, *Lincoln's Secretary*, 149; Randall, *Lincoln the President*, II, 87; McClellan, *McClellan's Own Story*, 149–184; Nicolay and Hay, *Lincoln*, V, 169; Powell, *Lincoln Day by Day*, III, 98–99; Lincoln, *CW*, V, 151. *Virginia* and *Monitor:* Nicolay's notes in Nicolay, *Lincoln's Secretary*, 136–137; Hay, *Diaries* (Dennett ed.), 37; Welles, *Diary*, I, 62–67; Browning, *Diary*, I, 532–533. Log guns and Manassas: Bates, *Diary*, 239, 240; Nicolay and Hay, *Lincoln*, V, 177–178; Williams, *Lincoln and His Generals*, 69–70; quotation ("scorn of world") in Francis Fessenden, *Life and Public Services of William Pitt Fessenden* (2 vols., Boston, 1907), I, 261; Hay, *Diaries* (Dennett ed.), 37; Thomas and Hyman, *Stanton*, 182–183; Lincoln's War Order #3, *CW*, V, 155. Peninsular plan: Williams, *Lincoln and His Generals*, 72–84; Powell, *Lincoln Day by Day*, III, 100–102; Lincoln, *CW*, V, 151n, 157, 182, 184; Thomas and Hyman, *Stanton*, 187–188, 208–209; Browning, *Diary*, I, 537–539; quotation ("against my better judgment") in Donald, *Sumner and the Rights of Man*, 59; Williams, *Lincoln Finds a General*, I, 152–186. For stout defenses of McClellan, turn to Hassler, *Commanders of the Army of the Potomac*, 37–43, and Randall, *Lincoln the President*, II, 74–89.

Pages 296–300

Quotation ("great movement of God") in Donald, *Sumner and the Rights of Man*, 50; Trueblood, *Lincoln*, 126–127; Lincoln's message, *CW*, V, 145–146; Hale, *Memories*, II, 193–196; *New York Tribune*, March 7, 1862; Lincoln, *CW*, V, 153. Meeting with border men: Nicolay memorandum, March 9, 1862, Nicolay Papers, Library of Congress, and Congressman J. W. Crisfield's account in Segal, *Conversations with Lincoln*, 165–168; quotation ("limb must be amputated") in Lincoln, *CW*, VII, 281–282. Quotation (entering "wedge") in *New York Herald*, April 5, 1862; Brooks, *Washington, D.C.*, 180–182. Hunter's proclamation: Lincoln, *CW*, IV, 219, 222–223, 225, 318; McPherson, *Struggle for Equality*, 107–112; Donald, *Sumner and the Rights of Man*, 62–67; Lincoln quotation ("only go as fast") in Nicolay memorandum, September 20, 1878, Nicolay Papers, Library of Congress; quotation ("I gave dissatisfaction") in Lincoln, *CW*, V, 318.

Pages 300–306

Quotation ("can't spare this man") in McClure, *Lincoln and Men of War-Times*, 193; correspondence between Lincoln and McClellan in Lincoln, *CW*, V, 182ff; Hay, *Diaries* (Dennett ed.), 39. Lincoln at Fort Monroe: Chase, *Inside Lincoln's Cabinet*, 74–87; Johnson and Buel, *Battles and Leaders*, II, 151–152. Peninsular operations: Lincoln, *CW*, V, 219–227; Nicolay, *Lincoln's Secretary*, 144; Williams, *Lincoln and His Generals*, 94–96. Jackson's Valley campaign: Lincoln, *CW*, V, 232, 235–237; McClellan, *McClellan's Own Story*, 346, 396–397; Lincoln to Frémont, McDowell, and Banks, *CW*, V, 243ff. Seven Pines and Seven Days: McClellan's report, June 1, 1862, *OR*, ser. I, vol. XI, pt. I, 749; quotation ("timid and irresolute") in Thomas, *Lincoln*, 325; Williams, *Lincoln and His Generals*, 106–129; Hassler, *Commanders of the Army of the Potomac*, 47–52; McClellan's report, June 25, 1862, *McClellan's Own Story*, 392–393; Lincoln to McClellan, *CW*, V, 286; Browning, *Diary*, I, 552; quotation ("wild with rumors") in Nicolay, *Lincoln's Secretary*, 146–147; quotation ("too many dead") in McClellan's report, June 28, 1862, *OR*, ser. I, vol. XI, pt. I, 61; Lincoln quotation ("maintain this contest") in Lincoln, *CW*, V, 291–292; Lincoln to McClellan, *ibid.*, 301, 305–306, 307; McClellan's reports, July 3 and 4, 1862, *OR*, ser. I, vol. XI, pt. III, 291–292, 294; Lincoln, *CW*, V, 307; Harrison's Landing letter in McClellan, *McClellan's Own Story*, 446, 487–489.

Pages 307–313

Quotation ("thrust at him") in Gideon Welles, "History of Emancipation," *Galaxy* (December, 1872), 842–843; quotation ("reconsecration of the day") in Donald, *Sumner and the Rights of Man*, 60; Trefousse, *Radical Republicans*, 216–223; Lincoln to the border men, *CW*, V, 317–319; quotation ("strong hand") in *ibid.*, VII, 281–282; quotation ("turn which way") in Welles, "Emancipation," *Galaxy* (December, 1872), 842–843. The carriage ride: *ibid.* (see also Segal, *Conversations with Lincoln*, 175–177); Welles, *Diary*, I, 70–71. Second confiscation act: Browning, *Diary*, I, 555, 558; Lincoln, *CW*, V, 328–331. Contrary to what many writers have claimed, Lincoln did enforce the act. See for example *ibid.*, 337n–338n, 341, VI, 45–46, 406, 408. Preliminary proclamation and Cabinet: Lincoln, *CW*, V, 336–337. In this draft of the proclamation, Lincoln announced that when Congress convened in December, he would push again for gradual, compensated emancipation in the loyal border states. Thus he would operate against slavery on two fronts, offering his old gradual plan to remove bondage in the border and threatening federally enforced military emancipation to eradicate it in the rebel states. This is discussed in the text in connection with the preliminary proclamation of September and Lincoln's December message to Congress. My account of the Cabinet meeting and postponement of the proclamation is based on Chase, *Inside Lincoln's Cabinet*, 98–99; Lincoln, *CW*, V, 337n; Nicolay and Hay, *Lincoln*, VI, 127; Van Deusen, *Seward*, 331; Thomas and Hyman, *Stanton*, 238–240; Carpenter, *Six Months in the White House* (Century House ed.), 14; Niven, *Welles,* 419–420; Sumner quotations in Donald, *Sumner and the Rights of Man,* 58, 80–81. Lincoln prepares the North: Voegeli, *Free But Not Equal,* 42, 97; Welles, *Diary,* I, 71, 150–153; Lincoln's address on colonization, *CW,* V, 371–375; McPherson, *Negro's Civil War,* 92–95; Lincoln to Louisiana Union men, New York financier, and Greeley, *CW,* V, 343, 346, 350, 388–389, and Voegeli's interpretation of the letter to Greeley, Voegeli, *Free But Not Equal,* 46.

Pages 313–317

Welles, *Diary,* I, 121, 331; Bates, *Diary,* 293; Stephen E. Ambrose, *Halleck, Lincoln's Chief of Staff* (Baton Rouge, 1962); Browning, *Diary,* I, 563; Williams, *Lincoln and His Generals,* 136–165. Second Bull Run: *ibid.;* Lincoln, *CW,* V, 395–402; Lincoln quotation ("whipped again") in Hay, *Diaries* (Dennett ed.), 45–47; Catton, *Terrible Swift Sword,* 428–437; Hassler, *Commanders of the Army of the Potomac,* 53–82, defends McClellan; Powell, *Lincoln Day by Day,* III, 137; Bates, *Diary,* 107; quotation ("wrung by bitterest anguish") in Lincoln, *CW,* V, 486n; Chase, *Inside Lincoln's Cabinet,* 94, 118–120; Lincoln quotation ("will of God prevails") in Lincoln, *CW,* V, 403–404. Lee's invasion: quotation ("first-rate clerk") in Hay, *Diaries* (Dennett ed.), 176; Welles, *Diary,* I, 119; Lincoln's vow to God about emancipation, *ibid.,* 143; Chase, *Inside Lincoln's Cabinet,* 150; Lincoln to Chicago ministers, *CW,* V, 419–425.

Pages 317–319

Antietam: Lincoln, *CW,* V, 426; McClellan's report, September 19, 1862, *OR,* ser. I, vol. XIX, pt. II, 330; McClellan, *McClellan's Own Story,* 613–614; Williams, *Lincoln and His Generals,* 167–169; Hassler, *Commanders of the Army of the Potomac,* 81–91. Cabinet meeting and proclamation: Lincoln, *CW,* V, 433–436; Artemus Ward's story in Segal, *Conversations with Lincoln,* 207; Welles, *Diary,* I, 143–149; Chase, *Inside Lincoln's Cabinet,* 149–153; Catton, *Terrible Swift Sword,* 461–463; Nicolay and Hay, *Lincoln,* VI, 162–163.

Pages 320–322

Response in the loyal states: McPherson, *Struggle for Equality,* 118–119; *Fred Douglass' Monthly* (October, 1862), 721–722; Emerson quotation in Mitgang, *Press Portrait,* 324–328; *Illinois State Journal,* September 24, 1862; Browning, *Diary,* I, 578; *Chicago Times,* September 24, 1862; *New York Herald,* September 27, 1862; Williams, *Lincoln and His Generals,* 170–171; *Louisville Journal* as quoted in the *Washington National Intelligencer,* October 8, 1862. Response in Confederacy: *Richmond Enquirer,* October 1, 1862; *Richmond Dispatch,* October 2, 1862; Davis, *Image of Lincoln in the South,* 82–85. Response in England: Adams, *Great Britain and the American Civil War,* II, 54–55; *London Star,* October 6, 1862; *London Times,* October 7 and 21, 1862. Lincoln quotation ("after full deliberation") in Lincoln, *CW,* V, 438; Hay, *Diaries* (Dennett ed.), 50.

Pages 322–323

Voegeli, *Free But Not Equal,* 54–64; Frank L. Klement, *The Copperheads in the Middle West* (Chicago, 1960), 11–39; Rice, *Reminiscences of Lincoln,* 271–272; Browning,

Diary, I, 611n; Lincoln, *CW,* V, 493–495, 503; N. Worth Brown and Randolph C. Downes (eds.), "A Conference with Abraham Lincoln: From the Diary of Reverend Nathan Brown," *Northwest Ohio Quarterly* (Spring, 1950), 58–60.

Pages 323–325

Lincoln, *CW,* V, 460–462; Halleck to Buell, October 23, 1862, *OR,* ser. I, vol. XVI, pt. II, 638; Williams, *Lincoln and His Generals,* 182–185; quotation ("horses of your army") in Lincoln, *CW,* V, 474–475, also 462n and 485; quotation ("hurt the enemy") in Hay, *Diaries* (Dennett ed.), 218–219; quotation ("sharp sticks") in Nicolay to Hay, October 26, 1862, and Nicolay to Therena Bates, October 26, 1862, Nicolay Papers, Library of Congress; Browning, *Diary,* I, 619; Hassler, *Commanders of the Army of the Potomac,* 92–101; Lincoln, *CW,* V, 505–506; Mary A. Livermore, *My Story of the War* (Hartford, 1889), 555–556. For McClellan, see Joseph L. Harsh, "On the McClellan-Go-Round," *Civil War History* (June, 1973), 101–118.

Pages 325–326

Lincoln, *CW,* V, 530–531, 534–537; Browning, *Diary,* I, 591.

Pages 327–330

Fredericksburg: Villard, *Memoirs,* I, 389–391; quotation ("Oh those men") in Johnson and Buel, *Battles and Leaders,* III, 138; Carpenter, *Six Months in the White House* (Century House ed.), 73; Donald, *Sumner and the Rights of Man,* 89; Fessenden, *Life of Fessenden,* I, 265–266; Strong, *Diary,* III, 256. Cabinet crisis: Welles, *Diary,* I, 79, 135–139; Chase, *Inside Lincoln's Cabinet,* 11, 16–17, 22–23, 94, 136, 175; Donald, *Lincoln Reconsidered,* 113–114; Van Deusen, *Seward,* 341–343; quotation ("to get rid of me") in Browning, *Diary,* I, 600–601, also 596–597; Fessenden, *Life of Fessenden,* I, 239–240; quotation ("earnest and sad") in Bates, *Diary,* 268–269; quotation ("Chase at the bottom") in Browning, *Diary,* I, 602–604; quotation ("certain men to say") in Hay, *Diaries* (Dennett ed.), 111–112; quotation ("new cabinet . . . new materials") in Bates, *Diary,* 268–269. December 19 meeting with the senators: Fessenden, *Life of Fessenden,* I, 243–246; Bates, *Diary,* 270. Quotation ("cuts Gordian knot") in Welles, *Diary,* I, 199–202; quotation ("damned plainly") in Hay, *Diaries* (Dennett ed.), 111–112.

Pages 330–333

Quotation ("is no hope") in Browning, *Diary,* I, 606–607. Colonization and Negro troops: Lincoln, *CW,* V, 371n; Voegeli, *Free But Not Equal,* 95–105; Thomas and Hyman, *Stanton,* 256; Nicolay and Hay, *Lincoln,* VI, 359–367; McPherson, *Negro's Civil War,* 96–97; quotation ("unavailed of force") in Lincoln, *CW,* VI, 149–150. New Year's Eve and Day: Thomas, *Lincoln,* 364; Lincoln's remark on being cast into a great civil war in undated clipping, Lincoln National Life Foundation, Fort Wayne, Ind., as cited in Warren, *Lincoln's Youth,* 225; Lincoln's final proclamation, *CW,* VI, 28–30; quotation ("Oh, Mr. French") in Benjamin French, "At the President's Reception," January 1, 1863, MS, in Brown University Library; Brooks, *Washington,*

D.C., 48–49; quotation ("If my name") in Segal, *Conversations with Lincoln*, 234–235; quotation ("a time of times") in McPherson, *Negro's Civil War*, 50. See also John Hope Franklin, *The Emancipation Proclamation* (paperback ed., Garden City, N.Y., 1965), 89–103.

PART NINE: THE NEW RECKONING
Pages 337–339

Lincoln, *CW*, VI, 31, 32–33, also 15, 22; Nicolay and Hay, *Lincoln*, VI, 31, 264–272; Williams, *Lincoln and His Generals*, 201–206; Hassler, *Commanders of the Army of the Potomac*, 120–130; Lincoln to Hooker, *CW*, VI, 78–79; quotation ("stuck in the mud") in Nicolay, *Lincoln's Secretary*, 166.

Pages 339–340

McPherson, *Struggle for Equality*, 120–122; Voegeli, *Free But Not Equal*, 76–90; *Chicago Times*, January 3, 1863; *New York World*, January 8, 1863; *New York Herald*, January 20, 1863; Bell I. Wiley, *Life of Billy Yank* (Indianapolis, 1951), 40–43; Forrest G. Wood, *Black Scare: the Racist Response to Emancipation and Reconstruction* (Berkeley and Los Angeles, 1968), 17–39; quotation ("These are dark hours") in Edward L. Pierce, *Memoir and Letters of Charles Sumner* (4 vols., Boston, 1887–1893), IV, 114; quotation ("upon brink of ruin") in Browning, *Diary*, I, 610–613, 616. Lincoln's defense of the proclamation: quotation ("slow walker") in Segal, *Conversations with Lincoln*, 215; quotation ("as much harm") in Rice, *Reminiscences of Lincoln*, 235–236; quotation ("fire in rear") in Pierce, as cited above; quotation ("broken eggs") in Lincoln, *CW*, VI, 48–49. England: Charles Francis Adams, Jr., *Charles Francis Adams* (Boston and N.Y., 1900), 299–300ff; Lincoln, *CW*, VI, 64; Mary Ellison, *Support for Secession: Lancashire and the American Civil War* (Chicago, 1973), 56–94; and Joseph M. Hernon, "British Sympathies in the American Civil War: A Reconsideration," *Journal of Southern History* (August, 1967), 356–367. See also Franklin, *Emancipation Proclamation*, 123–128, and Martin Duberman, *Charles Francis Adams, 1807–1886* (Stanford, Calif., 1960), 299ff.

Pages 340–342

Quotation ("bare sight") in Lincoln, *CW*, VI, 149–150; quotation ("imposing sight") in Thomas, *Lincoln*, 364. Negro recruitment: Dudley Taylor Cornish, *The Sable Arm: Negro Troops in the Union Army, 1861–1865* (paperback ed., N.Y., 1966), 94–131; Benjamin Quarles, *The Negro in the Civil War* (Boston, 1953), 183–213; McPherson, *Negro's Civil War*, 173–187. Quotation ("heartless bigotry") in *New York Tribune*, April 17, 1863. My account of the Union's refugee system draws heavily from Voegeli, *Free But Not Equal*, 95–112; Cornish, *Sable Arm*, 112–131; and Wiley, *Southern Negroes*, 199–259. See also Benjamin Quarles, *Lincoln and the Negro* (N.Y., 1962), 188–190. For the colonization fiasco in Haiti, see *ibid.*, 113–123, 191ff; McPherson, *Negro's Civil War*, 96–97; Lincoln, *CW*, VI, 178–179; Nicolay and Hay, *Lincoln*, VI, 359–367. For a different interpretation of Lincoln's racial views than that presented in this biography consult George M. Fredrickson, "A Man But Not a Brother: Abraham Lincoln and Racial Equality," *Journal of Southern History* (Feb., 1975), 39–58.

Pages 342–345

Thomas and Hyman, *Stanton*, 157–158, 245–281, 375; Hyman, *More Perfect Union*, 215–223; Voegeli, *Free But Not Equal*, 76–78; Randall, *Constitutional Problems*, 151; Lincoln quotation ("Must I shoot") and defense of martial law and internal security in Lincoln, *CW*, V, 260–269, 300–306; Welles, *Diary*, I, 306, 321; Bates, *Diary*, 306–307. See also Klement, *Copperheads*, chaps. I–IV, and Klement, *The Limits of Dissent: Clement L. Vallandigham and the Civil War* (Lexington, Ky., 1970), 102ff; G. R. Tredway, *Democratic Opposition to the Lincoln Administration in Indiana* (Indianapolis, 1973).

Pages 345–348

Vicksburg: Halleck to Grant, January 21, 1863, *OR*, ser. I, vol. XXIX, pt. I, 9; Madeleine Dahlgren, *Memoir of John A. Dahlgren* (N.Y., 1891), 389; Williams, *Lincoln and His Generals*, 224–230; quotation ("he is an ass") in Bruce Catton, *Grant Moves South* (Boston, 1960), 395, also 401–433; Lincoln, *CW*, VI, 326. Lincoln's visit with Hooker's army in Brooks, *Washington, D.C.,* 51–60. Chancellorsville: Lincoln, *CW*, VI, 189–190, 198–200; quotation ("My God!") in Brooks, *Washington, D.C.,* 61–63; Lincoln, *CW*, VI, 201, 217. See also Williams, *Lincoln and His Generals*, 232–242, and Hassler, *Commanders of the Army of the Potomac*, 134–150.

Pages 348–349

Quotation ("most brilliant in world") in Lincoln, *CW*, VI, 230; Williams, *Lincoln and His Generals*, 247–251; Catton, *Grant Moves South*, 435–462.

Pages 349–353

Gettysburg: quotation ("scary rumors") in Welles, *Diary*, I, 328; Lincoln, *CW*, VI, 273, 280–282; Welles, *Diary*, I, 332–333, 334, 347–350; Williams, *Lincoln and His Generals*, 255–270; quotation ("damned . . . turtle") in Catton, *Mr. Lincoln's Army*, 117; Powell, *Lincoln Day by Day*, III, 193–195; quotation ("restless solicitude") in Sandburg, *Lincoln, the War Years*, II, 343; Randall, *Mary Lincoln*, 324–325; Bates, *Lincoln in the Telegraph Office*, 155–156; Lincoln, *CW*, VI, 314, 318; quotation ("Is that all") in Rice, *Reminiscences of Lincoln*, 402; Hay, *Diaries* (Dennett ed.), 66–67; Welles, *Diary*, I, 363–365; quotation ("skill, toil, and blood") in Lincoln, *CW*, VI, 341, and Hay, *Diaries* (Dennett ed.), 69. Vicksburg: quotation ("do for the Secretary") in Welles, *Diary*, I, 364; quotation ("glorious old fourth") in Lincoln, *CW*, VI, 321, also 319; Nicolay, *Lincoln's Secretary*, 102. Lee escapes: quotation ("President is urgent") in *OR*, ser. I, vol. XXVII, pt. III, 605; quotations ("in our grasp" and "If I had gone up there") in Hay, *Diaries* (Dennett ed.), 67, 69; Brooks, *Washington, D.C.,* 80–82, 91–94; quotations ("bad faith" and "watching the enemy") in Welles, *Diary*, I, 369–371, 381; Lincoln to Meade, *CW*, IV, 328. See also Williams, *Lincoln Finds a General*, II, 672–759.

Pages 353–354

Lincoln, *CW,* VI, 326; B. H. Liddell Hart, "Sherman—Modern Warrior," *American Heritage* (August, 1962), 21–22. On Grant's drinking: Catton, *Grant Moves South,* 96–97, 396–397, 462–465, and John A. Carpenter, *Ulysses S. Grant* (N.Y., 1970), 17–18, 25, 37. Lincoln's remark on apocryphal story about Grant's type of whiskey in Chandler, "Lincoln and the Telegrapher" (ed. by Long), *American Heritage* (April, 1961), 34.

Pages 354–357

Quotation ("if vigorously applied") in Lincoln, *CW,* VI, 374, and Grant's reply August 23, 1863, RTL Coll.; quotation ("great river will bristle") in Hurlbut to Lincoln, August 15, 1863, *ibid.;* Lincoln, *CW,* VI, 357, 410; Cornish, *Sable Arm,* 151–156, 288; McPherson, *Negro's Civil War,* 183–191, 237–239. Douglass's interview with Lincoln: Douglass's article in the *New York Tribune,* July 5, 1885, and Douglass, *Life and Times* (reprint of revised 1892 ed., N.Y., 1962), 347–349, 484–486; Segal, *Conversations with Lincoln,* 276.

Pages 357–358

Nevins, *War for the Union,* III, 119–127; McPherson, *Negro's Civil War,* 69–75; Hay, *Diaries* (Dennett ed.), 71–72, 76; Lincoln quotation ("butcher drives bullocks") in Lincoln, CW, *VI,* 369–370, also 381, 390–391; quotation ("Are we degenerate") in *ibid.,* 444–449, also VII, 259; Welles, *Diary,* I, 432–434; Hay, *Diaries* (Dennett ed.), 76, and Nicolay, *Lincoln's Secretary,* 83. At the time he defended the draft in Cabinet (September 14 and 15, 1863), Lincoln was aggravated that "certain judges" tried to disrupt the draft by discharging conscripts through writs of habeas corpus. With Cabinet approval, Lincoln on September 15, 1863, suspended the writ for drafted men. See Lincoln, *CW,* VI, 451. For Lincoln's dealings with the governors, see William B. Hesseltine, *Lincoln and the War Governors* (N.Y., 1948).

Pages 358–360

Lincoln's letter, *CW,* VI, 406–410; Thomas and Hyman, *Stanton,* 294–295; quotation ("good deal of emotion") in Welles, *Diary,* I, 469–470; Voegeli, *Free But Not Equal,* 118–132; quotations ("Glory to God" and "puts his foot down") in Thomas, *Lincoln,* 398, 399; Donald, *Sumner and the Rights of Man,* 165–166; quotation ("my native depravity") in Lincoln, *CW,* VII, 24, also 50.

Pages 360–363

Lincoln, *CW,* VI, 373, 439, 466–467, 518–519; quotation ("same old story") in Welles, *Diary,* I, 439–440. Chickamauga and Chattanooga: *ibid.;* correspondence between Lincoln and Rosecrans and Lincoln and Burnside in Lincoln, *CW,* VI, 469–474, 480–481, 483–484, 486; quotation ("Damn Jonesboro") in Bates, *Lincoln in the Telegraph Office,* 202; Hay, *Diaries* (Dennett ed.), 92–93, 102; quotation ("duck hit on the head") in *ibid.,* 106; Williams, *Lincoln and His Generals,* 279–285; Bruce Catton, *Grant Takes Command* (Boston, 1969), 22–62.

Pages 363–367

Frank L. Klement, "These Honored Dead: David Wills and the Soldiers' Cemetery at Gettysburg," *Lincoln Herald* (Fall, 1972), 123–131; quotation ("short, short, short") in Brooks, *Washington, D.C.,* 252–253; Lincoln, *CW,* VII, 17–18; Turners, *Mary Lincoln,* 154–155; Randall, *Mary Lincoln,* 328–329; Powell, *Lincoln Day by Day,* III, 220. The train trip: *ibid.,* 220–221; Hay, *Diaries* (Dennett ed.), 119, 135. At Gettysburg: Klement's article as cited above; quotation ("from distant parts") in *Cincinnati Daily Commercial,* November 23, 1863; quotation ("notables and nondescripts") in Sandburg, *Lincoln, the War Years,* II, 463; quotation ("any foolish things") in Lincoln, *CW,* VII, 16–17; William E. Barton, *Lincoln at Gettysburg* (N.Y., 1930), 60–88; Allan Nevins, *Lincoln and the Gettysburg Address* (Urbana, Ill., 1954), 8–10; Nicolay, *Lincoln's Secretary,* 175–177; Hay, *Diaries* (Dennett ed.), 120–121. Lincoln made several drafts and copies of the Gettysburg Address (Lincoln, *CW,* VII, 17–23). I've quoted the final text.

Pages 367–372

Lincoln, *CW,* VII, 24–25; Hay, *Diaries* (Dennett ed.), 128; Lincoln's remarks to Nicolay in Nicolay, memorandum, December 7, 1863, Nicolay Papers, Library of Congress. Lincoln's message, *CW,* VII, 40, 47–52. Lincoln's early reconstruction efforts in Tennessee in *ibid.,* VI, 440, 469, and in Louisiana, *ibid.,* V, 504–505, VI, 365 (temporary arrangement of freed people), 440, 469, VII, 1–2; Trefousse, *Radical Republicans,* 280–282. Sumner and Blair: Donald, *Sumner and the Rights of Man,* 117–122, 137–141; quotation ("do nothing with Sumner") in Nicolay and Hay, *Lincoln,* X, 84–85; quotation ("little difference among loyal men") in Hay, *Diaries* (Dennett ed.), 112–113; Lincoln's plan of reconstruction in *CW,* VII, 50–56. Reactions to Lincoln's message and ten percent plan: Hay, *Diaries* (Dennett ed.), 131–135, 138; Brooks, *Washington, D.C.,* 150–151; Powell, *Lincoln Day by Day,* III, 226.

PART TEN: MIGHTY SCOURGE OF WAR
Pages 375–377

Lincoln, *CW,* VII, 63–64. Lincoln also wrote out a pardon for Emilie, conditional on her signing the oath of allegiance. According to Randall, *Mary Lincoln,* 336, 495, Emilie's "code of honor would not permit" her to sign the oath. See also Robert Gerald McMurtry, *Ben Hardin Helm* (Chicago, 1943), 54. The dialogues and quotations from Emilie's diary concerning her White House visit are from Helm, *Mary, Wife of Lincoln,* 222–232. The Lincolns' disagreement over Robert is from *ibid.,* 225–228; Keckley, *Behind the Scenes,* 121–122, reports a similar conversation. Nicolay, *Lincoln's Secretary,* 192–193; quotation ("Why should I sympathize") in Keckley, *Behind the Scenes,* 136.

Pages 377–379

Louisiana reconstruction: Lincoln, *CW,* VII, 66–67, 89–91, 95, 123–125, 161–162, 217–218, 269–270, VIII, 30–31, 106–107, 402; Jefferson Davis Bragg, *Louisiana in the Confederacy* (Baton Rouge, 1941), 212–213, 290–291; John D. Winters, *Civil War in Louisiana* (Baton Rouge, 1963), 394–395. Arkansas: Lincoln, *CW*, VII, 141–142, 144–145, 154–155, 189–191, 199. Congressional opposition: Donald, *Sumner and the Rights of Man*, 179–180; Nevins, *War for the Union*, III, 417–444. Lincoln and Negro suffrage: McPherson, *Negro's Civil War*, 275–281, 302–303; Lincoln quotation ("I barely suggest") in Lincoln, *CW*, VII, 243; Quarles, *Lincoln and the Negro*, 224–230; Krug, *Trumbull*, 222–223; Hyman, *More Perfect Union*, 209–215. See also the controversial Wadsworth letter in Lincoln, *CW*, VII, 101–102, and discussion in Trefousse, *Radical Republicans*, 286.

Pages 379–380

Quotation ("does not know how to help") in Donald, *Sumner and the Rights of Man*, 162–163; Lincoln, *CWS*, 243; Trefousse, *Radical Republicans*, 237–238; Quarles, *Lincoln and the Negro*, 123, 171–172; McPherson, *Negro's Civil War*, 245–270.

Pages 380–383

Quotation ("hear anything about it") in Hay, *Diaries* (Dennett ed.), 112; Welles, *Diary*, I, 501–502; quotation ("voting him a failure") in Brooks's letter in the *Sacramento Daily Union*, October 31, 1863; quotation ("eating my life out") in Bryan, *Great American Myth*, 127. Republican opposition to Lincoln: Chase, *Inside Lincoln's Cabinet*, 24–27; quotation ("national calamity") in Albert D. Richardson, *A Personal History of Ulysses S. Grant* (Hartford, Conn., 1868), 407, 434. Chase boom: Chase, *Inside Lincoln's Cabinet*, 5, 24–27, 32, 176, 183–184; Bates, *Diary*, 310; Randall, *Mary Lincoln*, 217, 242; quotation ("horsefly" and "devilish good joke") in Hay, *Diaries* (Dennett ed.), 54, 110, and Chase, *Inside Lincoln's Cabinet*, 24, 176–177; Ishbel Ross, *Proud Kate: Portrait of an Ambitious Woman* (N.Y., 1953), 158; Nicolay, *Lincoln's Secretary*, 189–191; Donald, *Sumner and the Rights of Man*, 165, 168–169. Pomeroy Circular: *New York Herald*, February 22 and 24, 1864; Lincoln, *CW*, VII, 200n–201n, 212–213; J. G. Randall and Richard C. Current, *Lincoln the President* (N.Y., 1955), IV, 85–110; Chase, *Inside Lincoln's Cabinet*, 210–211; Segal, *Conversations with Lincoln*, 310–311; Robert B. Warden, *Account of the Private Life and Public Services of Salmon Portland Chase* (Cincinnati, 1874), 576; quotation ("I claim not to have controlled events") in Lincoln, *CW*, VII, 282–283.

Pages 383–386

Carpenter, *Grant*, chaps. I, II, IV; Catton, *Grant Takes Command*, 141–158; Williams, *Lincoln and His Generals*, 299–310; Lincoln, *CW*, VII, 324; quotation ("those not skinning") and Lincoln in his nightshirt, Hay, *Diaries* (Dennett ed.), 179.

Pages 386–389

Wilderness: quotation ("Grant has gone into wilderness") in Porter, *Campaigning with Grant*, 98; Carpenter, *Six Months in the White House* (Century House ed.), 19; quotation ("I must have relief") in Schuyler Colfax, *Life and Principles of Abraham Lincoln* (Philadelphia, 1865), 11–12; quotation ("dogged pertinacity") in Hay, *Diaries* (Dennett ed.), 180; Catton, *Grant Takes Command*, 236–237; Brooks, *Washington,*

D.C., 137; quotation ("those poor fellows") in Isaac N. Arnold, *Life of Abraham Lincoln* (Chicago, 1885), 375; Welles, *Diary,* II, 44, 45; Williams, *Lincoln and His Generals,* 317–319. Renomination: quotations ("would be dangerous" and "little drinking") in Hay, *Diaries* (Dennett ed.), 183, 186, also 185; Brooks, *Washington, D.C.,* 142; Arnold, *Life of Lincoln,* 358; Lincoln, *CW,* VII, 376–377, 380–382; *New York Herald,* June 4 and 6, 1864; McPherson, *Struggle for Equality,* 260–261, 271–273; quotation ("you're the little woman") and Stowe's assessment of Lincoln in Mitgang, *Press Portrait,* 373–379.

Pages 390–393

Petersburg: Bates, *Diary,* 378; Welles, *Diary,* II, 54–55; Hay, *Diaries* (Dennett ed.), 195–196; Moore, *Rebellion Record,* XI, 333; Lincoln quotation ("chew and choke") in Lincoln, *CW,* VII, 499. Chase and Fessenden: Chase, *Inside Lincoln's Cabinet,* 30–31, 37–44, 223–235, 255; Lincoln, *CW,* VII, 419; Brooks, *Washington, D.C.,* 118–122; Hay, *Diaries* (Dennett ed.), 201–203. Wade-Davis bill and reaction: the bill and the Wade-Davis Manifesto in Commager, *Documents,* I, 436–440; Niven, *Welles,* 473–474; Trefousse, *Radical Republicans,* 286–289; Lincoln's remarks in the Capitol, July 4, in Hay, *Diaries* (Dennett ed.), 204–205; Brooks, *Washington, D.C.,* 153–154; Donald, *Sumner and the Rights of Man,* 183–184; Chase, *Inside Lincoln's Cabinet,* 232–233; Lincoln's proclamation, *CW,* VII, 433–434; Brodie, *Stevens,* 208; quotation ("To be wounded") in Brooks, *Washington, D.C.,* 156.

Pages 393–397

Early's raid: Brooks, *Washington, D.C.,* 160, 208–209; Lincoln, *CW,* VII, 437, 438; Hay, *Diaries* (Dennett ed.), 208–209; John H. Cramer, *Lincoln Under Enemy Fire* (Baton Rouge, 1948), chaps. I and II; Williams, *Lincoln and His Generals,* 326; Catton, *Grant Takes Command,* 336–345. Lincoln can't win: Greeley to Lincoln, July 7, 1864, and Weed to Seward, August 22, 1864, RTL Coll.; Chase, *Inside Lincoln's Cabinet,* 236–238; Randall and Current, *Lincoln the President,* IV, 198–232; Lincoln's memorandum on his probable failure, *CW,* VII, 514, and quotation ("General, the election has demonstrated") in Hay, *Diaries* (Dennett ed.), 238; quotation ("damned in time") in Lincoln, *CW,* VII, 507, also 499–501. Lincoln wavers: Philip S. Foner, *Life and Writings of Frederick Douglass* (4 vols., N.Y., 1950–1955), III, 422–424; Lincoln, *CW,* VII, 517–518; Nicolay and Hay, *Lincoln,* IX, 221. Campaign: quotation ("manly opponents") in *New York World* as quoted in *New York Herald,* August 2, 1864; Welles, *Diary,* II, 136; Voegeli, *Free But Not Equal,* 151–152; Randall and Current, *Lincoln the President,* IV, 237–249; Current, *Lincoln Nobody Knows,* 28; Wood, *Black Scare,* chap. IV; *La Crosse* (Wis.) *Democrat,* August 29, 1864.

Pages 397–399

Lincoln, *CW,* VII, 532, 533, VIII, 13; Bruce Catton, *Never Call Retreat* (Garden City, N.Y., 1965), 390; Lincoln quotation ("end of the struggle") in Lincoln, *CW,* VIII, 1–2, also 41, 52, 75; quotation ("occupy this big White House") in *ibid.,* VII, 512, also 528; Lincoln and patronage in Carman and Luthin, *Lincoln and the Patronage,*

282ff. Blair and Frémont: Trefousse, *Radical Republicans,* 295–296; Randall and Current, *Lincoln the President,* IV, 227–231; Hay, *Diaries* (Dennett ed.), 219–220; Nicolay and Hay, *Lincoln,* IX, 340; Lincoln, *CW,* VIII, 18; Welles, *Diary,* II, 156–159; Edward McPherson, *Political History of the United States . . . During the Great Rebellion* (Washington, D.C., 1865), 426. Republicans behind Lincoln: quotation ("fight like a savage") in Thomas, *Lincoln,* 447; Hay, *Diaries* (Dennett ed.), 226–230; Lincoln, *CW,* VIII, 46, 47, 58.

Pages 400–401

Mary's troubles: quotation ("Hundreds are getting rich") in Keckley, *Behind the Scenes,* 151; quotation ("almost a monomaniac") in Turners, *Mary Lincoln,* 180, also 163–165, 183; Brooks, *Washington, D.C.,* 195–197; Hay, *Diaries* (Dennett ed.), 232–236.

Pages 401–402

William Frank Zorow, *Lincoln and the Party Divided* (Norman, Okla., 1954), 200ff; McPherson, *Political History,* 623; Lincoln, *CW,* VIII, 100; quotation ("vote against McClellan") in Donald, *Sumner and the Rights of Man,* 190–191; quotation ("completely worn out") in Keckley, *Behind the Scenes,* 157, and Turners, *Mary Lincoln,* 183; quotation ("For God's sake") in Nicolay, *Lincoln's Secretary,* 217; Lincoln, *CW,* VIII, 100–102; Hay, *Diaries* (Dennett ed.), 239.

Pages 402–404

Quotation ("the bare thought") in Carpenter, *Six Months in the White House* (N.Y., 1866), 276. Sherman: quotation ("If people raise a howl") in James M. Merrill, *William Tecumseh Sherman* (Chicago, 1971), 258, 266; William Tecumseh Sherman, *Memoirs* (2 vols., N.Y., 1893), II, 111, 227–228; Stanton to Grant, October 12, 1864, *OR,* ser. I, vol. XXXIX, pt. III, 222; Williams, *Lincoln and His Generals,* 339–344; B. H. Liddell Hart, *Sherman* (N.Y., 1958), 308–355; also John Bennett Walters, *Merchant of Terror: General Sherman and Total War* (Indianapolis and N.Y., 1973), 87–183; quotation ("I know what hole") in McClure, *Lincoln and Men of War-Times,* 219–220; Eckert's description of Lincoln in Bates, *Lincoln in the Telegraph Office,* 317; Lincoln, *CW,* VIII, 181–182.

Pages 404–406

Lincoln quotations ("King's cure" and "at all events") in Lincoln, *CW,* VIII, 149, 254; quotation ("my chief hope") in Arnold, *Life of Lincoln,* 358–359, and Oldroyd, *Lincoln Memorial,* 491–494; Swett to Herndon, January 17, 1866, H-W Coll.; Rice, *Reminiscences,* 586; Lincoln's use of patronage in Albert G. Riddle, *Recollections of War Times* (N.Y., 1895), 323–324; quotation ("certain negotiations") in George W. Julian, *Political Recollections* (Chicago, 1884), 250; Trefousse, *Radical Republicans,* 298–300. Passage of the amendment: Brooks, *Washington, D.C.,* 185–187; Lincoln, *CW,* VIII, 254–255; quotation ("people over the river") in Segal, *Conversations with Lincoln,* 17, also 362–363.

Pages 406–407

Lincoln quotations ("better for poor black" and "sooner have the fowl") in Lincoln, *CW,* VIII, 106–108, 404; Sumner quotation ("eggs of crocodiles") in Sumner, *Works,* X, 44. My account of the compromise between Lincoln and Sumner and the debates over Louisiana draws from Donald, *Sumner and the Rights of Man,* 196–207.

Pages 408–409

Mary and Sumner: quotations ("make himself very agreeable," "extreme Republican," and "two schoolboys") in Turners, *Mary Lincoln,* 185, 186, 205–206; Brooks, *Washington, D.C.,* 227. Lincoln and Mary: quotation ("forbidden subjects") in Turners, *Mary Lincoln,* 183–184; quotation ("worst speech") in Sandburg and Angle, *Mary Lincoln,* 112; the scene over McManus in Turners, *Mary Lincoln,* 197–198. Robert: quotation ("deep waters") in *ibid.,* 189; quotation ("Robert in the army") in Luthin, *Lincoln,* 420; Lincoln, *CW,* VIII, 223–224.

Pages 409–412

Quotation ("looked badly") in Browning, *Diary,* II, 7–8; visit with Speed in Speed to Herndon, January 12, 1866, H-W Coll. Inauguration: quotation ("no disturbances") in Thomas and Hyman, *Stanton,* 349, 394; Dorothy M. and Philip B. Kunhardt, *Twenty Days* (N.Y., 1965), 33–35; Lincoln's inaugural address, *CW,* VIII, 332–333; Brooks, *Washington, D.C.,* 74, 195, 210–215; Douglass, *Life and Times,* 365–366.

PART ELEVEN: ''MOODY, TEARFUL NIGHT''
Pages 415–417

Lincoln, *CW,* VIII, 260–261, 284–285, 330–331; Nicolay, *Lincoln's Secretary,* 121; quotations ("Rebels would gain" and "Soon after I was nominated") in Carpenter, *Six Months in the White House* (Century House ed.), 74; gunshot story and Lamon's fears about Lincoln's safety in Lamon, *Recollections of Lincoln,* 263–281; Browning, *Diary,* II, 18–19; quotation ("not an American practice") in Seward, *Life of Seward,* II, 418; quotation ("crazy folks") in Carpenter, *Six Months* (Century House ed.), 75; quotation ("made up my mind") in Brooks, *Washington, D.C.,* 43–44. Stanton and Lincoln's protection: Bryan, *Great American Myth,* 63–72, 127, 136–137; Nicolay, *Lincoln's Secretary,* 121; quotation ("people know") in Cornelius Cole, *Memoirs* (N.Y., 1908), 214; quotation ("court-martialed and shot") in Thomas and Hyman, *Stanton,* 393–395.

Pages 417–422

At City Point: Lincoln, *CW,* VIII, 372–374; Catton, *Grant Takes Command,* 434–436; Lincoln's train trip to Petersburg in John S. Barnes, "With Lincoln from Washington to Richmond in 1865," *Appleton's Magazine* (May, 1907), 520–522; Richardson, *Grant,* 463; quotation ("appalling difficulties") in Porter, *Campaigning with Grant,*

406–409. Incident over Mrs. Ord: Adam Badeau, *Grant in Peace* (Hartford, 1887), 356–365; Barnes, as cited above; Turners, *Mary Lincoln,* 206–208, 211; Randall, *Mary Lincoln,* 372–374. Sherman's visit and Petersburg falls: David D. Porter, *Incidents and Anecdotes of the Civil War* (N.Y., 1885), 130–131; Williams, *Lincoln and His Generals,* 351–353; Catton, *Grant Takes Command,* 436–440; Hart, "Sherman," *American Heritage* (August, 1962), 21–23, 102–106. Lincoln, *CW,* VIII, 377; quotation ("pause by the army") in *ibid.,* 378; quotation ("something material") in Thomas, *Lincoln,* 510; Lincoln, *CW,* VIII, 381–385; Crook, "Lincoln's Last Day," *Harper's* (September, 1907), 519; quotation ("work was done") in Porter, *Campaigning with Grant,* 450–452; Lincoln, *CW,* VIII, 385. Richmond: Crook, "Lincoln's Last Day," *Harper's* (September, 1907), 520–522; account of *Boston Journal* correspondent in Mitgang, *Press Portrait,* 452–454; Charles A. Pineton (Marquis de Chambrun), *Impressions of Lincoln and the Civil War* (N.Y., 1952), 108; *New York Times,* April 8, 1865; Johnson and Buel, *Battles and Leaders,* IV, 727–728; Lincoln, *CW,* VIII, 386–389. Leaves City Point: quotation ("thing be pressed") in *ibid.,* 392, also 388; Bryan, *Great American Myth,* 133; Powell, *Lincoln Day by Day,* III, 325–326; Turners, *Mary Lincoln,* 212, 220; Pineton, *Impressions,* 82–86, and Pineton's (Marquis de Chambrun's) account in *Scribner's Magazine* (January, 1893), 34–35.

Pages 422–425

Lee's surrender: quotation ("back from Richmond") in Van Deusen, *Seward,* 411–412; quotations ("iron mask" and "been a good friend") in Thomas and Hyman, *Stanton,* 353, 354; quotation ("Guns are firing") in Welles, *Diary,* II, 278; Brooks, *Washington, D.C.,* 223–235; Lincoln, *CW,* VIII, 393–394. Lincoln's last speech: Brooks, *Washington, D.C.,* 225–227; Lincoln, *CW,* VIII, 399–405; Chase, *Inside Lincoln's Cabinet,* 262–263; Welles, *Diary,* II, 279–280; Lincoln, *CW,* VIII, 406–407; Thomas and Hyman, *Stanton,* 355–356, and Hyman, *More Perfect Union,* 281.

Pages 425–426

Lamon, *Recollections of Lincoln,* 115–118. Lamon said he wrote down Lincoln's description of his dream right after their conversation. Also see Arnold, *Life of Lincoln,* 433.

Pages 426–434

Bryan, *Great American Myth,* 145, 150; quotation ("playful note") in Turners, *Mary Lincoln,* 257; Randall, *Lincoln's Sons,* 208; Keckley, *Behind the Scenes,* 138. Cabinet meeting: Thomas and Hyman, *Stanton,* 357–358; Welles, *Diary,* II, 280–283; Frederick W. Seward, *Reminiscences of a War-Time Statesman and Diplomat* (N.Y., 1916), 254–257; quotation ("President so cheerful") in Hugh McCulloch, *Men and Measures of Half a Century* (N.Y., 1888), 222; quotation ("perhaps been too fast") in Speed's remarks to Chase, *Inside Lincoln's Cabinet,* 268; Welles, "Lincoln and Johnson," *The Galaxy* (April, 1872), 525–527. Afternoon: Bryan, *Great American Myth,* 159–161; Thomas and Hyman, *Stanton,* 395–396; John Russell Young, *Around the World with General Grant* (2 vols., N.Y., 1879), II, 356; Catton, *Grant Takes Command,* 474; Brooks, *Washington, D.C.,* 229. Carriage ride: Turners, *Mary Lincoln,* 218; Randall,

Mary Lincoln, 381; Arnold, *Life of Lincoln*, 429; Lamon, *Recollections of Lincoln*, 119–120; Duff, *Prairie Lawyer*, 370. Evening: quotation ("felt so unwilling") in Dr. Anson Henry to his wife, April 19, 1865, in Milton H. Shutes, *Lincoln and the Doctors: A Medical Study of the Life of Abraham Lincoln* (N.Y., 1933), 132–134; Bryan, *Great American Myth*, 162–163; Thomas and Hyman, *Stanton*, 395–396; Crook, "Lincoln's Last Day," *Harper's* (September, 1907), 525–530. At Ford's Theater: Bryan, *Great American Myth*, 180–182, 220–221; Kunhardts, *Twenty Days*, 28–37; G. A. Townsend, *The Life, Crime, and Capture of John Wilkes Booth* (N.Y., 1865), 7; Charles Sabin Taft, M.D., "Abraham Lincoln's Last Hours," *Century Magazine* (February, 1893), 634; Helen Truman's account in the *New York World*, February 17, 1924; quotation ("will Miss Harris think") in Dr. Henry to his wife, Shutes, *Lincoln and the Doctors*, 133, and Turners, *Mary Lincoln*, 222; Rathbone's affidavit in the *Washington Daily Morning Chronicle*, April 18, 1865; Benn Pitman (comp.), *The Assassination of the President and the Trial of the Conspirators* (Cincinnati, 1865); Keene's cry to the audience in the *New York Herald*, April 17, 1865; Ralph Borreson, *When Lincoln Died* (N.Y., 1965), 15–31, containing excerpts from eyewitnesses such as Dr. Leale.

The Petersen House: Turners, *Mary Lincoln*, 222–223; Mrs. Elizabeth Dixon to her sister, May 1, 1865, in *The Collector* (March, 1950), 49–50; Powell, *Lincoln Day by Day*, III, 330; Sumner quotation in Donald, *Sumner and the Rights of Man*, 216; Thomas and Hyman, *Stanton*, 396–400; Welles, *Diary*, II, 286–288; quotation ("dream was prophetic") in Arnold, *Life of Lincoln*, 433; quotation ("live but one moment") in Randall, *Lincoln's Sons*, 212–214; quotation ("given my husband to die") in Howard H. Peckham, "James Turner's Account of Lincoln's Death," *Abraham Lincoln Quarterly* (December, 1942), 176–183; Randall, *Mary Lincoln*, 383–384; Nicolay and Hay, *Lincoln*, X, 302; Bryan, *Great American Myth*, 186–189; Borreson, *When Lincoln Died*, 37–46. Bryan, *Great American Myth*, and Thomas and Hyman, *Stanton*, among other experts, have thoroughly destroyed the myth that Stanton and other men in the government were involved in the assassination plot.

Pages 434–436

Kunhardts, *Twenty Days*, 86–95, 119–132, 137, 140ff; Borreson, *When Lincoln Died*, 54–114; Whitman, "When Lilacs Last in the Dooryard Bloom'd," in Floyd Stovall (ed.), *Walt Whitman* (N.Y., 1934), 232–233. See also Victor Searcher, *The Farewell to Lincoln* (N.Y., 1965).

INDEX

About the Author

Stephen B. Oates is a professional biographer and historian. He has published nine books—among them, *With Malice Toward None* (1977); *The Fires of Jubilee: Nat Turner's Fierce Rebellion* (1975); *To Purge This Land with Blood: A Biography of John Brown* (1970); *Visions of Glory* (1970); *Rip Ford's Texas* (1963); and *Confederate Cavalry West of the River* (1961). In addition he has written numerous articles and shorter biographical studies for such journals as *American Heritage*, the *Nation*, the *American West, Civil War History*, and the *South Atlantic Quarterly*, and he has edited a two-volume anthology, *Portrait of America* (1973), dedicated to the proposition that biographical and historical writing can be literature. His own writings have been widely anthologized.

He was born in 1936, in the Texas Panhandle town of Pampa. He holds B.A., M.A., and Ph.D. degrees from the University of Texas at Austin and is an elected member of Phi Beta Kappa and the Texas Institute of Letters. His writing projects have taken him on extensive journeys into many parts of the South, the far Southwest, the Middle West, and the Northeast. He is currently a Professor of History at the University of Massachusetts, Amherst, where he teaches courses in antebellum and Civil War America and in the art and technique of biography.

In 1972 Mr. Oates received a Guggenheim Fellowship to facilitate his work on *With Malice Toward None*.